CURRENT FRONTIERS IN CRYOPRESERVATION

Edited by **Igor I. Katkov**

Current Frontiers in Cryopreservation
Edited by Igor I. Katkovi

Published by InTech

Janeza Trdine 9, 51000 Rijeka, Croatia

Publishing Process Manager Maja Bozicevic

Technical Editor Teodora Smiljanic

Cover Designer InTech Design Team

Additional hard copies can be obtained from orders@intechopen.com

Current Frontiers in Cryopreservation, Edited by Igor I. Katkov
p. cm.
ISBN 978-953-51-0302-8

For EU product safety concerns:
IN TECH d.o.o., Prolaz Marije Krucifikse Kozulić 3, 51000 Rijeka, Croatia,
info@intechopen.com or visit our website at intechopen.com.

Meet the editor

Igor I. Katkov, PhD, is a "seasoned" cryobiologist with more than 30 years of experience in the field. He received his MSc and PhD diplomas in Kharkov, Ukraine. He has collaborated with many prominent cryobiologists, including P. Mazur, F. Ostashko, V. Grischenko, the Isachenkos, and many others. Dr Katkov has more than 130 publications and is actively working in the fields of hyper-fast kinetic vitrification, osmotic modeling, cryopreservation of adherent cells, and many other fields related to fundamental cryobiology, assisted reproduction and regenerative medicine. Igor Ivanovich is a renowned scientist and a member of Editorial Boards of the "Cryo Letters" and "Problems of Cryobiology and Cryomedicine" journals. His hobbies are falconry, wildlife, sharp shooting, World history and "structure of vacuum".

Contents

Preface

Almost a decade has passed since the last textbook on the science of cryobiology and the most common methods of cryopreservation was published [Fuller *et al*, 2007], to which we will refer as *"the previous book"* here and below. When it was published, it became a useful guide for both "seasoned" cryobiologists and those who had just started their journey to this fascinating science.

However, there have been some serious tectonic shifts in cryobiology, which were perhaps not seen on the surface but may have a profound effect on both the future of cryobiology and on the development of new cryopreservation methods. We feel that it is time to revise the previous paradigms and dogmas, discuss the conceptually new cryobiological ideas and introduce the recently emerged practical methods of cryopreservation. The present books, *Current Frontiers in Cryobiology* [Katkov, 2012A] (referred here as *Book 1*) and *Current Frontiers in Cryopreservation* [Katkov, 2012B] (*Book 2*), will serve the purpose. These two books are not a substitute for *the previous book* but are rather complementary, so we highly recommend to all readers who want to know the background on which *the current books* were written to read *the previous book* as well.

Before we describe the current books, let us first briefly compare them to the previous book in retrospective. First of all, there were some very promising directions a decade ago that unfortunately did not meet the expectations. Molecular biology and genetics, particularly in regards to expression of stress proteins and other pathways related to the cell injury, have not introduced any serious breakthroughs except for the use of ROCK inhibitors for cryopreservation of human embryonic and induced pluripotent stem cells. The latter really was a revolutionary discovery, which however, was not made by cryobiologists; it was just "picked up" by them from the Watanabe's seminal work [Watanabe *et al*, 2007] (see the Chapter by Martin- Ibáñez in *Book 1* for details). In general, however, all those molecular biology tools have helped the solution but have not solved the cryopreservation problems *per se*. One of the backlashes of this new era is that the "traditional" cryobiologists now have little chances of getting a grant from many funding agencies such as NIH, whose panels are dominated by molecular biologists and geneticists, unless the applicant is willing to study those pathways and use of transcriptomics, proteomics, metabolomics, and other "omics". Yet, all those very expensive tools have so far added a little to the science of cryobiology, and especially to the practices of cryopreservation. Moreover, it is sad to see the how some new publications "rediscover the wheel", repeating many achievements of cryobiologists that had been done one or two decades before but were not referred as full size papers on PubMed, and these novel rediscoveries are often done at a much greater cost. We must agree with the author of *Foreword* of the previous book, who insightfully wrote *"I see now much of the early ground being replowed, often by equally empirical methods, albeit as far greater expenses...The concept of science as a community of colleagues engaged in public service ... has been eroded by the cost of research and the emergence of industry as not only a major source of research funding but as the ultimate exploiter of the results, and we have no choice but to play the game"* (Foreword in [Fuller *et al*, 2007] by H. Meryman). However, we hope that might change in the future and that an alliance between cryobiology and molecular and cellular biology will bring *real* practical fruits.

The slogan *"Let Us Learn from Mother Nature"*, while being attractive *per se* (it is actually imbedded in the title of our first Chapter by Katkov *et al* in *Book 1*), must be taken with a grain of salt. Yes, Mother Nature has liquid crystals in biological membranes, but LCD TV screens were invented by man. Yes, there are rotifers and other "molecular motors" in cells, but the wheel was discovered and built by the human race. And finally, there are TV, radio, internal combustion engines, and many other devices and apparatuses that have no close analogy in wildlife. Similarly, while some robust creatures are well adapted to survive for short time at up to -20°C, there is no place on Earth that cools down below the glass transition temperature of water (-136°C), and there is no place on Earth where liquid nitrogen is present. Ergo, practically no one natural biosystem can adapt to such low temperatures just by natural selection, it needs human help to be stabilized for infinite time at -196°C. Thus, while learning something from the natural phenomena, it is our strong opinion that we should not rely on them too much: the money for supplying an Antarctica deep lake drilling or a Mars expedition can be spent much more efficiently and *usefully for humankind* if channeled to the development of a new controllable freezer for cryopreservation of large tissues and organs, similar for instance, to the described in our *Book 1* by Butler and Pegg.

The next large area where the progress has been quite slower than it was expected a decade ago is lyophilization and desiccation of cells of vertebrates. So far, there is no compelling evidence that would convince us that there is a method of freeze-drying or desiccation that has produced *viable* mammalian cells that can be stored for sufficient amount of time (> 2 years) at temperature above +4°C despite the fact that the opposite was claimed many times in the last 50 years. We briefly explored this aspect in our *Chapter 1* of *Book 1*. We think that this field has remained to be trapped in a set of scientific misconceptions, such as the possibility of drying the sample to the glass transition T_g that is above the final temperature or drying $T_{dr,f}$, or the related misleading concept of the possibility of substantial movement of water in a sample below its glass transition. We think that such statements violate the laws of thermodynamics and the definition of the vitreous phase as the state with enormous viscosity, as well as that it is in contradiction with the Stokes-Einstein Law of diffusion. Numerous reports, which state T_g of the sample as high as + 60°C while drying was stopped at, say, $T_{dr,f}$ = +20°C are *incorrect*. Such overestimation of T_g (which in fact is 20-25 degrees lower than $T_{dr,f}$) lead to unsubstantiated expectations of long stability at supra-zero, even ambient temperatures at relatively *high* water content of the sample, which has never proven the case in thorough experiments (with some reservation to platelets as rather *"cell debris"* than true cells). All these data, if checked properly, are in fact either artifacts - the results of incorrect gravimetrical measurements or the use of inaccurate methods such as DSC. We are confident that the real viscosity, not those *mysterious* high temperature DSC peaks, should be measured for correct determination of the *biostabilization* T_g (defined as the point at which viscosity reaches 10 x $10^{13.6}$ Pa x sec).

And as *the last but not the least in our list* is the notion that all those ice-blockers, freeze and shock proteins, and other promising from a decade ago classes of molecules have not so far shown to be used in cryopreservation protocols alone but always in a concert with *ole good* permeable cryoprotectants and impermeable sugars, and other low molecular weight molecules, which have been around for decades. This is especially true in regards to vitrification, specifically of organs: *"the promise of the 21st century medicine"* has remained as far from the completion as it was 25+ years ago with the report on *equilibrium* vitrification of a kidney by Fahy and colleagues. On the other hand, the assisted reproduction cryobiology is rapidly moving toward *kinetic* vitrifcation, the very method of cryopreservation described by the most prominent pioneer of cryopreservation, Father Basil Luyett, more than 7 decades ago. We specifically dedicated our *Chapter 1* both to the memory of this brilliant scientist and to the detailed analysis of the situation, the difference between the two approaches to vitrifcation (*kinetic* vs. *equilibrium*), and to a quite opposite foreseeable future for them.

On the other hand, there are new directions (or the old ones, replowed with deeper and more thorough plowing techniques) on the horizon of cryobiology. Among them, we can mention the attractiveness of cryopreservation of *adherent* cells (often in monolayers) not only for the benefit of the cells *per se* (by avoiding anoikis triggered apoptosis, etc), but mainly for the convenience of the rapidly emerging field of cell based high throughput and high content analyses, where cells can be frozen, stored, shipped, and *ready-to-go* after thawing directly in multi-well plates (see Chapter by Martín-Ibáñez in *Book 1*).

The other serious breakthrough that was missed by many authors a decade ago is the kinetic vitrifcation of sperm and the emergence of what we call *"Race for the Pace"*, a set of new devices for ultra-fast cooling of samples such as Open Pulled Straws, Microdroplets, Vitrification on the Solid Surface, VitMaster (slush cooling), Cryogenic Oscillating Heat Pipes, Quartz Capillaries, and some others (see Chapter by Cipri *et al* in *Book 1*). We think that many of these devices are rather transient to a new generation of hyper-fast coolers and warmers, but yet, the rapid ascent of kinetic vitrification is the phenomenon that has been largely missed and often simply ignored by the "classical cryobiologists "at the end of the last century and the beginning of the current one. Our books dedicate a lot of space to those aspects and their future directions.

There are also some other differences between our books and *the previous book* published by CRC, which are mainly determined by the very nature of how the Open Access operates. To begin with, our books are closely related but yet are different. *Book 1* contains mainly reviews that were written by the leaders in the field and were solicited by the Editor. In contrast, *Book 2* (in general, with some exceptions) is dedicated mostly to the reports of concrete methods of cryopreservation, and its chapters are often written by young or emerging scientists who want to make their discoveries public as soon as possible. The Editor is well aware of how discouraging and often devastating the reviewing process in "standard' journals can be just because the reviewer(s) did not share innovative ideas proposed by the author, even though the experimental or mathematical aspects of the manuscript raised no questions. The Editor of these books has reviewed all submitted chapters, about a dozen of them has been rejected, and among 41 published chapters, many were revised one, two, or sometimes three times. But that was always regarding the quality of the manuscripts, not the quality of the author's science; if I sometimes disagreed with the author's opinion, I then *"let the cryo people go"* with their perception, not mine or the one of some external reviewer. In the revolutionary spirit of Open Access, let the common reader, not an elitist reviewer, be the judge in the end!

Another "democratic" aspect in our books in the times of globalization is that the contributions were made by people from 27 countries from *all* continents (except Antarctica). Editor *greatly* appreciates the *invaluable* contribution of the American, Australian, British, Canadian, and New Zealand scientists to the field, and 10 out of 42 Chapters in our books were contributed by authors from those countries. However, cryobiology has long existed in many other languages and cultures. We found the tone of some, especially "historical" reviews written by prominent cryobiologists that may make an impression that the scientists of *that* linguistic domain have predominantly contributed to cryobiology quite uncomfortable; in other words: *"If it is not published in English - it doesn't exist"* so to speak. In contrast to such biases, the seminal works of Luyett, Smirnov, Jahnel, Boutron, Milovanov, Cassou, Ostashko, Sumida, Kopeika, and many other scientists whose first language of publication and/or mother tongue were not English, but whose *pioneering* contribution to the theory and practice of low temperature stabilization has been recognized over the World, is also highly regarded in our books.

Yet another difference is that the chapters of our books are grouped into topics (Sections) that are "subject oriented" rather than loosely flocked to the "Themes" so none would wonder why one chapter on freezing of plants is in one section, while another one ends up in another. The sections are the same for both our *books*, the only difference is the type of the paper as described earlier, yet in many cases, this difference is rather vague: we do not consider chapters in *Book 2* as "second class" at all: *Books 1* and *2* are inseparable. The sections and chapters of the books are as follows:

Section *"Basic Cryobiology and Kinetic Vitrification"* opens *Book 1*. The first, and the two following chapters are dedicated to kinetic vitrification as the re-emerging method of cryopreservation. Chapter 1 by Katkov and colleagues reiterates the idea that basically all methods of long-term stabilization of cells are in fact different ways (the authors identify 5 of them) of vitrifcation of the intracellular milieu. The chapter gives a detailed thermodynamical description and analysis of the methods. The second part of the chapter is dedicated to the kinetic vitrification of human and animal spermatozoa, the concept of the *"Universal Cryopreservation Protocol"* and what the author called *"Race for the Pace"*, though the last one needs a separate chapter and is only mentioned briefly as one of the future directions. The chapter by *Isachenko and colleagues* tells the story of successful vitrification of human and animal spermatozoa, and its emerging as a valuable tool applied to the assisted reproduction technologies. The third chapter, by a Canadian group (Moskovtsev *et al*) is an *independent* report of the success of vitrifcation of human sperm without permeable (and potentially toxic) cryoprotectants (vitrificants) with certain modifications of the Isachenko's method. The chapter by Gao & Zhou is dedicated to the basic cryobiology of osmotic effects, prevention of the osmotic injury, as well as to the equipment for the optimal addition and elution of osmotically active permeable cryoprotective agents (CPAs).

Section *"Stem Cells and Cryopreservation in Regenerative Medicine"* in *Book 1* is presented by a review by Martín-Ibáñez on cryopreservation of human *pluripotent* stem cells; it is the cutting edge of the contemporary cryobiological science where major discoveries have been made very recently. Cryopreservation of *adult* rat mesenchymal stem cells by vitrification is the theme of the chapter by Bahadori *et al* in *Book 2*. It is the one of the chapters when the Editor disagrees with the evaluation of the convenience of cryopreservation of stem cells by vitrification in small containers such OPS, but as we said before, we judged the experimental science, not the concept, and the former one was self-evidently good. The chapter by Campbell & Brockbank reports very interesting results on cryopreservation of adherent smooth muscle and endothelial cells, a direction that, as we mentioned before, may bring about some interesting practical applications. The other two chapters in *Book 2* are the one by U. Santos and colleagues, dedicated to cryopreservation of musculoskeletal cells and tissues, and the other application of regenerative medicine - cryopreservation of allograft for knee ligament construction is the theme of the Chapter by Bitar *et al.*

Section *"Human Assisted Reproduction Techniques (ART)"* opens with a review by Liebermann on vitrifcation of embryos and oocytes, a fast developing method of ART. Juergen and Michael Tucker have edited an excellent book dedicated to the use of vitrifcation in human ART [Liebermann & Tucker, 2007] that we highly recommend for reading to the specialists in the field. This review in *Book 1* summarizes the latest achievements in the area. Another chapter in *Book 1* (Bigelow & Copperman) is also dedicated to cryopreservation of human oocytes, and altogether, both chapters provide a good glance at the comparative advantages of slow freezing vs. vitrifcation in cryopreservation of human eggs. As the background in cryopreservation of human spermatozoa is extensively covered by three chapters of the first section in *Book 1*, the third chapter of this section written by Honaramooz discusses the recent advances in cryopreservation of testicular tissues. The chapter in *Book 2* by Criado covers a very "hot" topic of contamination associated with vitrifcation in the so called "open systems", in which there is a direct contact (or a possibility of such) of the vitrified sample with liquid nitrogen. The chapter also provides a comprehensive review of current containers used for vitrifcation in human ART.

The section *"Farm / Pet/ Laboratory Animal ART"* is generically related to the previous section, but with an emphasis on animal reproductive cells and tissues. The first chapter in *Book 1* by Rodriguez-Martinez covers the cryopreservation of porcine (pig) gametes, embryos and genital tissues. It is followed by a chapter on cryopreservation of embryos of model animals, written by Tsang & Chow. *Book 2* contains a series of reports and mini-reviews on cryopreservation of boar (Kaeoket) and rat (Yamashiro & Sato) spermatozoa, cryopreservation of genetic diversity (sperm, oocytes, embryos, somatic cells) of rabbit species by Jolly *et al*, cryopreservation of ovarian tissues of large domestic animals (cow, pig and sheep) and non-primates (macaque) by Milenkovic and colleagues, and cryopreservation of reproductive cells of domestic animals (Neto *et al*). While there is a certain overlap in the coverage among those chapters, we feel that such diversity enriches the *Book 2* as different points of view are considered.

Section titled *"Cryopreservation of Wildlife Genome"*, particularly of terrestrial vertebrate species, is comprehensively covered by Saragusty and is supplemented by a review on cryopreservation of genome of wild *Felidae* by Paz in *Book 1*.

Section *"Cryopreservation of Aquatic Species"* lacks a general review but several aspects are covered in a variety of chapters. Zilli & Vilella (*Book 1*) discuss the effect of cryopreservation on bio-chemical parameters, DNA integrity, protein profile and phosphorylation state of proteins of seawater fish spermatozoa, with a similar topic, but at a different angle and with their own recent experimental results, covered by Li and colleagues in *Book 2*. This book also contains several experimental reports on cryopreservation of sperm of freshwater species, such as European pikeperch and catfish (Bokor *et al*), a variety of Malaysian freshwater species (Chew), brown trout and koi carp (Bozkurt *et al*) and African giant catfish (Omitogun *et al*).

The section titled *"Cryopreservation in Plants"* is the most *populous* and is represented by four chapters in *Book 1* and five chapters in *Book 2*. Two extensive reviews by Kaczmarczyk *et al* and by Kami cover general aspects of plant cryopreservation. Again, while those two chapters overlap in many aspects, they are rather complementary. The third chapter in *Book 1,* written by Babu and colleagues, is dedicated to cryopreservation of species of spices plants, while the review by Quain *et al* discusses the current advances in cryopreservation of vegetatively propagated tropical crops. Similar subject (vegetatively propagated crops), but with an emphasis on the thermal analysis of the cryopreservation methods using DSC, is covered by Zámečník *et al* in *Book 2*. In the same book, Martinez-Montero and colleagues cover current frontiers in cryopreservation of sugarcane and pineapple, C. Santos reports the results of cryopreservation of cork oak, and Radha *et al* discuss cryopreservation of a medicinal Indian plant of *Icacinaceae* species. The Chapter by Burritt covers an interesting topic of action of proline as a "natural" multi-functional cryoprotectant that is accumulated in higher plants under stress, and can be considered as an attractive CPA candidate for cryopreservation in general.

Section *"Equipment and Assays"* is the last but definitely not the least important section of these two books, as the entire progress of cryobiology and cryopreservation depends on the development of devices and containers for cryopreservation, and proper and adequate assays of cryopreserved cells after resuscitation. The first chapter in *Book 1*, by Butler & Pegg, covers the precision in and control of cryopreservation, the pivotal components in the modern cryopreservation technologies. While this chapter covers mostly slow (equilibrium) programmed freezing, it is supplemented by a review by Cipri *et al*, which discusses some novel equipment and carriers, particularly for vitrification. To some extent it complements the last sub-section of our chapter 1, but we have to emphasize that none of the devices described in the Cipri's chapter can achieve the very rapid rates of cooling as many of the inventors claim. For example, the notion that *VitMaster* can achieve as high as 135,000 ^{o}C/min is largely overestimated even for very small samples, as slush freezing does not completely eliminate the Leidenfrost effect. In regard to assays, Partyka and colleagues review the methods of assessment of viability of cryopreserved sperm, many of which can be adapted to other types of cells as well. Finally, Pérez Campos *et al* present some interesting ideas on using X-ray diffraction for the assessment of quality of cryopreserved tissues in tissue banks in *Book 2*.

In conclusion, *Books 1* and *2* cover a vast variety of topics regarding the current development of both fundamental cryobiology and practical aspects of cryopreservation, and we hope they will help the researches to grasp the background, state of the art, and future of this captivating and very important field of Life Sciences.

Igor I. Katkov, Ph.D.
Head of Cryobiology and Biostabilization
Stem Cell Center
Burnham-Sanford Institute for Medical Research
and
Chief Scientific Officer of CELLTRONIX
San Diego, California,
USA

References

Fuller B. J., Lane N., & Benson E.E.., Eds. (2004). *Life in the Frozen State*. CRC Press, 701 p., ISBN 978-041-524-700-9

Katkov I.I.., Ed. (2012A). *Current Frontiers in Cryobiology*. InTech Open Access Books, ISBN 979-953-307-300-1.

Katkov I.I., Ed. (2012B). *Current Frontiers in Cryopreservation*. InTech Open Access Books, ISBN 979-953-307-743-6.

Tucker M. J. & Liebermann J., Eds. (2007). *Vitrification in Assisted Reproduction: A User's Manual and Troubleshooting Guide*. London, UK, Infroma Healthcare, 320 p., ISBN: 978-041-540-882-0.

Watanabe K., *et al* (2007). "A ROCK inhibitor permits survival of human embryonic stem cells." Nat Biotechnol 25: 681-6.

Part 1

Stem Cells and Cryopreservation in Regenerative Medicine

1

Cryopreservation of Rat Bone Marrow Derived Mesenchymal Stem Cells by Two Conventional and Open-Pulled Straw Vitrification Methods

Mohammad Hadi Bahadori*
Cellular and Molecular Biology Research Center, Faculty of Medical Science, Guilan University of Medical Sciences, Rasht, Iran

1. Introduction

Bone marrow (BM) is a complex tissue containing populations of progenitor and stem cells (1). One type, hematopoietic stem cells (HSCs), can renew circulating blood elements such as red blood cells, monocytes, platelets, granulocytes and lymphocytes. The other is mesenchymal stem cells (MSCs), which possess two important properties of long-term self renewal and differentiate into osteoblasts, chondroblasts, adipocytes and hematopoiesis supporting stroma (2, 3). Their mesenchymal differentiation potential is retained even after repeated subcultivation *in vitro* (4, 5). Besides originating the forming mesenchymal tissue, many studies have demonstrated that MSCs could differentiate into various non-mesenchymal tissue lineages under appropriate experimental conditions *in vitro* and *in vivo*, such as hepatocytes (6, 7), cardiomyocytes (8, 9), lung alveolar epithelium (10), olfactory epithelium (11), inner hair cells (12), neurons and neuroglia (1, 4, 13). MSCs are spindle shaped fibroblast-like cells that are easily isolated, cultured and expanded *in vitro* due to their adherent characteristics, and not associated with any ethical debate (14). Thus, MSCs may be used in the treatment of a diverse variety of clinical conditions (15) such as engraftment of various organs (16). The long-term cultivation of MSCs may fail for many reasons: genotypic drift, senescence, transformation, phenotypic instability, and contamination or incubator failure. The inability to cultivate MSCs will result in the lack of MSCs for both experimental and clinical use (17). Therefore, it is necessary to cryopreserve MSCs as cell seeds. Although increasing telomerase expression of cells may overcome cell senescence (18), cryopreservation of hMSCs may be more practical in order to save time and culture materials (16, 19). Resuscitated MSCs can be subcultivated for many passages without a noticeable loss of viability and capability of osteogenic differentiation (20-22).

Formulating a cryopreservation protocol for hMSCs is required because these cells cannot survive for long periods under *in vitro* culture conditions. Slow rate cooling methods using dimethylsulfoxide (DMSO) as a cryoprotectant have been used for a wide variety of MSC lines established from bone marrow (23, 24), umbilical cord blood (23-25), hematopoietic progenitor cells (26) and mouse ES cell lines (27). In most protocols, cells are suspended in

* Corresponding Author

freezing medium containing DMSO at 5-20%, transferred into glass or plastic cryovials and then frozen by cooling at 1.0 to 2.0 °C/min (28). Slow freezing reduces ice crystal formation and eliminates toxic and osmotic damage to cells through exposure to low concentrations of cryoprotectants while slowly decreasing temperatures (29). However, it is difficult to completely eliminate injury by intracellular ice formation. Damage by ice crystal formation in the cytoplasm during the freezing process is one of the possible causes of cell death; such conventional methods, are not applicable to hMSC cells because many of these cells die immediately after thawing (28). Alternately, vitrification, a rapid cooling method using a high concentration of cryoprotectant, could also be used. Vitrification can completely eliminate damage caused by ice crystal formation in the cytoplasm of cells during freezing (29, 30). It is also advantageous because the procedure takes a relatively short time and a programmable temperature decreasing container is not required (31).

Vitrification has been used for the cryopreservation of oocytes, fertilized eggs and embryos of several mammalian species including humans in order to prevent ice crystal formation (32). There have been some reports demonstrating that embryonic stem (ES) cells could be successfully cryopreserved by vitrification in recent years (27, 33,34). Moon et al. tested vitrification of the human amnion-derived mesenchymal stem cells (HAMs) by using a two-step exposure to equilibration and vitrification solutions (21). They used an EG-based cryoprotectant and their findings were in line with previous reports that showed the superiority of EG. EG has been proven to be less toxic on fibroblast and other somatic cells in comparison with permeating agents such as DMSO and propylene glycol (PROH) that have been used on murine and human embryos (35). However, as a long-term preservation method for HAMs, a well-defined protocol of cryopreservation needs to be established for a human bone marrow derived mesenchymal stem cell bank. In the present study, to confirm the proliferative capability and pluridifferentiation of cryopreserved adult hMSCs; we chose ethylene ficoll sucrose (EFS) 40 that contained 40% v/v EG for the vitrification solution. hMSCs that were cryopreserved for two months were resuscitated and cultivated for 15 passages. An analysis of their expansion, morphological and pluridifferentiation characteristics was undertaken. Finally, under induction conditions, adipogenic and osteogenic potentials have been discussed.

2. Materials and methods

2.1 Preparation and culture of MSCs

For isolation of rat MSCs; female Sprague-Dawley rats (weighing 200-250 g) with the approval from the Institute for Animal Care were obtained from the Animal Center, Faculty of Medicine, Guilan University of Medical Sciences. Rats were killed by intraperitoneal administration of a lethal dose of sodium pentobarbital. The femurs and tibias were carefully dissected away from attached soft tissue as previously reported with modification (1). The ends of the bones were cut, and the bone marrow was aseptically extruded with 5 ml PBS solution by using a syringe with a 21G needle and flushing the shaft ten times. The marrow tissue was dissociated by pipetting. The cell suspension was then centrifuged at 500 × g for 5 minutes and the supernatant was discarded. Bone marrow mesenchymal stem cells (BMSCs) were then mechanically dispersed into a single-cell suspension so that the density of BMSCs reached 10^6 cells/ml. At this point, marrow cells were plated in a 25 cm^2 plastic flask in Dulbecco's modified eagle medium (DMEM) containing 20% fetal bovine serum

(FBS), 100 U/ml penicillin, and 100 mg/ml streptomycin. All cells were incubated at 37 °C, in an atmosphere of 5 % humidified CO2. After 48 hours incubation, the nonadherent cell populations were removed and the medium was added and replaced every three or four days for about two weeks. When the cells grew to 80% confluency they were harvested with 0.25% trypsin and 1 mM EDTA (Gibco, UK) for 5 minutes at 37°C, replated and diluted 1:3 on a 25 cm^2 plastic flask, again cultured to the next confluency and harvested. Prior to their use in inducing differentiation and vitrification MSCs that were passaged approximatly 15 times were morphologically evaluated.

2.2 Cryopreservation of MSCs

MSCs at passage 4 of pre-cryopreservation were harvested and centrifuged at 400 × g for 15 minutes as mentioned above. Approximately 1 $\times 10^6$ cells/ml of randomly selected batches were cryopreserved by using the vitrification method or OPS vitrification.

2.3 Vitrification and thawing procedure

MSCs were cryopreserved by using a two-step exposure to the equilibration and vitrification solutions (34). The equilibration solution was 20% ethylene glycol (EG; Sigma) and the vitrification solution was composed of 40% EG, 18% Ficoll - 70 (Sigma) and 0.3 M sucrose (Sigma). All solutions were based on PBS (Sigma) containing 20% FBS. A pellet of ~1 $\times 10^6$ MSCs (~10 µl) was first suspended in 50 µl equilibration solution for 5 minutes and then mixed with 500 µl vitrification solution for 40 seconds. Suspended MSCs were immediately transferred to 1.2 ml cryovials (Nunc) and plunged directly into liquid nitrogen. The OPS vitrification method was carried out according to Reubinoff et al. (33). For OPC, a pellet of ~1 $\times 10^6$ MSCs (~10 µl) was first suspended in 50 µl equilibration solution for 5 minutes and then mixed with 500 µl vitrification solution for 40 seconds. Suspended MSCs were at once transferred to 0.25 ml plastic straws (IMV, L'Aigle, France). Immediately afterwards, the straws were immersed in liquid nitrogen for two months. Following storage, the cells were thawed by rapidly immersing the vials and straws in a water bath at 37 °C. After warming for about 7 seconds, (at approx. 1800 °C/min) the contents of the vials and straws were suspended serially in 0.5, 0.25 and 0 M sucrose in PBS containing 20% FBS. After thawing, the survival rate was evaluated by the trypan blue staining method. After removing some of the cell pellet and adding 0.4% trypan blue (Sigma), the cells were plated onto a slide and unstained cells were counted as live cells (26). The remaining cells were centrifuged at 200 × g for 10 minutes and washed three times with DMEM medium supplemented with 20% FBS, 100 U/ml penicillin and 100 mg/ml streptomycin. Cells were immediately plated at a density of 1 $\times 10^6$ cells/ml in a 25 cm^2 culture flask and subcultured over seven days in the above described condition.

2.4 Evaluation of the differentiation potential of cryopreserved MSCs

2.4.1 Adipogenic induction

Pre and post-cryopreserved MSCs were seeded on coverslips in a six-well plate and cultured in DMEM with 10% FBS. Cells with nearly 80% confluency were exposed to DMEM supplemented with 5µg/ml insulin, 1 µM dexamethasone, 100 nM indomethacine, 0.5 mM methylisobutylxanthine (Sigma), and 10% FBS for 48 hours. Cells were then incubated in the same medium without dexamethasone. For control, cells were cultured in regular medium as above. The medium was changed every third or fourth day. One week after induction, adipogenic differentiation was evaluated by the cellular accumulation of neutral lipid vacuoles that were stained with oil-red O (Sigma) and observed under an inverted microscope (17). Briefly, after fixation in 5% metanol, induced MSCs were stained in filtered oil red O for 2-3 hours and then rinsed with 60% isopropyl alcohol.

2.4.2 Osteogenic induction

To identify osteogenic differentiation, thawed and non-cryopreserved MSCs were cultured in 100 nM dexamethasone, 10 mM ß-glycerol phosphate and 50 µM ascorbic acid-2-phosphate in 400 µl DMEM-LG supplemented with 10% FBS on coverslips in a six-well plate for subsequent staining. During the culture period, the medium was changed once per week. After 14 days, osteogenic differentiation was evaluated by staining the coverslips with fresh 0.5% alizarin red solution (1).

2.4.3 Colony-forming unit assays

For these assays, both thawed and non-cryopreserved cells were plated at 1×10^6 cells per ml and cultured for 14 days in 25 cm^2 tissue culture flasks. After 14 days, the cultures were stained with giemsa for 5 minutes. The formations of colonies were considered acceptable until passage 15 (P15) and those less than 2 mm in diameter or faintly stained were excluded.

2.4.4 Statistical analysis

Statistical analysis for comparison of the postthaw survival rate was performed using the χ^2 test. Statistically significant values were defined as $p<0.05$. All experiments were conducted in triplicate.

3. Results

The growth and morphology of MSCs appeared rather heterogeneous in primary culture as seen in Fig 1A. Under a phase contrast microscope, the cells appeared fibroblast-like, elongated and spindle shaped with a single nucleus (Fig 1B). These cells showed the ability to form colonies with the occasional cell sphere formation giving an impression of embryoid bodies (Fig 1C). However, they progressively showed homogenous fibroblast-like features following subsequent subculture (Fig 1D).

3.1 Morphology and growth of vitrified-thawed MSCs

The duration of storage in frozen state for MSCs was two months. Post-cryopreserved MSCs from both the vitrification method and OPS vitrification had similar cellular morphology and colony-formation. Resuscitated MSCs first grew as isolated colonies after initial plating. Subsequently these adherent cells grew as typically fibroblastic or spindle shaped.

As the cells approached confluence, they assumed a more spindle-shaped, fibroblastic morphology (Fig 2A, B). The thawed and non-cryopreserved MSCs were subcultured until P15. Until the P9, fibroblast-like morphology was consistently observed in both the thawed and non-cryopreserved MSCs. At the P10, cells in both cultures became large and flat, suggesting senescence.

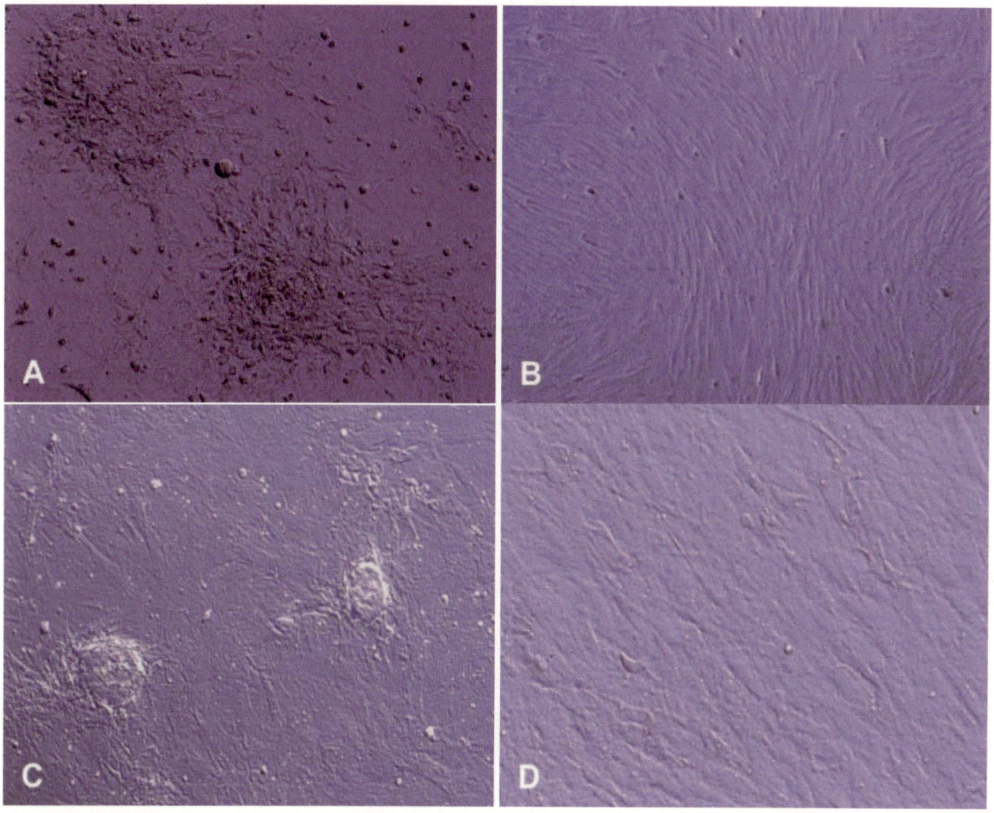

Fig. 1. Morphology and growth of MSCs. (A) Primary (x 40), (B) Passage 4 (14 days, x 100), (C) Passage 4 (35 days, x 100) and (D) Passage 4 (40 days, x 100). In primary culture, cell growth was scattered with some colony formation. Followin subsequent subculture, the cells changed into spindle-like fibroblasts.

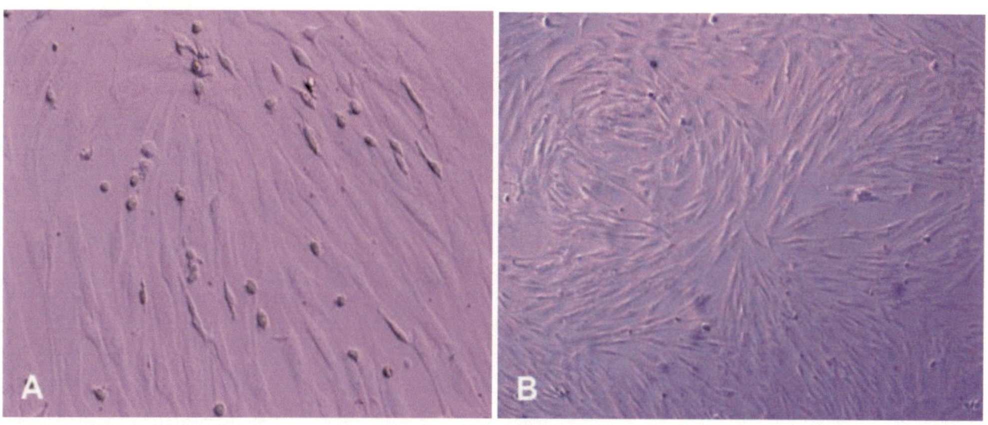

Fig. 2. Morphology and growth of vitrified-thawed MSCs. Phase contrast images of MSCs two months after thawing from: (A) vitrification method (x 100) and (B) OPS vitrification (x 40). MSCs had a similar morphology to fibroblasts and were indistinguishable from non-cryopreserved MSCs.

4. Viability of vitrified-thawed MSCs

Live/dead viability of MSCs was determined by the trypan blue staining test. The number of MSCs was counted and compared to that of the control. After thawing, the viability rates were 81.33% ± 6.83 for the vitrification method and 80.83% ± 6.4 for the OPS vitrification, while values with the pre-vitrification control group were 88.16% ± 6.3, respectively (Table 1). There were no differences in viability between them.

No.	Before vitrification (%)	Vitrification method (%)	OPS method (%)
1	85	81	82
2	93	92	90
3	78	77	70
4	96	81	79
5	89	72	81
6	88	85	83
Mean ± SD	88.16(SD ± 6.30)	81.33 (SD ± 6.83)	80.83(SD ± 6.40)

OPS: open-pulled straw, student t-test ($p<0.05$).

Table 1. Percentage of cell viability of non-cryopreserved and two different cryopreserved vitrification methods by the trypan blue staining test.

5. Colony forming unit assay

For these assays, cells of both thawed and noncryopreserved MSCs or cultures were plated at 1×10^6 cells/ml and MSCs in culture showed a colony formation consisting of 40-80 cells in 25 cm^2 flasks for 14 days (Fig 3). After P9, the cells showed no colony forming ability which illustrated that colony forming ability decreases with increasing passages.

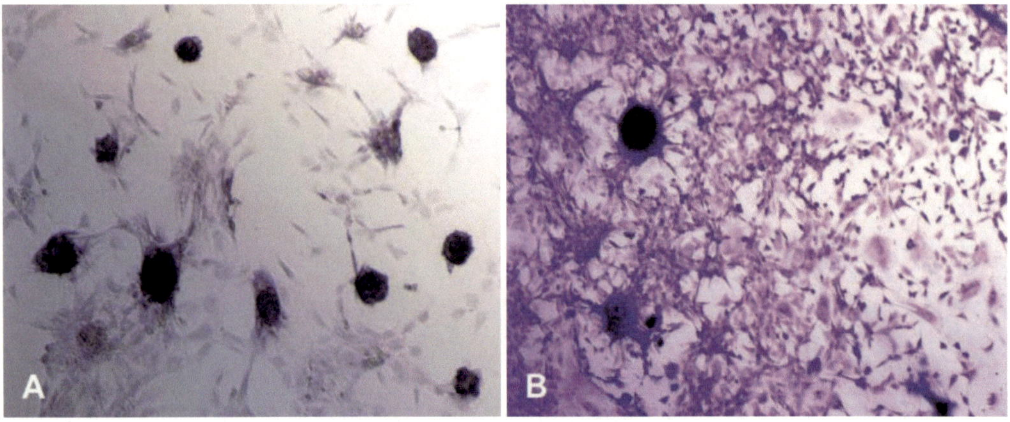

Fig 3. Phase contrast images of typical MSCs colony morphology before (A) and after (B) cryopreservation. Cell sphere formation in the MSCs culture produced colony formation units that stained with giemsa (A and B: x 40).

6. Differentiation of post-cryopreserved MSCs

After culturing for adipogenic differentiation, the accumulations of numerous neutral lipid vacuoles were detectable in the cytoplasm of vitrified-thawed cells (Fig 4A). Following three weeks of induction, oil Red O staining showed the lipid droplets with orange red color, which demonstrated the committed differentiation of MSCs into adipocytes (Fig 4B). The control cells showed no detectable lipid vacuoles. Under culture with osteogenic induction medium, resuscitated MSCs detached and floated in the medium. After three weeks, mineral accumulations were observed by alizarin red staining (Fig 5A, B). Similar results were observed in the group of non- cryopreserved MSCs when osteogenically induced under the same condition (Fig 5C, D).

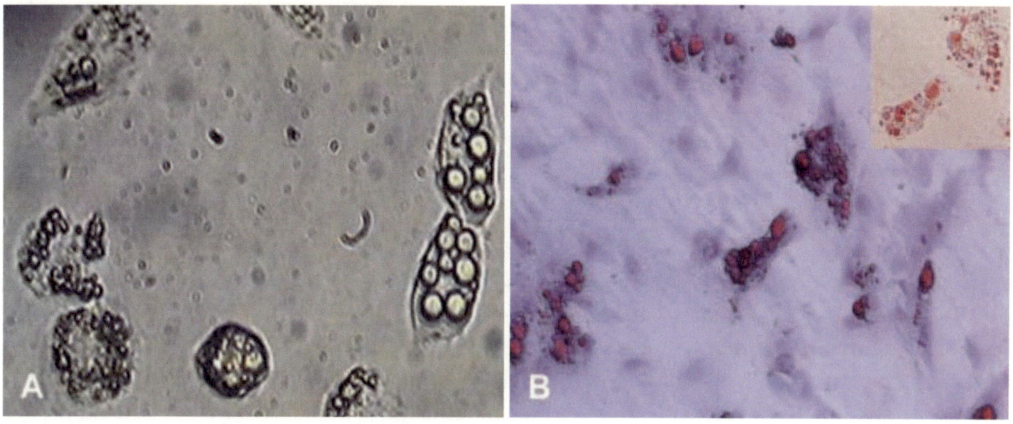

Fig 4. Evaluation of adipogenesis potential of MSCs under a phase-contrast microscope. Both the non cryopreserved MSCs and vitrified-thawed MSCs after treatment by adipogenic medium showed numerous neutral lipid vacuoles which accumulated in the cytoplasm. (A) Confirmed by oil red O staining (oil red O + hematoxiline, which one is the top right one: oil red without hematoxiline, x 100), (B).

7. Discussion

Cryopreservation is an important method to maintain cells for biological research and medical applications such as tissue engineering, gene therapy, cell transplantation, pharmacological testing and future therapeutic indications (17, 28). A study on the long-term storage of BM-derived MSCs is of critical importance (1). The objective of the current investigation was to test the possibility that vitrification could be a useful method for the cryopreservation of MSCs. Thus, in the present study, we isolated MSCs from bone marrow of adult female rats. In culture; MSCs are characterized by their capacity to adhere to a plastic culture surface and form a fibroblast-like shape (Fig 1). Our data corroborated previous findings from other groups which showed homogenous fusiform features with oval vesicular nuclei (36) and the colony forming ability of MSCs, which decreased with increasing passages (37). The achievement of pure fibroblastic clones from murine bone marrow was first reported by Wang and Wolf (38).

Furthermore, Eslaminejad et al. obtained an average of 15-17 clones, each one consisting of several fibroblastic cells per 24-well plate (39). This approach yielded both the number and

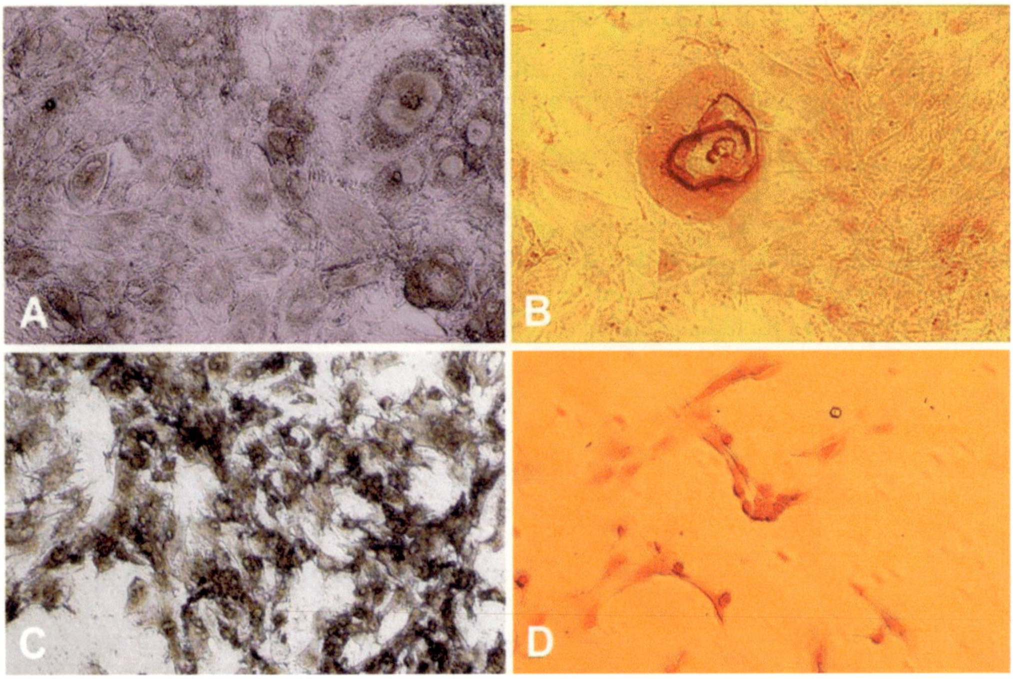

Fig 5. Differentiation potency of nonvitrified and vitrified-thawed MSCs observed under a phase-contrast microscope. (A) Nonvitrified MSCs after osteogenic induction. (B) Alizarin red staining, x 100. (C) Vitrified MSCs after osteogenic induction. (D): Alizarin red staining, x 100. After induction of differentiation for 2-3 weeks in respective induction media, the mineralized extracellular matrix of pre and post-vitrification MSCs stained positively with alizarin red (arrows).

cellular densities of colonies that were dependent upon the number of the cells plated per culture dish and was the method employed in this investigation. We also successfully vitrified MSCs and confirmed the morphology, viability rate and differentiation capacity of post-vitrification MSCs as compared with non-cryopreserved controls. Generally, most comprehensive studies on the cryopreservation of MSCs were carried out by using slow-rate cooling methods (17, 22-24). Slow-rate cooling methods using dimethylsulfoxide (DMSO) as a cryoprotectant are effective for a wide variety of cell lines, including ES cell lines (27, 28). This method by decreasing temperature slowly with a low concentration of cryoprotectant is used to balance damage caused by various factors including ice crystal formation, fracture, toxic and osmotic damage (29).

It is difficult to completely eliminate injuries from intracellular ice formation, which is the main source of fracture and damage to the cytoplasm (29). It is also a time-consuming procedure and requires an expensive programmable freezer (31). On the other hand, there are some reports that stem cells are highly sensitive to cryoinjury and the vitrification method is a better choice for HES cell cryopreservation than conventional slow freezing and rapid thawing (28, 33). This method has been previously applied for cryopreservation of oocytes, fertilized eggs and embryos of several mammalian species, including humans (32).

In vitrification method, the concentrations of cryoprotectants seem to be dangerously high at the final phases, it happens at low temperatures, where the real toxic effect is minimal. Moreover, the high cooling and warming rates applied at vitrification provide an unique benefit compared to the traditional freezing (40).

Umbilical cord blood-derived mesenchymal stem cells (UCB-derived MSCs) were vitrified by vitrification and by programmed freezing without dimethyl sulfoxid (DMSO) by Wang et al. (2011). Their results showed that the viability of thawed UCB-derived MSCs was enhanced from 71.2% to 95.4% in the presence polyvinyl alcohol (PVA) for vitrification, but only < 10% to 45% of viability was found for programmed freezing (41). While, Kim et al. reported that Post-thaw colony-formation of embryonic stem cells (ESCs) was detected only after a slow freezing using DMSO by stepwise placement of a freezing container into a -80°C deep freezer and subsequently into -196°C liquid nitrogen, while no proliferation was detected after vitrification (42). Also, hMSCs from pre- and post-cryopreservation by slow freezing had similar colony-formation and cellular morphology similar to our resuls (17). Low survival of human ESCs has been also reported when they are frozen slowly with DMSO (43). There have been several reports to demonstrate the superiority of vitrification to other freezing programs for human ESCs, because it is able to avoid cell injury resulted from ice crystal formation (42, 44).Carvalho et al. (2008) reported that viability of frozen BM-MSCs by slow freezing which was to added 5% DMSO was 94.76% and 90.58% viability before versus after cryopreservation (45). Also, the high survival rate (81.8%) is obtained after cryopreservation of human ESCs by programmed freezing (46). Our result showed 81.33% and 80.83% cell viability of two different cryopreserved vitrification methods versus 88.16% viability of non-cryopreserved cells. Nevertheless, we are not able to deny the feasibility of vitrification program for effective cryopreservation of SCs (42).

Also Fujioka et al. vitrified ESCs using EFS40; ethylene glycol, 18% ficoll 70,000 MW and 0.3 M sucrose (similar to this work) and slow-freezed in freezing medium containing 10% DMSO. Their results indicated that the vitrification methods yielded higher cell recovery and survival rates than did the slow-rate freezing methods (28).

DMSO has been well known as a toxic agent for stem and progenitor cells, and particularly for human embryonic stem cells. It is also known as a powerful differentiation agent that may interact with the chromatin structure (47).

Therefore, the selection of suitable cryoprotectants is essential. Cryopreservation procedure ought to become different according to cell type and cellular characteristics. So, all components consisting of freezing and thawing procedures should specifically be determined in each case of cryopreservation (42).

This study was DMSO-free vitrification of MSCs using cryovial and straw. Straws, preferably thin straws, even in sealed form, can be cooled safely with an increased cooling rate (40). There was any report about vitrification of MSCs without DMSO.

So that in the present experiment, we have observed the viability and proliferation capability as well as differentiation potential of cryopreserved MSCs *in vitro*. Inverted microscope findings showed that, after culturing for seven days, numerous MSCs adhered well to the surface culture dish (Fig 1). In a study by Moon et al. on human amnion-derived mesenchymal stem cells (HAMs), they observed that slow freezing resulted in a lower survival rate compared with vitrification, indicating a high efficiency of the vitrification procedure (21).

Moreover, Heng (48) demonstrated that 39.8 ± 0.9% of the hMSCs could be recovered after cryopreservation using a conventional slow freezing method which was lower than that our result (Table 1). Here, resuscitated MSCs kept a high proliferative potential. They first grew as clones after limiting dilution and then expanded rapidly with the typical features of spindle-shaped cell bodies and confluence after a lag phase of 6-14 days. There were no differences among the pre and post-cryopreservation of colony formation at the same passage (Fig 2). In addition, the results showed that the passage procedure was selective for MSCs and it could be inferred that the passaging of resuscitated MSCs increased cellular homogeneity.

Ji et al. demonstrated that cryopreservation of encapsulated HES cells offers better cellular viability, higher colony recovery, and less differentiation than the slow-freezing techniques most commonly used to preserve HES colonies. Therefore, this difference in recovery may be due to differences in cell lines, freezing and thawing protocols, or growth substrate (49).

On the other hand, previous studies have shown that cyropreservation had no effect on either the proliferation or osteogenic and adipogenic differentiation of human MSCs *in vitro* (5, 22). In agreement with these reports, Liu et al. using slow cooling with Me2SO as a cryoprotectant and rapid thawing demonstrated that thawed cryopreserved human MSCs had higher survival rates in comparison with non-cryopreserved MSCs and differentiated into osteoblasts when cultured in osteogenic media. Also they found that cryopreserved hMSCs could not differentiate into osteoblasts spontaneously when cultured in basic culture media (50). In addition to the characteristics described above, our present study demonstrated that post-cryopreserved MSCs from bone marrow were still pluripotential and differentiated into osteoblasts and adipocyte under appropriate culture conditions (Fig 3, 4). These observations suggest that the "memory" of proliferation and differentiation in MSCs is not affected by the process of vitrification. The ability of frozen BM-MSCs by slow freezing to differentiate into mesenchymal derivatives (such as osteogenic and adipogenic) reported by Lee et al. (4). In this study, we established a two-step vitrification protocol for MSCs using EFS containing 40% v/v EG for the vitrification solution, which is widely used for successful vitrification of mouse embryos (30), human blastocysts (32) and ESCs (34). Our findings are in line with the reports by Gajda et al. who used the same methods for somatic cells which have been proven to be less toxic on bovine skin fibroblast and cumulus cells (35).

EG is the most commonly used cryoprotectant for vitrification due to its low molecular weight and low toxicity (35). In addition, additives with high molecular weights, such as sucrose, can significantly reduce toxicity by decreasing the concentration of permeating agents required for the vitrification solution. We also used ficoll as a macromolecule to promote permeation by cryoprotectants, which seems to have the advantages of lower toxicity, higher solubility and lower viscosity (30). In a study by Moon et al. on HAMs, they observed that the combination of EG with either PROH or DMSO resulted in a very low survival rate of HAMs as compared with EFS alone (21). Also, Kuleshova and Lopata ascertained the advantages of vitrification when compared with earlier applied cryopreservation techniques (51). These advantages include the control of solute penetration and dehydration rates, prevention of prolonged temperature shock and damage from ice formation, and inexpensive equipment and running costs. Vitrification is a process where glass-like solidification of a solution occurs without the formation of ice crystals inside living cells, by exposure to a high concentration of cryoprotectant with a higher cooling rate (52). This procedure is a simple method to circumvent the obstacles of slow freezing without the need for a freezing container to modulate the reduction of temperature in a deep freezer before storing in liquid nitrogen (21).

8. Conclusion

In the present experiment, it was shown that vitrification can be an efficient storage method for MSCs without losing their activity and usual properties. Such a system will be exceedingly helpful for both experimental research and medical applications.

9. Acknowledgments

This research was a part of proposal supported by a grant (3/132/8162 /P) funded by the Guilan University of Medical Sciences and was done in the Cellular and Molecular Research Center at the Faculty of Medicine. The authors wish to thank both centers. There is no conflict of interest in this article.

10. References

[1] Eslaminejad MB, Nazarian H, Taghiyar L. Mesenchymal stem cells isolation from removed medium of rat's bone marrow primary culture and their differentiation into skeletal cell lineage. Yakhteh. 2008; 10(1): 62-72.

[2] Bianco P, Riminucci M, Gronthos S, Robey PG. Bone marrow stromal stem cells: nature, biology, and potential applications. Stem Cells. 2001; 19(3): 180-192.

[3] Wu QY, Li J, Feng ZT, Wang TH. Bone marrow stromal cells of transgenic mice can improve the cognitive ability of an Alzheimer's disease rat model. Neurosci Lett. 2007; 417(3): 281-285.

[4] Lee MW, Choi J, Yang MS, Moon YJ, Park JS, Kim HC, et al. Mesenchymal stem cells from cryopreserved human umbilical cord blood. Biochem Biophys Res Commun. 2004; 320(1): 273-278.

[5] Bruder SP, Jaiswal N, Haynesworth SE, Growth kinetics, self renewal, and the osteogenic potential of purified human mesenchymal stem cells during extensive subcultivation and following cryopreservation. J Cell Biochem. 1997; 64(2): 278-294.

[6] Kang XQ, Zang WJ, Song TS. Xu XL, Yu XJ, Li DL, et al. Rat Bone marrow mesenchymal stem cells differentiation into hepatocytes *in vitro*. World J Gastroenterol. 2005; 11 (22): 3479-3484.

[7] Jiang Y, Jahagirdar BN, Reinhardt RL, Schwartz RE, Keene CD, Ortiz- Gonzalez XR, et al. Pluripotency of mesenchymal stem cells derived from adult marrow. Nature. 2002; 418: 41-49.

[8] Baharvand H, Mathae KI. Culture condition difference for stablishment of new embryonic stem cell lines from the C57 BL/6 and BALB/c mouse strains. *In Vitro* Cell Dev Biol Anim. 2004; 40(3-4): 76-81.

[9] Toma C, Pittenger MF, Cahill KS, Byrne BJ, Kessler PD. Human mesenchymal stem cells differentiate to a cardiomyocyte phenotype in the adult murine heart. Circulation. 2002; 105: 93-98.
[10] Kotton DN, Ma BY, Cardoso WV, Sanderson EA, Summer RS, Williams MC, et al. Bone marrow-derived cells as progenitors of lung alveolar epithelium. Development. 2001; 128: 5181-5188.
[11] Tsujigiwa H, Nishizaki K, Teshima T, Takeda Y, Yoshinobu J, Takeuchi A, et al. The engraftment of transplanted bone marrow-derived cells into the olfactory epithelium. Brain Res. 2005; 1052(1): 10-15.
[12] Jeon SJ, Oshima K, Heller S, Edge AS. Bone marrow mesenchymal stem cells are progenitors *in vitro* for inner ear hair cells. Mol Cell Neurosci. 2007; 34(1): 59-68.
[13] Suzuki H, Taguchi T, Tanaka H, Kataoka H, Li Z, Muramatsu K, et al. Neurospheres induced from bone marrow stromal cells are multipotent for differentiation into neuron, astrocyte, and oligodendrocyte phenotypes. Biochem Biophys Res Commun. 2004; 322(3): 918-922.
[14] Alviano F, Fossati V, Marchionni C, Arpinati M, Bonsi L, Franchina M, et al. Term Amniotic membrane is a high throughout source for multipotent mesenchymal stem cells with the ability to differentiate into endothelial cells *in vitro*. BMC Dev Biol. 2007; 7:11.
[15] Korbling M, Estrov Z. Adult stem cells for tissue repair. A new therapeutic concept? N Engl J Med. 2003; 349(6): 570-582.
[16] Liechty KW, MacKenzie TC, Shaaban AF, Radu A, Moseley AM, Deans R, et al. Human mesenchymal stem cells engraft and demonstrate site-specific differentiation after in utero transplantation in sheep. Nat Med. 2000; 6(11): 1282-1286.
[17] Xiang Y, Zheng Q, Jia BB, Huang GP, Xu YL, Wang JF, et al. Ex vivo expansion, adipogenesis and neurogenesis of cryopreserved human bone marrow mesenchymal stem cells. Cell Biol Int. 2007; 31(5): 444-450.
[18] Poh M, Boyer M, Solan A, Dahl SL, Pedrotty D, Banik SS, et al. Blood vessels engineered from human cells. Lancet. 2005; 365(9477): 2122-2124.
[19] Guy JA, Wouters R, Successful vitrification of umbilical cord tissue to harvest stem cell for therapies in the future. J Cryobiol/abstract. 2008; 57: 329.
[20] Perry BC, Zhou D, Wu X, Yang FC, Byers MA, Chu TM, et al. Collection, cryopreservation, and characterization of human dental pulp derived mesenchymal stem cells for banking and clinical use. Tissue Eng Part C Methods. 2008; 14(2): 149-156.
[21] Moon JH, Lee JR, Jee BC, Suh CS, Kim SH, Lim HJ, et al. Successful vitrification of human amnionderived mesenchymal stem cells. Hum Reprod 2008; 23(8): 1760-1770.
[22] Kotobuki N, Hirose M, Machida H, Katou Y, Muraki K, Takakura Y, et al. Viability and osteogenic potential of cryopreserved human bone marrow derived mesenchymal cells. Tissue Eng. 2005; 11(5-6): 663-673.
[23] Haack- Sorensen M, Bindslev L, Mortensen S, Friis T, Kastrup J. The influence of freezing and storage on the characteristics and functions of human mesenchymal stromal cells isolated for clinical use. Cytotherapy 2007; 9(4): 328-337.

[24] Lee RH, Kim B, Choi I, Kim H, Choi HS, Suh K, et al. Characterization and expression analysis of mesenchymal stem cells from human bone marrow and adipose tissue. Cell Physiol Biochem. 2004; 14: 311- 324.

[25] Romanov YA, Svintsitskaya VA, Smirnov VN. Searching for alternative sources of postnatal human mesenchymal stem cells: candidate MSC-like cells from umbilical cord. Stem Cells. 2003; 21(1): 105-110.

[26] Zhao J, Hao HN, Thomas RL, Lyman WD. An efficient method for the cryopreservation of fetal human liver hematopoeitic progenitor cells. Stem Cells. 2001; 19(3): 212-218.

[27] Yang PF, Hua TC, Wu J, Chang ZH, Tsung HC, Cao YL. Cryopreservation of human embryonic stem cells: a protocol by programmed cooling. Cryo Letters. 2006; 27(6): 361-368.

[28] Fujioka T, Yasuchika K, Nakamura Y, Nakasuji N, Suemori H. A simple and efficient cryopreservation method for primate embryonic stem cells. Int J Dev Biol. 2004; 48(10): 1149-1154.

[29] Vajta G, Nagy Z. Are programmable freezers still needed in the embryo laboratory? Review on vitrification. Reprod Biomed Online. 2006; 12(6): 779-796.

[30] Kasai M, Mukaida T. Cryopreservation of animal and human embryos by vitrification. Reprod Biomed Online. 2004; 9(2): 164-170.

[31] Karlsson JO. Cryopreservation: freezing and vitrification. Science. 2002; 296(5568): 655-656.

[32] Cho HJ, Son WY, Yoon SH, Lee SW, Lim JH. An improved protocol for dilution of cryoprotectants from vitrified human blastocysts. Hum Reprod. 2002;17(9): 2419-2422.

[33] Reubinoff BE, Pera MF, Vajta G, Trounson AO. Effective cryopreservation of human embryonic stem cells by the open pulled straw vitrification method. Hum Reprod. 2001; 16(10): 2187-2194.

[34] Nie Y, Bergendahl V, Hei DJ, Jones JM, Palecek SP. Scalable culture and cryopreservation of human embryonic stem cells on microcarriers. Biotechnol Prog. 2009; 25(1): 20-31.

[35] Gajda B, Katska- Ksiazkiewicz L, Rynska B, Bochenek M, Smorag Z. Survival of bovine fibroblasts and cumulus cells after vitrification. Cryo letters. 2007; 28(4): 271-279.

[36] Stock P, Staege MS, Muller LP, Sgodda M, Volker A, Volkmer I, et al. Hepatocyte Derived from adult stem cells. Transplant Proc. 2008; 40(2): 620-623.

[37] Polisetty N, Fatima A, Madhira SL, Sangwan VS, Vemuganti GK. Mesenchymal cells from limbal stroma of human eye. Mol Vis. 2008; 14: 431-442.

[38] Wang QR, Wolf NS. Dissecting the hematopoietic microenvironment. Clonal isolation and identification of cell types in murine CFU-F colonies by limiting dilution. Exp Hematol. 1990; 18(4): 355-359.

[39] Eslaminejad MB, Nikmahzar A, Taghiyar L, Nadri S, Massumi M. Murine mesenchymal stem cells isolated by low density primary culture system. Dev Growth differ. 2006; 48(6): 361-370.

[40] Vajta G, Kuwayama M. Improving cryopreservation systems. Theriogenology. 2006; 65: 236–244.

[41] Wang HY, Lun ZR, Lu SS. Cryopreservation of umbilical cord blood-derived mesenchymal stem cells without dimethyl sulfoxide. Cryo Letters. 2011; 32(1): 81-88.

[42] Kim GA, Lee ST, Lee EJ, Choi JK, Lim JM. Simplified slow freezing program established for effective banking of embryonic stem cells. Asian-Aust J Anim Sci. 2009; 22(3): 343-349.

[43] Katkov MS, Kim R, Bajpai YS, Altman M, Mercola JF, Loring AV, et al. Cryopreservation by slow cooling with DMSO diminished production of Oct-4 pluripotency marker in human embryonic stem cells. Cryobiol. 2006; 53:194-205.

[44] Suemori H, Yasuchika K, Hasegawa K, Fujioka T, Tsuneyoshi N, Nakatsuji N. Efficient establishment of human embryonic stem cell lines and long-term maintenance with stable karyotype by enzymatic bulk passage. Biochem Biophys Res Commun. 2006; 345:926-932.

[45] Carvalho KAT, Cury CC, Oliveira L, Cattaned RII, Malvezzi M, Francisco JC, et al. Evaluation of Bone Marrow Mesenchymal Stem Cell Standard Cryopreservation Procedure Efficiency. Transplant Proc. 2008; 40: 839–841.

[46] Yang PF, Hua TC, Tsung HC, Cheng QK, Cao YL. Effective cryopreservation of human embryonic stem cells by programmed freezing. Conf Proc IEEE Eng Med Biol Soc. 2005; 1: 482-485.

[47] Katkov II, Kan NG, Cimadamore F, Nelson B, Snyder EY, Terskikh AV. DMSO-Free Programmed Cryopreservation of Fully Dissociated and Adherent Human Induced Pluripotent Stem Cells. Stem Cells Int. 2011;2011:981606.

[48] Heng BC. Effect of Rho-associated kinase (ROCK) inhibitor Y-27632 on the post-thaw viability of cryopreserved human bone marrow- derived mesenchymal stem cells. Tissue Cell. 2009;41(5):376-380.

[49] Ji L, de Pablo JJ, Palecek SP. Cryopreservation of adherent human embryonic stem cells. Biotechnol Bioeng. 2004; 88(3): 299-312.

[50] Liu G, Shu C, Cui L, Liu W, Cao Y. Tissue-engineered bone formation with cryopreserved human bone marrow mesenchymal stem cells. Cryobiology. 2008; 56(3): 209-215.

[51] Kuleshova L, Lopata A. Vitrification can be more favorable them slow cooling. Fertil Steril. 2002; 78: 449-454.

[52] Chian RC, Kuwayama M, Tan L, Tan J, Kato O, Nagai T. High survival rate of bovine oocytes matured *in vitro* following vitrification. J Reprod Dev. 2004; 50: 685-696.

2

Cryopreservation of Adherent Smooth Muscle and Endothelial Cells with Disaccharides

Lia H. Campbell[1] and Kelvin G.M. Brockbank[1,2,3]
[1]*Cell & Tissue Systems, Inc., North Charleston, SC*
[2]*Georgia Tech / Emory Center for the Engineering of Living Tissues, The Parker H. Petit Institute for Bioengineering and Bioscience, Georgia Institute of Technology, Atlanta, GA*
[3]*Medical University of South Carolina, Department of Regenerative Medicine and Cell Biology, Charleston, SC*
USA

1. Introduction

There is a need for mammalian cell cryopreservation methods that either avoid or improve upon outcomes employing dimethyl sulfoxide (DMSO) as a cryoprotectant. DMSO was the second effective cryoprotectant to be discovered (Lovelock, 1959). Cell cryopreservation usually involves slow rate freezing with DMSO in culture medium and storage below -135°C for later use. Typically as long as there are enough cells surviving to start an expanding proliferating culture the yield of viable cells after thawing is not an important consideration. However, there are instances where cell yield and viability can be very important. Examples include minimization of expensive delays when starting cultures for bioreactor protein manufacturing runs and cellular therapies that involve administering cells into patients for treatment of various diseases, such as cancer. While some cells, for example fibroblasts, are easily cryopreserved other cell types like keratinocytes, hepatocytes, and cardiac myocytes do not freeze well and cell yields are often <50%. Furthermore, current opinion is that DMSO should be removed before cells are infused into patients (Caselli et al., 2009; Junior et al., 2008; Mueller et al., 2007; Otrock et al., 2008; Schlegel et al., 2009). The mechanism for DMSO cytotoxicity has not been determined, however, it is thought to modify membrane fluidity, induce cell differentiation, cause cytoplasmic microtubule changes and metal complexes (Barnett 1978; Katsuda et al., 1984, 1987; Miranda et al., 1978). DMSO also decreases expression of collagen mRNAs in a dose-dependent manner (Zeng et al., 2010).

One strategy for finding interesting new cryoprotectants and cryopreservation strategies is by evaluating what happens in nature (Brockbank et al., 2011). No examples of organisms synthesizing DMSO to survive freezing conditions have been found to date, however several creatures have been found that employ glycerol (Brockbank et al., 2011) the first effective cryoprotectant to be discovered (Polge, 1949). Nature has developed a wide variety of organisms and animals that tolerate low temperatures and dehydration stress by accumulation of large amounts of disaccharides, particularly trehalose, including plant seeds, bacteria, insects, yeast, brine shrimp, fungi and their spores, cysts of certain crustaceans, and some soil-dwelling animals. While the cryoprotective capabilities of sucrose and trehalose has been known for years, conventional cryopreservation protocols have generally not employed them even though early work with them demonstrated their ability to protect proteins and membrane vesicles during freezing (Rudolf & Crowe, 1985; Crowe et al., 1990). Trehalose has both major advantages and disadvantages for potential preservation of mammalian cells. On the negative side mammalian cells do not have an active trehalose transport system for uptake of trehalose from the extracellular environment, while on the plus side if you can get it in mammalian cells it is not metabolized giving the opportunity for trehalose to be accumulated to potentially effective preservation concentrations. The purpose of the studies presented here were; 1) to assess or review alternative strategies for delivery of trehalose into mammalian cells, and; 2) to determine whether the benefits were specific to trehalose by investigating alternative sugars employing the same loading strategies.

2. Materials and methods

2.1 Cell culture

Cells used in these studies are described in Table 1. All were grown and maintained at 37°C in 5% carbon dioxide.

Description	Acronym	Culture conditions
Rat aortic myofibroblast cells	A10 (ATCC# CRL-1476)	DMEM* (4.5 g/L) 10% FCS 1.0 mM sodium pyruvate
Bovine calf pulmonary artery endothelial cells	CPAE (ATCC# CCL-209)	EMEM** 10% FCS 1 mM sodium pyruvate 2 mM glutamine 1X non-essential amino acids
Rat aortic smooth muscle cells	A7R5 (ATCC# CRL-1444)	DMEM (4.5 g/L) 10% FCS 1.0 mM sodium pyruvate
Bovine corneal endothelial cells	BCE (ATCC# CRL-2048)	DMEM (4.5 g/L) 10% FCS 1 mM sodium pyruvate 4 mM glutamine

*Dulbecco's Modified Eagle's Medium
**Eagle's Minimum Essential Medium

Table 1. Cell types

2.2 Cell poration with H5

The pore-forming protein H5 was obtained from the lab of Hagan Bayley (Bayley, 1994). It is derived from the bacterial toxin α-hemolysin, which forms constitutively opened pores in cell membranes. The modified bacterial toxin has been engineered to form pores in the membrane that can be opened and closed by the addition of Zn^{+}. More specific details are presented in the discussion. Cells were plated at 10,000-20,000 cells/well the night before in 96 well microtiter plates. The next day, the cells were washed with DMEM containing 1 mM EDTA for 2 minutes and then again with DMEM to remove the EDTA. 0.2M trehalose was added and incubated for 20 minutes at 37°C followed by the appropriate concentration of H5 for the respective cell type. Cells were porated and loaded with trehalose for 1 hour at 37°C before addition of DMEM with 25 μM $ZnSO_4$ or 10% serum to close the pores. Trehalose in DMEM was then added to the wells followed by cryopreservation using a controlled rate freezer (Planar) at ~-1.0°C/min from 4°C to -80°C with a programmed nucleation step at -5.0°C. Cryopreserved cells were stored overnight at <-135°C. The next day, the cells were placed at -20°C for 30 minutes followed by rapid thawing at 37°C (Campbell et al., 2003; Taylor et al., 2001). The cell cultures were washed twice and then placed at 37°C for 1 hour to recover under normothermic cell culture conditions before assessment of cell viability.

2.3 Pretreatment (Incubation) with trehalose

Cells were plated at 10,000-20,000 cells/well and placed in culture. The next day, the culture medium was replaced with EMEM or DMEM containing trehalose (0-0.6M) and cultured at 37°C for varying periods of time. After culture, the solution was replaced with fresh medium containing trehalose (0-0.6M) and the cells were cryopreserved using a controlled rate freezer as described for H5 above.

2.4 Cell poration with ATP

Cells were plated at 10,000-20,000 cells/well and placed in culture. The next day, the cells were washed with poration buffer (phosphate-buffered saline [PBS] with 1X essential amino acids, 1X Vitastock, 5.5 mM glucose) designed to optimize binding of ATP^{4-} to the receptor and facilitate formation of the pore. The cells were then placed in 50 μl poration buffer, pH of 7.45, with 0.2M trehalose. A stock solution of 100 mM ATP^{4-}, pH of 7.45, was made fresh and added to each well to achieve a final concentration of 5 mM. After addition of the ATP^{4-}, the cells were left at 37°C for 1 hour to allow sugar uptake. Following incubation, 200 μl of DMEM plus 1 mM $MgCl_2$ was then added to the cells at 37°C to close the pores. After 1 hour of recovery from the loading procedure cryopreservation was initiated.

2.5 Assessment of cell viability

Cell viability was determined using the non-invasive metabolic indicator alamarBlue™ (Trek Diagnostics). A volume of 20 μl was added to cells in 200 μl of DMEM (10%FCS) and the plate was incubated at 37°C for 3 hours. Plates were read using a fluorescent microplate reader (Molecular Dynamics) at an excitation wavelength of 544 nm and an emission wavelength of 590 nm. Viability was measured before and after sugar loading, immediately after thawing and at several later time points post-thaw.

2.6 Statistical methods

All experiments were repeated at least four times with four replicates in each experiment. Statistical differences were assessed by two way analysis of variance. P-values <0.05 were regarded as significant.

3. Results

3.1 H5 poration

One of the first strategies for utilizing disaccharide sugars as cryoprotectants involved the use of a modified pore forming complex. In our initial studies we evaluated the H5 mutant α-hemolysin (Bayley, 1994) using two adherent cell lines, A10 and CPAE. The earlier studies had been done with cells in suspension (Eroglu et al., 2000). We also evaluated sucrose, another disaccharide sugar that is commonly found in nature, for its potential usefulness as a cryoprotectant. Using the protocol of Eroglu et al as a starting point, a protocol was established for adherent cells. Several parameters were evaluated and included the H5 concentration, time of poration, concentration of trehalose loaded, and time for loading trehalose. Conditions that worked best with adherent cells included 20 minutes for poration followed by 60 minutes for trehalose loading. The highest concentration of trehalose that caused the least drop in cell viability was 0.2M. The optimum H5 concentration varied according to cell type. The A10 smooth muscle cells were porated with 12.5 µg/mL of H5 while the endothelial CPAE cells were porated with 50 µg/mL. In contrast, the fibroblasts and keratinocytes in the literature were porated with 25 µg/mL (Eroglu et al., 2000). Other changes to the protocol were made that benefited viability for adherent cells specifically and included addition of trehalose prior to the addition of H5, the base solution used for poration, and the amount of EDTA (1 mM versus 10 mM) for the removal of Zn^+ prior to poration. After cryopreservation, however, poor viability was obtained with both cell types. A10 cells demonstrated a viability of 5.57±0.17%. The endothelial cells demonstrated similar viabilities. These values were not as good as those observed when suspended cells were cryopreserved with sugars in the literature. However, it is our experience that adherent cells are generally more difficult to cryopreserve regardless of the cryoprotectant used.

3.2 Trehalose exposure without poration

When we started adding the trehalose to cells in the H5 experiments, control cells were exposed to trehalose by addition to the culture medium prior to cryopreservation. An unanticipated observation of cell survival was made with slow rate cryopreserved CPAE cells prompting further investigation. The cells exposed to trehalose overnight were observed to develop vacuoles (Fig 1) suggesting a possible pinocytotic uptake mechanism.

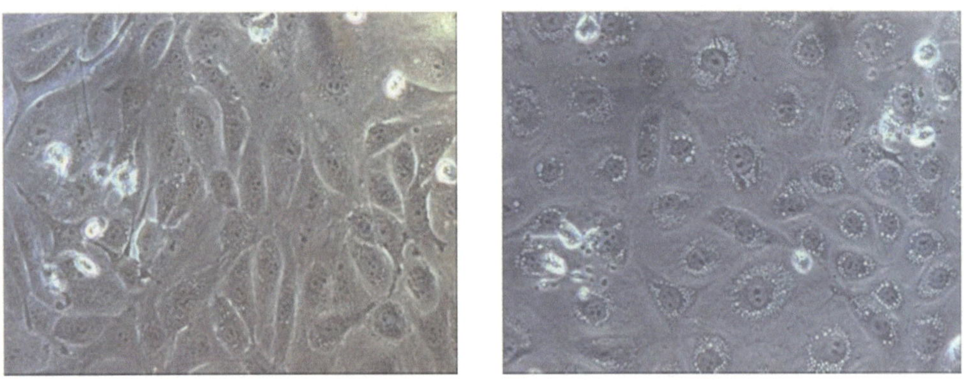

Fig. 1. CPAE cells after exposure to trehalose. Left: no sugar, Right: 0.1M trehalose. 40X magnification

After these observations were made, further experiments were designed to examine cell viability after extended trehalose exposure. CPAE cells were exposed to 0.2M trehalose in Dulbecco's Modified Eagle's Medium (DMEM) buffered with 25 mM Hepes for 0-72 hours at 37°C. After exposure the cells were left in 0.2M trehalose and cryopreserved at ~-1.0°C/min (Fig. 2). CPAE cell viability was observed immediately after thawing. An exposure time of 24 hours provided the best overall cell survival. Extracellular exposure alone during cryopreservation failed to produce any cell survival. In contrast, A10 smooth muscle cells generally did not survive cryopreservation after trehalose exposure as well as the CPAE endothelial cells. Examination of optimal concentrations of trehalose during incubation and during cryopreservation showed that a concentration of 0.1-0.2M trehalose for incubation produced the best viability with a similar concentration being required during the freezing process.

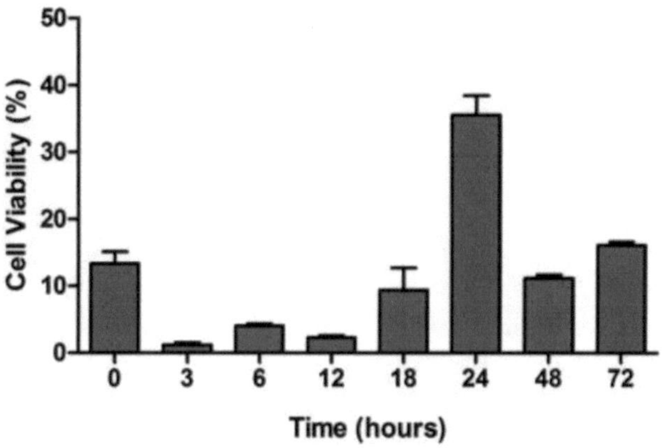

Fig. 2. Impact of cell culture time with trehalose on cryopreservation. CPAE cells were cultured with 0.2M trehalose for up to 72 hours followed by cryopreservation with 0.2M trehalose. Percent viability was calculated based on the pre-cryopreservation controls. ($p<0.05$)

Several other parameters were also examined to further improve cell viability. Other studies have shown that not only the concentration and choice of cryoprotectant but also the vehicle solution for the cryoprotectant can have a significant impact on cell viability after cryopreservation (Campbell & Brockbank, 2007; Mathew et al., 2004; Sosef et al., 2005). Initial experiments were performed using DMEM, however, it was observed that CPAE cells, which are grown in EMEM medium, actually preferred exposure to trehalose in EMEM medium. Further experiments examined the buffers used to maintain the pH of the system. Four cell lines were evaluated. CPAE cells demonstrated decreased viability when the zwitteronic buffer, 4-(2-hydroxyethyl)-1-piperazineethanesulfonic acid (HEPES) was used while the other 3 cell lines did not show decreased viability. Rather a combination of HEPES and sodium bicarbonate was preferred by the CPAE cells. This unusual choice of buffer prompted examination of solution pH during incubation, a pH of 7.4 was optimal for all the 4 cell lines tested. Once loaded with sugar, the cells could either be left in the extracellular sugar at another concentration or an alternative cryoprotectant for preservation.

These studies were then extended to include other sugars, sucrose, raffinose, and stachyose (Fig. 3). The potential cryoprotective benefits of these sugars were evaluated and it was found that stachyose was as good as trehalose using an identical protocol, sucrose was not quite as good and raffinose had very little benefit. All cell lines showed evidence of some cell survival days after cryopreservation and thawing. The second smooth muscle cell line, A7R5, demonstrated low levels of viability with stachyose. Both endothelial cell lines, CPAE and BCE, showed good viability after exposure and freezing with sucrose. Overall, the CPAE cell line had the best viability in these experiments. Use of an optimized protocol with trehalose produced excellent post cryopreservation results with 10-14mM intracellular trehalose (Campbell, 2011). Conditions included 24 hours of cell culture with 0.2M trehalose followed by cryopreservation with 0.2-0.4M trehalose in sodium bicarbonate buffered EMEM at pH 7.4 resulting in ~75% post-preservation cell viability (Campbell et al., 2011).These experiments confirmed that this technique is more effective for endothelial cells than smooth muscle cells and demonstrated that stachyose is effective for cryopreservation.

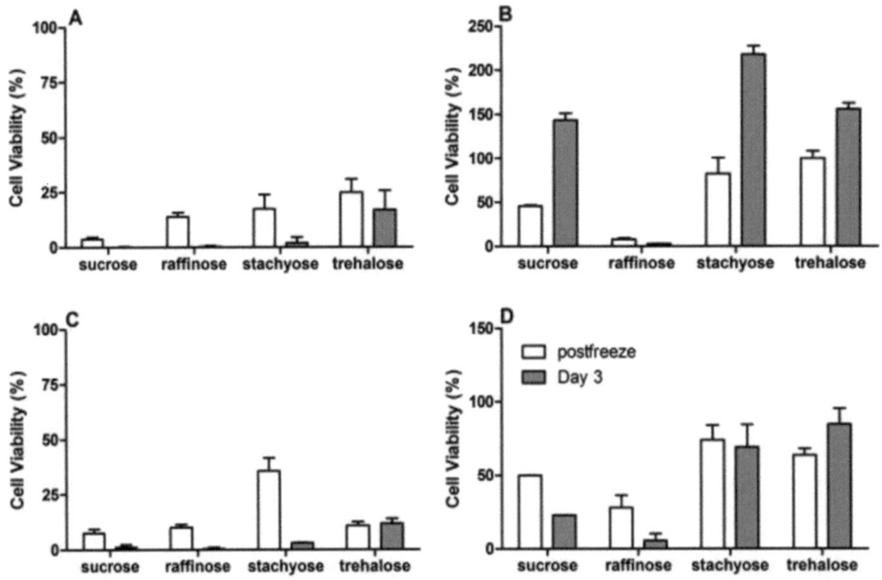

Fig. 3. Cell viability for A10 (A), CPAE (B), A7R5 (C) and BCE (D) cells after exposure and freezing in the presence of sugars.

3.3 ATP poration

In addition to the H5 mutant α-hemolysin poration strategy, we sought other poration techniques that could be used to permeate mammalian cells with disaccharides. Cells expressing the $P2_{X7}$ purinergic cell surface receptor, also known as the $P2_z$ receptor, may be permeabilized by the formation of a channel/pore that allows passage of molecules into and out of the cell when the active form of ATP (ATP^{4-}) binds to the receptor. Our initial studies focused on determination of whether or not the $P2_{X7}$ was expressed on smooth muscle and endothelial cells. Experiments using the ELICA assay demonstrated the presence of the $P2_{X7}$ receptor on both endothelial and smooth muscle cell lines to varying degrees (Fig 4). The smooth muscle cell lines demonstrated the greatest density of the receptor.

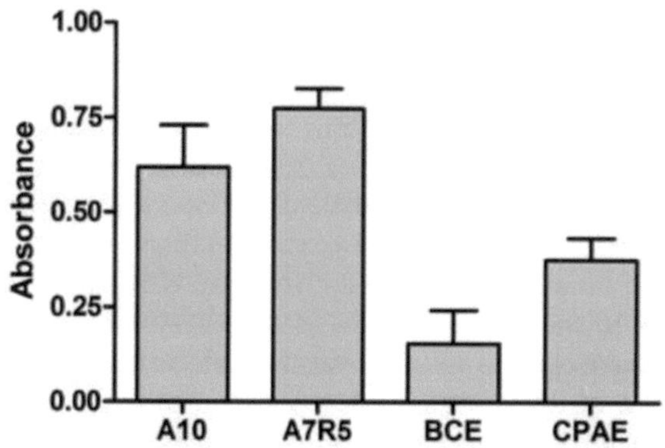

Fig. 4. Detection of the $P2_{X7}$ receptor by ELICA. Cells were probed for the presence of the $P2_{X7}$ receptor using antiserum specific for the receptor at a dilution of 1:25. The graph represents the average absorbance (±SEM) for 10 replicates.

ATP-permeabilized cells retained better viability than untreated cells both immediately after thawing and five days later (Fig 5). Immediate metabolic activity in A7R5 and CPAE cells demonstrated dependence upon increasing ATP concentrations, while for A10 and BCE cells immediate metabolic activity was increased at all ATP concentrations with only slight improvement at the higher concentrations tested. However, survival at five days demonstrated that intermediate concentrations of ATP (0.5-2.5mM) were best. Further cryopreservation studies were performed to optimize cell survival resulting in at least 25% cell survival for both endothelial cell lines but only low levels of survival for the smooth muscle cells.

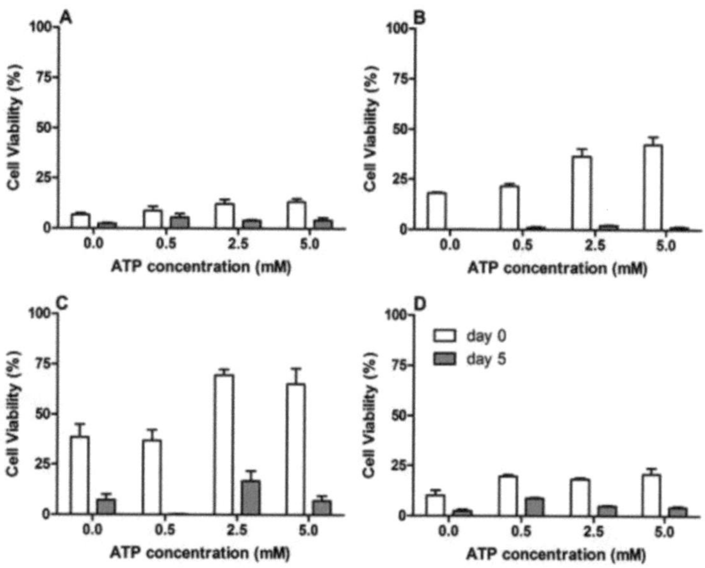

Fig. 5. Cell viability after poration and freezing with increasing concentrations of ATP. Cells were loaded with 0.2M trehalose using the $P2_{X7}$ receptor and the indicated concentrations of ATP. After poration and cryopreservation, cell viability was evaluated by alamarBlue. (A) A10, (B) A7R5, (C) CPAE, (D) BCE.

4. Discussion

As cryopreservation has been applied to cells and tissues for clinical use, concerns about toxicity relating to the various cryoprotectants being used, particularly DMSO, have developed. Because of this, there has been renewed interest in finding less toxic cryoprotectants. The cryoprotective capabilities of some sugars, disaccharide sugars in particular, has been known for years and early work with them demonstrated their ability to protect proteins and membrane vesicles during freezing (Crowe et al., 1990; Rudolph & Crowe, 1985). Coupled with these early studies are observations made in nature regarding organisms that can survive extremes in temperature and desiccation due to their ability to accumulate large amounts of disaccharide sugars, specifically trehalose and sucrose, until more favorable conditions are available. The protective effects of trehalose and sucrose have been determined and may be classified under two general mechanisms: (1) "the water replacement hypothesis" or stabilization of biological membranes and proteins by direct interaction of sugars with polar residues through hydrogen bonding, and (2) stable glass formation (vitrification) by sugars in the dry state (Crowe et al., 1987, 1988, 1998, 2001; Slade & Levine, 1991).

Two primary stresses that destabilize membranes have been defined, fusion and lipid phase transition. Studies have shown that when the water that hydrates the phospholipid molecules of the membrane is removed, packing of the head groups increases. The result is an increase in van der Waals interactions and a dramatic increase in the phase transition temperature (T_m) (Crowe et al., 1987, 1988, 1990, 1991). At the phase transition the phospholipid bilayer shifts from a gel phase to a liquid crystalline phase, the state normally observed in fully hydrated cells. For example, the T_m of a cell membrane might be -10°C when fully hydrated but when water is removed the T_m increases to over 100°C. Thus, the membrane is in the gel phase at room temperature. As the membrane shifts between the gel phase and the liquid crystalline phase it becomes transiently leaky allowing its intracellular contents to leak out. Therefore it would be advantageous to avoid the lipid phase transition as this can compromise the health of a rehydrated cell. Addition of disaccharide sugars, in particular trehalose, depresses T_m allowing the membrane to remain in the liquid crystalline state even when dried, so that upon rehydration no phase transition takes place and no transient leaking. During cryopreservation water is not necessarily lost, but it undergoes a phase change forming ice as the temperature drops and depending upon the rate of cooling, the cells become more or less dehydrated rendering the cells vulnerable to damage by mechanisms similar to those proposed for desiccated cells.

The stabilizing effect of these sugars has been shown in a number of model systems including liposomes, membranes, viral particles, and proteins. The mechanism by which disaccharide sugars are able to decrease the T_m for a given bilayer has been elucidated. Interactions take place between the sugars and the -OH groups of the phosphate in the phospholipid membrane preventing interaction or fusion of the head groups as the structural water is removed (Crowe et al., 1986, 1988, 1989a, 1989b). Although not as well understood, a similar mechanism of action stabilizes proteins during drying (Carpenter et al., 1986, 1987a, 1987b, 1989). Despite their protective qualities, the use of these sugars in mammalian cells has been somewhat limited mainly because mammalian cell membranes are impermeable to disaccharides or larger sugars and there is strong evidence that sugars need to be present on both sides of the cell membrane in order to be effective (Crowe et al.,

2001; Eroglu et al., 2000; Beattie et al., 1997). This is why, in addition to loading sugars, we added sugars to the cryopreservation solution just before initiating cooling.

In addition to trehalose and sucrose, we were interested in other sugars that could be used as cryoprotectants avoiding monosacharides that would likely be degraded in the cell. Larger more complex sugars such as disaccharides or larger would be less likely to be degraded and utilized inside cells and might therefore be more stable as cryoprotectants. The comparative structures of the sugars we considered for preservation of mammalian cells are illustrated in Figure 6. Three other sugars were evaluated besides trehalose and included sucrose, raffinose and stachyose. Sucrose and trehalose are both non-reducing sugars, so they do not react with amino acids or proteins and should be relatively stable under low pH conditions and at temperature extremes. Raffinose is a trisaccharide and stachyose is a tetrasaccharide.

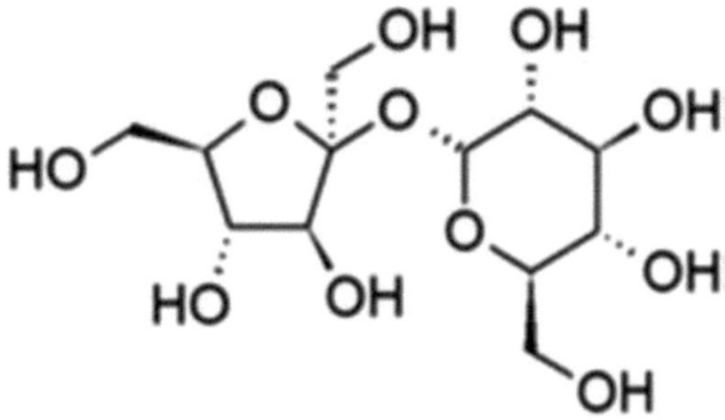

Sucrose (mw: 342.30)

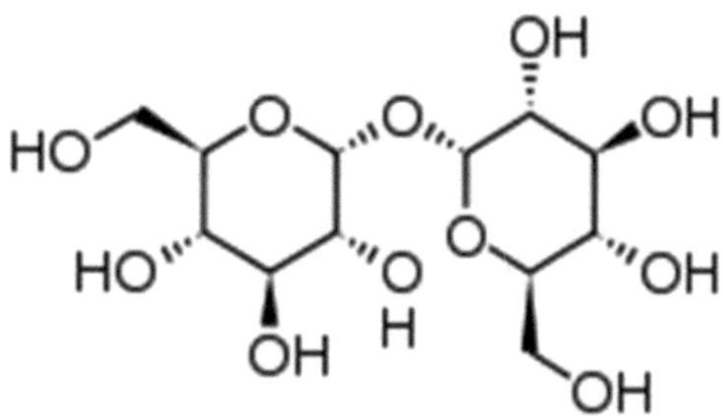

Trehalose (mw: 342.30)

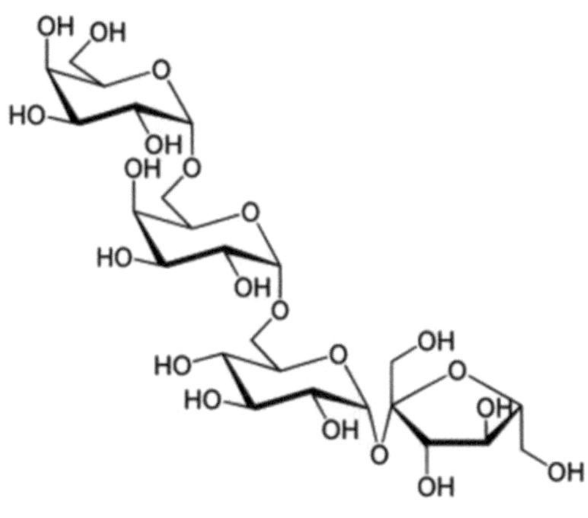

Stachyose (mw: 684.59)

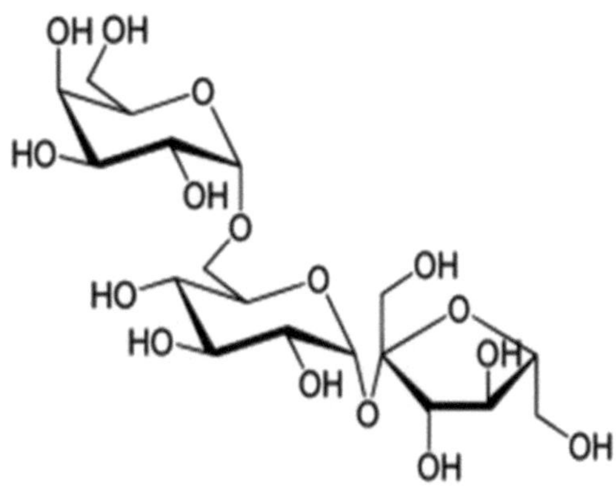

Raffinose (mw: 504.44)

Fig. 6. Sugar structure

Before going further, it is important to point out that the cells we have employed were cryopreserved and thawed while adherent in 96-well plates using cooling and warming conditions defined and reported at the turn of the century (Campbell et al., 2003; Taylor et al., 2001). We have since used these conditions to cryopreserve several adherent cell types (Campbell et al., 2007, 2010, 2011). Our rationale for using this adherent model was two-fold. First, due to our interest in regenerative medicine we thought that adherent cells more closely mimicked cells in tissue engineered devices. Second, we thought there might be a market for cells cryopreserved on plates for research and cytotoxicity testing, CryoPlate™. More recently another group has been using adherent cells for investigation of preservation by vitrification and drying and have reported on cryopreservation of adherent pluripotent stem cells (Katkov et al., 2006; Katkov et al., 2011;). Katkov et al. presented results for preservation of human embryonic stem cells in 4-well plates and pointed out several advantages of cryopreservation in adherent mode. These included elimination of possible bias due to selective pressure within a pluripotent stem cell line after cryopreservation and distribution of multiwell plates for immediate use for embryotoxicity and drug screening in pluripotent stem cell-based toxicology in vitro kits (Katkov et al., 2011).

There are several methods in the literature that could be employed for intracellular delivery of these sugars including those already discussed (Campbell et al., 2011; Table 2). Many drugs, therapeutic proteins and small molecules have unfavorable pharmacokinetic properties and do not readily cross cell membranes or other natural physiological barriers within the body. This has resulted in the search for and discovery of alternative methods to transport materials, like sugars, across mammalian cell membranes.

Some of these strategies have been presented in depth in the results sections. The first involved the use of the *Staphylococcus aureus* toxin, α-hemolysin. This toxin is produced as a monomer by the bacteria. It then oligomerizes to form pores on mammalian cell membranes. Hagan Bayley and his group modified the wild type α-hemolysin protein by replacing 4 native residues with histidines, termed H5. In addition to pore formation on cell membranes, the H5 mutant also enabled it to be opened and closed at will. When inserted into the membrane, it is open and molecules up to 3000 daltons are able to pass through. Then the pores are closed in the presence of Zn^+. To reopen the pore, addition of a chelating agent such as EDTA will remove the Zn^+ and the pore is ready to be used again (Bayley, 1994; Walker et al., 1995). Early studies showed that H5 could create pores in mammalian cell membranes and that they could be used for efficient intracellular loading of trehalose (Eroglu et al, 2000; Acker et al., 2003). Our experiments with H5 worked well initially using adherent cells. The results demonstrated good poration and loading of trehalose into cells. However, after adherent cells were cryopreserved, their viability was not very good (<6%). At this point in our studies, several issues arose that prevented further studies using H5. First, the H5 pore was derived from the bacterial toxin α-hemolysin so there were concerns raised whether regulatory approval could be obtained if it was ever to be used clinically with human cells and tissues. There were some indications during these studies that the pores were shed from the membrane over time. However, H5 was still detectable in picogram quantities after 7 days in culture. Finally, as new batches of H5 were delivered the activity varied greatly and more H5 was required to achieve the same level of poration compared with earlier batches. Ultimately the batch variation was attributed to a protein stability issue. When these issues were not resolved other strategies for introducing trehalose into cells were explored.

An unexpected outcome of our H5 experiments was the development of a new, simple strategy for introduction of trehalose into cells which involved incubating cells in sugar for extended periods of time at physiological temperature (Brockbank, 2007). One possible mechanism to explain this observation was that the trehalose is substituting for water molecules in the cell membranes keeping the membrane stable and preventing it from going through a phase transition (Crowe et al., 1988, 1989). A second mechanism is most likely an active uptake mechanism involving endocytosis similar to that proposed for loading of trehalose by Oliver et al (Oliver et al., 2004). Her results suggested that human MSCs are capable of loading trehalose from the extracellular space by a clathrin-dependent fluid-phase endocytotic mechanism that is microtubule-dependent but actin-independent (Oliver et al., 2004). Further research is required to elucidate the mechanism by which culture in the presence of trehalose facilitates cell cryopreservation and determine the degree of cell viability retention under different storage conditions.

The last method presented was poration using the $P2_z$ receptor and ATP. This was a somewhat unique strategy in that it took advantage of the cell's own machinery. It was shown that cells expressing the $P2_{X7}$ purinergic cell surface receptor, also known as the $P2_z$ receptor, could be permeabilized when the receptor binds to ATP^{4-}. The interaction with ATP resulted in the formation of a non-selective pore that allows molecules up to ~900 Daltons to pass through (Nihei et al., 2000). The $P2_{X7}$ receptor selectively binds to only ATP^{4-} whose presence in solution is dependent on temperature, pH, and the concentration of divalent cations such as Mg^{2+}. Closure of the pore after activation by ATP is achieved by simply removing ATP from the system or adding exogenous Mg^{2+} that has a high affinity for the active form of ATP, ATP^{4-}. The $P2_z$ receptor is found on a number of cell types including cells of hematopoietic origin (Nihei et al, 2000). There were several factors that likely affected the cell viability and survival of cells after ATP poration. First, is the density of the receptor on the cells which directly affects the amount of trehalose that can be loaded into the cells and how long it takes. Another factor is that poration with ATP tends to promote the detachment of adherent cells from their substrate. Part of the protocol requires a recovery period of 1 hour at 37°C to allow cells that may have been perturbed by the poration process the chance to settle back onto their substrate. Finally, cell loss is at least in part due to apoptosis. There is evidence in the literature that poration with ATP induces apoptosis in some cell types (Murgia et al., 1992).

In marked contrast the human stem cell line, TF-1, demonstrated excellent post cryopreservation survival (Buchanan et al., 2004; 2005). We have exposed TF-1 cells to ATP with trehalose for 1 hour followed by a 10-fold dilution of the ATP and inactivation of the active form of ATP (ATP^{4-}) by the addition of 1 mM $MgCl_2$ followed by a 1-hour recovery period at 37°C (Brockbank et al., 2011). When the cells were compared to cells cryopreserved with 10% DMSO, the DMSO group demonstrated greater initial viability close to 100% that steadily declined over days in culture post thaw. However by day 4 of culture post-cryopreservation cells cryopreserved in disaccharides were similar to the viability of cells cryopreserved in DMSO. Similarly colony forming assays with TF-1 cells demonstrated similar outcomes compared with DMSO. Furthermore, the use of disaccharides, trehalose and sucrose, appeared to result in similar results at both slow (1°C/min) and rapid (100°C/min) cooling rates. Buchanan et al (Buchanan et al., 2010) have extended these studies obtaining excellent TF-1 cell line and cord blood-derived multipotential hematopoietic progenitor cell survival after freeze drying and storage at room temperature for 4 weeks! It is studies such as Buchanan's that keep us optimistic that disaccharide introduction/preservation strategies can be developed for preservation of other mammalian cell types. Further development work is required with the cell culture and $P2_{x7}$ methods with the promise of preservation by freezing and freeze-drying.

Existing Techniques	Description	Pitfalls	References
H5	Derived from α-hemolysin, which normally forms a constitutively opened pore in the membrane. Engineered to close in the presence of Zn^{+} or serum.	Derived from a bacterial toxin. Batch to batch variation and instability.	Acker et al. 2003 Bayley, 1994 Eroglu et al. 2000
ATP	The naturally occurring $p2_{x7}$ receptor forms a non-specific pore upon binding of ATP^{4-} able to allow molecules <900 daltons to pass through.	$P2_{x7}$ receptor found on some but not all cell types.	Buchanan et al. 2005
Culture methods	1) Prolonged incubation of cells in the presence of disaccharide sugars at 37°C. 2) Fluid phase endocytosis: disaccharide sugars are taken up by cells via a clathrin dependent endocytotic mechanism.	Works better with some cells but not others.	Brockbank et al. 2007 Oliver et al. 2004
Temperature manipulation	A shift in temperature can cause a lipid phase transition which temporarily changes the membrane permeability and allows molecules to pass through.	Has been demonstrated with pancreatic islets and kidney cells. Requires optimization by cell type.	Beattie et al. 1997 Mondal 2009

Table 2. Strategies for Loading Disaccharide Sugars

There are still other methods in the literature that could lead to intracellular delivery of disaccharides in addition to those already discussed (Campbell et al., 2011; Table 2). One method takes advantage of the lipid phase transition described above when the cell membrane is exposed to changes in temperature. As the membrane changes from the liquid crystalline phase to the gel phase it becomes leaky providing an opportunity to introduce molecules into the cell that would not normally cross like trehalose. Beattie used this method to cryopreserve pancreatic islets by introducing DMSO and trehalose into the islets during the thermotropic phase transition between 5 and 9°C. The islets were then cryopreserved in combination with DMSO and the viability of the islets after thawing was greater than when DMSO alone was used, 94% versus 58% (Beattie et al., 1997). In a related study, Mondal et al, cryopreserved kidney cells (MDBK) using 264 mM trehalose. The cells were suspended in trehalose with 20% fetal bovine serum in culture medium then incubated at 40°C for 1 hour before slow rate cooling for storage at -80°C. Viability was measured using Trypan Blue exclusion at 74% upon thawing (Mondal, 2009).

In another variation for loading molecules into cells, a number of proteins have been discovered that possess the ability to cross the cell membrane. These protein transduction domains (PTDs) generally correspond to portions of native proteins. Examples of PTDs include the Tat protein from the human immunodeficiency virus type I, the envelope glycoprotein E^{rns} from the pestivirus and the DNA binding domains of leucine zipper proteins such as c-fos, c-jun and yeast transcription factor GCN4 (Futaki et al., 2001, 2004; Langedijk, 2002; Langedijk et al., 2004; Lindgren et al., 2000; Richard et al., 2003; Vives et al., 1997). These PTDs are short cationic peptides that cross the cell membrane in a concentration-dependent manner that is independent of specific receptors or other transporters. The exact mechanism of translocation has not been defined. Enrichment of basic amino acids, particularly arginine and in some instances lysine, have been shown to be important for the translocation activity (Futaki et al., 2001, 2004; Vives et al., 1997). Some studies have suggested that endocytosis is involved (Lundberg & Johansson, 2002; Richard et al., 2003), however, the current theory includes interaction with glycosaminoglycans and uptake by a non-endocytotic mechanism that may involve the charged heads of the phospholipid groups within the cell membrane. (Langedijk, 2002; Langedijk et al., 2004; Mai et al., 2002).

While most of these peptides need to be cross linked to the molecule of interest, there are peptides that can move proteins and other peptides across the membrane without the requirement for cross-linking. Examples include Pep-1, a 21-residue peptide which contains three domains; a tryptophan rich region (5 residues) for targeting the membrane and forming hydrophobic interactions; a lysine rich domain to improve intracellular delivery whose design was taken from other nuclear localization sequences from other proteins like the simian virus 40 large T antigen, and, a spacer region with proline that provides flexibility and maintenance of the other two regions. When mixed with other peptides or proteins, Pep-1 rapidly associates and forms a complex with the protein of interest by noncovalent hydrophobic interactions to form a stable complex. Once in the cytoplasm the peptide dissociates from the protein that has been carried across the membrane causing little if any interference regarding the protein's final destination or function. The process occurs by an endocytosis independent mechanism (Morris et al., 1999, 2001). We anticipate that such peptides may eventually lead to methods for introduction of disaccharides into mammalian cells (Campbell et al., 2011).

Another alternative method is electroporation, also called electropermeabilization, which involves the application of an electric pulse that briefly permeabilizes the cell membrane. Since its introduction in the 1980's it has been primarily used to transfect mammalian cells and bacteria with genetic material. Initially electroporation tended to kill most cells. However, further work and development of the electroporation process, such as alternate electrical pulses like the square wave pulse, have refined the process so that better permeabilization and cell viability can be achieved (Gehl, 2003; Hapala, 1997; Heiser, 2000). The formation of pores, their size and the recovery of the membrane are important factors that influence the success of an electroporation protocol (Gehl, 2003; Hapala, 1997; Heiser, 2000). Most importantly, electroporation is applicable to all cell types.

It was hypothesized that trehalose provided protection during electropermeabilization in a manner similar to chelating agents such as EDTA or lipids like cholesterol (Katkov, 2002;

Mussauer et al., 2001). Effective electroporation protocols are a balance between how much material can be loaded into the cells and cell survival after membrane permeabilization. So, while it cannot be predicted how well certain cell types will respond to electroporation, there is ample evidence that electroporation can be used with a reasonably certainty of success. A short culture period may be all that is required to permit restabilization of membranes post-electroporation. Additionally, like trehalose which interacts with membranes under stressful conditions such as drying, other compounds, such as cholesterol and unsaturated fatty acids, can also interact with membranes and may facilitate resealing of the membranes increasing overall cell survival (Katkov, 2002). Efficient resealing of cell membranes after permeabilization is thought to be essential for promoting cell recovery (Gehl et al., 1999) and compounds such as Poloxamer 188 facilitate membrane resealing (Lee et al., 1992).

5. Conclusion

In conclusion, there are multiple potential ways to introduce trehalose into mammalian cells and in some cases excellent cell preservation can be achieved. However, it is clear that methods for each cell type will need to be diligently developed and many years of work remain before we can replace DMSO as the lead cryoprotectant. In the mean time, we must not forget that there are other relatively low molecular weight sugars available. Preliminary evidence suggests that with further work sucrose and stachyose may, in some cases, be equally effective for cell preservation.

6. Acknowledgements

We would like to thank Elizabeth Greene for her assistance in the preparation of this manuscript. This work was supported by a cooperative agreement (No. 70NANB1H3008) between the U.S. Department of Commerce, National Institute of Standards and Technology— Advanced Technology Program, and Organ Recovery Systems, Inc

7. References

Acker, J.P., Lu, X.M., Young, V., Cheley, S., Bayley, H., Fowler, A. & Toner M. (2003). Measurement of trehalose loading of mammalian cells porated with a metal-actuated switchable pore. *Biotechnology and Bioengineering*, Vol.82, No.5, (June 2003), pp. 525-32, PMID: 12652476.

Barnett, R. (1978). The effects of dimethyl sulfoxide and glycerol on Na+. K+-ATPase and membrane structure. *Cryobiology*, Vol.15, (April 1978), p. 227, PMID: 149651..

Bayley, H. (1994). Triggers and Switches in a Self-Assembling Pore-Forming Protein. *Journal of Cellular Biochemistry*, Vol.56, (October 1994), pp. 177-82, PMID: 7829577.

Beattie, G.M., Crowe, J.H., Lopez A.D., Cirulle, V., Ricordi, C. & Hayek, A. (1997). Trehalose: A cryoprotectant that enhances recovery and preserves function of human pancreatic islets after long-term storage, *Diabetes*, Vol.46, (March 1997), pp. 519-23, PMID: 9032112.

Brockbank, K.G.M., Campbell, L.H., Ratcliff, Kelly M. & Sarver, K.A. (2007, 2011). Method for treatment of cellular materials with sugars prior to preservation. U.S. Patents #7,270,946, #8,017,311.

Brockbank, K.G.M., Campbell, L.H., Greene, E.D., Brockbank, M.C.G. & Duman, J.G. (2011). Lessons Learned from Nature for Preservation of Mammalian Cells, Tissues and Organs, *In Vitro Cellular & Developmental Biology–Animal*, Vol.47, No.3, (March 2011), pp. 210-217, PMID: 21191664.

Buchanan, S.S., Gross, S.A., Acker, J.P., Toner, M., Carpenter, J.F., & Pyatt, D.W. (2004). Cryopreservation of Stem Cells Using Trehalose: Evaluation of the Method Using a Human Hematopoietic Cell Line, *Stem Cells and Development*, Vol.13, (June 2004), pp. 295–305, PMID: 15186725.

Buchanan, S.S.; Menze, M.A.; Hand, S.C.; Pyatt, D.W.& Carpenter, J.F. (2005). Cryopreservation of Human Hematopoietic Stem and Progenitor Cells Loaded with Trehalose: Transient Permeabilization via the ATP-Dependent P2Z Receptor Channel, *Cell Preservation Technology*, Vol. 3, No.4, (2005), pp. 212–22.

Buchanan, S.S., Pyatt, D.W. & Carpenter J.F. (2010). Preservation of Differentiation and Clonogenic Potential of Human Hematopoietic Stem and Progenitor Cells during Lyophilization and Ambient Storage, *PLoS ONE*, Vol.5, No.9, (September 2010), pp. e12518, PMID: 20824143.

Campbell, L.H., Taylor, M.J. & Brockbank, K.G.M. (2003) Two stage method for thawing cryopreserved cells, U.S. Patent #6,596,531.

Campbell, L.H., & Brockbank, K.G.M. (2007). Serum free solutions for the cryopreservation of cells, *In Vitro Cellular & Developmental Biology–Animal*, Vol.43, No.8-9, (September 2007), pp. 269-275, PMID: 17879124.

Campbell, L.H., Brockbank, K.G.M. (2011). Comparison of Electroporation and Chariot™ for Delivery of β-galactosidase into Mammalian Cells: Strategies to Use Trehalose in Cell Preservation, *In Vitro Cellular & Developmental Biology–Animal*, Vol.47, No.3, (March 2011), pp.195-199, PMID: 21184200.

Campbell, L.H., Sarver, K.A., Hylton, K.R., Leman, B.B., & Brockbank, K.G.M. (2011). Culturing with trehalose produces viable endothelial cells after cryopreservation, *Cryobiology*, Submitted, 2011.

Carpenter, J.F., Hand, S.C., Crowe, L.M., & Crowe, J.H. (1986). Cryoprotection of Phosphofructokinase with Organic Solutes: Characterization of Enhanced Protection in the Presence of Divalent Cations, *Archives of Biochemistry and Biophysics*, Vol.250, No.2, (November 1986), pp. 505-1, PMID: 2946263.

Carpenter, J.F., Crowe, L.M .& Crowe JH. (1987a). Stabilization of Phosphofructokinase with Sugars During Freeze-Drying: Characterization of Enhanced Protection in the Presence of Divalent Cations, *Biochimica et Biophysica Acta*, Vol.923, No.1, (January 1987), pp. 109-15, PMID: 2948571.

Carpenter, J.F., Martin, B., Crowe, L.M. & Crowe, J.H. (1987b). Stabilization of Phosphofructokinase During Air-Drying with Sugars and Sugar/Transition Metal Mixtures, *Cryobiology*, Vol.24, No.5, (October 1987), pp. 455-64, PMID: 2958239.

Carpenter, J.F. & Crowe, J.H. (1989). An Infrared Spectroscopic Study of the Interactions of Carbohydrates with Dried Proteins, *Biochemistry,* Vol.28, No.9, (May 1989), pp. 3916-22, PMID: 2526652.

Caselli, D., Tintori, V., Messeri, A., Frenos, S., Bambi, F. & Arico, M. (2009). Respiratory depression and somnolence in children receiving dimethyl sulfoxide and morphine during hematopoietic stem cell transplantation. *Haematologica,* Vol.94. No. 1, (January 2009), pp. 152–153, PMID: 19001279.

Crowe, L.M., Womersley, C., Crowe, J.H., Reid, D., Appel, L. & Rudolph, A. (1986). Prevention of Fusion and Leakage in Freeze-Dried Liposomes by Carbohydrates". *Biochimica et Biophysica Acta,* Vol.861, (1986), pp. 131-40.

Crowe, J.H., Crowe, L.M., Carpenter, J.F. & Aurell Wistrom, C. (1987). Stabilization of Dry Phospholipid Bilayers and Proteins by Sugars, *The Biochemical Journal,* Vol.242, No.1, (February 1987), pp. 1-10, PMID: 2954537.

Crowe, J.H., Crowe, L.M., Carpenter, J.F., Rudolph, A.S. & Aurell Wistrom, C., Spargo, B.J. & Anchordoquy, T.J. (1988). Interactions of Sugars with Membranes, *Biochimica et Biophysica Acta,* Vol.947, No.2, (June 1988), pp. 367-84, PMID: 3285894.

Crowe, J.H., Crowe, L.M. & Hoekstra, F.A. (1989a). Phase Transitions and Permeability Changes in Dry Membranes During Rehydration, *Journal of Bioenergetics & Biomembranes,* Vol.21, No.1, (February 1989), pp. 77-91.

Crowe, J.H., McKersie, B.B. & Crowe, L.M. (1989b). Effects of Free Fatty Acids and Transition. Temperature on the Stability of Dry Liposomes, *Biochimica et Biophysica Acta,* Vol.979, No.1, (February 1989), pp. 7-10, PMID: 2917168.

Crowe, J.H., Carpenter, J.F., Crowe, L.M. & Anchordoguy, T.J. (1990). Are Freezing and Dehydration Similar Stress Vectors? A Comparison of Modes of Interaction of Stabilizing Solutes with Biomolecules, *Cryobiology,* Vol.27, No.3, (June 1990), pp. 219-31.

Crowe, J.H. & Crowe, L.M. (1991). Preservation of Liposomes by Freeze Drying, In: *Liposome Technology,* G. Gregoriadis, CRC Press, ISBN 0849340764, Boca Raton, FL.

Crowe, J.H., Carpenter, J.F. & Crowe, L.M. "(1998). The role of Vitrification in Anhydrobiosis, *Annual Review of Physiology,* Vol.60, (1998), pp. 73-103, PMID: 9558455.

Crowe, J.H., Crowe, L.M., Oliver, A.E., Tsvetkova, N., Wolkers, W. & Tablin, F. (2001). The Trehalose Myth Revisited: Introduction to a Symposium on Stabilization of Cells in the Dry State, *Cryobiology,* Vol.43, No.2, (September 2001), pp. 89-105, PMID: 11846464.

Eroglu, A., Russo, M.J., Bieganski, R., Fowler, A., Cheley, S., Bayley, H. & Toner, M. (2000). Intracellular trehalose improves the survival of cryopreserved mammalian cells, *Nature Biotechnology,* Vol.18, No.2, (February 2001), pp. 163-67, PMID: 10657121.

Futaki, S., Suzuki, T., Ohashi, W., Yagami, T., Tanaka, S., Ueda. K., Sugiura, Y. (2001). Arginine-rich Peptides: An Abundant Source of Membrane-Permeable Peptides having Potential as Carriers for Intracellular Protein Deliver, *Journal of Biological Chemistry,* Vol. 276, No.8, (February 2000), pp. 5836-40, PMID: 11084031.

Futaki, S., Goto, S., Suzuki, T., Nakase, I. & Sugiura, Y. (2004). Structural Variety of Membrane Permeable Peptides, *Current Protein & Peptide Science*, Vol.4, No.2, (April 2004), pp. 87-96, PMID: 12678848.

Gehl, J., Sorensen, T.H. & Nielsen, K. (1999). In Vivo Electroporation of Skeletal Muscle: Threshold, Efficacy and Relation to Electric Field Distribution, *Biochimica et Biophysica Acta*, Vol.1428, No.2, (August 1999), pp. 233-40, PMID: 10434041

Gehl, L. (2003). Electroporation: theory and methods, perspectives for drug delivery, gene therapy and research, *Acta Physiologica Scandinavica*, Vol.177, No.4, (April 2003), pp. 437-47, PMID: 12648161

Hapala, I. (1997). Breaking the Barrier: Methods of reversible permeabilization of cellular membranes, *Critical Reviews in Biotechnology*, Vol.17, No.2, (1997), pp. 105-22, PMID: 9192473.

Heiser, W.C. (2000). Optimizing electroporation conditions for the transformation of mammalian cells, *Methods in Molecular Biology*, Vol.130, (2000), pp. 117-34, 2000. PMID: 10589426

Junior, A.M., Arrais, C.A., Saboya, R., Velasques, R.D., Junqueira, P.L. & Dulley, F.L. (2008). Neurotoxicity associated with dimethyl sulfoxide-preserved hematopoietic progenitor cell infusion. *Bone Marrow Transplantation*, Vol.41, No. 1, (January 2008), pp. 95–6, PMID: 17934528.

Katkov, I. (2002). Electroporation of cells in applications to cryobiology: summary of 20-year experience, *Problems of Cryobiology*, Vol.2, (2002), pp. 3-8.

Katkov, I., Isachenko, V., Isachenko, E., Kim, M., Lulat, A., Mackay, A. & Levine, F. (2006). Low- and high-temperature vitrification as a new approach to biostabilization of reproductive and progenitor cells, *International Journal of Refrigeration*, Vol.29, (February 2006), pp. 346-357.

Katkov, I., Kan, N., Cimadamore, F., Nelson, B., Snyder, E., Terskikh, A. (2011). DMSO-Free programmed cryopreservation of fully dissociated and adherent human induced pluripotent stem cells, *Stem Cells International*, Vol.2011, (June 2011), pp.1-8, PMID: 21716669.

Katsuda, S., Okada, Y., Nakanishi, I. & Tanaka, J. (1984). The influence of dimethyl sulfoxide on cell growth and ultrastructural features of cultured smooth muscle cells, *Journal Electron Microscopy*, Vol.33, No. 3, (July 1984), pp. 239-241, PMID: 6533232.

Katsuda, S., Okada, Y. & Nakanishi, I. (1987). Dimethyl sulfoxide induces microtubule formation in cultured arterial smooth muscle cells. *Cell Biology International Reports*, Vol.11, (February 1987), pp. 103-110, PMID: 3549003.

Langedijk, J.P.M. (2002), Translocation Activity of C-terminal Domain of Pestivirus E^{rns} and Ribotoxin L3 Loop, *Journal of Biological Chemistry*, Vol.277, No.7, (February 2002), pp. 5308-14, PMID: 11673454.

Langedijk, J.P.M., Olijhoek, T., Schut, D., Autar, R. & Meloen, R.H. (2004). New Transport Peptides Broaden the Horizon of Applications for Peptidic Pharmaceuticals, Molecular Diversity, Vol. 8, No.2, (2004), pp. 101-11, PMID: 15209161

Lee, R.C., River, L.P., Pan, F.S. & Ji, L. (1992). Surfactant-Induced Sealing of Electropermeabilized Skeletal Muscle Membranes In Vivo, *Proceedings of the*

National Academy of Sciences USA, Vol.89, No.10, (May 1992), pp. 4524-28, PMID: 1584787

Lindgren, M., Hällbrink, M., Prochiantz, A., Langel, Ü. (2000). Cell-penetrating Peptides, *Trends in Pharmacological Sciences*, Vol.21, No.3, (March 2000), pp. 99-103, PMID: 10689363.

Lovelock, J. & Bishop, M., (1959). Prevention of freezing damage to living cells by dimethyl sulfoxide. *Nature*, Vol.183, (May 1959), p. 1394, ISSN: 0022-202X.

Lundberg, M. & Johansson, M. (2002). Positively Charged DNA-Binding Proteins Cause Apparent Cell Membrane Translocation, *Biochemical and Biophysical Research Communications*, Vol.291, No.1, (February 2002), pp. 367-71, PMID: 11846414.

Mai, J.C., Shen, H., Watkins, S.C., Cheng, T. & Robbins, P.D. (2002). Efficiency of Protein Transduction is Cell Type-dependent and is Enhanced by Dextran Sulfate, *Journal of Biological Chemistry*, Vol.277, No.33, (August 2002), pp. 30208-18, PMID: 12034749.

Mathew, A.J., Baust, J.M., Van Buskirk, R.G. & Baust, J.G. (2004). Cell preservation in reparative and regenerative medicine: evolution of individualized solution composition, *Tissue Engineering*, Vol.10, No. 11, (November 2004), pp. 1662-71, PMID: 15684675.

Miranda, A.F., Nette, G., Khan, S., Brockbank, K.G.M. & Schonberg, M. (1978). Alteration of myoblast phenotype by dimethyl sulfoxide, *Proceedings of the National Academy of Sciences USA*, Vol.75, (August 1978), pp. 3826-3830, PMID: 278996.

Mondal, B. (2009). A simple method for cryopreservation of MDBK cells using trehalose and storage at −80°C, *Cell and Tissue Banking*, Vol.10, No.4, (November 2009), pp. 341-344, PMID: 19381873.

Morris, M.C., Robert-Hebmann, V., Chaloin, L., Mery, J., Heitz, F., Devaux, C., Goody, R.S. & Divita G. (1999). A New Potent HIV-1 reverse transcriptase Inhibitor. A Synthetic Peptide Derived from the Interface Subunit Domain, *Journal of Biological Chemistry*, Vol.274, No.35, (August 1999), pp. 24941-6, PMID: 10455170.

Morris, M.C., Depollier, J., Mery, J., Heitz, F., Divita, G. (2001). A Peptide Carrier for the Delivery of Biologically Active Proteins into Mammalian Cells, *Nature Biotechnology*, Vol.19, No. 12, (December 2001), pp. 1173-76, PMID: 11731788.

Mueller, L.P., Theurich, S., Christopeit, M., Grothe, W., Muetherig, A., Weber T, Guenther, S. & Behre, G. (2007). Neurotoxicity upon infusion of dimethyl sulfoxide-cryopreserved peripheral blood stem cells in patients with and without pre-existing cerebral disease. *European Journal of Haematology*, Vol.78, No.6, (June 2007), pp. 527–31, PMID: 17509106.

Murgia, M., Pizzo, P., Steinberg, T.H. & Di Virgilio, F. (1992). Characterization of the Cytotoxic Effect of Extracellular ATP in J774 Mouse Macrophages, *The Biochemical Journal*, Vol.288, (December 1992), pp. 897-901, PMID: 1472003.

Mussauer, H., Sukhorukov, V.L. & Zimmermann, U. (2001). Trehalose Improves Survival of Electrotransfected Mammalian Cells, *Cytometry*, Vol.45. No.3, (November 2001), pp. 161-9, PMID: 11746084

Nihei, O.K., Savino, W. & Alves, L.A. (2000). Procedures to Characterize and Study $P2_Z/P2_{X7}$ Purinoceptor: Flow Cytometry as a Promising Practical, Reliable Tool,

Memorias do Instituto Oswaldo Cruz, Vol.95, No.3, (May 2000), pp. 415-28, 2000, PMID: 10800201.

Oliver, A.E., Jamil, K., Crowe, J.H. & Tablin, F. (2004) Loading Human Mesenchymal Stem Cells with Trehalose by Fluid-Phase Endocytosis, *Cell Preservation Technology,* Vol.2, No.1, (2004) pp. 35-49.

Otrock, Z.K., Beydoun, A., Barada, W.M., Masroujeh, R., Hourani, R. & Bazarbachi A. (2008). Transient global amnesia associated with the infusion of DMSO-cryopreserved autologous blood stem cells. *Haematologica,* Vol.93, No.3, (March 2008), pp. 36-7, PMID: 18310533.

Polge, C., Smith, A.U. & Parkes A. (1949). Revival of spermatozoa after vitrification and dehydration at low temperatures. *Nature.* Vol.164, (October 1949), p. 666, ISSN: 0028-0836

Richard, J.P., Melikov, K., Vives, E., Ramos, C.,Birgit, V., Gait, M.J., Chernomordik, L.V. & Lebleu, B. (2003). Cell-penetrating Peptides: A Reevaluation of the Mechanism of Cellular Uptake, *Journal of Biological Chemistry,* Vol. 278, No.1, (January 2003), pp. 585-90, PMID: 12411431.

Rudolph, A.S. & Crowe, J.H. (1985). Membrane Stabilization During Freezing: The Role of Two Natural Cryoprotectants, Trehalose and Proline, *Cryobiology,* Vol.22, No.4, (August 1985), pp. 367-77, PMID: 4028782.

Schlegel, P.G., Wölfl, M., Schick, J., Winkler, B. & Eyrich, M. (2009). Transient loss of consciousness in pediatric recipients of dimethyl sulfoxide (DMSO)-cryopreserved peripheral blood stem cells independent of morphine co-medication, *Haematologica,* Vol.94, No.10, (October 2009), pp. 1473-5, PMID: 19608681.

Slade, L. & Levine, H. (1991). A food polymer science approach to structure-property relationships in aqueous food systems: non-equilibrium behavior of carbohydrate-water systems, *Advances in Experimental Medicine and Biology,* Vol.302, (1991), pp. 29-101, PMID: 17416335.

Sosef, M.N., Baust, J.M., Sugimachi, K., Fowler, A., Tompkins, R.G. & Toner, M. (2005) Cryopreservation of isolated primary rate hepatocytes: enhanced survival and long-term hepatospecific function, *Annals of Surgery,*.Vol.241, No.1, (January 2005), pp. 125-33, PMID: 15622000.

Taylor, M., Campbell, L., Rutledge, R., Brockbank, K. (2001). Comparison of Unisol with Euro-collins solution as a vehicle solution for cryoprotectants, *Transplantation Proceedings,* Vol.33, (February 2001), pp. 667-9, PMID : 11267013.

Vives, E., Brodin, P. & Lebleu, B. (1997). A Truncated HIV-1 Tat Protein Basic Domain Rapidly Translocates through the Plasma Membrane and Accumulates in the Cell Nucleus, *Journal of Biological Chemistry,* Vol.272, No.25, (June 1997), pp. 16010-17, PMID: 9188504.

Walker, B., Braha, O., Cheley, S. & Bayley, H.(1995). An intermediate in the assembly of a pore-forming protein trapped with a genetically-engineered switch. *Chemistry & Biology,* Vol.2, No,2, (February 1995), pp. 99-105, PMID: 9383410.

Zeng, X., Zhao, C., Wang, H., Li, S., Deng, Y. & Li, Z. (2010). Dimethyl Sulfoxide Decrease Type-I and -III Collagen Synthesis in Human Hepatic Stellate Cells and Human

Foreskin Fibroblasts. *Advanced Science Letters*, Vol.3, No. 4, (December 2010), pp. 496–499.

3

Cryopreserved Musculoskeletal Tissue Bank in Dentistry: State of the Art and Perspectives

Luiz Augusto U. Santos[1], Alberto T. Croci[2], Nilson Roberto Armentano[3], Zeffer Gueno de Oliveira[4], Arlete M.M. Giovani[5], Ana Cristina Ferreira Bassit[6], Graziela Guidoni Maragni[7], Thais Queiróz Santolin[7] and Lucas da Silva C. Pereira[8]

1. Introduction

Maxillary and mandibular bone loss has long been a challenge to dental surgeons who seek to reconstruct these lost segments. These lesions lead to deformation of some maxillary and mandibular areas which interferes in the functional rehabilitation process of these structures. The most common cause of these lesions is prolonged use of total prostheses in a large part of the Brazilian population and the searches for surgical techniques and bone substitutes are today proposed and studied by the academic class. In this context, Brazil is starting to distribute allogeneic tissue obtained, processed and qualified by musculoskeletal tissue banks. Such banks already have experience in dispensing tissue to the orthopedic area, which has been using reconstructive techniques with allografts for many years. The first studies proposing the use of bone substitutes for replacement of these faulty parts commenced in the decades subsequent to 1860. (Carrel, 1912;Groves, 1917; Sharrard, Collins, 1961; Urist, 1965; Fischer, 1998; Tomford, 2000).

After the verification of the disadvantages in the use of autologous tissues for this purpose, such as the increase in donor morbidity, greater risk of nerve lesion and of infection inherent to the second surgical procedure and limitation in the availability of the tissue in quantity and variety, the use of homologous tissue became another option that was gradually indicated (Cunningham, Reddi, 1992; Tomford, 2000).

[1]*Institute of Orthopedics and Traumatology, Hospital das Clínicas of the School of Medicine of the University of Sao Paulo, dentist, Tissue Bank Technical Responsible and. Sao Paulo/SP, Brazil*

[2]*Institute of Orthopedics and Traumatology, Hospital das Clínicas of the University of São Paulo school of Medicine, Professor and Tissue Bank Director - Sao Paulo/SP, Brazil*

[3]*School of Dentistry of the University of Santo Amaro- São Paulo, Brazil*

[4]*Orthopedic Nurse Specialist. São Paulo/SP, Brazil*

[5]*Nurse, Institute of Orthopedics and Traumatology, Hospital das Clínicas of the University of São Paulo school of Medicine, Tissue Bank Coordinator - São Paulo/SP, Brazil*

[6]*Veterinarian, Tissue Bank Researcher, University of Florida, Gainesville, FL – Flórida- US*

[7]*Nurse, Institute of Orthopedics and Traumatology, Hospital das Clínicas of the University of São Paulo school of Medicine, Tissue Bank Team - São Paulo/SP, Brazil*

[8]*Dental Student, Institute of Orthopedics and Traumatology, Hospital das Clínicas of the University of São Paulo school of Medicine, Tissue Bank Team - São Paulo/SP, Brazil*

The good results with the clinical application of allografts in dentistry motivate their use on an increasing scale, until in 2005 Dentistry came into the scene with the use of tissues in maxillary and mandibular pre-prosthetic surgery. A consensus between the National Transplant System and the Federal Board of Dentistry allows the use of allografts by specialists in the areas of Implant Dentistry, Periodontics and Oral and Maxillofacial. The tissue banks, in turn, prepare a tissue processing line geared toward dental needs with a focus on quality control and traceability.

Thus usage has become both abrupt and a tendency in the last 5 years (RBT, 2006-2010). In spite of a significant number of bone transplants in the dental area with good clinical results, the dental profession is still lacking information about activities that involve the area of tissue banks, particularly in the rigid quality control and traceability. Such activities are founded on international standards[2], literature[3] and legislation[4] and implemented according to Good Manufacturing Practice- GMP.

We consider it very important to gather epidemiological data on bone transplants in dentistry, elucidating the size and the limits of this type of treatment that is already considered a tendency in our field. In addition, to report on our perspectives of investigation into the efficacy and safety of the use of allografts, with tests that can enable us to expand our knowledge about the osseointegration of allografts. In other words, knowledge that allows us to reach what we consider most important in dental treatments: the predictability of treatment.

2. Bone tissue

Bone tissue is composed of two portions: 1. Organic, consisting of intrinsic bone cells (osteoblasts, osteoclasts and osteocytes) and the organic matrix synthesized thereby; 2. Inorganic, consisting of hydroxyapatite, deposited amorphously in an initial phase and that in a short space of time is converted into another crystalline hydroxyapatite. Organic matrix corresponds to 35% of the bone volume and inorganic matrix to 65%.

In spite of the resistance and hardness, bone tissue is very plastic and has a high capacity to remodel through various situations to which it is submitted, such as fractures, lesions and bone loss. The bone tissue regeneration process starts from important biological reactions, triggered by the actual tissue lesion. Grafting triggers a mechanism of migration of the bone cells belonging to the receptor bed to the inside of the graft, with the purpose of resorbing it and replacing it with neoformed bone.

---

[2]European Association of Tissue Banks. Common Standards for Tissues and Cells Banking: Berlin: European Association of Tissue Banks; 2004.
American Association of Tissue Banks. Standards for Tissue Banking. 11th ed. McLean : American Association of Tissue Banks; 2007
[3]Phillips GO, Strong DM, Versen RV, Nather A. Advances in Tissue Banking. Vol. 4. World Scientific . New Jersey, 2000.
Bancroft JD, Stevens A. Theory and practice of histological techniques. Fourth Edition. Churchill Livingstone. United Kingdom, 1999.
[4]Law n.9434 of February 5, 1997; Decree n.2268 of June 30, 1997;Administrative Ruling n.1686 of September 20, 2002; Resolution n. 220 of December 27, 2006; Administrative Ruling n. 2600 of October 21, 2009.

The cells belonging to the bone tissue are the osteoblasts, osteocytes and osteoclasts.

The osteoblasts are cuboid, elongated cells of mesenchymal origin that are located in the bone margins; their function is to produce the organic matrix of the bone tissue. In reduced activity these cells assume a more slender shape. The osteocytes are encapsulated osteoblasts, which after maturation became imprisoned inside the mineralized matrix, but that still maintain contact with other cells through cytoplasmic ramifications, thus maintaining physiologic functionality of the tissue (Junqueira, Carneiro, 1999; Davies 2000). This contact with surface cells such as the osteoblasts and lining cells is related bone structure maintenance and to the physiological responses that lead to tissue formation or resorption (Aubin et al., 2006).

The osteoclasts are giant cells with multiple nuclei and their function is related to resorption. In synergy with the osteoblasts they promote bone remodeling.

The interaction and the synergism among bone cells is called creeping substitution, and this occurs through three essential cellular events: osteogenesis (cellular event that favors the synthesis of bone matrix by the osteoblasts), osteoinduction (ability to induce the migration of mesenchymal cells and their differentiation into osteoblasts) and osteoconduction (ability of the tissue to serve as a mold or guide for the cellular processes involved in bone tissue repair).

Moreover, as is the case with others, the bone cells pass through the stages of the cell cycle, which range from formation to cell division (mitosis). Mitosis is susceptible to external interferences, and the cell can either enter a state of rest or continue to split cyclically (Urist, 1965; Enneking et al., 1975; Junqueira, Carneiro, 1999; Perren, Claes, 2002).

3. Bone transplantation

The term transplant is not widely used by the dental community to refer to the use of bone tissue. The common term is bone grafting. The bone graft can receive a nomenclature and be classified according to the origin of its obtainment and on the implant site (Table 1).

Autologous graft or autograft	Graft of tissue from one site to another in the same individual
Isograft	Graft between people with the same genotype (homozygotes; e.g. identical twins)
Allograft or homologous graft	Graft between individuals of the same species with disparate genotype
Xenograft or heterologous graft	Graft from one individual of a species onto a different species

Table 1. Classification of grafts according to their nature. Source: Drumond, 2000

4. Clinical application of grafts over the years

The use of bone tissue for replacement of bone losses is not a recent procedure. Since last century there have been accounts of the use of these tissues in humans and in experiments with animals as a means of assessing their efficacy.

In these studies, many treatments with the use of bone grafts of autologous and homologous nature have been proposed over the years. The first homologous bone transplant is described by William MacEwen in 1878 (Giovani, 2005). At this time the treatment of osteomyelitis was performed by means of surgical resections of the infected segments. Homologous tibial segments (obtained from patients submitted to osteotomies) were used in the reconstruction of a bone defect caused by the resection of part of the humerus of a young man suffering from osteomyelitis (Tomford, 2000). Encouraging results of the consolidation of the bone graft with the receptor bed (MacEwen, 1909) motivated researchers back then. In this period, there was no consensus about which bones can specifically be used for transplantation. Tissues were obtained randomly from donors that were victims of fractures, resections and amputations. For the storage and processing of these tissues, professionals used protocols and equipment that today are not the best suited to this purpose (Tomford, Mankin, 1999; Tomford, 2000). Nowadays a large portion of these studies has only historical value with respect to the pioneer spirit of these researchers.

The clinical use of allografts during this period was in low demand, and occurred on an experimental basis at some centers from all around the world. With the availability of antibiotic therapy, changes occurred in the indication of these tissues, and patients with osteomyelitis were then submitted to pharmacological treatments, to the detriment of surgical procedures (Tomford, 2000). Thus surgical resections of infected segments cease to be a priority, and allografts are then used in the reconstructions of bone defects caused by tumor resections. This change causes the studies from the time to evolve as well, and to give more detailed accounts of the cellular processes involved in the osseointegration of grafts. The consensuses of these first studies serve as guidelines for the first grafts performed at that time.

The use of cryopreserved allografts presents some advantages over autologous tissue, such as the availability of the necessary quantity of tissue and a decrease in postoperative morbidity. As regards morphology, there are some differences in the vascularization process of the cortical and spongy bone grafts. In the cortical bone, the repair is started by the action of osteoclasts and in the spongy bone, by the osteoblasts. Another difference lies in the revascularization time, which is slower for cortical bone and faster for spongy bone.

At that time, Urist (1965) was already describing that the osteoprogenitor cells responsible for the bone repair process are derived from monocytes, present in an elevated number in the repair zone coming from the bone marrow, and that the osseointegration of grafts is achieved, since primitive cells not yet differentiated can differentiate into viable osteoblasts from osteoinductive substances. Perivascular mesenchymal cells disaggregate and migrate to the grafting area, where they reaggregate, proliferate and differentiate to form new bone. (Urist et al, 1983)

Some substances secreted by certain cell types interfere in or even modulate the cellular processes of osseointegration, today known as Cytokines or growth factors.

Urist made important discoveries in this area back in 1965, even proposing a new bone processing procedure aimed at the removal of a calcified layer (demineralization) from the matrix, making this graft more osteoinductive. At the time this kind of tissue was called Demineralized Bone Matrix (DBM), and the name is still in use today. Later, it was understood that part of this capacity is the responsibility of the superfamilies of proteins (TGF), including the morphogenetic protein (BMP) present in the bone matrix. (Malafaya et al, 2002).

The decade of 1980 was marked by the advent of the acquired immunodeficiency syndrome (AIDS) that gives rise to discussions on the safety of the clinical use of homologous tissues. The biological risk of disease transmission between tissue donors and recipients is the topic of greatest relevance and importance in the period. The tissue banks existing at that time were encouraged to review donor selection protocols with the objective of avoiding the transmission of these diseases. This encouragement was provided mainly by the international public health regulatory agencies, such as the FDA (Food and Drug Administration) and other institutions related to the Haemovigilance and tissue transplantation systems. The result was a standardization of the internal processes of banks with respective preparation of standards by the main global tissue bank associations (American Associating of Tissue Banks - AATB and European Association of Tissue Banks-EATB), which contributes to the gain of quality of tissue made available by these services (Nather, 1991; Galea, Kearney, 2005; Santos, 2007).

With the availability of more reliable grafts by the tissue banks, their use, biological behavior and indication by surgeons become a viable treatment option. (Santos, 2011).

Similar to the repair process in fractures and in the development of the musculoskeletal system, the osseointegration of grafts takes place after a selection of primordial cells that are differentiated into osteoblasts under the influence of osteogenic factors. (Thies et al., 1992). The main objective expected in the use of grafts is the ability to selectively induce the primordial events of the integration process, such as osteoinduction, osteoconduction and osteogenesis (Lindhe et al., 1997).

According to Tomford and Mankin (1999), for cortical bone grafts to be incorporated into the receptor bed, there must be revascularization of this bed. When this process does not occur, the repair area loses balance in resorption and the graft might suffer fatigue fractures. Spongy grafts are consolidated more quickly.

Boldt et al. (2001) evaluate the use of frozen bone graft in 173 acetabular reconstructions and 79 femoral reconstructions in humans. Femoral heads obtained from a local bone bank are particulated and impacted in the faults. After a mean follow-up period of four years, they report acetabular clinical stability in 97.2 % of the cases, graft incorporation in 74% in the acetabula and 61% in the femurs, according to radiological analysis. They conclude that the results obtained with the use of impacted grafts are promising, except for the reconstruction of type III acetabular defects, where a reinforcement cage is recommended.

Janssen et al. (2001) studied the use of homologous cortical rings obtained from femoral diaphysis for reconstruction of intervertebral discs of the lumbar spine in 137 patients. The results show that arthrodesis was achieved in 94% of the cases and they do not report signs of resorption.

Weyts et al. (2003) analyze samples of femoral heads collected after the primary arthroplasty of two human donors. The tissues are cryopreserved at - 80 °C for a minimum period of six months, according to the protocols of the American and European Tissue Bank Associations. After this period, these tissues are biopsied and submitted to cell culture (for survival observation) and PCR (polymerase chain reaction) for genetic screening. The results show the presence of live cells belonging to the donors in the analyzed samples.

Lavernia et al. (2004) researched the adoption and use of allografts by orthopedic surgeons in 340 U.S. hospitals. The frozen graft from tissue banks certified by the American Association of Tissue Banks - AATB is used the most often by orthopedists for the treatment of knee and hip bone loss.

Schreurs et al. (2005) conduct a study to evaluate the use of homologous bone graft from a tissue bank in the reconstruction of 33 femoral defects. Bone graft impaction precedes the fixation of the femoral nail. After a minimum follow-up of eight years there is functional improvement of the joint according to the Harris hip score (from 49 to 85 points from pre- to postoperative period) and good survival according to the Kaplan-Meier method. Although four patients have had femoral fractures, the authors conclude that the graft impaction technique and use of cemented femoral nail results in excellent survival for eight to thirteen years.

Cabrita (2007) studies the treatment of infected hip arthroplasties with and without the use of the antibiotic-impregnated cement spacer. For the reconstruction of bone stock, Cabrita uses the massive particulated homologous graft in 60.9% of the patients treated and does not report complications related to their use.

In a review of concepts, Giannoudis et al. (2005) emphasize the advantages of the use of bone grafts in the area of orthopedics and traumatology. They describe the cellular events present in literature that involves the osseointegration process of autologous, homologous grafts and of biocompatible synthetic substitutes. They stress the osteoinductive characteristic of fresh grafts and the osteoconductive characteristic of frozen and lyophilized grafts. Heyligers and Kleim (2005) verify the presence of live cells with growth potential in samples of femoral heads cryopreserved at - 80 °C, over a minimum period of six months. The authors stress the importance of discussing the osteoconductive potential of grafts and highlight the need to investigate the role of these surviving cells from the frozen tissue in the bone formation process after its implantation. Besides the bone cells, other cells and inflammatory factors play a vital role in the bone repair process. The macrophages and substances such as interleukin one, six, eleven (IL-1, IL-6, IL-11), RANKL and osteoprotegerin (OPG) are found during the first three days after the lesion (Gerstenfeld, Einhorn, 2006). As far as macrophages are concerned, Knighton et al. (1982) explain that these cells are present in the repair zone and are capable of producing growth factors, which, in turn, stimulate the neovascularization, proliferation and migration of other cell types, such as the fibroblasts.

5. Evolution of musculoskeletal tissue banks in Brazil

Considering the satisfactory results in the use of allografts obtained from multiple donors of organs and tissues with brain death besides studies showing revascularization, osteointegration and bone formation at sites of graft (Barros Filho, et al., 1989; Croci et al., 2003, Zhang et al., 2004; Dallari et al., 2006, Bitar et al, 2010), an increasing number of orthopedic surgeons and dentists currently opt to use of homologous grafts in our country. This fact is corroborated by the disadvantages already known in the use of autologous tissues, such as the increase in postoperative morbidity, greater risk of infection inherent to the second surgical procedure required for their obtainment, risk of nerve lesion and limitation of the quantity and variety of graft obtained (Smith et al. 1984; Cunningham, Reddi, 1992; Drumond, 2000).

This tendency, combined with the growing number of patients with bone loss that seek specialized orthopedic and dental services, leverages the creation of some Tissue Banks in the country in an experimental manner.

In Brazil, the standardization of Musculoskeletal Tissue Banks is linked to specific laws of our country. We have a General Coordination Office for the National Transplant System – SNT of the Ministry of Health that discusses and prepares, together with technical chambers, legislations involving the use of tissue by the medical and dental community. The guidelines are based on protocols already developed by some centers and on the standards of international associations (European Association of Tissue Banks - EATB and American Association of Tissue Banks-AATB). The creation of the first reference centers in the large-scale capture, processing and distribution of musculoskeletal tissue and some of these experiences are described. (Amatuzzi et al., 2000, Amatuzzi et al., 2004). Tissue Banks in Brazil are controlled by the General Management of Blood, other Tissues, Cells and Organs - GGSTO of the National Health Surveillance Agency - ANVISA, an institution similar to the U.S. FDA. This agency focuses its activities on health surveillance and on quality control, traceability, appraisal of risks and of adverse effects involving tissue transplants in the country and its guidelines are published in the form of legislation[5] that also defines Musculoskeletal Tissue Bank as "the service that, with physical facilities, equipment, human resources and adequate techniques, has as its duties the performance of clinical, laboratory and serological triage of tissue donors, the removal, identification, transportation, processing, storage and delivery of bones, soft tissues (cartilage, fasciae, serous membranes, muscle tissue, ligaments and tendons) and their derivatives, of human origin for therapeutic purposes, research and teaching".

6. Activities of a musculoskeletal tissue bank

The description of activities of a musculoskeletal tissue bank is summarized in the algorithm below (**Illustration 01**).

Every activity related to the bank should also be based on the ethical principles inherent to the activities of any organ transplantation. They are:

- **Autonomy and self-determination:** The recipient of tissue from the musculoskeletal system should be provided with information in accessible language about the entire tissue obtainment process, the risks and the chances of success or failure of the treatment. The following stage is the patient's decision, after their evaluation of the information received, set out in an informed consent form.
  Professionals with specific training provide all the required information, using language exempt from complex or technical terms, enabling the patient to achieve easy understand in order to make the final decision. For this document to be authentic, the consent must be free, that is, not caused by coercion. The professionals of a musculoskeletal tissue bank should be objective and impartial while providing guidance to recipients. Every process is recorded in the Recipient's Form.

[5]Administrative Ruling no. 211, of March 24, 2003

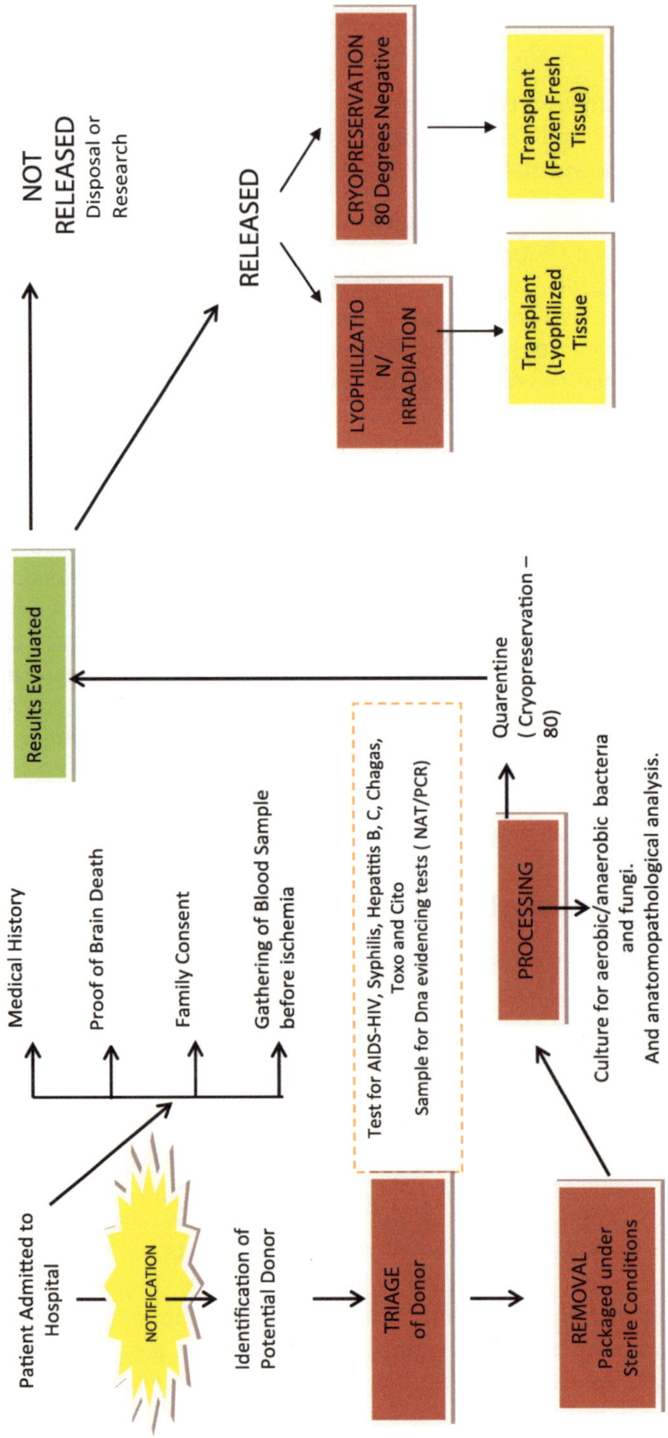

Illustration 01. Algorithm of the Musculoskeletal Tissue Donation and Transplantation Process

- **Justice** - The principle of equal opportunities for the use of available tissues. The definition of ethical parameters in distribution is imperative.
- **Symbolism of the body** - This principle is employed mainly in the reconstruction of the deceased donor's body after removal, which should be performed carefully, with the apparent anatomical parameters respected, thus ensuring that the family receives a body in adequate conditions.

7. Obtaining musculoskeletal tissues

The main source of musculoskeletal tissues is the notification of deceased donors to the Transplant Centers, Organ Service Services, and Hospital Transplant Departments. The teams that receive the organs and tissues are only notified after a series of procedures and exams has been carried out to ascertain brain death and obtain the family's consent for the process of organ and tissue donation. Brain death is initially verified by a neurologist, using techniques of physical and imaging (doppler) exams, which are repeated after six hours in the presence of a family member of the potential donor. Once there is no doubt as to the irreversible diagnosis of brain death, the family members are asked whether they would consider donating their loved one's organs. The family interview is done by trained members belonging to an intra-hospital committee, or by an organization that looks for organs. The entire donation process should be recorded and legally signed before the teams are notified to remove each organ (heart, liver, kidney, pancreas, lung, intestine) and tissues (osteochondral and fascial-ligamentous, skin, vessels, cornea, heart valves). Each team should have clearly-defined criteria for selecting, and at the time of notification, accepting or refusing the donor in question.

For donors of musculoskeletal tissue, the selection follows a rigorous control process, with serological tests for antigens and HIV antibodies, Hepatitis A, B and C, HTLV-1 and 2, Syphilis, Chagas disease, Toxoplasmosis and Cytomegalovirus, as well as state-of-the-art tests for evidenciation of D (Nucleic Acid Amplification - NAT) HIV and Hepatitis B and C, bone marrow aspirate smear of the sternum and iliac crest sample, both for histopathological investigation.

Donors with the following criteria were excluded: orthopedic pathologies, such as osteoporosis, osteonecrosis, rheumatoid arthritis, lupus erythematosus, neoplasias, age group that compromises that characteristics of the tissues, blood transfusions, tattoos or piercings within the period of the immunological window, use of illegal drugs, travel to endemic zones, generalized or localized infections, fractures, open sores on the limbs from which the musculoskeletal tissues are to be removed, or any other situation that places in doubt the quality of these tissues, pursuant to the Brazilian legislation.

The whole procedure is carried out in totally antiseptic conditions, just as in surgery. Special gowns made from synthetic material are used, and all the surgical stages of antisepsis are followed.

The removed tissues are immediately packaged in triple packaging, hermetically sealed, and delivered, under refrigeration (-4°C) to the Tissue Bank.

A very important stage of the capture process is donor reconstruction. The body should be delivered to the family free from any deformation and as close as possible to its appearance before the tissue removal. This is because the fear of deformation has been one of the main causes of refusals of bone donation by the family. For a perfect reconstruction the professionals use prostheses especially developed for this purpose, plaster, sutured and gauze. **(Illustration 02)**. This rigorous reconstruction is the most laborious stage in the removal procedure. Areas possibly visible during the funeral (face, anterior side of arm, anterior side of shoulder, etc.) are not approached. All the anatomical parameters are respected, and therefore donor deformation does not occur.

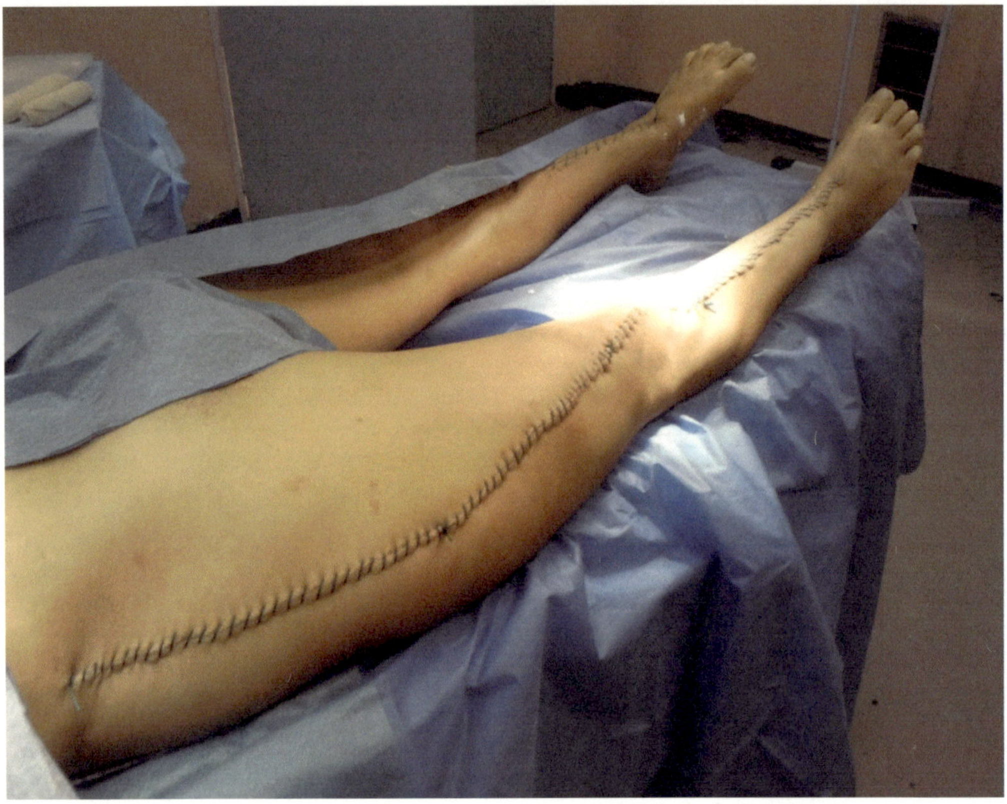

Illustration 02.Limbs reconstructed with prostheses after tissue removal.

8. Processing the musculoskeletal tissues

Once the tissues and organs have been obtained, they are delivered to the BTME in portable refrigerators, with temperature monitoring throughout the transport process. The processing stage is preceded by a planning of the activities necessary to accomplish it, such as provision of materials and instruments, summoning the processing team, defining the preparation and dimensions, according to the needs of the service (waiting list) and requests for orthopaedic and odontological surgeons. This stage is done in a special, classified operating theater (class 100 or ISO 5) equipped with a laminar flow module. (Illustration 03). A Class 100 room means it has purity of 100 particules per m3 of air. For the purposes of comparison, an operating theater should have 10,000 particles/m3 of air.

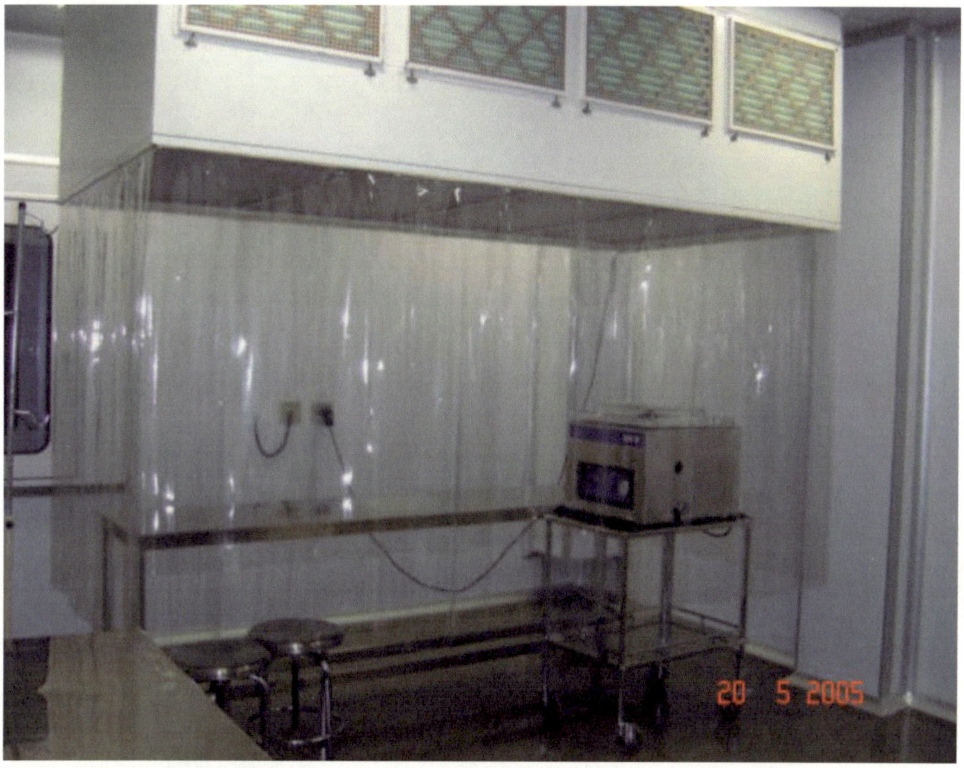

Illustration 03. Classified processing room (100 particles/m3 air: Class 100/ISO 5; HEPA Filters 99.9 %)

The room also has a pass-through anti-chamber, and all the environments have rigorous control of air particles and positive pressure, to ensure the quality of the tissues processed in it. Specific gowning of the professional team is also necessary, using only non-fabric clothing (Spunbond - Meltblown – Spunbond - SMS) to prevent the dispersion of particles given off by conventional cotton clothing. (**Illustration 04**)

In addition to non-fabric gowning; the team must also adopt certain behaviors. For example, sudden movements, use of cosmetic products and exposure of the skin should be avoided while in this room. Adequate conduct is ensured through special training, not only for the processing team that actually carries out the procedure, but also for other professionals who enter the environment (e.g. for cleaning and maintenance purposes).

A BTME carries out various types of tissue, for use in orthopedic and odontological surgery, and each procedure requires careful planning. (**Illustrations 05 ,06 and 07**)

For the processing of fresh, frozen tissues, a process called mechanical processing is carried out, i.e. removal of the adventitious tissues such as the blood, periosteum, subcutaneous tissue, muscles, fasciae, and fibrotic tissue.

The fragments are then submitted **chemical processing**, where they are immersed in hydrogen peroxide based emulsifying solutions and alcoholic solutions under ultrasound stirring. (**Illustrations 08 and 09**)

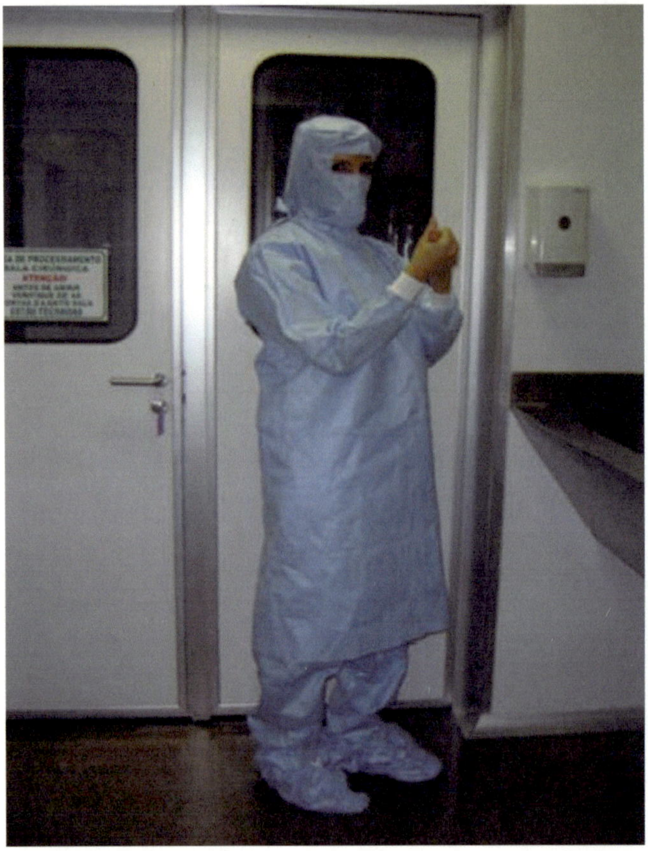

Illustration 04. Specific gowning to work in the classified room.

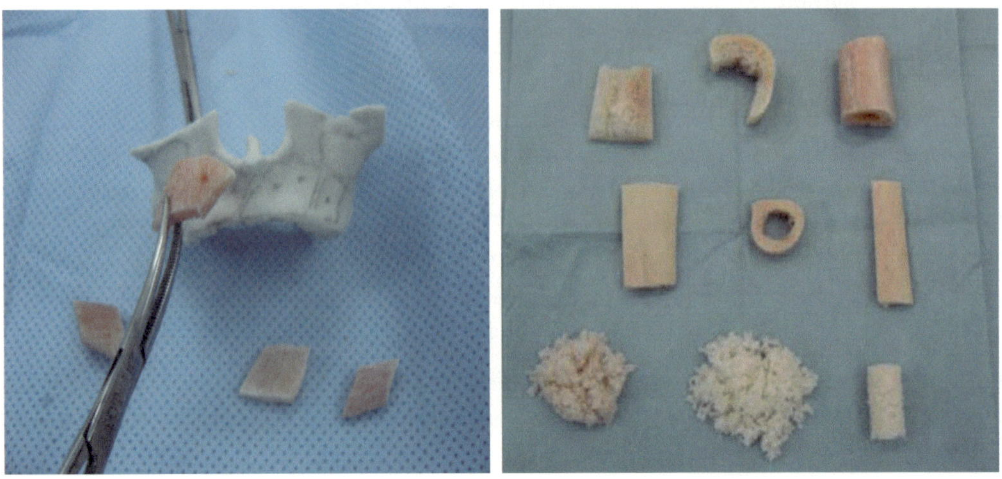

Illustration 05 and 06. Processing of grafts for use in odontological surgery. The photograph on the left shows modeling of maxillary bone defects based on a resin prototype. The one on the right shows fragments of allografts for use in odontological surgery.

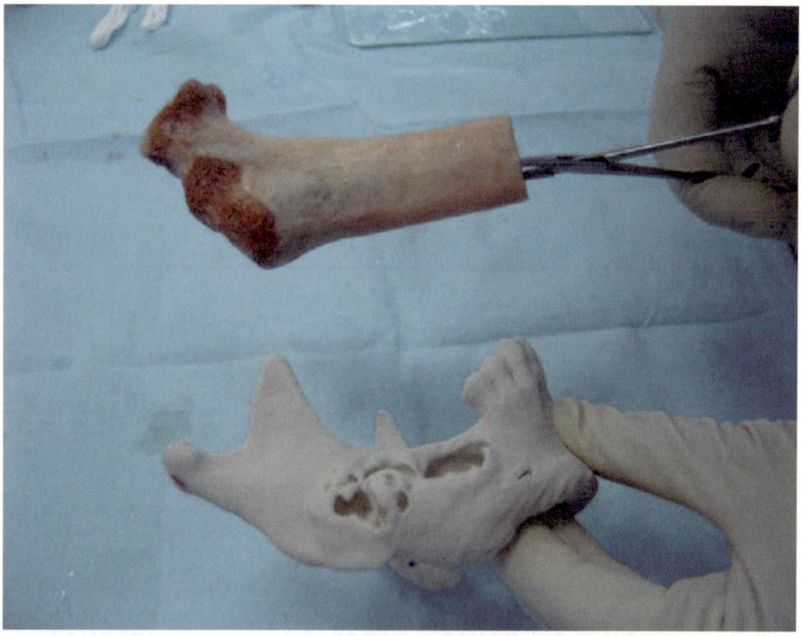

Illustration 07. Modeling of a jaw bone from a segment of proximal femur.

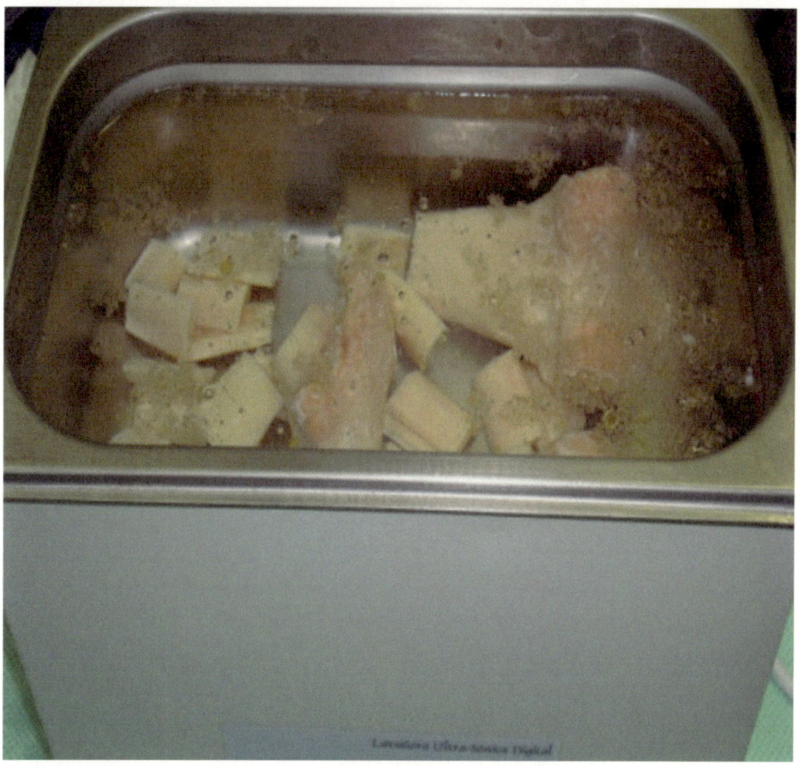

Illustration 08. Ultrasound washing in emulsifying solution.

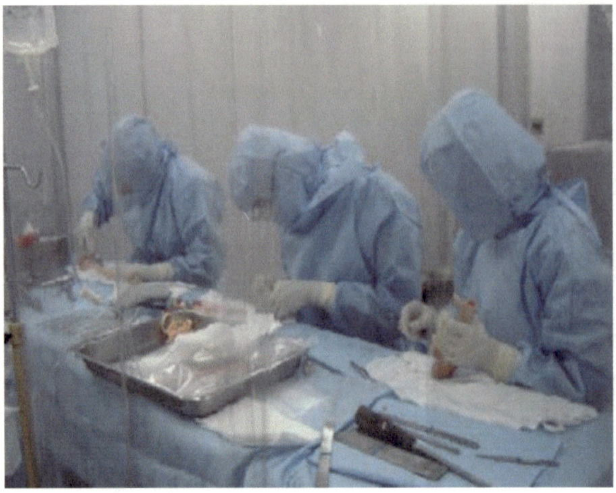

Illustration 09. Processing Team at work.

Immediately afterwards, samples of bone marrow from the long bones and fragments of each tissue submitted to processing are collected from these resulting solutions, and submitted to microbiological processing (general culture, anaerobic and fungal culture) using the direct inoculation technique. Samples are also obtained for histopathological analysis.

Finally, the packaging procedure is begun for all the grafts processed, which are measured (length, height, diameter, weight volume, perimeter), packed in sterile, triple packaging, vacuum sealed, and duly labeled as analysis tissue. (Figure 8). The tissues are labeled with the following information: donor, exams carried out, batch number, item, validity period, type of conservation, and bar code.

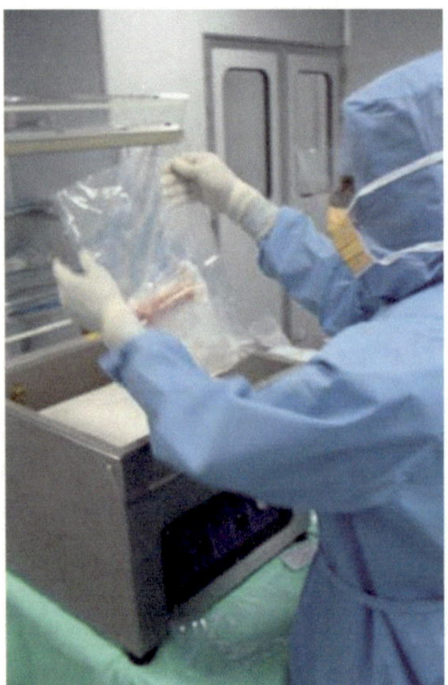

Illustration 10. Vacuum sealing and labeling.

Once they have been labeled, the tissues are submitted to radiography at the BTME, then sent for cryopreservation.

9. Cryopreservation of the musculoskeletal tissues

The musculoskeletal tissue cryopreservation process at - 80◦C is described in chapter 1 (VALIDATION OF PRIMARY PACKAGING FOR CRYOPRESERVED MUSCULOSKELETAL TISSUES)

10. Lyophilization of musculoskeletal tissue

The bones can also be processed in their lyophilized form. The lyophilization process should be validated and be, like all the tissue handling procedures, in conformity with the Manual of Good Manufacturing Practice - GMP and in accordance with international standards[6], literature[7] and legislation. The procedure involves the use of an automated lyophilization system composed of a Labconco® (**Illustration 11**) freeze drying, or lyophilization chamber with Condensation Chamber/Vacuum. During lyophilization the tissues remain frozen for the prevention of ice crystal liquefaction inside the matrix. The sublimation process should be validated by analyses of the Residual Moisture by automated thermogravimetric method. The lyophilization process is divided into 2 stages: Primary and Secondary Drying. In the primary phase, the largest fraction of water present in the matrix in its solid state (ice crystals) is removed by sublimation induction and gaseous migration. This induction is achieved by the driving force resulting from the difference of pressure gradient between the lyophilization chamber and condenser. The heat generated by this gaseous transportation should be controlled continually by digital sensors strategically positioned inside the lyophilization chamber.

At the end of the sublimation, the aliquot of unfrozen water linked to the organic components of the matrix (proteins) is then removed in the secondary phase, with an increase of pressure in the lyophilization chamber followed by the gradual increase of temperature at positive levels.

The analysis and control of residual moisture are essential to ensure the integrity of the protein matrix. The residual moisture is determined by thermogravimetric method, using an **Ohaus® (Illustration 12,13)** moisture analyzer. This method analyzes the initial weight of the sample on precision scales, followed by the promotion of heating with continuous recording of evaporation and weight. The percentage of residual moisture (**RM**), solid mass (**SM**), initial weight (**IW**) and final weight (**FW**) are analyzed in this method. The limit of RM is < **6** % as described by literature (Phillips, 2000).

---

[6]European Association of Tissue Banks. Common Standards for Tissues and Cells Banking: Berlin: European Association of Tissue Banks; 2004.
American Association of Tissue Banks. Standards for Tissue Banking. 11th ed. McLean : American Association of Tissue Banks; 2007
[7] Phillips GO, Strong DM, Versen RV, Nather A. Advances in Tissue Banking. Vol. 4. World Scientific . New Jersey, 2000.
Bancroft JD, Stevens A. Theory and practice of histological techniques. Fourth Edition. Churchill Livingstone. United Kingdom, 1999.

Illustration 11. Lyophilization chamber: Labconco® equipment from the Tissue Bank of Hospital das Clínicas - Universidade de São Paulo.

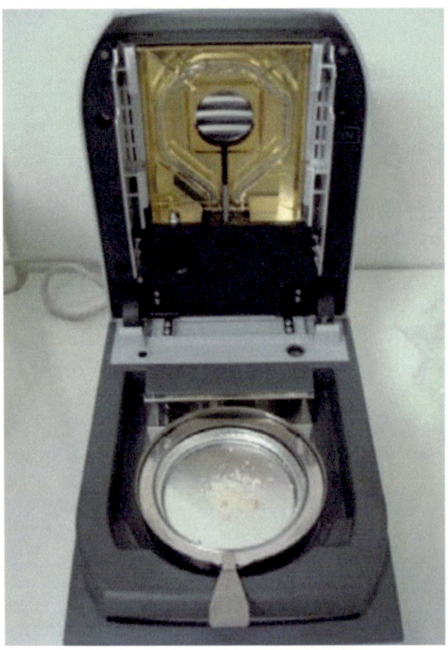

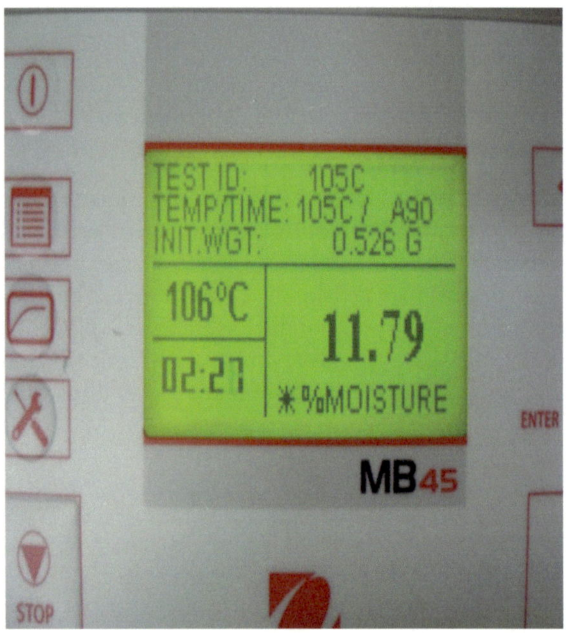

Illustrations 12,13. Residual Moisture Analyzer: Ohaus® equipment belonging to the Tissue Bank of Hospital das Clínicas - Universidade de São Paulo.

The result of the lyophilization is a dry tissue, conservable at room temperature and that should receive final sterilization by radiation. The lyophilized tissue radiation process is carried out in a Multi-purpose Irradiator, with gamma radiation from sources of ^{60}CO. The appropriate radiation dose is 25 kGy, with a dose rate of approximately 7 kGy/h. In order to avoid temperature variation, the samples are radiated in the presence of cooling elements, keeping the temperature between 4 and 8 °C. Red Perspex dosimeters are used for dose control.

Time Interval	Process	Indicators	Sample Temperature
2 hours	Pre-freezing of the lyophilization chamber	Temp. Chamber: - 40 °C	- 80 °C
10 minutes	Vacuum	Chamber Temp.: - 30 °C Chamber Vacuum : 8 x 10^{-3} mBar Condenser Vacuum : 2 x 10^{-3} mBar	- 65 °C
6 hours	Primary Drying	Chamber Temp.: - 30 °C Chamber Vacuum: 8 x 10^{-3} mBar Condenser Vacuum : 1 x 10^{-3} mBar **Analysis Final Moisture of Phase 01** RM: 6.26% Initial Weight: 0.527g Final Weight: 0.494g Solid: 93.74% **Analysis Temperature: 105°C**	- 40 °C
4 hours	Secondary Drying	Chamber Temp.: up to + 5 °C Chamber Vacuum: 7 x 10^{-3} mBar Condenser Vacuum: 1 x 10^{-3} mBar **Analysis Final Humidity of Phase 02** RM: 2.32% Initial Weight: 0.732g Final Weight: 0.715g Solid: 97.68% Analysis Temperature: 105°C	+ 5 °C

Table 1. Primary and Secondary Drying Process

11. Storage, distribution and quality control of tissue

At the end of any processing mode, the procedures should be documented in a specific file. This file is used to contain all the documentation related to the process, including records of the participant team, inputs used, documents evidencing sterility of instruments and other records that are part of the quality control program and that provide subsidies for traceability of the processes. This stage is imperative in the legislation and recommended by the Standards and Quality Control Manuals.

The tissue stock can be kept both frozen and at room temperature (lyophilized) following the same standards used by the global tissue banks and associations. Other processing methods have been investigated with the purpose of reducing the costs related to banking and maintenance. The glycerolization of bone tissue is presented as a processing methodology able to maintain the viability of the matrix and to prevent bacterial growth, besides enabling storage at room temperature (Giovani, 2006).

Until the transplantation occurs, all the processed tissues should be submitted to rigid quality assurance criteria. It is necessary to have an evaluation of all the data pertaining to the donor, results of exams, maintenance and control of equipment, material and instruments used in all the phases of each procedure. Management software can be used to record all the stages, allowing the fast retrieval of information such as the particulars of the donor, lot, shelf life, exams and status of the tissue (analysis, released, excluded, used) making it easier to trace each graft processed and made available, especially in the presence of evidence of an adverse effect and implementation of corrective and preventive actions.

For a lot of grafts under analysis to be released for use, the qualified technical professional from the Tissue Bank must analyze the results of all the exams performed: NAT or PCR serology for HIV, HBC and HCV, General Culture, Anaerobic Culture, Fungal Culture, Anatomopathological Exam and radiology reports. These exam reports are ultimately evaluated and released by the Qualified Clinical Professional of the service.

Besides exams, it is necessary to consider an evaluation of the printed records of temperature during the banking period. The service should have equipment that detects temperature oscillations even at a distance (satellite monitoring system).

Moreover, the installation of buzzers at strategic points of the hospital as well as Co2 backups ensures the reliability of the system.

After the release of each lot, there should also be a final inspection of each tissue, besides the substitution of tags of tissues under analysis by replaced. The banking logistics of the tissues in the ultra-low temperature (ULT) freezers considers the tissue type and search agility.

Services that execute rigorous quality control use annotation systems featuring checklists with double checking and consent.

All the data pertaining to the donor and to the lot, in compliance with the legislation, should be kept in single folders and stored in specific files of the musculoskeletal tissue bank for a minimum period of 25 years.

A serum bank with samples of donor plasma should also be made available by the musculoskeletal tissue bank in case of the need for counterproof exams.

As soon as the quality criteria have been evaluated and approved, the tissues are made available for use.

The tissues are distributed to the various specialties (Hip, Knee, Shoulder, Tumors and Dentistry) according to the availability of and requests for grafts.

The transplanter (physician or dentist) places the order for the tissue through a discussion of cases and by sending a specific form. The tissue reservation takes into account the demand for each type of transplant, waiting list and stock. The waiting list for transplantations performed within the Unified Health System - SUS complies with the prevailing legislation and today is organized and managed by the musculoskeletal tissue banks themselves, observing an order by date of inclusion. Urgent cases appointed by the medical team, such as malignant tumors and situations with a risk of severe complications, are communicated to the musculoskeletal tissue bank through an Emergency Form, for immediate response.

In dentistry, transplants have evolved differently and their distribution features some particularities that will be described further on.

12. Global data on tissue transplantation activities in Brazil

Today it can be seen that tissue transplantations in general are on the rise. Events such as officialization in the legislation and the creation of public promotion policies corroborate this evolution. Considering bone transplants alone, 30x growth has been observed in the last 5 years (Brazilian Transplantation Register, 2010), a fact motivated by the start of large-scale distribution of tissues for dental surgery. Although statistics show the number of tissues to be growing, the quantity of donors is still a concern. The vast majority of donors in Brazil are still for the removal of perfused solid organs (heart, kidney, liver, etc. A minority (6% on an average) accept the donation of musculoskeletal tissues. Of these, just 8% on average, become effective donors and the rest are discarded due to the presence of exclusion criteria such as infections, blood transfusion, and inadequate profile (**Graph 1**).

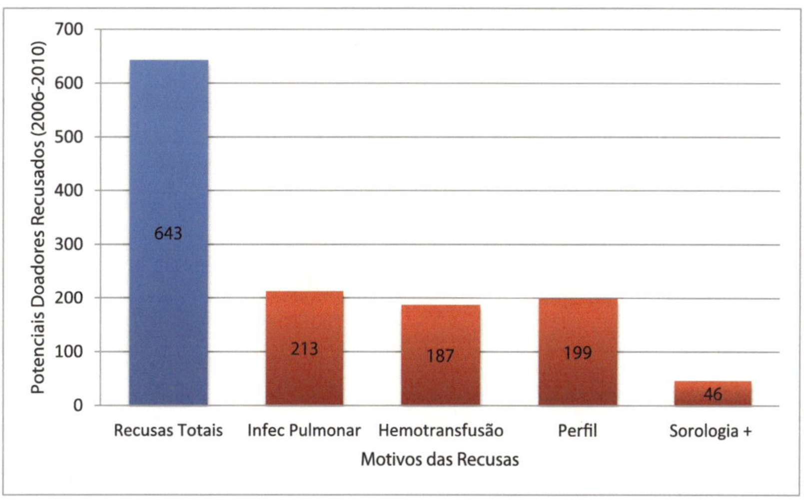

Graph 1. Reasons for refusals of bone donors between 2006-2010.
(Source: File BTME-HC-USP)

13. Homologous tissue transplants in dentistry

Dentistry has also sought biomaterials usable in the replacement of mandibular and maxillary bone loss over the years. The transplantation of bone portions taken from the iliac crest and menton (autologous transplant) has been broadcast in recent years, yet similarly to what happened in the orthopedic area, the disadvantages related to donor morbidity and the good results observed in the use of allografts in orthopedic patients, motivated professionals and patients to adopt this form of treatment.

In Brazil the use of bone transplants in dentistry started in 2005 after a consensus regarding the need for use and administrative mobilizations with the National Transplant System, Federal Council of Dentistry and Musculoskeletal Tissue Banks. This consensus serves as a starting point for the definition of criteria for requests for these tissues at the existing banks, where the professional must be a specialist in the areas of Implant Dentistry, Periodontics or Oral and Maxillofacial.

Rules related to logistics and traceability were incorporated into the activities of existing tissue banks, which then started to implement the bone tissue processing and distribution programs for dental purposes. In this program it is crucial to record the entire process with a focus on health control, including adverse deeds and traceability from the request to the actual transplantation. Specific forms are used for this purpose, including the Request Form, Terminated Transplantation Form, Non-Conformity Term and Adverse Effect Form.

Once standardized, this type of transplant is initiated in the country with widespread adoption by dentists, as observed in the Brazilian transplantation records. (RBT, 2006-2010)

Naturally, the Maxilla and Mandible are today the main bone tissue receptor areas in dentistry that have very distinctive characteristics when submitted to the osteolysis processes, which, in turn, require distinctive techniques during the bone transplant. The size and shape of the bones are influenced by several factors, which range from the genetic conditions of the individual to the environment in which they live. In other words, age, sex, physical characteristics, health, diet, race and place of residence are aspects to be considered (Moore, 1990).

Maxillary and Mandibular development and growth are determined by the appearance of teeth from the first months of life. It is interesting to note that the mandibular and maxillary bone tissue responds to intrinsic and extrinsic factors throughout the lifetime of an individual, and, therefore is very plastic, which counteracts its rigid and inert appearance.

Pathophysiology of Bone Loss
Metabolic Factors: • Age; • Gender; • Hormone balance; • Osteoporosis; • Nutritional disorders. *Mechanical Factors:* • Functional (force applied to edge (pressure, compression, tension, shearing)):

• Frequency; • Direction; • Quantity. *Prosthetic Factors:* • Type of prosthesis base; • Shape and type of teeth.

Table 2. Factors related to the bone loss process (Source: Fonseca & Davis, 1995)

One of the factors most closely related to the indication of bone reconstructions in dentistry are maxillary and mandibular resorptions due to lack of the dental element.

The resorption of the alveolar edge is a chronic, progressive, irreversible and cumulative alteration. This condition, observed in the toothless individual, becomes faster in the first six months after exodontias or dental extractions. Once the function of providing support to the teeth has been lost, the alveolar process tends to undergo resorption due to disuse (Mecall & Rosenfeld, 1991). And this resorption can be exacerbated by local factors (traumatism, infections and pathologies), systemic factors (osteopenia, osteoporosis, osteomalacia, endocrine and nutritional alterations), systemic health problems, prosthetic treatments and others (Fonseca & Davis, 1995; Gassen et al, 2008)

Projections by the Brazilian Institute of Geography and Statistics (IBGE) show that the elderly population in Brazil is set to increase considerably in future years. Life expectancy in 2020 is estimated at 71.2 years (men) and 74.7 years (women) and will represent 13% of the population (IBGE, 2011). Data from the Epidemiological Survey indicate the elderly age bracket as having the highest rates of edentulism and of prosthesis use for prolonged periods. (Ministério da Saúde, 2003)

The previous use not only of total prostheses but also of removable partial prostheses is identified as a predisposing factor of tissue resorption (odds ratio = 2.4), and the flaccid tissue from the edge is related to the severity of resorption (odds ratio = 2.4). (Watzek, 1996, Xie et al, 1997)

In this context, it is noted that the majority of atrophic edentulous cases (total or partial) has increasingly resorted to the adoption of dental implants and for this reason, bone grafts appear as an option of biomaterial used in pre-prosthetic surgery (Galea, 2005).

Frozen homologous bone tissue is biocompatible and can be used successfully in treatments that require maxillary sinus lifting. Its use favors bone neoformation, integration, and absence of inflammatory infiltrate as well as an increase in the percentage of bone volume (Stacchi et al, 2008). In the long term, it is possible to observe the formation of viable and mature bone tissue, providing adequate reconstruction techniques are adopted. (Contar et al, 2009)

Some studies in the dental area have evaluated the efficacy of the use of allografts through an analysis of the biomechanics of implants placed in the grafting zone. This is possible through a resonance frequency analysis (RFA) and removal torque (RT) analysis. The results show that there is no difference in stability between implants installed with autogenous and allogeneic grafts. (Ribeiro, 2009) It is also possible to evaluate the osseointegration process through an evaluation of the neoformed bone volume. Lima (2010) studied homologous bone grafts processed in a tissue bank with different methods (lyophilized, demineralized and radiated (ALD); mineralized frozen (ACM) besides autogenous grafts (AT) and blood clot (CG). In the Guided Bone Regeneration (GBR) technique, samples of the groups of grafts were placed in 32 cylinders fastened to the calvaria of 08 animals. After 13 weeks the cylinder fill rates (bone volume of the ALD group) were similar to the ACM and superior to the autogenous graft). Bone neoformation also occurs during the use of homologous grafts in maxillary sinus lifting surgery, besides affording lower morbidity levels (Viscioni et al, 2010). Hence it should be considered a valid alternative for the replacement of autologous grafts in patients submitted to implant therapy.

14. Prospects in the application of osseointegration investigation methodologies in grafting areas

Some techniques can be used today in the analysis of interface sites between receptor and donor bone, which generate information capable of providing subsidies for a greater understanding of the osseointegration process.

14.1 Histomorphometry

Histomorphometry aims to analyze bone morphology and its components (measurements of volume, area, perimeter etc.). This technique is developed primarily for rock analysis, and is currently employed to analyze cellular behaviors of tissues starting from their structural conformity, expressed in thin slices on a slide. The histological reading and the definition of elements that compose the bone microtexture is the main goal of this procedure. The method can be manual, semiautomatic and automatic. A microscope coupled to a micrometer ruler is used in the first. In the semiautomatic version the microscope is coupled to a computer, which in turn uses software that allows users to record and to quantitatively analyze the images of a slide, seen through the microscope lenses, which are projected onto a digitizing board and drawn manually. The definition of each histological structure is performed by a professional. Finally, the automatic technique allows users to capture the images from the microscope using video cameras and the definition of each structure is determined by the actual computer, which automatically analyzes the coloration of each structure. Although the latter is the fastest technique, it is also the least sensitive. (Jorgetti, 2003; Aaron, Shore, 2003)

This technique permits the analysis of primary or static parameters (extension, number and distance) and the derived parameters, which are divided into structural (analyze bone structure) and kinetic (analyze the dynamics of the bone tissue).

The histomorphometric parameters are measured at 125x zoom. The choice of the analysis area should consider the contact surface of the receptor bone with the allograft.

A microscope equipped with objective, micrometer ruler, cursor, digitizing board and image analysis software is used to quantify the structural static, bone tissue formation and resorption histomorphometric parameters. **(Illustration 22)**

The static histomorphometric parameters are classified as:

STRUCTURAL PARAMETERS

- Total area (**T.Ar** mm^2): total area measured;

- Bone Volume (**BV/TV** %): percentage of bone tissue formed by mineralized or unmineralized trabecular bone;
- Thickness of the woven bone (**Tb.Th** μm): thickness of the bone trabeculae expressed in micra;
- Trabecular number (**Tb.N**/mm): the number of bone trabeculae, by millimeter of tissue, which is also an index that expresses trabecular density;
- Trabecular separation (**Tb.Sp** μm): the distance between the bone trabeculae expressed in micra;
- Number of osteocytes (**N.Ot**): number of osteocytes present in the area of the bone tissue evaluated.

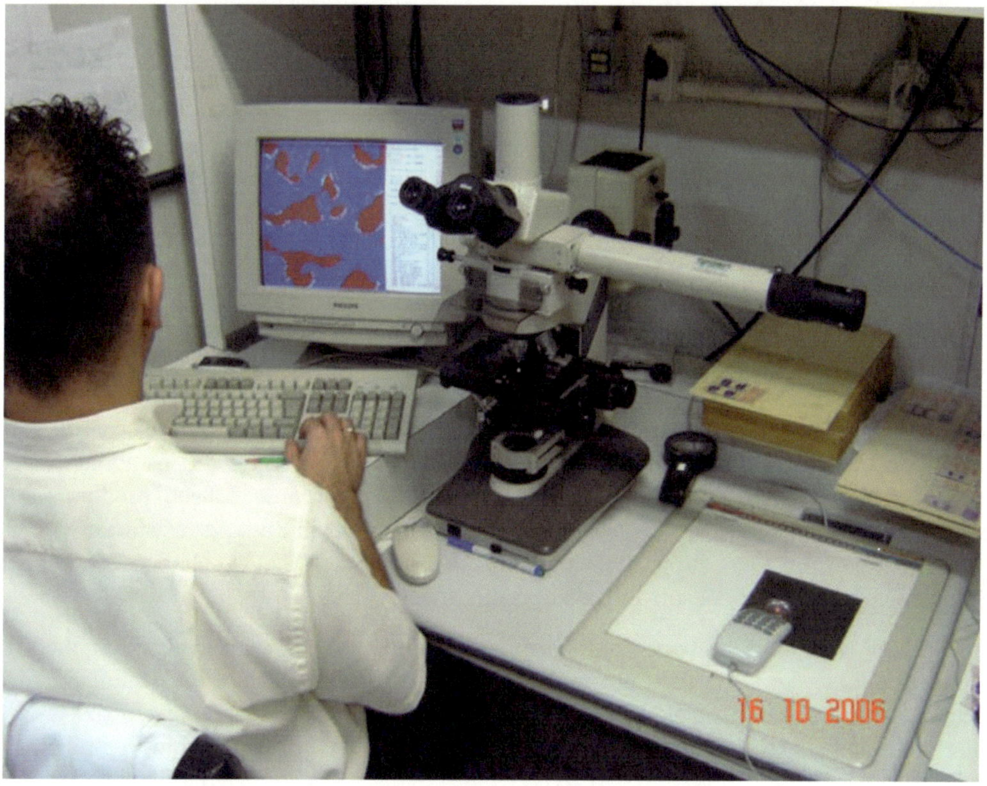

Illustration 14. Equipment used for histomorphometric analysis (microscope coupled to digitizing board and Osteomeasure® software)

FORMATION PARAMETER

- Osteoid volume (**OV/BV** %): percentage of osteoid matrix in relation to the trabecular bone;
- Osteoid surface (**OS/BS** %): percentage of trabecular surface covered by osteoid matrix;
- Osteoblast surface (**Ob.S/BS** %): percentage of the trabecular surface that presents osteoblasts;
- Osteoid thickness (**O.Th** μm): the thickness of the osteoid matrix deposited on the bone trabeculae, expressed in micra;

- Number of osteoblasts (**N.Ob**): absolute number of osteoblasts present in the measured area;
- Number of osteoblasts by tissue area (**N.Ob/T.Ar**): number of osteoblasts by tissue area analyzed, expressed in square millimeters;
- Number of osteoblasts by bone perimeter (**N.Ob/B.Pm**): number of osteoblasts by bone perimeter analyzed, expressed in millimeters.

RESORPTION PARAMETER

- Osteoclast surface (**Oc.S/BS**%): percentage of trabecular surface that presents osteoclasts;
- Number of osteoclasts (**N.Oc**): number of osteoclasts present in the area of the bone tissue evaluated;
- Resorption surface (**ES/BS** %): percentage of surface that presents bone resorption lacunae with or without the presence of osteoclasts.

The histomorphometric parameters adopted in this technique should follow a universal nomenclature agreed upon by the American Society of Bone and Mineral Research - ASBMR (Parfitt et al., 1987).

14.2 Immunohistochemistry

With the advent of osseointegrated dental implants it is extremely important for us to know the cellular processes involved in the osseointegration of allografts, as already mentioned, and this is possible with the immunohistochemistry technique. In vivo mineralization on the implant surface is directly related to the components of the extracellular bone matrix, both collagen and non-collagen components. The bone structure is composed of about 70% of inorganic matter (hydroxyapatite), while the rest is the organic matrix, predominantly consisting of collagen (95%), which is responsible for its flexibility and resistance whereas type I collagen provides support to the mineral structure. The non-collagen components include osteocalcin, osteonectin and osteopontin.

Osteocalcin is related to bone matrix mineralization, and is exclusively expressed by osteoblasts (Ducy et al. 1995). Osteonectin influences the synthesis of extracellular matrix components (Bradshaw et al. 2003) and the calcification of the organic matrix as it is selectively linked to hydroxyapatite and type I collagen fibrils (Termine et al. 1981). Osteopontin is strongly linked to hydroxyapatite crystals and is involved in the anchoring of the osteoclasts in the inorganic bone matrix, controlling crystal nucleation and growth.

Hence the evaluation of these proteins in the osseointegration process can provide us with important information about how the allogeneic bone graft used can corroborate for better interaction of the cellular mechanisms between donor and recipient focusing on the maintenance of the bone tissue after the implant installation.

Moreover, the remodeling of this matrix needs specialized enzymes. Matrix metalloproteinases (MMPs) are an important family of endopeptidases, with 25 known human members, and represent the largest class of enzymes that, collectively, are responsible for the degradation or resorption of extracellular matrix components, and are the only enzymes capable of cleaving fibrillar collagens (Curran et al. 1999), several proteins from the cellular surface and from the pericellular environment and some intracellular proteins.

This technique makes it possible to evaluate bone matrix proteins (type I collagen, osteocalcin, osteopontin and osteonectin) and the MMPs during the osseointegration process of bone grafts in patient to be submitted to future dental implants.

14.3 Peripheral quantitative computed bone tomography (pQCT)

Peripheral quantitative computed tomography - pQCT is a method used to measure the appendicular skeleton that, besides furnishing a broad range of data, also makes it possible to evaluate the cortical and trabecular bones separately.

The obtainment of images is performed by a scanner (ex Stratec XCT Research M instrument scanner (Norland Medical Systems, Fort.) specifically for the analysis of small grafting sites.

The parameters evaluated in this technique are:

- total BMC – total bone mineral content, mg/mm;
- total BMD - total bone mineral density, mg/cm3;
- trabecular BMC - trabecular mineral content, mg/mm;
- trabecular BMD - trabecular mineral density, mg/cm3;
- cortical content, mg/mm;
- cortical density, mg/cm3;
- total area, mm2;
- trabecular area, mm2.
- periosteal circumference, mm;
- endosteal circumference, mm.

15. Discussion

There have been many differences in bone tissue preparation techniques since the first tests in the 30s, yet the desire for the recovery of lost bone tissue among physicians and dentists still remains. The Ministry of Health has recently expressed interest in increasing and promoting this kind of transplant in the country. It is a fact that an increasing number of patients with bone loss for various reasons have approached specialized orthopedics and dentistry services in pursuit of reconstructive solutions. This context includes procedures involving the use of allografts made available by musculoskeletal tissue banks.

In dentistry, the officialization of allograft use arrives very late in 2005. Dental transplanters were curtailed in their use of this kind of treatment for more than a decade. Public policies of oral health were also very late in arriving. The result is perceptible in the statistics that evidence a high rate of edentulism in our population. (Ministerio da Saúde, 2003). In response to this situation, we can also observe high rates of prosthetic rehabilitations in the Brazilian population.

Data from the Epidemiological Oral Health Survey in Brazil (2003) shows that about 67% of the population between 65 and 74 years of age use some kind of prosthesis as a form of edentulism treatment. We know that edentulism and the prolonged use of these prostheses lead to maxillary and mandibular bone resorption over the years (Davis, 1995; Gassen et al,

2008). The high rate of individuals with this necessity, the greater economic access of the population to treatments with osseointegrated implants and to the start of authorization of allograft use in dentistry, corroborated the abrupt growth of bone transplants in dentistry at around **600%** in these last 5 years (**Graph 2**).

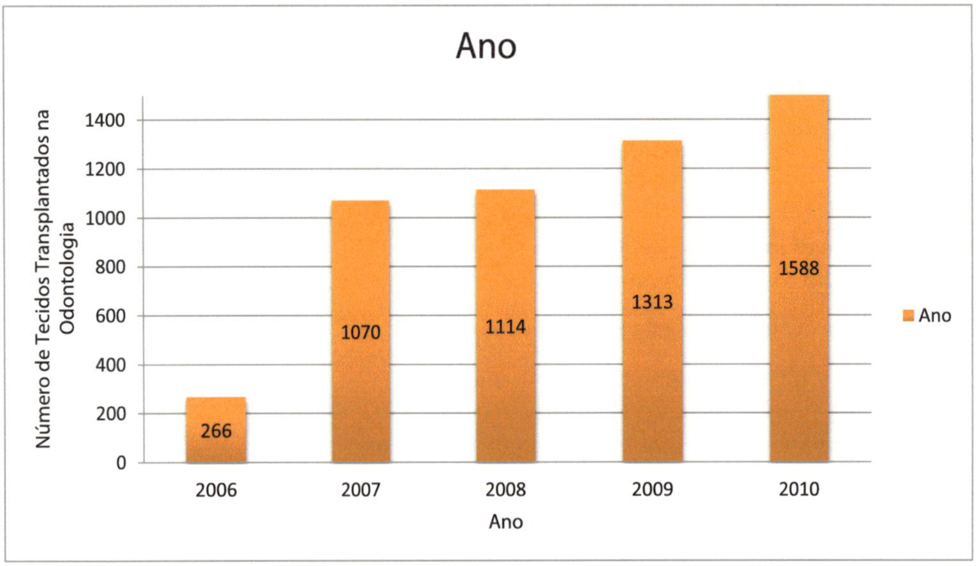

Graph 2. Annual Distribution of homologous tissues made available by a tissue bank that serves as a national reference for Dental Transplants (2006 to 2010).

Such a situation was also observed in our study in relation to the number of transplanting dentists, whose number climbed steeply from **22** to **3585** also in 5 years (**Graph 3**).

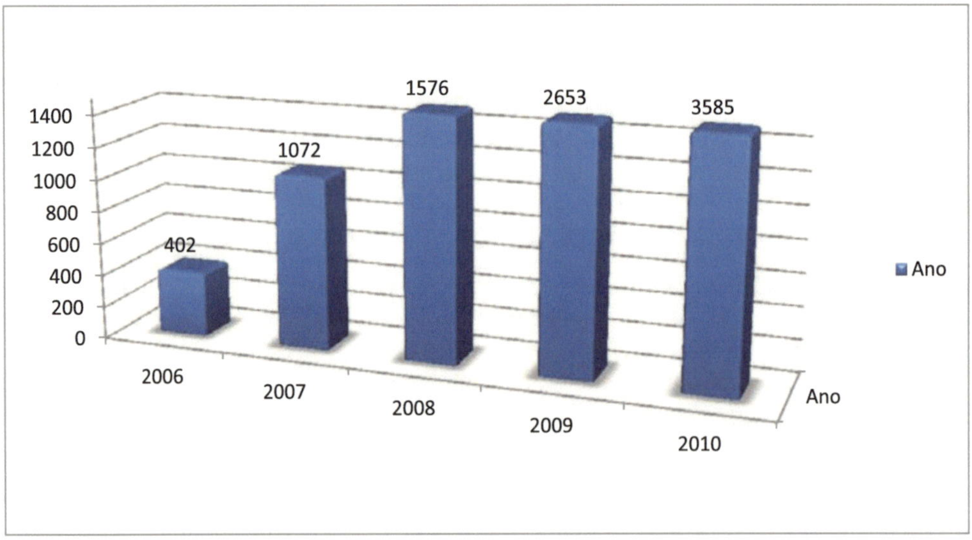

Graph 3. Absolute Frequency of dentists accredited for transplants over the years (2006-2010)

In the next few years we will observe growth of the elderly population in our country (around 13% in 2020), motivated by the increase in life expectancy. It is estimated that in 2020 this expectancy will be 71.2 years for men and 74.7 years for women (IBGE, 2011). Considering that the elderly population is exposed to factors accumulated during their lifetimes that lead to the need for pre-prosthetic reconstructive bone interventions (edentulism, prolonged use of prostheses, osteoporosis, hypothyroidism) we can reflect here on a possible influence over prosthetic rehabilitations of the elderly population in future years, and that in some of them the use of biomaterials (including allograft) may be indicated.

What then has become an obvious trend gives rise to the need to discuss the future of dental transplants. Public health surveillance policies seek, together with trade associations (Regional Federal Council of Dentistry) and tissue banks, the assurance of traceability in procedures of this nature. This is undoubtedly the major challenge for tissue banks. Unlike other kinds of transplants, where the data are compiled and made available by hospitals, in dental transplantations there is the need for total involvement of the transplanting professional in disclosing and informing the adverse effects detected in the treatment. Following the example of the national policy implemented by ANVISA in the area of hemoderivates transfusion, the intention is to create a similar system for the surveillance of tissue transplants in the country, which will include dental transplants. In this model, the adverse effects can be notified nationwide (electronically), enabling the construction of a database that will serve as a source for the definition of corrective and preventive actions of complications in dental surgeries with the use of allografts. Such information is extremely valuable to banks, particularly in the quality control of the tissues produced thereby.

After a more detailed analysis of dental transplants, we detected that approximately 76% are related to the use of cortical tissues, which in our understanding is expected due to the greater availability of this type of tissue at banks. While a spongy segment of a long bone such as the femur and tibia (metaphyseal region) yields 10 to 15 units of spongy blocks, 50 to 60 units of cortical blocks are processed from a diaphysis. It should also be considered that due to the lesser availability of spongy tissues, surgeons have opted in particular for cortical blocks, especially for use in sinus lifting surgery preceding implant installation.

GRAFT MORPHOLOGY	Relative Frequency (%)
Cortical	75.64 %
Spongy	15.86 %
Cortico-Spongy	8.5 %
TOTAL	**100.00**

Table 3. Distribution of Grafts in Terms of Morphology. BTM (musculoskeletal tissue bank) - HCFMUSP, 2006 to 2010.

The most frequent surgical site was undoubtedly the maxilla (98.2%- Graph 4) and we hereby emphasize the main reasons for this finding. Firstly, edentulism in the Brazilian population has appeared to affect the maxilla more than the mandible which certainly leads to the greater use of upper prostheses (57.9%) versus lower prostheses (34.18%) by the population aged between 65 and 74 years (Ministério da Saúde, 2003). Secondly, the actual anatomy of the maxilla due to the presence of maxillary sinuses that are grafted during the sinus lifting technique. This technique is very frequent in the branch in implant dentistry. Thirdly, the lower bone quality of the maxilla in comparison to the mandible, focusing on the primary stability of implants and propensity for resorption after the prolonged use of prostheses. (Mezzomo et al, 2010) It is also worth emphasizing that the maxilla appears as a more esthetic area that demands more attention from individuals concerned about its rehabilitation.

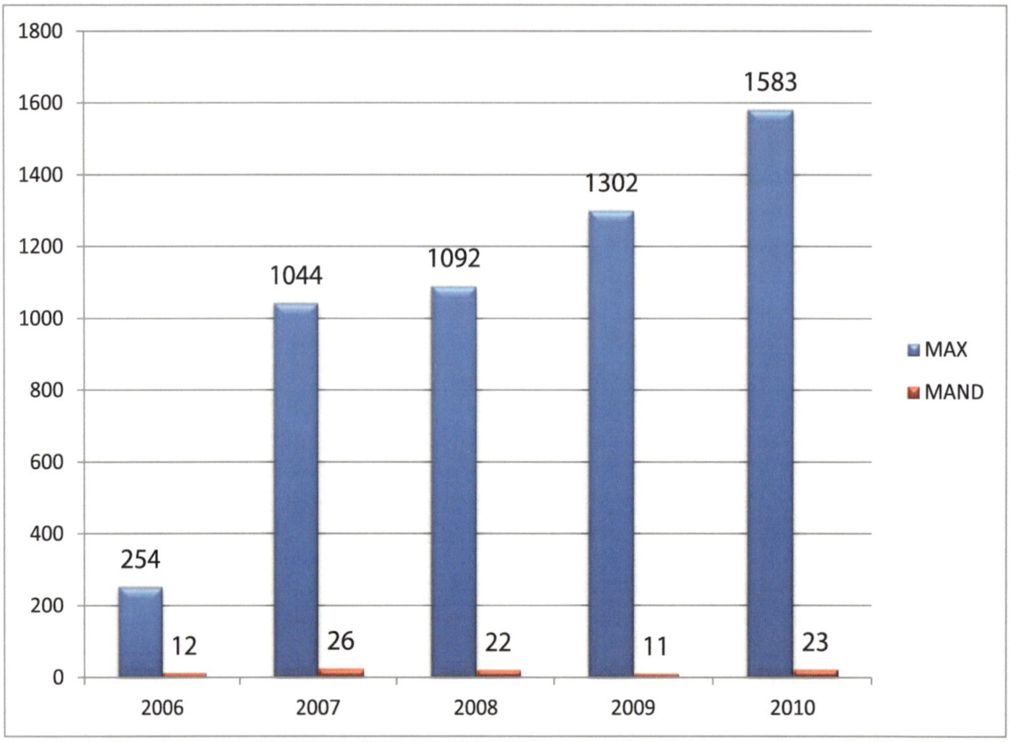

Graph 4. Distribution of Type and Surgical Site of use of the tissues made available by a bank that serves as a national reference from 2006 to 2010.

Although epidemiological data shows the increasing use of homologous tissues in dental surgery, few studies in our field assess their efficacy, hence the need to establish consensuses on their applicability and more importantly, predictability of treatments with the use of allografts. We take predictability to mean knowledge of the responses of bone tissue to the interventions performed in our areas of specialty. Thus in this study we have highlighted a chapter on possible lines of investigation enforceable in our field, which can help us in our search for answers related to the osseointegration process of allografts at sites of dental interest.

The analysis of grafting area by histomorphometry allows us to accurately quantify the structure of the grafted tissue (area, volume) and how much tissue was formed after interaction with the recipient's body (osteoid volume, osteoid surface), to judge the participation of each bone cell (number of osteoblasts and osteocytes) and also to quantify resorptions that have occurred (resorption surface) as a result of this interaction. This data is supplemented by peripheral quantitative computed tomography - pQCT, which unlike histomorphometry, allows us to evaluate the density of the tissue before and after the grafting period. Data of extreme importance, in view of the need for assurance of mechanical stability of dental implants, when implanted in the newly formed bone.

And finally, immunohistochemistry solves some mysteries related to the cellular processes involved in osseointegration. The technique allows us to evaluate, by means of reaction by selected antibodies, the expression of proteins related to the mineralization process (osteocalcin, collagen) and remodeling of bone grafts (osteopontin, osteonectin).

As the basis for all these cellular events, the use of specific antibodies (Tartrate-resistant Acid Phosphatase - TRAP, Osteoprotegerin, Rank-L and Cd34) also allows us to evaluate the inflammation and the vascularization of grafting areas throughout the osseointegration period.

Nowadays in Brazil we are experiencing a time of transition in the indication of these transplants, and studies with this information content are certainly necessary for better knowledge and indication of allografts in dentistry, as well as the criteria and methodologies that can be used to analyze osseointegration.

16. References

[1] Aaron JE, Shore PA. Bone histomorphometry: concepts and commom techniques. In: An YH, Martin KL. Handbook of histology methods for bone and cartilage. New Jersey: Humana Press; 2003. p. 331-51.

[2] Amatuzzi MM, Croci AT, Giovani AMM, Santos LAU, Maragni GG, Shinzato JC. Banco de tecidos:estruturação e normatização. In: Amatuzzi, Joelho:articulação dos membros inferiores. São Paulo:Roca; 2004. p.687-99.

[3] Amatuzzi MM, Croci AT, Giovani AMM, Santos LAU. Banco de tecidos:estruturação e normatização. *Rev Bras Ortop.* 2000; 35(5):165-72.

[4] American Association of Tissue Banks. Standards for Tissue Banking. 11th ed. McLean : American Association of Tissue Banks; 2007.

[5] Aubin JE, Lian JB, Stein GS. Bone formation: maturation and functional activities of osteoblast lineage cells. In: American Society for Bone and Mineral Research. Primer on the metabolic bone diseases and disorders of mineral metabolism. 6th ed. Washington: ASBMR; 2006. p.20-9.

[6] Bancroft JD, Stevens A. Theory and practice of histological techniques.Fouth Edition. Churchill Livingstone. United Kingdom, 1999.

[7] Barros Filho TEP, Rossi JDMBA, Rodrigues CJ, Lage LA, Reis PR. Poder osteogênico dos enxertos ósseos: estudo experimental comparativo entre enxertos autólogo, homólogo irradiado e homólogo AAA. *Rev Bras Ortop.* 1989; 24(1-2):36-40.

[8] Bitar AC, Santos LAU, Croci AT, Pereira JARM, França Bisneto EN, Giovani AMM, Oliveira CRGCM. Histological Study of Fresh versus Frozen Semitendinous Muscle Tendon Allografts. CLINICS 2010;65(3):297-303

[9] Boldt JG, Dilawari P, Agarwal S, Drabu KJ. Revision total hip arthroplasty using impaction bone grafting with cemented nonpolished stems and charnley cups. *J Arthroplasty.* 2001; 16(8):943-52.

[10] Bradshaw, A.D., Graves, D.C., Motamed, K., & Sage, E.H. 2003. SPARC-null mice exhibit increased adiposity without significant differences in overall body weight . *Proc.Natl.Acad.Sci.U.S.A*, 100, (10) 6045-6050 available from: PM:12721366

[11] Brasil, Leis etc. Decreto n.2268 de 30 de junho de 1997. Dispõe sobre a remoção de órgãos, tecidos e partes do corpo humano para fins de transplantes e tratamento. *Diário Oficial da União*, Brasília (DF). 1997 30 jun; seção 1:1.

[12] Brasil, Leis etc. Lei n.9434 de 5 de fevereiro de 1997. Dispõe sobre a remoção de órgãos, tecidos e partes do corpo humano para fins de transplantes e tratamento. *Diário Oficial da União,* Brasília (DF). 1997 5 fev; seção 1:25.

[13] Brasil, Leis etc. Portaria n.1686 de 20 de setembro de 2002. Dispõe sobre a regulamentação para funcionamento de banco de tecidos músculo esqueléticos. *Diário Oficial da União,* Brasília (DF). 2002 24 jul; seção 1:1.

[14] Brasil, Leis etc. PORTARIA Nº 2.600, DE 21 DE OUTUBRO DE 2009. Aprova o Regulamento Técnico do Sistema Nacional de Transplantes. *Diário Oficial da União,* Brasília (DF). 2009 21out.

[15] Brasil, Leis etc. Resolução n. 220 de 27 de dezembro de 2006. Dispõe sobre o Regulamento Técnico para o Funcionamento de Bancos de Tecidos Musculoesqueléticos e de Bancos de Pele de origem humana. *Diário Oficial da União,* Brasília (DF). 2006 29 dez.

[16] Cabrita, H.B., et al., Prospective study of the treatment of infected hip arthroplasties with or without the use of an antibiotic-loaded cement spacer. Clinics (Sao Paulo), 2007. 62(2): p. 99-108.

[17] Carrel A. The preservation of tissues and its applications in surgery. 1912. *Clin Orthop Relat Res.* 1992; 278:2-8

[18] Contar CM, Sarot JR, da Costa MB, Bordini J, de Lima AA, Alanis LR, Trevilatto PC, Machado MÂ. Fresh-frozen bone allografts in maxillary ridge augmentation: histologic analysis. J Oral Implantol. 2011 Apr;37(2):223-31.

[19] Croci AT, Camargo OP, Oliveira CRGC, Baptista AD, Sorrilha A. Estudo histológico dos enxertos ósseos homólogos humanos. *Acta Ortop Bras.* 2003b; 11(4):220-4.

[20] Cunningham N, Reddi AH. Biologic principles of bone induction: application to bone grafts. In: Habal MB, Reddi AH. Bone grafts and bone substitutes.Philadelphia: W.B . Saunders C; 1992. p.93-8.

[21] Curran, S. & Murray, G.I. 1999. Matrix metalloproteinases in tumour invasion and metastasis. *J.Pathol.,* 189, (3) 300-308 available from: PM:10547590

[22] Drumond SN. Transplantes ósseos. In: Pereira WA. *Manual de transplantes de órgãos e tecidos.* 2a ed. Rio de Janeiro: Medsi; 2000. p.359-80.

[23] Ducy, P. & Karsenty, G. 1995. Two distinct osteoblast-specific cis-acting elements control expression of a mouse osteocalcin gene *Mol.Cell Biol.,* 15, (4) 1858-1869 available from: PM:7891679

[24] Enneking WF, Burchardt H, Puhl JL, Piotrowski G. Physical and biological aspects of repair in dog cortical-bone transplants. *J Bone Joint Surg Am.* 1975; 57:237-52.

[25] European Association of Tissue Banks. Common Standards for Tissues and Cells Banking: Berlin: European Association of Tissue Banks; 2004.

[26] Fischer LP, Fischer-Athiel C, Fischer BS. One hundred years of bone surgery in the Lyons Teaching Hospitals (1897-1997). *Ann Chir.* 1998; 52(3):264-78.

[27] Fonseca RJ, Davis WH .Reconstructive Preprosthetic Oral and Maxillofacial Surgery, ed 2. Philadelphia, WB Saunders, 1995, pp 498-510

[28] Galea G, Kearney. Clinical effectiveness of processed and unprocessed bone. *Transfus Med.* 2005; 15(3):165-74.

[29] Gassen, Humberto Thomazi; Filho, Rolf Muner; Siqueira, Bianca Munari de; Oliveira, Samia Bohm; Junior, Aurelício Novaes Silva. Reconstrução óssea de maxila atrófica utilizando enxerto de ramo mandibular. Stomatos, v.14, n.26, jan./jun. 2008.

[30] Gerstenfeld LC, Einhorn TA. Fracture healing: the biology of bone repair and regeneration. In: American Society for Bone and Mineral Research. Primer on the

metabolic bone diseases and disorders of mineral metabolism. 6th ed. Washington: ASBMR; 2006. p.42-48.

[31] Giannoudis PV, Dinapoulos H, Tsiridis E. Bone substitutes: an update. Injury. 2005; 36S:S20-27.

[32] Giovani AMM, Croci AT, Oliveira CRGCM, Filippi RZ, Santos LAU, Maragni GG. Comparative study of cryopreservedbone tissue and tissue preserved in a 98% glycerol solution. Clinics. 2006;61(6):565-70.

[33]Giovani AMM. *Estudo comparativo entre o tecido ósseo criopreservado e o conservado em glicerol a 98%* [dissertação]. São Paulo: Faculdade de Medicina, Universidade de São Paulo; 2005.

[34] Groves EWH. Methods and results of transplantation of bone in the repair of defects caused by injury or disease. *Br J Surg*. 1917; 5(18):185-242.

[35] Heyligers IC, Klein-Nulend J. Detection so living cells in non-processed but deep-frozen bone allografts. *Cell Tissue Bank*. 2005; 6 :25-31.

[36] IBGE: Instituto Brasilerio de Geografia e Estatística: http://www.ibge.gov.br/home/ . Acessado em 15 de julho de 2011.

[37] Janssen ME, Lam C, Beckham R. Outcomes of allogenic cages in anterior and posterior lumbar interbody fusion. *Eur Spine J*. 2001; 10(Suppl 2):158-68.

[38] Jorgetti V. Biópsia óssea e análise histomorfométrica. Manual Fleury de diagnóstico de doenças ósteo-metabólicas. 2003. Disponível em: http://www.fleury.com.br/htmls/cdrom/capitulo5.htm.

[39] Junqueira LC, Carneiro J. Histologia básica. 9a ed. Rio de Janeiro :Guanabara-Koogan; 1999. Cap.8, p.111-28: Tecido ósseo.

[40] Knighton DR, Hunt TK, Thakral KK, Goodson WH 3rd. Role of platelets and fibrin in the healing sequence: an in vivo study of angiogenesis and collagen synthesis. *Ann Surg*. 1982; 196 (4): 379-88.

[41] Lavernia CJ, Malinin TI, Temple HT, Moreyra CE. Bone and tissue allograft use by orthopaedic surgeons. *J Arthroplasty*. 2004; 19(4):430-35.

[42] Lindhe J, Karring T, Lang NP.Tratamento periodontal regenerativo. Tratado de periodontia clínica e implantologia oral. 3a ed. Rio de Janeiro: Guanabara Koogan;1997. p.428-62.

[43] Malafaya, PB, G. A. Silva, E. T. Baran, R. L. Reis, Curr.Opin. Solid State Mater. Sci. 2002, 6, 283.

[44] Mecall RA, Rosenfeld AL. The influence of residual ridge resorption patterns on implant fixture placement and tooth position. Part I. Int J Periodontics Restorative Dent. 1991; 11(1): 8-23

[45] Mezzomo RJ, Garbin CA, Schuh C, Rigo L. Critical analysis of studies of immediately loaded implants supporting fixed prostheses in the edentulous maxxila. Implantnews 7 (4), 2010.

[46] Ministério da Saúde : Levantamento das condições de saúde bucal da população brasileira; 2003. Disponível em : http://dab.saude.gov.br/CNSB/vigilancia.php

[47] Moore KL. Anatomia Orientada para a Clínica. 2ª ed. Rio de Janeiro - RJ: Editora Guanabara. 1990.

[48] Nather A. Organisation, operational aspects and clinical experience of National University of Singapore Bone B. *Ann Acad Med Singapore*. 1991; 20(4):453-7.

[49] Parfitt AM, Drezner MK, Glorieux FH, Kanis JA, Malluche H, Meunier PJ, et al. Bone histomorphometry: standardization of nomenclature, symbols, and units. Report of the ASBMR Histomorphometry Nomenclature Committee. *J Bone Miner Res*. 1987; 2(6):595-610

[50] Perren SM, Claes L. Biologia e biomecânica no manejo de fraturas. In: Rüed TP, Murphy WM. *Princípios AO do tratamento de fraturas*. Tradução de Jacques Vissoky.Porto Alegre: Artmed; 2002. p.7-30.

[51] Phillips GO, Strong DM, Versen RV, Nather A. Advances in Tissue Banking. Vol. 4. Worl Scientific . New Jersey, 2000.

[52] RBT- Registro Brasileiro de Transplantes [on line] São Paulo: Associação Brasileira de Transplantes de Órgãos; 2006-2010. Disponível em: http://www.abto.org.br/profissionais/profissionais.asp.

[53] Santos LAU. Efeito da utilização de plasma rico em plaquetas na osteointegração dos enxertos ósseos homólogos criopreservados : estudo histomorfométrico em coelhos [dissertação]. São Paulo: Faculdade de Medicina, Universidade de São Paulo; 2007. Disponível em: http://www.teses.usp.br/teses/disponiveis/5/5140/tde-16082007-160750/

[54] Santos, LAU et al .Banco de Tecidos Musculoesquelético: Da captação ao uso clínico. In: Mancussi e Faro, AC. Emergencias ortopédicas. São Paulo. Editora Manole.304p.2011

[55] Schreurs BW, Busch VJ, Welten ML, Verdonschot N, Slooff TJ, Gardeniers JW. Acetabular reconstruction with impaction bone-grafting and a cemented cup in patients younger than fifty years old. *J Bone Joint Surg Am*. 2004; 86(11):2385-92.

[56] Sharrard WJW, Collins DH. The fate of human decalcified Bone grafts. *Proc R Soc Med*. 1961; 54:1101-2.

[57] Smith SE, DeLee JC, Ramamurthy S. Ilioinguinal neuralgia following iliac bone-grafting. Report of two cases and review of the literature. *J Bone Joint Surg Am*.1984;66(8):1306-8.

[58] Stacchi C, Orsini G, Di Iorio D, Breschi L, Di Lenarda R. Clinical, histologic, and histomorphometric analyses of regenerated bone in maxillary sinus augmentation using fresh frozen human bone allografts. J Periodontol. 2008 Sep;79(9):1789-96.

[59] Thies RS, Bauduy M, Ashton BA, Kurtzberg L, Wozney JM, Rosen V. Recombinant human bone morphogenetic protein-2 induces osteoblastic differentiation in w-20-17 stromal cells. *Endocrinology*. 1992; 130(3):1318-24.

[60] Tomford WW, Mankin HJ. Bone Banking: Update on methods and materials. *Orthop Clin North Am*. 1999; 30:565-70.

[61] Tomford WW. Bone allografts: past, present and future. *Cell Tissue Bank*. 2000; 1:105-9.

[62] Tuominen T. *Native bovine bone morphogenetic protein in the healing of segmental long bone defects* [dissertação]. Oulu: Division of Orthopaedic and Trauma Surgery, University of Oulu; 2001.

[63] Urist MR, DeLange RJ, Finerman GA. Bone cell differentiation and growth factors. *Science*.1983; 220(4598):680-6.

[64] Urist MR. Bone formation by auto induction. *Science*.1965; 50:893-99.

[65] Viscioni A, Dalla Rosa J, Paolin A, Franco M. Fresh-frozen bone: case series of a new grafting material for sinus lift and immediate implants. J Ir Dent Assoc. 2010 Aug-Sep;56(4):186-91

[66]Watzek G. Endosseous implants:scientific and clinical aspects. Chicago; Quintessence; 1996.

[67] Weyts FA, Bos PK, Dinjens WN, van Doorn WJ, van Biezen FC, Weinans H, Verhaar JA. Living cells in 1 of 2 frozen femoral heads. *Acta Orthop Scand*. 2003; 74(6): 661-4.

[68] Xie Q, Narthi T. Oral status and prosthetic factors related to residual ridge resorption in elderly subjects. Acta Odonto Scand 1997; 55:306-13.

4

Validation of Primary Packaging for Cryopreserved Musculoskeletal Tissues

Luiz Augusto Ubirajara Santos[1], Rosa Maria Vercelino Alves[2], Alberto T. Croci[1], Fábio Gomes Teixeira[2], Paulo Henrique Kiyataka[2], Marisa Padula[2], Mary Ângela Fávaro Perez[2], Monica Beatriz Mathor[3], Renata Miranda Parca[6], Arlete M.M . Giovani[1], Cesar Augusto Martins Pereira[4], Graziela Guidoni Maragni[1], Thais Queiróz Santolin[1] and Lucas da Silva Pereira[5]

1. Introduction

Bones and tendons are obtained from donors who have been pronounced brain dead after a rigid and extensive screening process. The tissues obtained are sent to a Tissue Bank and submitted to processing steps, packaging and cryopreservation at -80°C. It is vital to maintain sterility and integrity, so that no tissue is discarded.These precautions also extend to the packaging, which should promote containment and protection. Although minimum technical criteria have already been defined for the food industry, this has still not been regularized in Brazil. We emphasize, therefore, the need to study this subject, focusing on maintaining the quality of the musculoskeletal tissues produced by tissue banks. All procedures developed byTissue Banks currently present in the country have rigid control over the quality and the traceability of tissues made available, based on international standards (EATB, 2004; AATB, 2007) on the legislation (BRASIL, 1997-2006) and in conformity with Good Manufacturing Practices - GMP.

2. Cryopreservation of musculoskeletal tissues

In the cryopreservation room, the tissues are stored according to their status in the process. Thus, there are designated areas for tissues under analysis or in quarantine (where they remain for around 60 days before the result of all the exams) and for those that have been liberated for use. both areas are equipped with ultra-low temperature freezers, with temperatures ranging from minus 80 to minus 100 degrees Celsius.

---

[1]*Institute of Orthopedics and Traumatology, Hospital das Clínicas of the University of São Paulo School of Medicine, Sao Paulo, SP, Brazil*

[2]*Packaging Technology Center – Institute of Food Technology, Campinas, SP, Brazil*

[3]*Nuclear and Energy Research Institute - IPEN/CNEN-SP – São Paulo, Brazil*

[4]*Biomechanics Laboratory - Institute of Orthopedics and Traumatology, Hospital das Clínicas of the University of São Paulo school of Medicine, Brazil*

[5]*Dental Student Institute of Orthopedics and Traumatology, Hospital das Clínicas of the University of São Paulo school of Medicine, São Paulo, SP, Brazil*

[6]*National Health Surveillance Agency- ANVISA*

The room is also equipped with an air conditioning system, its own energy generator, and carbon gas backup as protection against defrosting, as well as a rigorous temperature monitoring system that generates 24-hour printed temperature recordings, and a satellite alarm system, which ensures adequate maintenance of the temperature and early detection of any flaws.

Depending on the results of the analyses, the tissues are transferred to the room designated for materials liberated for use. The maximum cryopreservation period is 5 years for bone tissue and 2 years for soft or tendinous tissues.

It is vital to maintain sterility and integrity, in order to avoid disposal of any material. This is the purpose of **packaging**, which is aimed at containment and protection. In Brazil, there are no specific regulations for packaging of sterile, cryopreserved tissues for transplants, and in this chapter, we present our experience in the definition and validation of a type of packaging used for this purpose. Our proposal is to characterize a coextruded plastic film structure used for packaging of musculoskeletal tissues at low temperatures, in regard to the aspects sterility, cytotoxicity, migration, mechanical resistance and oxygen permeability.

Fig. 1. Cryopreservation of tissues at -80 °C in drawers, separated by batches.

3. Validation of packaging for use in tissue cryopreservation.

3.1 Description of our sample

A roll of transparent, unprinted seven layers, coextruded plastic:

Layer 01: LLDPE - 1-octene comonomer Linear Low-Density Polyethylene
Layer 02: LLDPE - 1-octene comonomer Linear Low-Density Polyethylene and 1-octene comonomer Linear Low-Density Polyethylene modified with anhydride maleic
Layer 03: PA - hexamethylenediamine, adipic acid e caprolactam copolyamide
Layer 04: LLDPE - 1-octene comonomer Linear Low-Density Polyethylene and 1-octene comonomer Linear Low-Density Polyethylene modified with anhydride maleic
Layer 05: PA - hexamethylenediamine, adipic acid e caprolactamcopolyamide
Layer 06: LLDPE - 1-octene comonomer Linear Low-Density Polyethylene and 1-octene comonomer Linear Low-Density Polyethylene modified with anhydride maleic
Layer 07: LLDPE - 1-octene comonomer Linear Low-Density Polyethylene

In relation to the physicomechanical and barrier properties characteristics, we investigated the alterations that occurred in the packages after 30, 60, 90, 120, and 150 days of cryopreservation at -80°C. For the cytotoxicity and sterility analysis, two groups (before and after sterilization by ethylene oxide) were analysed.

3.2 Physicomechanical and barrier properties tests

A) Characterization of coex film in relation to its thickness and water vapor transmission rate (WVTR)

The thickness of each layer of plastic material in the film sample was determined through images captured by a Metaval inverted microscope operating at a magnification of 200x, using the image analysis system Axio Vision (Zeiss®). The cross-section of the sample was obtained using a Leica microtome, model RM2245, with section thickness set to 40 µm. To facilitate visualization of the barrier layer, 2% iodine solution was used as a contrasting agent. Five cross-section samples of the material were obtained; five measurements were performed for each sample, totalling 25 thickness measurements. The test was carried out at a temperature of around 23°C, after packaging of the sample in a controlled environment at 23°C ± 2°C and (50 ± 3)% relative humidity for a minimum period of 48 hours.

The water vapor transmission rate was determined using a MOCON PERMATRAN-W 3/31 device, following the procedure described in regulation ASTM F1249-06 - Standard test methods for water vapor transmission rate through plastic film and sheeting using a modulated infrared sensor. In this test, the water vapor that passes through the film is carried to the infrared sensor by ultra-dry nitrogen flow. The sensor measures the fraction of energy absorbed by the water vapor, and emits an electrical signal with amplitude proportional to the concentration of water vapor. The range of this signal is compared with that of the signal produced by the water vapor that passes through a calibration film with a known water vapour transmission rate. The effective permeation area of each sample was $50 cm^2$. The assay was performed at 38°C/100%RH and in this condition, the calibration standard showed water vapour transmission rate of 4.54g water.m^{-2}.day^{-1}. The water vapor transmission rate of the sample was corrected for the condition 38°C/90%RH, multiplying the results by a factor of 0.9.

The total thickness of each layer of coextruded film is shown in Table 1. Figure 10 shows an example of a cross-section image of the sample obtained for the assay.

Determinations	Thickness (μm)		
	Mean	Variation Interval	Coefficient of variation (%)
Total	91.1	90.0 - 92.9	0.9
LLDPE (A)	27.3	26.2 - 28.5	2.4
LLDPE blend (B)	7.6	7.0 - 8.0	3.8
PA (C)	5.6	5.1 - 5.9	4.8
LLDPE blend (D)	6.8	5.3 - 7.8	11
PA (C)	5.6	5.0 - 6.0	5.6
LLDPE blend (F)	7.7	7.1 - 8.6	5.9
LLDPE (G)	31.0	30.1 - 32.1	1.9

Values for 25 measurements
A-G: Text for visualization in Figure 10

Table 1. Total thickness of each layer of sample of coextruded plastic film.

Thus, by optical microscopy, it was observed that it is a coextruded material with seven layers, with two intermediate layers of PA of approximately 6μm each. The other LLDPE layers totalled 80μm.

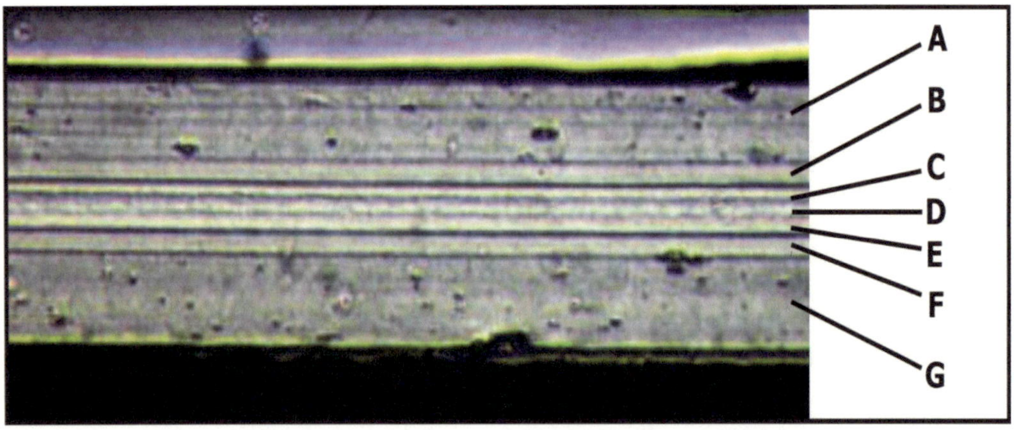

Fig. 2. Example of cross-section image of the sample obtained with the microscope operating at a magnification of 200x.

The characterization of the film in relation to water vapor transmission rate is shown in Table 2.

Sample	WVTR (g water . m-2.day-1)		
	Mean	VI	CV (%)
Coextruded film LLDPE / LLDPE / PA / LLDPE / PA / LLDPE / LLDPE	3.30	3.2 - 3.4	4.4

Values relate to four measurements
VI - variation interval; CV - coefficient of variation

Table 2. Water vapor transmission rate (WVTR) of the coextruded plastic film at 38°C/90%RH - Permatran-W 3/31 method.

The WVTR depends on the thickness of the LLDPE layers of the coex film (80μm).

B) Penetration resistance

The penetration resistance of the coextruded film was determined based on Standard ASTM F 1306-90 (2008) e1 - Standard test method for slow rate penetration resistance of flexible barrier films and laminates, on an Instron 5500R universal testing machine, using load cells of 100 N. The speed of penetration, performed with a spherical-tipped metal probe with diameter of approximately 3.2 mm, was 25 mm/min. The penetration was performed from the inner surface to the outer surface of the material. The test was conducted in an environment of 23°C ± 2°C and (50 ± 3)% relative humidity, after leaving the packaged samples for a minimum period of 48 hours in this environment.

The evaluation of penetration resistance was carried out on samples in their original or non-frozen condition (0) and 30, 60, 120 and 150 days after freezing at -80 °C. The data are shown in Table 3 and Figure 3.

Penetration resistance		Evaluation periods (days)					
		0	30	60	90	120	150
Force at break (N)	Mean	12.1 ^{ab}	12.6 ^{bc}	12.9 ^{c}	12.6 ^{bc}	11.1 ^{d}	11.8 ^{a}
	VI	11.5-12.7	11.8- 13.5	11.9- 14.2	11.2- 14.1	9.4 - 12.3	10.9- 12.4
	CV (%)	3.1	3.6	4.7	6.9	9.1	3.7
Deformation at break (mm)	Mean	9.4 ^{a}	11.0 ^{b}	10.0 ^{c}	9.8 ^{c}	8.3 ^{d}	9.0 ^{e}
	VI	9.2 - 9.8	10.7- 11.4	9.5 - 10.5	8.9 - 10.6	7.4 - 9.1	8.2 - 9.5
	CV (%)	2.3	2.3	2.6	5.2	5.5	3.7
Energy at break (mJ)	Mean	70 ^{ad}	76 ^{b}	78 ^{b}	73 ^{ab}	62 ^{c}	69 ^{d}
	VI	64 - 74	70 - 84	72 - 85	64 - 86	52 - 73	62 - 73
	CV (%)	4.6	5.5	5.4	8.6	11	5.6

Values relate to 10 measurements: 1 N = 0.102 kgf
VI/CV: Variation interval/Coefficient of variation
a,b,c,d: for an analyzed property, mean values on the same line accompanied by the same superscript letter did not show any difference between them in the least significant difference (LSD) test, at a level of error of 5%.

Table 3. Penetration resistance of the sample during storage at -80°C.

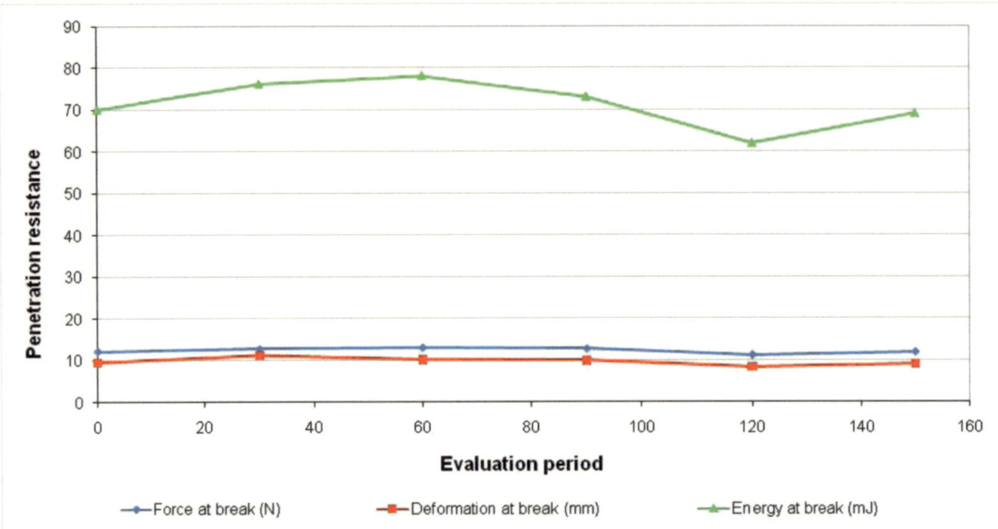

Fig. 3. Penetration resistance of the sample during the period of storage at -80°C (days).

The results shown in Table 3 and Figure 3 demonstrate that the process of sterilization and packaging of human bone, and storage for 150 days at -80°C, did not alter the penetration properties of the coextruded film. Variations were observed in the results of the three properties evaluated, including statistical differences in some periods/properties, but as these variations were small, without any clearly-defined trend, no alteration is expected in penetration resistance of the film during packaging and storage of human tissue for a period of 150 days at -80 °C.

C) Seal strength

The seal strength of the packaging was determined according to ASTM F 88/F 88M-09 - Standard test method for seal strength of flexible barrier materials. Samples of 25.4 mm in width were submitted to tensile test in a 5500R Instron universal testing machine, operating with load cells of 50N and 100N, at a speed of 300mm/min. The distance between the fixing clamps and the sample was 25mm. The test was conducted in an environment of 23 °C ± 2 °C and (50 ± 3)% relative humidity, after leaving the pre-prepared, packaged samples in this same environment for at least 48 hours.

The strength of the side seal is different from that of the top seal, because the top seal of the packaging was made after the bone tissue packaging process, while the side seals were made by the package manufacturer.

Despite this, there was no statistical difference between the results obtained for both seals. Thus, the sterilization and packaging process of human tissue, and storage for 150 days at -80°C did not alter the seal strength.

Maximum seal strength (kgf/25.4 mm)		Evaluation periods (days)					
		0	30	60	90	120	150
Side seal	Mean	3.60 ab	3.65 b	3.60 ab	3.73 b	3.73 b	3.44 a
	VI	3.36 – 3.95	3.41 – 3.83	3.38 – 3.80	3.34 – 3.97	3.43 – 3.96	3.12 – 3.78
	CV (%)	4.4	**4.0**	**4.4**	**5.7**	**4.5**	7.0
Top seal	Mean	-	4.69 a	4.75 a	5.07 a	5.09 a	4.84 a
	VI	-	3.65 – 5.20	3.20 – 5.85	4.31 – 5.95	4.68 – 5.54	4.48 – 5.44
	CV (%)	-	9.6	15	9.6	5.1	6.6

Values relate to ten determinations: 1 kgf/25.4 mm = 386.1 N/m
VI/CV: Variation interval/Coefficient of variation
a,b: for a seal type, mean values on the same line accompanied by the same superscript letter did not show any difference between them in the least significant difference (LSD) test, at a level of error of 5%.

Table 4. Seal strength during storage at -80°C.

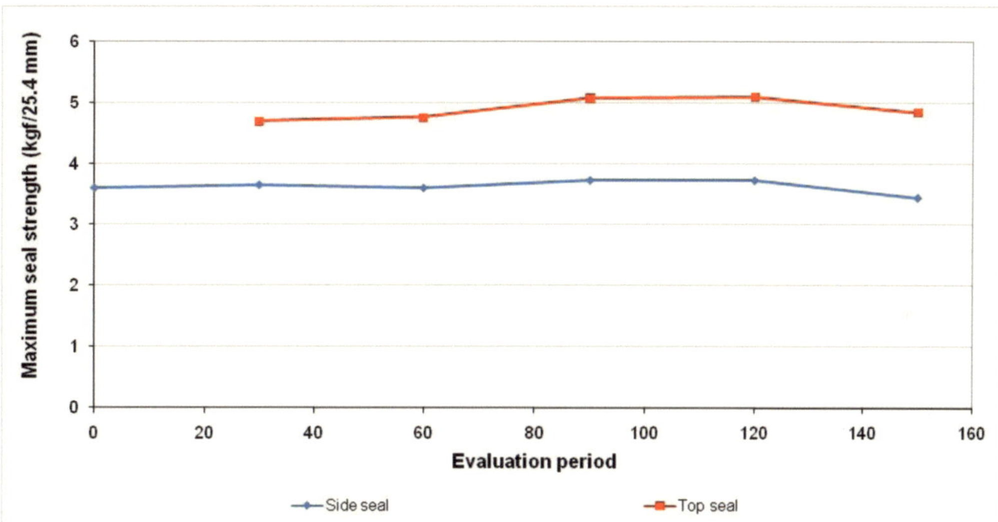

Fig. 4. Maximum seal strength during the period of storage at -80°C (days).

D) Oxygen transmission rate

The oxygen transmission rate at humid was determined by the coulometric method, according to the procedure described in ASTM F 1927 - Standard test method for determination of oxygen gas transmission rate, permeability and permeance at controlled relative humidity through barrier materials using a coulometric detector, on MOCON OXTRAN device, model 2/20, operating with pure oxygen as permanent gas. The tests were carried out at 23°C to 75%RH, with the samples packaged for 88 to 112 hours in a temperature-controlled room at 25°C and 75%RH. The effective area of permeation of each sample was 50cm². The result obtained was corrected for 1 atm of partial pressure gradient of oxygen.

O_2TR mL (STP).m-2.day-1 at 23°C/75%RH	Evaluation periods (days)					
	0	30	60	90	120	150
Mean	69.75 a	75.95 a	74.97 a	82.57 a	76.98 a	113.67 b
VI	68.63–70.87	71.77-80.14	74.48-75.46	79.34-85.80	73.83-80.13	105.05-122.29
CV (%)	2.3	7.8	0.9	5.5	5.8	10.7

values relate to two determinations
VI – variation interval; CV – coefficient of variation
a,b: for a seal type, mean values on the same line accompanied by the same superscript letter did not show any difference between them in the least significant difference (LSD) test, at a level of error of 5%.

Table 5. Oxygen transmission rates (O_2TR) during storage at -80°C.

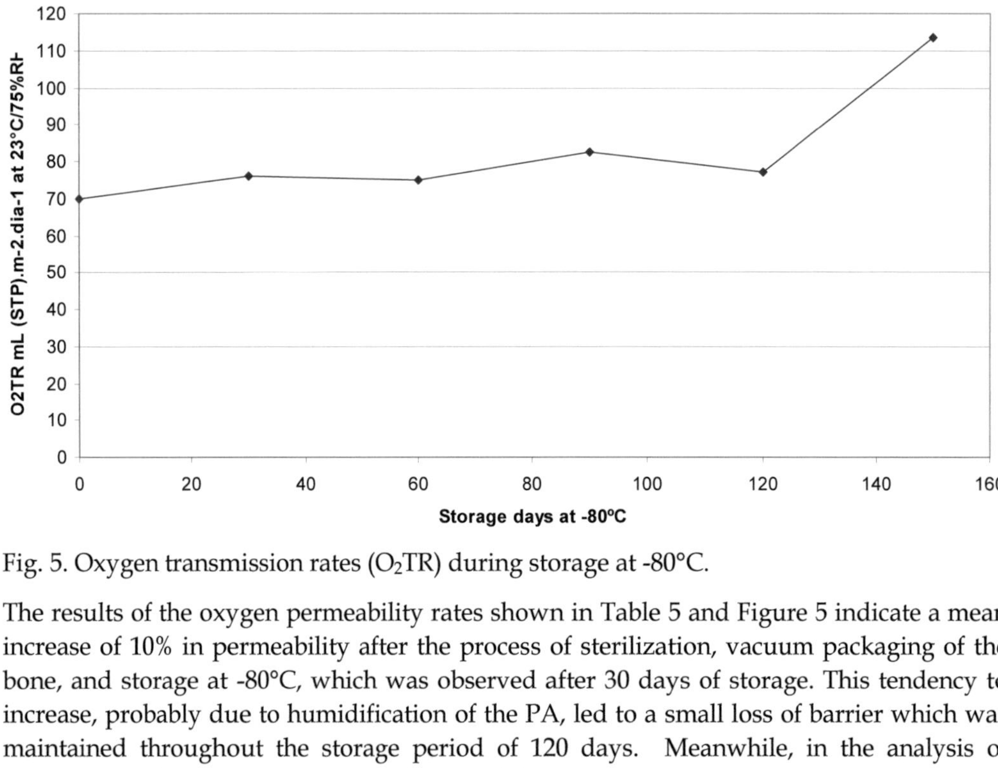

Fig. 5. Oxygen transmission rates (O_2TR) during storage at -80°C.

The results of the oxygen permeability rates shown in Table 5 and Figure 5 indicate a mean increase of 10% in permeability after the process of sterilization, vacuum packaging of the bone, and storage at -80°C, which was observed after 30 days of storage. This tendency to increase, probably due to humidification of the PA, led to a small loss of barrier which was maintained throughout the storage period of 120 days. Meanwhile, in the analysis of samples from 30 days of storage at -80°C, a high oxygen transmission rate of was observed, which was not expected, and as we did not have any more stored samples, it was not possible to re-evaluate this result. This higher O2TR may be the result of some variation in thickness of the PA in the samples evaluated in this period. In any case, the level of oxygen transmission rate of the film did not lead to loss of vacuum in the samples stored in the Tissue Bank for 150 days at -80°C.

E) Overall Migration

The evaluations of overall migration were performed according to Resolution 51 of 26 November 2010 published by *Agência Nacional de Vigilância Sanitária do Ministério da Saúde* (National Agency of Sanitary Surveillance of Health Ministry) in Diário Oficial da União (official journal of the Brazilian Government) on 30 November 2010. This Resolution internalizes Mercosul Technical Regulation GMC 32/10.

The methodology to quantify the overall migration was according to the method of the European Standard EN 1186-1: materials and articles in contact with foodstuffs. Plastics. Part 1: guide to the selection of conditions and test methods for overall migration, EN 1186-3: Materials and articles in contact with foodstuffs. Plastics. Test methods for overall migration into aqueous food simulants by total immersion and EN 1186-14: materials and articles in contact with foodstuffs. Plastics. Part 1: Part 14: test methods for "substitute tests" for overall migration from plastics intended to come into contact with fatty foodstuffs using test media isooctane and 95% ethanol and consists of the sample contact with extraction solutions in certain periods and temperatures that simulate their actual condition of use. The residues of overall migration were determined by the weight difference after the contact and evaporation of the solutions through an analytical scale with 0.01mg of accuracy. The sample was evaluated under the contact conditions shown in Table 6.

Model Solution	**Contact condition**
Ultra purified water	40°C/10 days
Acetic acid solution in Ultra purified water at 3% (w/v)	40°C/10 days
Isooctane	40°C/10 days

Table 6. Conditions of time and temperature used in the overall migration.

The results of the overall migration tests performed on the transparent coextruded plastic film, obtained using the model solutions and the specific contact conditions are shown in Table 7.

Model Solution/Contact Condition	**Maximum limit of overall migration**	**Sample**	**Mean**	**Standard Deviation**	**Variation interval**
Ultra purified water/ 40°C/10 days	8.0	Before EtO	≤ 0.72	0.21	≤ 0.50(2) - 1.00
		After EtO	≤ 0.51	0.02	≤ 0.50(2) - 0.55
3% Acetic acid solution (w/v)/ 40°C/10 days	8.0	Before EtO	≤ 0.54	0.09	≤ 0.50(2) – 0.67
		After EtO	0.92	0.19	0.72 – 1.12
Isooctane/ 20°C/48 hours	8.0	Before EtO	≤ 1.21	0.64	≤ 0.50(2) – 1.85
		After EtO	≤ 1.50	0.72	≤ 0.50(2) – 2.05

(1) Result of four determinations.
(2) Limit of quantification of the method in the analytical conditions used.
(3) Not applicable.

Table 7. Residues of overall migration obtained for the transparent coextruded plastic film, before and after the application of EtO, in mg/dm^2 (1).

The maximum limit of overall migration provided by Resolution nº105/99 is of 8mg of residue per dm^2 of contact plastic material, with an analytical tolerance of 10%. Therefore 8.8mg/dm^2 is the maximum tolerable value.

The overall migration values found in the samples analyzed, in the analytical conditions used, were below the established limit. There was no statistical significance with sterilization with ethylene oxide (EtO).

E 1.) Physicochemical tests - According to USP 33

The physicochemical tests were conducted based in the methodology describe in the Chapter <661> Containers - Plastics - Physicochemical Tests of the **United States Pharmacopeia (USP 33)**.

In accordance to the United States Pharmacopeia, physicochemical tests are designed to determine physical and chemical properties of plastic materials. The extracts methodology consists of the sample contact with a extraction solution (deionized water) at 70ºC during 24 hours, maintaining the ratio area / volume of 120 cm^2 total surface area of plastic material for each 20 mL of extraction solution.

The analyzed sample was received in the form of the film cut into strips with dimensions of 5.0 cm long, 0.3 cm wide and thickness less than 0.1 cm. In this case, the thickness of the material to determine the total area was not considered and was maintained the ratio of 120 cm^2 for each 20 mL of extraction solution. Water was used as extraction solution.

After the contact, the extraction solution and blank reagent were analyzed by the following tests:

Buffering Capacity: Titrate 20 mL of the extraction solution potentiometrically to a pH of 7.0, using 0.01 N sodium hydroxide. Treat a 20.0 mL portion of the blank reagent similarly. The difference between the two volumes can not be greater than 10.0 mL.

Nonvolatile Residue: 50 mL of the extraction solution were evaporated on a hot plate, after the residue was dried at 105ºC for 1 hour on a oven and finally the nonvolatile residue was weighted through an analytical balance with 0.01mg of accuracy. Treat a 50.0 mL portion of the blank reagent similarly. The difference between the two volumes can not be greater than 15.0 mg.

Residue on Ignition: in the residues obtained in nonvolatile residues test, add sulfuric acid and burn on the muffle furnace until constant weight. Treat the blank reagent similarly. The difference between the two volumes should not be greater than 5.0 mg. It is not necessary to perform this test when the nonvolatile residue test result does not exceed 5.0 mg.

Heavy Metal, as lead: an aliquot of extraction solution has been transferred to a volumetric flask and acidified with nitric acid and the volume was completed with the extraction solution. After treating the sample, the lead content was quantified by atomic emission spectrometry induced by plasma, with an optical detector, in a Perkin Elmer equipment, model OPTIMA 2000DV, using appropriate calibration curves for the analyses. This test was conducted in replacement to the heavy metal test stablished by American Pharmacopeia, whose result is expressed as lead (and is based on the colour comparison among the test solution and a solution of lead with concentration of 1.0 mg / kg).

Physicochemical Assay	Limit based on USP	Sample	Mean	Standard deviation	Variation interval
Buffering capacity (mL)	10.0 mL	Before EtO	≤ 0.5(2)	(3)	(3)
		After EtO	≤ 0.5(2)	(3)	(3)
Non-volatile residue: (mg)	15 mg	Before EtO	≤ 1.0(2)	(3)	(3)
		After EtO	≤ 1.0(2)	(3)	(3)
Heavy Metals (as lead) (mg/kg (ppm))	1 mg/kg (ppm)	Before EtO	≤ 0.05(2)	(3)	(3)
		After EtO	≤ 0.05(2)	(3)	(3)

(1) Results of three determinations.
(2) Corresponds to the limit of quantification of the method in the analytical conditions used.
(3) Values no applicable

Table 8. Physicochemical assays of the analyzed samples(1).

E 2.) Assays According to the European Pharmacopoeia

The tests acidity or alkalinity, sulfated ash, absorbance, extractable aluminum, chromium, titanium, vanadium, zinc, zirconium and extractables from heavy metals, expressed as lead were conducted based in the methodology described in the European Pharmacopoeia, Chapters "3.1.3 Polyolefines", "3.1.4 Polyethylene without Additives for Containers for Parenteral Preparations and for Ophthalmic Preparations" and "3.1.5 Polyethylene with Additives for Containers for Parenteral Preparations and for Ophthalmic Preparations".

Extractables of aluminium, chromium, titanium, vanadium, zinc and zirconium

The methodology for quantification of extractables aluminum, chromium, titanium, vanadium, zinc and zirconium involved contact of 100 grams of sample with a solution of 0.1 M hydrochloric acid for one hour at the reflux temperature. After treating the sample, the metals contents were quantified by atomic emission spectrometry induced by plasma, with an optical detector, in a Perkin Elmer equipment, model OPTIMA 2000DV, using appropriate calibration curves for the analyses.

Extractables of Heavy Metals, expressed as lead

The same procedure described for quantification of extractable aluminum, chromium, titanium, vanadium, zinc and zirconium was used. The method for quantification of extractable lead was used instead of the colorimetric method of determination of heavy metals established by the European Pharmacopoeia, which is based on color comparison between the extracting solution and a solution of lead at a concentration of 2.5 mg/kg.

Alkalinity or Acidity, and Absorbance

The method for quantification of alkalinity or acidity and absorbance involved contact of 12.5 grams of sample with 250 mL of deionized water for five hours at the reflux temperature.

After the extraction time, both the extracting solution in contact with the samples and a blank solution (reference) were assessed as to the following tests:

Alkalinity and Acidity: measurement of pH in a Micronal pHmeter, model B 474 and titration of 100 mL of extracting solution with sodium hydroxide 0.01M or hydrochloric acid 0.01M up to pH 7.0.

Absorbance: absorbance of the extracting solution was measured in the 220 nm to 340 nm wavelength range, using a quartz cuvette with 10 mm pathlenght in a UV / VIS spectrophotometer, using an Analytik Jena instrument, model Specord 210.

Sulphated Ash

Sulfated ash were assessed per requirements in European Pharmacopoeia 6.0, Chapter 2.4.14 - Sulfated Ash. The method for quantification of sulfated ash consisted in weighing 5.00 g ± 0.01 g of sample on an analytical scale with 10^{-5} g resolution and incineration at a temperature of 600 °C ± 20 °C using a Milestone microwave heating furnace, model Pyro. After incinerating the sample, the ash were determined gravimetrically, using an analytical scale with 10^{-5} g resolution. There was not need to use sulfuric acid.

Extractable aluminum, chromium, titanium, vanadium, zinc and zirconium

The results of the extractable aluminum, chromium, titanium, vanadium, zinc and zirconium tests for the analyzed sample are shown in Table 9.

Extractables	Limit based on European Pharmacopeia 6.0	Sample	Mean	Standard Deviation	Variation interval
Al	1.0	Before EtO	0.71	0.08	0.62 - 0.78
		After EtO	0.74	0.06	0.70 - 0.81
Cr	0.05	Before EtO	0.03	0.00	0.02 - 0.03
		After EtO	0.33	0.05	0.27 - 0.37
Ti	1.0	Before EtO	≤ 0.10(2)	(3)	(3)
		After EtO	≤ 0.10(2)	(3)	(3)
V	0.10	Before EtO	≤ 0.10(2)	(3)	(3)
		After EtO	≤ 0.10(2)	(3)	(3)
Zn	1.0	Before EtO	0.05	0.00	0.05 - 0.06
		After EtO	0.05	0.01	0.05 - 0.06
Zr	0.10	Before EtO	≤ 0.05(2)	(3)	(3)
		After EtO	≤ 0.05(2)	(3)	(3)

(1) Result of four determinations.
(2) Corresponds to the limit of quantification of the equipment in the analytical conditions used.
(3) Not applicable.

Table 9. Extractables of the metals aluminum (Al), chromium (Cr), titanium (Ti), vanadium (V), zinc (Zn) and zirconium (Zr), in mg/kg(1).

Extractable Heavy Metal, Expressed as Lead

The results of the extractable heavy metals, expressed as lead, acidity, absorbance and sulphated ash tests, for the analyzed sample are shown in Table 10.

Physicochemical Assay	Limit based on European Pharmacopeia 6.0	Sample	Mean	Standard Deviation	Variation interval
Extractables of Heavy Metals, expressed as lead (ppm)	2.5mg/kg (ppm)	Before EtO	≤ 0.1(2)	(3)	(3)
		After EtO	≤ 0.1(2)	(3)	(3)
Acidity (mL of NaOH 0.01 M)	1.5 mL de sodium hydroxide 0.01 M	Before EtO	≤ 0.5(2)	(3)	(3)
		After EtO	≤ 0.5(2)	(3)	(3)
Absorbance (UA)	0.2 UA	Before EtO	0.23	0.01	0.22 - 0.23
		After EtO	0.24	0.04	0.22 - 0.26
Sulphated Ash (%)	0.02% and 1.00%(4)	Before EtO	0.29	0.01	0.28 - 0.30
		After EtO	0.25	0.12	0.11 - 0.33

UA- Unit of Absorbance.
(1) Results of three determinations.
(2) Corresponds to the limit of quantification of the method in the analytical conditions used.
(3) Values no applicable
(4) Varies according to the presence of additives

Table 10. Results for heavy metals, expressed as lead, acidity, absorbance and ash of the samples analyzed (1).

The lead values, volume of 0.1 M sodium hydroxide used, and sulphated ash values found in the samples analyzed were below the maximum limits established in the European Pharmacopeia 6.0. In relation to absorbance, the values obtained in the two samples analyzed, before and after the application of sterilization with ethylene oxide (EtO), were slightly higher than the maximum limits established.

A038-2/11 - Packaging for human tissue - Final Report 17/20 This means that some substance of the coextruded film may have migrated to the extraction solution in contact with the sample (deionized water), a fact that requires further investigation.

It should be emphasized that the methodology of the American and European Pharmacopeias apply to single-layer packaging, and that substances from internal layers of the film analyzed may have been extracted, slightly increasing the absorbance of the extraction solution.

E 3.) Specific migration

The extruded film before and after sterilization with ethylene oxide was evaluated in relation to specific migration of 1-octene and ε-caprolactam. The evaluations of specific migrations of 1-octene, ε-caprolactam and hexamethylenediamine were carried out according to the Brazilian legislation.

E 3.1) Specific migration of 1-octene

The quantification of specific migration of 1-octene was evaluated based on Standard **CEN/ TS 13130-26**: materials and articles in contact with foodstuffs - Plastics substances subject to limitation - Part 26:

Determination of 1-octene and tetrahydrofuran in food simulants, and consists of contact of the sample with solutions of extraction with times and temperatures that simulate its real condition of use.

The internal sides of the samples were placed in contact with the simulants, obeying an area:volume ratio of 600 cm2 to 1000 mL. The same was evaluated in the contact conditions shown in Table 11.

Simulants/Contact Condition	Maximum limit of specific migration of 1-octene	Sample	Mean (1)	Standard Deviation	Variation Interval
Ultra purified water/ 40°C/10 days	15	Before EtO	≤2.8(2)	(3)	(3)
		After EtO	≤2.8(2)	(3)	(3)
3% Acetic acid solution in ultra purified water (w/v)/ 40°C/10 days	15	Before EtO	≤1.6(2)	(3)	(3)
		After EtO	≤1.6(2)	(3)	(3)
Olive oil/ 40°C/10 days	15	Before EtO	≤8.7(2)	(3)	(3)
		After EtO	≤8.7(2)	(3)	(3)

(1) Result of three determinations.
(2) Quantification Limit of the method under the analytical conditions.
(3) Not applicable.

Table 11. Specific migration of 1-octene obtained for transparent coextruded plastic film, before and after the application of EtO, in mg/dm^2.

The specific limit of monomer migration of 1-octene established in Resolution 105/99 of the National

Health Surveillance Agency - ANVISA of 19 May 1999 is 15 mg/kg of simulant. The values for specific migration of 1-octene found in the samples analyzed, in the analytical conditions used, were below the established limit. Sterilization with ethylene oxide (EtO) did not affect the monomer migration potential of 1-octene.

E 3.2) Specific migration of ε-caprolactam

The quantification of specific ε-caprolactam was evaluated based on Standard **CEN/ TS 13130-16**: materials and articles in contact with foodstuffs - Plastics substances subject to limitation - Part 16: Determination of caprolactam and caprolactam salt in food simulants.

The internal surfaces of the samples were placed in contact with the simulants, obeying an area:volume ratio of 600 cm2/1000 mL. The samples were evaluated under the contact conditions shown in Table 12.

The specific limit of monomer migration of ε-caprolactam established in Resolution 105/99 of the National Health Surveillance Agency - ANVISA of 19 May 1999 is 15 mg/kg of simulant. The values for specific migration of ε-caprolactam found in the samples analyzed, in the analytical conditions used, were below the established limit. Sterilization with ethylene oxide (EtO) did not affect the potential monomer migration of ε-caprolactam for the fatty simulant, but was significantly lower (probability of 95% confidence - Tukey Test) for the ultrapure water simulants and 3% acetic acid solution (m/v) in ultrapure water after sterilization with ethylene oxide.

Simulants/Contact Condition	**Maximum limit of specific migration of** ε-caprolactam	**Sample**	**Mean** (1)	**Standard Deviation**	**Variation Interval**
Ultra purified water/ 40ºC/10 days	15	Before EtO	2.4	0.3	1.9 - 2.7
		After EtO	1.7	0.2	1.3 - 1.9
3% Acetic acid solution in ultra purified water (w/v)/ 40ºC/10 days	15	Before EtO	2.7	0.1	2.6 - 2.8
		After EtO	2.2	0.3	1.8 - 2.6
Olive oil/ 40ºC/10 days	15	Before EtO	3.4	0.4	3.0 - 4.0
		After EtO	3.2	0.2	2.9 - 3.5

(1) Result of three determinations.

Table 12. Specific migration of ε-caprolactam obtained for transparent coextruded plastic film, before and after the application of EtO, in mg/dm^2.

E 3.3) Specific migration of hexamethylenediamine

The quantification of specific hexamethylenediamine migration was evaluated based on Standard **CEN/ TS 13130-21**: materials and articles in contact with foodstuffs - Plastics substances subject to limitation - Part 21: Determination of ethylenediamine and hexamethylenediamine in food simulants. The sample of hexamethylenediamine, adipic acid e caprolactam copolyamide in the form of a film, was placed in contact with the simulants, obeying an area:volume ratio of 600 cm2 to 1000 mL. The sample was evaluated under the contact conditions shown in Table 13.

Simulants/Contact Condition	**Maximum limit of specific migration of** hexamethylenediamine	**Mean** (1)	**Standard Deviation**	**Variation Interval**
Ultra purified water/ 100 ºC/30 minutes + 40 ºC/10 days	2,4	≤1.2(2)	(3)	(3)
3% Acetic acid solution in ultra purified water (w/v)/ 100 ºC/30 minutes + 40 ºC/10 days	2,4	≤1.1(2)	(3)	(3)
Olive oil/ 100 ºC/30 minutes + 40 ºC/10 days	2,4	≤2.0(2)	(3)	(3)

(1) Result of three determinations.
(2) Quantification Limit of the method under the analytical conditions.
(3) Not applicable.

Table 13. Specific migration of hexamethylenediamine obtained for a 42 μm film copolyamide, in mg/kg.

The specific limit of monomer migration of hexamethylenediamine established in Resolution 105/99 of the National Health Surveillance Agency - ANVISA of 19 May 1999 is 2.4 mg/kg of simulant. The values for specific migration of hexamethylenediamine found in the analysis of the 42 μm film copolyamide, under the analytical conditions used, were below the established limit.

3.3 Cytotoxicity assay

Assay performed in accordance with Standard ISO 10993-5: Biological evaluation of medical devices - Part 5: Cytotoxicity Assays: *in vitro* methods in samples of coextruded plastic film.

3.3.1 Cytotoxicity

Definitions of cytotoxicity vary, depending on the nature of the study and whether cells are killed or simply have their metabolism altered. Cytotoxicity is the toxicological effect that a substance can cause *in vitro*, at cellular level (Freshney, 2000).

As defined in ISO 10993, "the numerous methods used and end-points measured in cytotoxicity determination can be grouped into categories of evaluation type, like assessments of cell damage by morphological means; measurements of cell damage; measurements of cell growth and measurements of specific aspects of cellular metabolism" [IOS, 2010]. The cells can be exposed to the samples or their extracts.

3.3.2 The protocols

Cell culture

Chinese hamster ovary cell line (CHO-k1) was standardized for cytotoxicity and genotoxicity tests. Cells were maintained in RPMI medium supplemented with antibiotics and antimycotics (100 units/mL penicillin, 100 μg/mL streptomycin and 0.025 μg/mL amphotericin), 2mM glutamine, and 10% calf serum, at 37° C in a humidified 5% CO_2 atmosphere until they reached confluence. For subculturing and for experiments, cells were harvested using 0.05% trypsin and 0.02% EDTA in phosphate-buffered saline, pH 7.4.

Extract preparation

Samples of packaging, before and after sterilization, were submitted to this assay. The samples were immersed separately in RPMI culture medium at a final concentration of 1 cm^2 / mL and left in an incubator at 37°C for 72 hours to fulfil the extraction condition. The first concentration was sterilized by filtration and the subsequent dilutions were performed in sterile RPMI medium at a ratio of 1:2.

Cytotoxicity test

A colorimetric method that uses the tetrazolium compound MTS was used to determine the number of viable cells in proliferation (Cory et al, 1991). 96-well microplates were prepared with 50 μL of extract diluted from 100 to 6.25% in RPMI medium in quadruplicates. The positive control was a phenol solution (0.5% v/v) as 100% concentration and the negative control was a high density polyethylene (HDPE) extract. The 100% concentration was the non-extract well A suspension of CHO-k1 (from second to fourth passages after thawing) with 6 x 10^4 cell/mL was prepared and 50 μL/well was pipetted into the microplates. The microplates were incubated for 72 hours at 37°C in a humidified 5% CO_2 atmosphere. Blank and controls of the cells were also prepared. Cell viability was measured by adding 20 μL of MTS/PMS (20:1) solution to the humidified 5% CO_2 incubator, followed by incubation for 2 hours at 37°C. The microplates were read in a spectrophotometer reader at 490 nm.

Cell viability was calculated by the equation:

$$CV\% = \frac{OD\,sample}{OD\,non\,extract} \times 100$$

Where: CV% = cell viability, OD sample = optical density at 490 nm of the extract dilution, OD non extract = optical density at 490 nm of the well without extract.

The results consider the following parameters:

a. Controls (positive and negative)

Positive control: 0.5% Phenol solution

Negative control: HDPE (high-density polyethylene) extract.

b. Observations: definitions of some terms.
   - Positive control: material which, when tested according to Standard ISO 10993-5, promotes a cytotoxic response.
   - Negative control: material which, when tested according to Standard ISO 10993-5, does not promote a cytotoxic response.
   - $IC_{50(\%)}$: cytotoxicity index 50%, concentration of extract that kills 50% of the viable cell population.

The results of the assay showed that the samples of packaging material, before and after sterilization by ethylene oxide (EtO), resulted in viability of over 90%, and therefore do not become cytotoxic.

3.4 Sterility assay and ethylene oxide residues

The samples were classified according to their position (0 to 5a; 0 to 5b) during exposure to EtO, for the penetration analysis. The outermost or surface position corresponds to the number 5, and the innermost position to the number 0. (Figure 6)

Sterility tests were carried out through the analyses of two biological indicators (bi 3M - ATTEST -™ Bacillus atrophaeus and Terragene - Bionova BT40 - Bacillus atrophaeus). Incubation time was 48 h at a temperature of 35°C ± 1.5.

Samples of packaging were also submitted to direct incubation for 7 days with TSB liquid culture at a temperature of 35°C ± 1.5. The methodology used is in accordance with the Brazilian Pharmacopoeia. The analyses of the three sterility tests (biological indicator Bionova BT40, 3Mattest and direct incubation) confirm the sterility of the packaging material after being submitted to ethylene oxide gas.

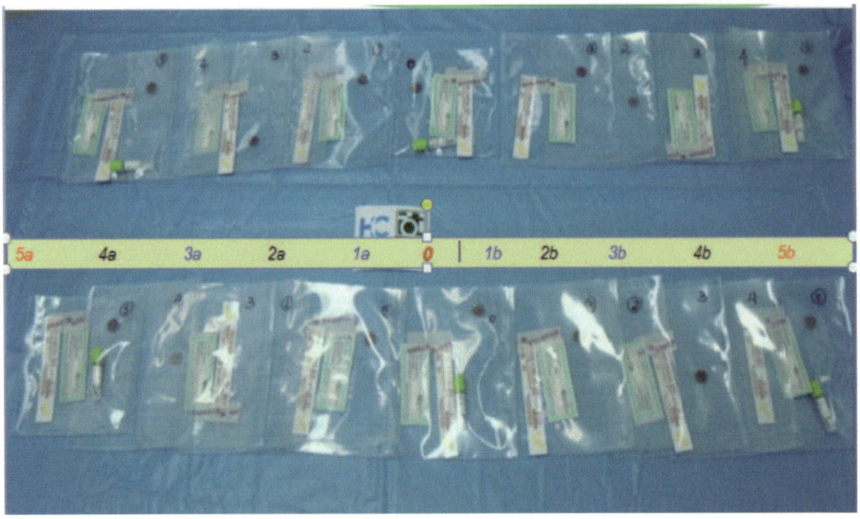

Fig. 6. Samples of packaging positioned from 0 to 5 for EtO penetration analysis.

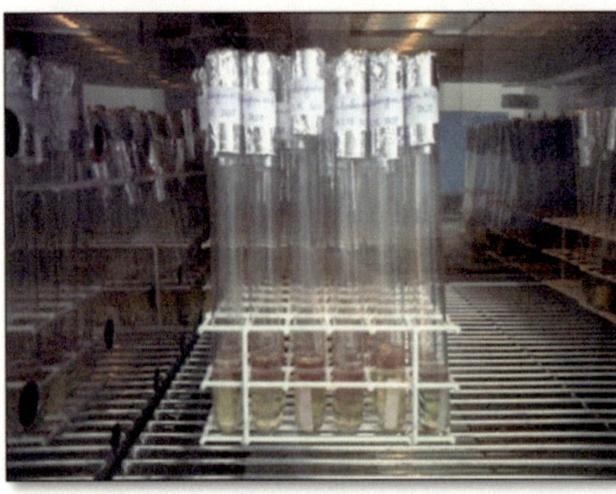

Fig. 7 and 8. Performance of the sterility test by direct incubation in samples submitted to ethylene oxide.

The analysis of ethylene oxide residues was performed by the Gas Chromatography test, determining the levels of Ethylene chlorohydrin and Ethylene glycol. The data are shown in **Figure 9** and show that the levels are within the limits accepted by our legislation.

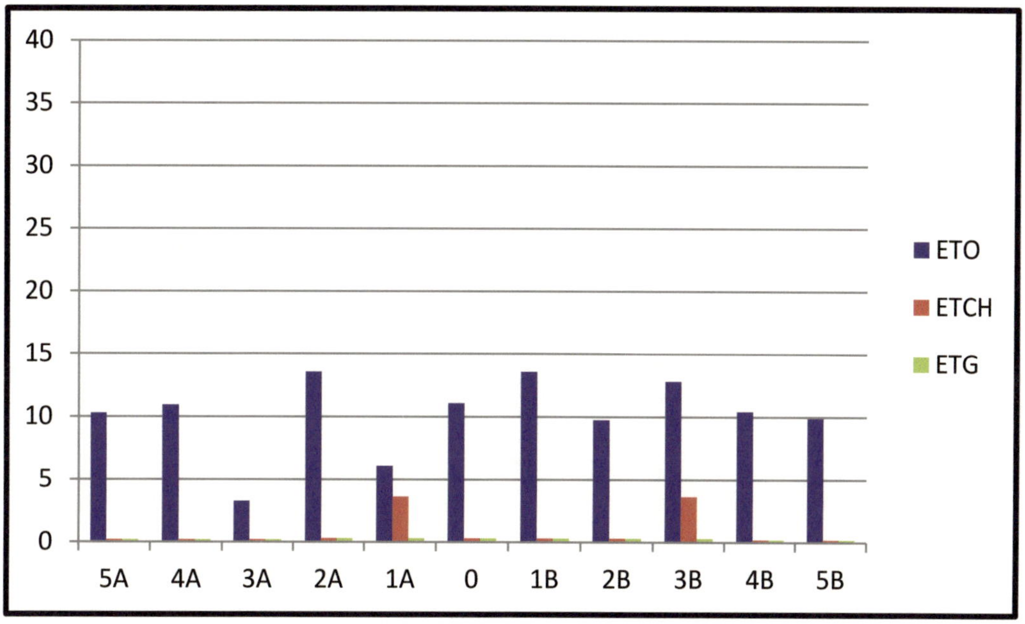

Note. Maximum limits according to the Brazilian legislation. [ETO up to 25/ ETCH up to 25/ETG up to 250]

Fig. 9. Residues (ETO, ETCH, ETG) found in the samples submitted to sterilization.

4. Final considerations

Analyses of the packaging used in this study demonstrated that it is a good option for cryopreservation of tissues at a temperature of - 80°C.

Our experience with assays to validate coextruded polyethylene and polyamide plastic film shows that the mechanical properties of this material are not altered by cryopreservation and sterilization. Penetration resistance of the thermal seal remained unaltered after all the processes carried out in a tissue bank, such as sterilization and cryopreservation.

We found a loss of barrier due to increased oxygen permeability of around 10% after sterilization and cryopreservation, which can be explained by the humidification of the polyamide. However, this slight alteration in oxygen permeability does not compromise the inner vacuum of the packaging, and does not place at risk the tissue packaged in it.

In relation to total migration, we did not observe any alterations in the assays, i.e. once again, sterilization and cryopreservation did not lead to monomer migration at levels above those required by our legislation. This is also valid for specific migrations of 1-octene, ε-caprolactam and hexamethylenediamine.

The application of ethylene oxide is safe for sterilization of this type of packaging, as it results in good penetrability and safe levels of Eto, Etch and Etg residues at the end of the procedure.

In the cytotoxicity test, we observed levels of cell viability of over 90%, therefore they do not become cytotoxic.

Thus, analyses of coextruded plastic polyethylene and polyamide film used in this study proved to be a good option for cryopreservation of tissues at temperatures of -80°C, even for prolonged periods of 150 days.

5. References

American Association of Tissue Banks. Standards for Tissue Banking. 11th ed. Mclean : American Association of Tissue Banks; 2007.

ASTM INTERNATIONAL. ASTM F1927-07: standard test method for determination of oxygen gas transmission rate, permeability and permeance at controlled humidity through barrier materials using a coulometric detector. Philadelphia, 2007. 6 p.

ASTM INTERNATIONAL. F 1306-90 (2008) e1: standard test method for slow rate penetration resistance of flexible barrier films and laminates. Philadelphia, 1990. 5p.

ASTM INTERNATIONAL. F 88/F 88M-09: standard test method for seal strength of flexible barrier materials. Philadelphia, 2007. 11p.

BRASIL, Leis etc. Decreto n.2268 de 30 de junho de 1997. Dispõe sobre a remoção de órgãos, tecidos e partes do corpo humano para fins de transplantes e tratamento. *Diário Oficial da União,* Brasília (DF). 1997 30 jun; seção 1:1.

BRASIL, Leis etc. Lei n.9434 de 5 de fevereiro de 1997. Dispõe sobre a remoção de órgãos, tecidos e partes do corpo humano para fins de transplantes e tratamento. *Diário Oficial da União,* Brasília (DF). 1997 5 fev; seção 1:25.

BRASIL, Leis etc. Portaria n.1686 de 20 de setembro de 2002. Dispõe sobre a regulamentação para funcionamento de banco de tecidos músculo esqueléticos. *Diário Oficial da União,* Brasília (DF). 2002 24 jul; seção 1:1.

BRASIL, Leis etc. PORTARIA N° 2.600, DE 21 DE OUTUBRO DE 2009. Aprova o Regulamento Técnico do Sistema Nacional de Transplantes. *Diário Oficial da União,* Brasília (DF). 2009 21out.

BRASIL, Leis etc. Resolução n. 220 de 27 de dezembro de 2006. Dispõe sobre o Regulamento Técnico para o Funcionamento de Bancos de Tecidos Musculoesqueléticos e de Bancos de Pele de origem humana. *Diário Oficial da União,* Brasília (DF). 2006 29 dez.

BRASIL. Agência Nacional de Vigilância Sanitária.ANVISA. Resolução n° 105 de 19 de maio de 1999. Aprova o regulamento técnico sobre disposições gerais para embalagens e quipamentos plásticos em contato com alimentos. Diário Oficial

[da] República Federativa do Brasil, Brasília, DF, 20 maio, 1999a. N. 95, Seção 1,p.21-34.

BRASIL. Ministério da Saúde - Agência Nacional de Vigilância Sanitária. Resolução - RDC n° 51 de 26 de novembro de 2010. Dispõe sobre migração em materiais, embalagens equipamentos plásticos destinados a entrar em contato com alimentos. Diário Oficial da República Federativa do Brasil, Brasília, DF, 22 dez. 2010. Seção 1, n. 244, p. 75-79.

Cory,AH; Owen,TC; Barltrop,JA; Cory, JG. (1991). Use of an aqueous soluble tetrazolium/formazan assay for cell growth assays in culture. Cancer Comm., 3, 207-212.

European Association of Tissue Banks. Common Standards for Tissues and Cells Banking: Berlin: European Association of Tissue Banks; 2004.

EUROPEAN COMMITEE FOR STANDARDIZATION - CEN/TS 13130-16: materials and articles in contact with foodstuffs - Plastics substances subject to limitation - Part 16: Determination of caprolactam and caprolactam salt in food simulants. February 2005. 15p.

EUROPEAN COMMITEE FOR STANDARDIZATION - CEN/TS 13130-21: materials and articles in contact with foodstuffs - Plastics substances subject to limitation - Part 21: Determination of of ethylenediamine and hexamethylenediamine in food simulants. February 2005. 17p.

EUROPEAN COMMITEE FOR STANDARDIZATION - CEN/TS 13130-26: materials and articles in contact with foodstuffs - Plastics substances subject to limitation - Part 26: Determination of 1-octene and tetrahydrofuran in food simulants. February 2005. 25p.

EUROPEAN COMMITTEE FOR STANDARDIZATION. EN 1186-1: materials and articles in contact with foodstuffs - plastics. Part 1: guide to the selection of conditions and test methods for overall migration. Brussels, 2002. 49 p.

EUROPEAN COMMITTEE FOR STANDARDIZATION. EN 1186-14: materials and articles in contact with foodstuffs - plastics. Part 14: test methods for "substitute tests" for overall migration from plastics intended to come into contact with fatty foodstuffs using test media iso-octane and 95% ethanol. Brussels, 2002. 20 p.

EUROPEAN COMMITTEE FOR STANDARDIZATION. EN 1186-3: materials and articles in contact with foodstuffs - plastics Part 3: test methods for overall migration into aqueous food simulants by total immersion. Brussels, 2002. 17 p.

EUROPEAN PHARMACOPOEIA. 6. ed. Strasbourg, France: European Department for the Quality of Medicines, 2008.

Freshney, RI (2000). Culture of animal cells: a manual of basic technique. 4.ed. New York: Wiley, 577p.

GRUPO MERCADO COMÚN. Mercosur//GMC/Res. n° 32/10. Regulamento técnico Mercosur sobre migración en materiales, envases y equipamientos plásticos destinados a estar en contato con alimentos. Disponível em: <http://www.puntofocal.gov.ar/doc/r_gmc_32-10.pdf>. Acesso em: 22 ago. 2011

International Organization For Standardization. (2010), ISO 10993-5; Biological evaluation of medical devices – Part 5: Tests for cytotoxicity: in vitro methods. Switzerland

Rockville: The United States Pharmacopeial Convention, Inc., 2010.

THE UNITED STATES PHARMACOPEIAL CONVENTION. The United States Pharmacopeia. USP 33; NF 28.

5

Cryopreservation – A Viable Alternative in Preparation for Use of Allografts in Knee Ligament Reconstruction

Alexandre C. Bitar et al.*
Vita Institute
Brazil

1. Introduction

The use of allogenic tissues is growing in orthopedic practice, as well as the number of studies on methods for processing, sterilization and cryopreservation that interfere as little as possible with the original physiological properties of the tissues (Nutton et al., 1999).

In addition to bone tissue, other tissues of the locomotor system can be captured, processed and stored in tissue banks with the purpose of transplantation. Therefore a strict quality control must be implemented and set after discussions compiled by international organizations such as AATB e EATB.

The first report of the use of allografts in humans dates back to 1881. The first tissue bank of bone grafts was created in 1940 in the United States and the initial clinical results were published in 1942 by Inclan, 1942. Since then a series of regulations and studies has emerged relating to the use of grafts in orthopedic practice[19]. Currently, tendon allografts are used in knee surgeries, in elbow ligament reconstructions and in revisions of the acromioclavicular joint (Costic et al., 2004).

In our country we have few tissue banks. The tissue bank (BTME) from the Institute of Orthopedics and Traumatology (IOT), Hospital das Clínicas da Faculdade de Medicina da Universidade de Sao Paulo was the first and is the biggest bank in activity nowadays. It thas been in operation since 1999 and is governed by local legislation (Amatuzzi et al., 2000). With the restructuring of our service in 2005, we initiated a new technique aimed at the provision of the tendon with a well-structured quality program in line with other centers of excellence nationally and internationally. Today our service is provided by a series of tendons (tibial tendon, Achilles, patellar and peroneal) taken from different regions with very specific applications. Thus, we can then follow the technological trend in the use of

*Caio Oliveira D'Elia[1], Antônio Guilherme P. Garofo[1], Wagner Castropil[1], Luiz Augusto U. Santos[2], Marco K. Demange[2], José Ricardo Pécora[2] and Alberto T. Crocci[2]
[1]*Vita Institute, Brazil*
[2]*Universidade de São Paulo; Institute of Orthopedics and Traumatology, Hospital das Clinicas da Faculdade de Medicina da Universidade de São Paulo, SP, Brazil*

tissues already practiced in other global centers of excellence in orthopedics and traumatology.

In our medical service, Vita Institute (private use), and in the Institute of Orthopedics and Traumatology (public and private use), allografts are used mainly in knee surgeries, ACL reconstruction, multiple ligament reconstructions, ligament surgery in skeletally immature patients and with double bundle reconstruction.

2. Importance of allografts in knee ligament reconstructions

Grafts are used in various procedures in different branches of orthopedics, including ligament reconstruction. The literature shows the importance of the use of allografts in knee surgery, especially in revision surgeries, multiple ligament reconstructions and, more recently, surgery for primary reconstruction of the anterior cruciate ligament (ACL) in active patients and in those aged over 40 years (Marrale et al., 2007; Sherman & Banffy, 2004). There have been at least 11 clinical studies comparing the use of auto and allografts in the reconstruction of the ACL (Chang et al., 2003; Marrale et al., 2007). Most of these show little difference between the two techniques with respect to long-term results. However, there has been few prospective randomized studies , and the comparison methods (scores), types of graft, as well as methods of preparing and fixing the graft are highly variable.

Furthermore, most studies use the patellar tendon graft; therefore, it may not be possible to generalize the conclusions of these studies to the flexor tendon. Some authors, such as Lawhorn and Howell, suggest the use of allografts without a bone plug because of the potential for slower incorporation of the bone due to immunogenicity and smaller cross-sectional area of transplants with bone plugs (Lawhorn & Howell, 2003). Recently, Sun et al. (2011) published a prospective randomized study comparing non-irradiated allograft with flexors autograft showing similar results between groups in terms of subjective clinical scores, goals, rate of return to sports and incidence of complications (Sun et al., 2011).

However, compared with autologous transplants, allografts do have some advantages. For example, they do not increase morbidity for the donor, they require a shorter surgery time, and they are available without restriction on size and morphology. In ligament reconstruction surgeries, the possibility exists of an immune response from the recipient tunnel enlargement, and delayed incorporation of the allograft (Marrale et al., 2007). The risk of disease transmission and the potential for immunogenicity are the major disadvantages of allografts, but these complications can be controlled (Albert et al., 2006; Barrios et al., 1994, Urabe et al., 2007).

3. Controversy of methods of preparation of allografts

The increased frequency of the use of allografts in traumato-orthopedics requires the adoption of storage techniques that interfere as little as possible in the quality of the parts (Vangsness et al., 2003). Allografts can be stored in different ways; they can be chilled in residential mechanical freezers at temperatures of + 2°C to - 4°C for up to five days. In freezers with temperatures of -20°C to -40°C, they can be stored for up to six months[14]. At these temperatures, the enzymes present in the tissue are still active and can destroy the tissue. Therefore, storage periods of longer than a few months are not recommended. The methods of sterilization used at low temperatures are effective against fungi and do not seem to change the mechanical characteristics of the grafts. The period of 40 days chosen in our service for cryopreservation coincides with the period of incubation necessary for microbiological investigations for bacteria and fungi (Vangsness et al., 2003). The deep-freezing process enables storage for up to five years, and this is the method we use in our service[19].

Many services prefer to carry out the manipulation of tissues under aseptic conditions from acquisition through clinical use, and the samples are discarded when microbiological assays show positive bacterial cultures (20 to 30%) (Zimmerman et al., 1994). Sterilization methods, therefore, are not completely safe. They can alter the biomechanical characteristics of tissues or fail to penetrate tissue layers, resulting in the protection of microorganisms rather than their destruction. Irradiation with gamma rays is the most common method of sterilization (Sterling et al., 1995). However, to achieve safer sterilization in frozen tissues, high-dose irradiation is necessary, which can alter the biomechanical properties of the tissue in a dose-dependent manner (Curran et al., 2004; Fideler et al., 1995).

Doses as low as 2 Mrad resulted in a statistically significant reduction in biomechanical properties, outcomes, or physical examination measures. Rappe et al. (2007) studied the effect of irradiation on clinical outcomes of ACL reconstruction, they found the irradiated group had an unacceptable higher rate of failure (33%) than the non- irradiated group (2.4%). Fideler et al. (1995) found that the dose of 2.5 Mrad, which was a dose commonly used by tissue banks for sterilization, was just bacteriocidal but ineffective in eliminating viruses such as human immunodeficiency virus (HIV) (Sterling et al., 1995). Doses of 3 to 4 Mrad were necessary to inactivate the virus. Grieb et al. also proved that lower levels of radiation may be inadequate to kill hepatitis and HIV viruses, with a dose of 5 Mrad being necessary (Grieb et al., 2006). When dosage is increased, its clinical implications increase correspondingly. We must question the use of gamma irradiation as there are so many adverse effects and it fails to sterilize the allograft as required.

Also, the sterilization effectiveness against viruses is low (Vangsness et al. 2003). Ethylene oxide sterilization requires strict control of the levels of waste gas in contact with the allograft and is no longer used by tissue banks, due to the possibility of toxic effects for the recipient (dissolution of the graft and articular inflammatory reactions) (Vangsness et al. 2003). The processing techniques used in the preparation and preservation of grafts have been questioned as potentially altering the initial resistance and mechanical properties of the graft prior to implantation.

Two studies carried out in Brazil address the biomechanical properties of patellar and calcaneus tendons of cadavers with the same preparation method as that used in our study, comparing fresh and cryopreserved allografts (Giovani et al., 2006; Reiff et al., 2007). They found no differences. A study on metric measurements and attachment levels of the medial patellofemoral ligament shows this to be a distinct structure (Zimmerman et al., 1994). Although there have been studies on the biomechanical behavior of tendons, the literature does not address histological changes of tendons cryopreserved at -80°C under aseptic conditions (Pearsall et al., 2003). During cryopreservation at -80°C, thedestruction of the allograft enzyme appears to be minimal and at least one enzyme, collagenase, which can destroy the tissue, is inactive (Tomford, 1997). Furthermore, with cryopreservation there is no intracellular free water, which is thought to be necessary for enzymatic activity, bacterial proliferation and lipid oxidation (Galea & Keamey, 2005; Laitinien et al., 2006). Lipid oxidation inside the tissues induces apoptosis and inhibits cell differentiation; such oxidation can be minimized or avoided with cryopreservation at temperatures of at least -70° C (Laitinien et al., 2006). The literature refers to histological changes due to cryopreservation only in cartilage (one of the most commonly used grafts in surgical practice), concluding that during freezing, the vitality of the cells is threatened (Schchar & McGann, 1986). Other injuries may also occur, such as the formation of extracellular ice crystals, intracellular ice nucleation, collapse of the matrix, and breakage of intercellular bridges.

In our study, the histological study of one tendon (not cartilage) was carried out, and none of these histological phenomena were observed with cryopreservation at -80ºC. Freezing with liquid nitrogen at -179ºC has also been used as a storage method with similar results but higher cost (Zimmerman et al., 1994). Another widespread storage method is lyophilization. Cryopreservation and lyophilization have been related to a reduction in allograft antigenicity (Jackson et al., 1990). The use of chilled saline solution is not a guaranteed method because the stock can only be kept safely for short periods (Zimmerman et al., 1994). Treatment with paraformaldehyde and fixation with glutaraldehyde are no longer recommended because of the toxicity of these solutions to the recipient tissue.

We recently published a study in which we proved the histological properties of the flexor tendons of the knee from cadavers subjected to cryopreservation and experience with the use of allografts of the Knee Group from IOT (Bitar et al., 2010; Damasceno et al., 2009).

4. Cryopreservation: Our method

4.1 Tendinous tissue removal

The attainment of the musculoskeletal tissues has as its source deceased donors with brain death reported by the Committees Intra-hospital - CIHDOTs, Organ Procurement Organizations - OPOS and by the twenty-three central of notification and collection of organs and tissues - CNCDOs, logistically spread throughout country. Notifications for teams pickups are made after the execution of a series of procedures and tests that aim beyond the evidence of brain death, family consent of the donation of organs and tissues.

The donor selection follows a rigorous research with control antigen and antibody serology for HIV, Hepatitis A, B and C, HTLV-1 and 2, syphilis, Chagas disease, toxoplasmosis and cytomegalovirus in addition to testing of last generation for evidencing of DNA (Nucleic Acid Amplification - NAT) for HIV and Hepatitis B and C, required of musculoskeletal tissues. The capture of musculoskeletal tissues (bone and tendons) is performed after the initial screening of donors of multiple organs and tissues (heart, kidney, liver, pancreas, lung, cornea, etc.).

In our specific field, we follow a protocol of evaluation of the donor that counts with a written anamnesis of a term to capture and physical examination. Are excluded donors with orthopedic disorders such as osteoporosis, osteonecrosis, rheumatoid arthritis, lupus erythematosus, malignancy, age that compromises characteristic of tissues, blood transfusion, tattoos or adornments (piercings) within the window period, users of illicit drugs, permanence in endemic areas, generalized or localized infections, fractures, bruises on the limbs which are absorbed in the musculoskeletal tissues or any other situation that would call into doubt the quality of these tissues, as arranged in the existing laws. The tissues removed are immediately packed in triple enclosures, hermetically sealed and sent under refrigeration (- 4 ° C) to the Tissue Bank.

A very important step of the process of capture is the reconstruction of the donor and for this matter we use prosthesis specially designed using plaster, wire suture, gauze. This reconstruction is done rigorously and is characterized as the most laborious phase of the procedure. All anatomical parameters are respected, and therefore the deformation of the donor does not occur (Figures 1 and 2).

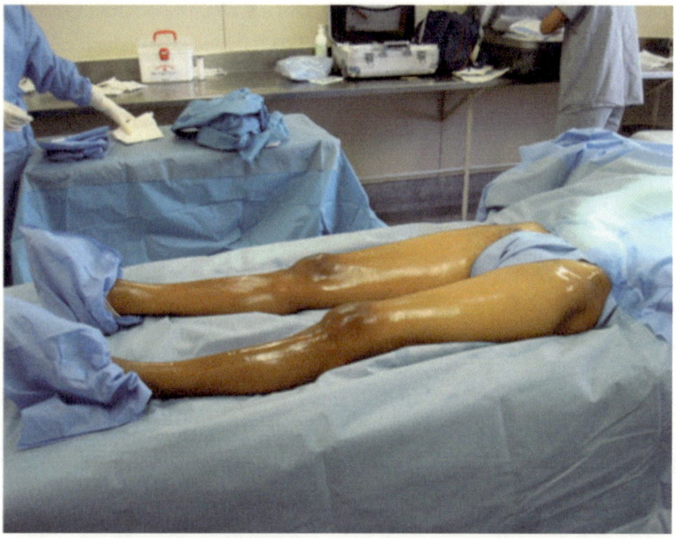

Fig. 1. Pre-operative preparation of the potential musculoskeletal tissues doner.

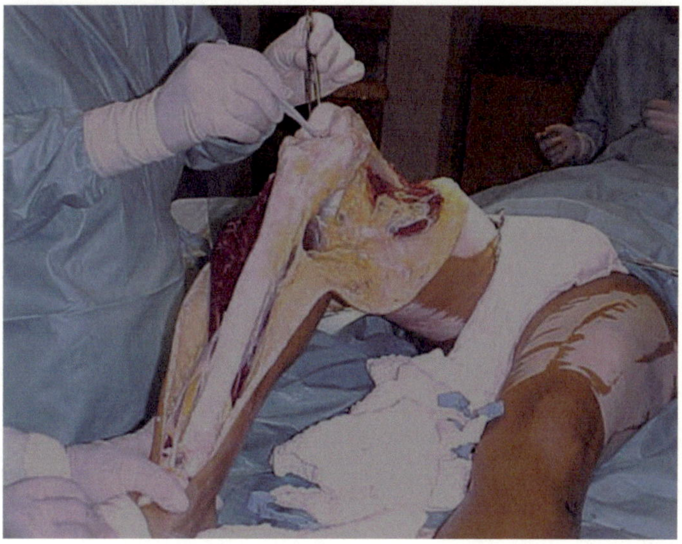

Fig. 2. Tissue removal: bone and tendon dissecation under asseptic conditions.

4.2 Processing and cryopreservation of musculoskeletal tissues

At the end of the uptake, the tissues are sent to BTME chilled in coolers with temperature monitoring throughout the period of transportation. The processing step is preceded by a planning of activities needed for its implementation, such as provision of materials and instruments, convocation of the processing team, definition of preparation and dimensioning according to the need for service (queue) requests from orthopedic and dental surgeons. This step is performed in the operating room properly rated (class 100 or ISO 5) equipped with integrated laminar flow (Figure 3). The room also has an antechamber and pass-through and all environments have strict control of air particles and positive pressure for quality assurance of tissues processed there.

Fig. 3. Tissue cryopreservation área. Ultrafreezer with a temperature of -80ºC.

In addition, specific attire is required of the professional team that should only use non-woven clothes to avoid dispersion of particles that emit the cotton clothes (Figure 4). Not only the non-woven attire is required but the team's behavior should be differentiated. Thus, sudden movements, use of cosmetics and hair exposure should be avoided during the permanence in this room. Ensuring an appropriate approach is not only a result of training of the nursing staff that performs the procedure, but also other professionals who access the environment (cleaning maintenance).

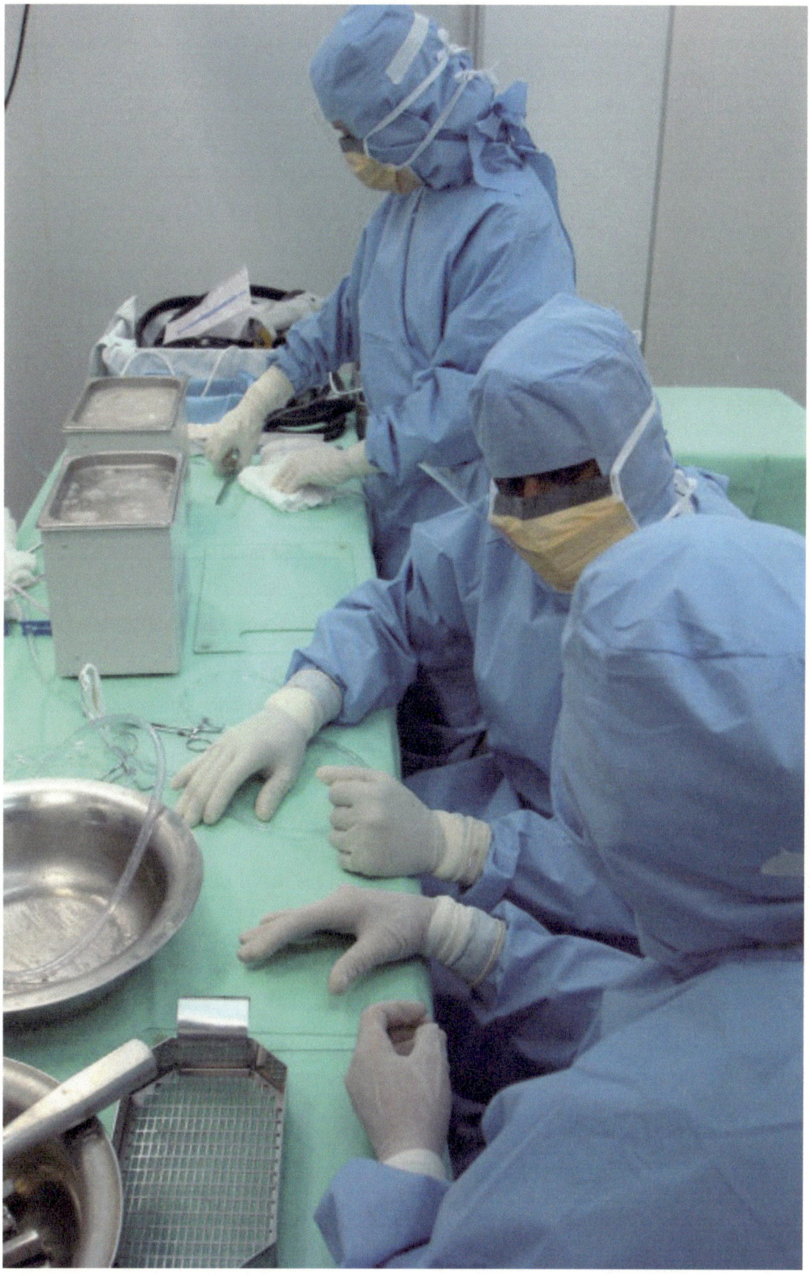

Fig. 4. Processing team in activity in the controlled área (ISO 5: Class 100).

The BTME conducts various types of processing of these tissues with the purpose to use in orthopedic and dental surgeries, each of which requires a specific plan. For the processing of fresh frozen tissues it is performed what we call mechanical processing, ie removal of adventitial tissue such as blood, periosteum, subcutaneous tissue, muscle, fascia and fibrotic tissue. Then, these tissues are immersed in emulsifying solutions based on hydrogen peroxide and alcohol under ultrasonic agitation (Figure 5).

Fig. 5. Tissue's chemical processing. Ultrasonic clining with emulsifying solution.

Then, a sampling of these resultant solutions, of bone marrow of long bones and fragments of each tissue processing are subjected to microbiological examination (General Knowledge, and Culture of Anaerobic Fungi). Furthermore, it is also obtained samples to histopathological analysis.

Finally starts the procedure of packaging of all the processed grafts which are measured (length, height, weight, volume, perimeter) and kept in sterile triple wrappers, vacuum sealed and properly identified as tissue in analysis. The label contains information from the donor, examination, lot number, item, expiration date, type of conservation and barcode.

Once all the tissues are identified, they are x-rayed at the very BTME and referred to cryopreservation.

The bones can also be processed in its lyophilized form, where all water is removed with the tissue still frozen. The process involves placing the tissue in a lyophilizer chamber where ice crystals sublimate by the action of the high pressure, not passing through the liquid phase and thus maintaining the viability of bone matrix. The result is a dry tissue, conservable at room temperature that must receive final sterilization by irradiation.

At the end of the processing it is performed the documentation of the procedure in the Processing and archived in the donor's chart. The stock of tissues can be kept either frozen or dried, if necessary, according to the same standards used by the Global Association of Tissue Banks . Other forms of processing have been investigated in order to reduce costs related to storage and maintenance. The glicerolization of bone tissue is presented as a processing methodology capable of maintaining the viability of the matrix and prevent bacterial growth, and allows storage at room temperature (Giovani et al., 2006).

4.3 Tissue cryopreservation

In the room of cryopreservation tissues are stored according to their status in the process. Thus, there is a space for tissues in analysis or in quarantine (where they remain for about 60 days until the results of all examinations) and those already released for use. Both rooms are equipped with ultrafreezers with temperatures ranging from 85 to 110 degrees below zero (Amatuzzi et al., 2000).

The room is also equipped with air-conditioning system, own power generator and the unfreezing protection of carbon dioxide (CO2 Backup), and a rigorous system of monitoring the temperature, with printed record of temperature for 24 hours and alarm system via satellite, which guarantees the right temperature and early detection of complications.

Depending on the outcome of the analysis, the tissues are transferred to the room of material released for use. The maximum period of cryopreservation is five years to bone tissue and two years for soft tissues and tendons.

4.4 Quality control and distribution

By the time of transplantation, all tissues processed are subjected to rigorous quality assurance criteria. It requires the evaluation of all data pertinent to the donor, test results, maintenance and control of equipment, materials and instruments used in all phases of each procedure (Figures 6 and 7).

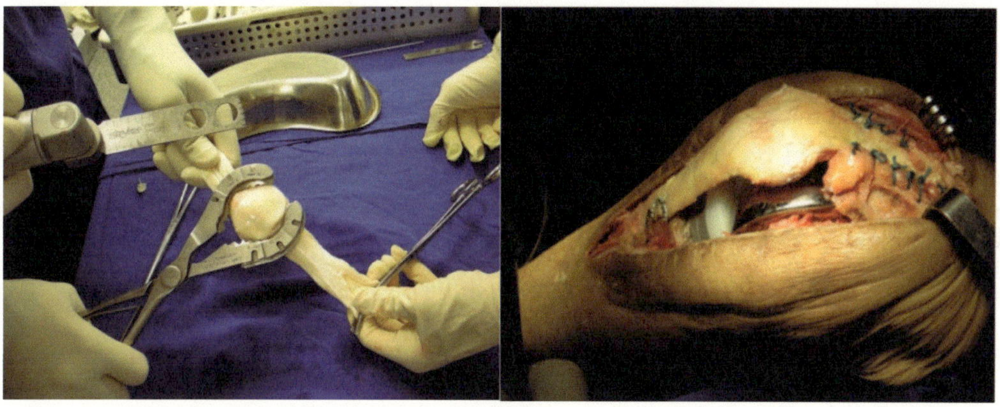

Fig. 6. and 7. Patellar tendon allograft transplant.

All processes are computerized through the System Manager of the Tissue Bank, a program designed to record all the steps which allows the rescue and traceability of each graft processed and delivered. Through a coding is possible to identify the donor, lot, expiration, and status of the tissue examination (under review, released, deleted and used).

Given the need of retrieval of information, as the evidence of an adverse effect, you can quickly and safely obtain the information and implementation of corrective and preventive actions.

For a lot of graft in analysis to be released for use nurses must analyze the results of all tests performed: NAT serology or PCR for HIV, HBC and HCV, General Culture, Anaerobic Culture and Culture of Fungi, pathology reports and radiographic findings. These reports of examinations are assessed and ultimately released by the Technical Director of the Tissue Bank.

Besides examinations, evaluation of the printed record of temperature during the storage period is considered. The temperature oscillations are quickly detected and reported to the team members from the BTME even remotely by cell phones. In addition, audible alarms at strategic points in the hospital and the presence of CO2 backups, ensure system reliability.

After the release of each lot, the nurses carry a detailed examination of integrity of each tissue during the replacement of tissues in analysis labels to released labels and posterior transfer of the sector. The logistics of storage of tissues in ultrafreezers considers the type of tissue and speed up the search.

We emphasize that for a rigorous quality control all steps of each procedure are carried out through check-lists with double checking and approval.All relevant data of the donor or lot records are filed in a single file and stored at the BTME for a minimum of 25 years.

A serum bank with plasma samples of donors are offered by BTME if necessary examination of counterproof.

5. Our experience

In the last five years, we have performed 35 knee ligament reconstructions, including multiple ligament, and isolated ACL and PCL reconstructions with single and double bundle techniques, and ACL reviews. Twenty seven men and eight women underwent surgery with the employment of the following grafts from the tissue bank: anterior tibial tendon (48 units), patellar tendon (4 units), quadriceptal tendon (5 units), semitendinosus tendon (4 units), calcaneus tendon (1 units) and fibular tendon (1 units). The patients follow-up range from 6 to 57 months and we are still collecting data from this cases. Our first results showed that there no viral or bacterial infection associated to the use of allografts in any of the cases or other complications, and clinical outcome of these patients has shown good results with the use of tendons from the tissue bank.

A study conducted in another service, in which we colaborate, revised the records of 46 patients who were submitted to ligament reconstructions between 1999 and 2007 using grafts supplied by the same tissue bank (Damasceno et al., 2009). Thirty-four male patients and 12 female patients were reviewed, with follow-up time ranging from 10 months to 9 years (mean: 3.1 years).The surgical procedures used 9 units of patellar tendons, 9 units of anterior tibial tendons, 8 units of calcaneal tendons, 6 units quadriceptal tendons and 1 unit of fibular tendon, mainly for multiple ligamentar reconstructions and ACL reviews[41]. There were also no viral or bacterial infection cases[41].

The decrease in morbidity and postoperative complications allied to good results obtained in our samples reinforces the idea that the use of allografts is a good and safe option in knee ligament reconstruction.

6. References

Albert, A; Leemrijse, T; Druez, V; Delloye, C; Cornu, O. (2006) Are bone autografts still necessary in 2006? A three-year retrospective study of bone grafting. *Acta Orthop Belg*, 72, 734-40. ISSN 0001-6462.

Amatuzzi, MM; Croci, AT; Giovani, AMM; Santos, LAU. (2000) Banco de tecidos: estruturação e normatização. [Tissue bank: structure and organization]. *Rev Bras Ortop*, 35, 165-72. ISSN 0102-3616.

Barrios, RH; Leyes, M; Amillo, S; Oteiza, C. (1994) Bacterial contamination of allografts. *Acta Orthop Belg*, 60, 293-5.

Bitar, AC; Santos, LAU; Croci, AT; Pereira, JARM; França Bisneto, EM; Giovani, AMM; Oliveira, CRGCM. (2010) Histological Study of Fresh versus Frozen Semitendinous Muscle Tendon Allografts. *CLINICS*, 65(3),297-303 ISSN 1807-5932.

Chang, SK; Egami, DK; Shaieb, MD; Kan, DM; Richardson, AB. (2003) Anterior cruciate ligament reconstruction: allograft versus autograft. *Arthroscopy*, 19, 453-62. ISSN 0749-8063.

Costic, RS; Labriola, JE; Rodosky, MW; Debski, RE. (2004) Biomechanical rationale for development of anatomical reconstructions of coracoclavicular ligaments after complete acromioclavicular joint dislocations. *Am J Sports Med*, 32, 1929-36. ISSN 1552-3365.

Curran, AR; Adams, DJ; Gill, JL; Steiner, ME; Scheller, AD. (2004) The biomechanical effects of low-dose irradiation on bone-patellar tendon-bone allografts. *Am J Sports Med*, 32, 1131-1135. ISSN 1552-3365.

Damasceno, ML; Ferreira, TF; D'Elia, CO; Demange, MK; Pécora, JR; Hernandez, AJ; Camanho, GL; Croci, AT; Santos, L; Helito, CP. (2009) Use of Allograft in Ligamentar Reconstruction of Knee. *Acta Ortop Bras*, 17(5), 265-8 ISSN 2176-7521.

Fideler, BM; Vangsness, CT Jr; Lu, B; Orlando, C; Moore T. (1995) Gamma irradiation: effects on biomechanical properties of human bone-patellar tendon-bone allografts. *Am J Sports Med*, 23, 643-646. ISSN 1552-3365.

Flahiff, CM; Brooks, AT; Hollis, JM; Vander Schilden, JL; Nicholas, RW. (1995) Biomechanical analysis of patellar tendon allografts as a function of donor age. *Am J Sports Med*, 23, 354-8. ISSN 1552-3365.

Galea, G; Kearney, JN. (2005) Clinical effectiveness of processed and unprocessed bone. *Transfus Med*. 15:165-74.

Giovani, AM; Croci, AT; Oliveira, CR; Filippi, RZ; Santos, LA; Maragni, GG. (2006) Comparative study of cryopreserved bone tissue and tissue preserved in a 98% glycerol solution. *CLINICS*, 61, 565-70. ISSN 1807-5932.

Grieb, TA; Forng, RY; Bogdansky, S. (2006) High-dose gamma irradiation for soft tissue allografts: high margin of safety with biomechanical integrity. *J Orthop Res*, 24, 1011-1018. ISSN 0736-0266.

Inclan A. (1942) The use of preserved bone graft in orthopaedic surgery. *J Bone Joint Surg Am*, 24, 81-96.

Jackson, DW; Corsetti, J; Simon TM. (1996) Biologic incorporation of allograft anterior cruciate ligament replacements. *Clin Orthop*, 324, 126-33.

Jackson, DW; Windler, GE; Simon, TM. (1990) Intraarticular reaction associated with the use of freeze-dried, ethylene oxide-sterilized bone-patella tendon-bone allografts in the reconstruction of the anterior cruciate ligament. *Am J Sports Med*, 18, 1-10. ISSN 1552-3365.

Laitinen, M; Kivikari, R; Hirn M. (2006) Lipid oxidation may reduce the quality of a fresh-frozen bone allograft. Is the approved storage temperature too high? *Acta Orthop*, 77, 418-21. ISSN 1745-3674.

Lawhorn, KW; Howell, SM. (2003) Scientific justification and technique for anterior cruciate ligament reconstruction using autogenous and allogenic soft-tissue grafts. *Orthop Clin North Am*, 34, 19-30. ISSN 0030-5898.

Levitt, RL; Malinin, T; Posada, A; Michalow, A. (1994) Reconstruction of anterior cruciate ligaments with bone-patellar tendon-bone and achilles tendon allografts. *Clin Orthop Relat Res,* 303, 67-78. ISSN 0009-921X.

Marrale, J; Morrissey, MC; Haddad, FS. (2007)A literature review of autograft and allograft anterior cruciate ligament reconstruction. *Knee Surg Sports Traumatol Arthrosc,* 15, 690-704. ISSN 0942-2056.

Nutton, RW; McLean, I; Melville, E. (1999) Tendon allografts in knee ligament surgery. *J R Coll Surg Edinb,* 44, 236-40.

Papandrea, P; Vulpiani , MC; Ferretti, A; Conteduca, F. (2000) Regeneration of the semitendinosus tendon harvested for anterior cruciate ligament reconstruction. Evaluation using ultrasonography. *Am J Sports Med,* 28, 556-61. ISSN 1552-3365.

Pearsall, AW; Hollis, JM; Russell, GV Jr; Scheer, Z. (2003) A biomechanical comparison of three lower extremity tendons for ligamentous reconstruction about the knee. *Arthroscopy,* 19, 1091-6. ISSN 0749-8063.

Rappe, M; Horodyski, M; Meister, K; Indelicato, PA. (2007) Nonirradiated versus irradiated Achilles allograft: in vivo failure comparison. *Am J Sports Med,* 35, 1653-1658. ISSN 1552-3365.

Reiff, RBM; Croci, AT; Bolliger Neto, R; Pereira, CAM. (2007) Estudo comparativo de propriedades biomecânicas da porção central do tendão calcâneo congelado e a fresco. [Comparative study on biomechanical properties of the central portion of frozen and fresh calcaneus tendon]. *Acta Ortop Bras,* 15, 6-8. ISSN 2176-7521.

Schachar, NS; McGann, LE. (1986) Investigations of low-temperature storage of articular cartilage for transplantation. *Clin Orthop Relat Res,* 146- 50. ISSN 0009-921X.

Sherman, OH; Banffy, MB. (2004) Anterior cruciate ligament reconstruction: which graft is best? *Arthroscopy,* 20, 974-80. ISSN 0749-8063.

Sterling, JC; Meyers, MC; Calvo RD. (1995) Allograft failure in cruciate ligament reconstruction. Follow-up evaluation of eighteen patients. *Am J Sports Med,* 23, 173-8. ISSN 1552-3365.

Sun, K; Zhang, J; Wang, Y; Xia, C; Zhang, C; Yu, T; Tian, S. (2011) Arthroscopic Reconstruction of the Anterior Cruciate Ligament With Hamstring Tendon Autograft and Fresh-Frozen Allograft A Prospective, Randomized Controlled Study. *Am J Sports Med.* [Epub ahead of print] ISSN 1552-3365.

Tomford, W. Transmission of disease through musculoskeletal transplantation. *Portland Bone Symposium.* Portland: Oregon Health Sciences University, 1997.

Urabe, K; Itoman, M; Toyama, Y; Yanase, Y; Iwamoto, Y; Ohgushi, H. (2007) Current trends in bone grafting and the issue of banked bone allografts based on the fourth nationwide survey of bone grafting status from 2000 to 2004. *J Orthop Sci,* 12, 520-525. ISSN 1436-2023.

Vangsness, CT Jr; Garcia, IA; Mills, CR; Kainer, MA; Roberts, MR; Moore, TM. (2003)Allograft transplantation in the knee: tissue regulation, procurement, processing, and sterilization. *Am J Sports Med,* 31,474-481. ISSN 1552-3365.

Zimmerman, MC; Contiliano, JH; Parsons, JR; Prewett, A; Billotti, J. (1994) The biomechanics and histopatology of chemically processed patellar tendon allografts for anterior cruciate ligament replacement. *Am J Sports Med,* 22, 378-386. ISSN 1552-3365.

Part 2

Human Asisited Reproduction Techniques (ART)

6

The Problem of Contamination: Open *vs.* Closed *vs.* Semi-Closed Vitrification Systems

Enrique Criado Scholz
CERAM: Centre for assisted reproduction in Marbella
Spain

1. Introduction

The Development of cryopreservation techniques, the increase in demand for cryopreserved cells or tissues and the use of these techniques in cells or tissues from patients with infectious diseases, has forced us to reduce the risk of contamination during the freezing process and the risk of cross-contamination during the storage of this material. Recent publications that demonstrate the survival of pathogens at low temperatures and possible contamination of the cells or tissues stored have changed the laws of each country and the customs and protocols used so far in the cryopreservation.

To understand the problem of contamination in cryopreservation we need to have an overview of the current problem in which all researchers are concerned about, seeking a cryopreservation protocol with good results but without contamination problems. Discussing the cryopreservation's different techniques such as slow freezing, vitrification, kinetic vitrification (extra-, hyper-, super-, ultra-fast vitrification) and the various components that help us understand the difficult balance between technique, device used and the risk of contamination. We need to use new products and new protocols' to have good results ensuring biological samples and patient safety.

Dr. Katkov's idea of find the *"Universal crypreservation protocol"* (see the other Chapter in the Book) by Katkov at al. that we can use worldwide with all the possible biological samples would lower considerably the price of cryopreservation process and we would have better results because everybody would work with the same protocol and the same results.

We have to comprehend the difference between open, closed or semi-closed devices and the importance of choosing one device or another both in morphological survival post thaw cell or tissue as on non-contamination of these samples. The device used, the protocol used and the cooling solution used can change the outcome of cryopreservation and therefore we have to find a protocol for cryopreservation with a cooling solution and a secure device to provide us good results free of contamination.

2. Contamination and cross-contamination

The first thing we must learn is to differentiate their respective importance are the concepts of contamination and cross contamination of samples. The first relates to the contamination of the sample by freezing or by direct contact with the cooling solution and the second refers to the contamination of the sample within the common container which is in contact with all cryopreserved samples, some samples may be contaminated or the liquid nitrogen (LN2) might be contaminated producing a possible cross-contamination. The potential for disease transmission and pathogen survival through contaminated LN2 has been proposed by many authors (1-3), and the evidence of contamination in human patients has been described for different pathogens (4-10). It has to be stated that none of the reported infections after insemination or ET in humans and domestic animals can be clearly attributed to the applied cryopreservation and storage procedure but the use of safe cryopreservation protocol is very important to avoid human cell contamination or cross-contamination in common LN2 tanks.

Although cryopreservation had a boom in the mid 70's and early 80's with the opening of the first sperm banks in America and Europe, it was not until the mid-80's when we saw the need for biological samples cryopreserved in quarantine and the lack of screening leads to infection of several recipients that had been inseminated with semen samples from donors HIV+ those unaware of their disease (11). In these cases it was found that samples stored in the same containers with frozen HIV+ samples were not contaminated, otherwise in 1995, six patients undergoing cytotoxic treatments hermetic problems developed an outbreak of acute hepatitis B after undergoing an autologous cryopreserved material that had been stored in the same cryogenic container as other patients infected with hepatitis B (12).

2.1 Cells and tissue contamination

In the field of assisted reproduction, although it hasn't been detected any contamination in the cryopreservation of gametes and embryos, the probability and the occurrence is low, the risk is not zero so it is recommended to follow the rules in biosecurity manuals for both the physical and chemical risk as well as the risk of contamination and cross contamination of samples.

The case in 1985 where there was infection with hepatitis B in the cryopreserved samples (12) the infection was due to an error in packaging and storage of samples. With time a deterioration of the bags containing infectious material causing the infection of the LN2 and other samples was observed.

Further studies have shown that the storage of samples is decisive. There is evidence that frozen samples in hermetically sealed straws are not contaminated even if they are in contaminated containers with contaminated LN2 and LN2 does not contaminate infective biological samples that were frozen in a sealed container (13,14)

During the cryopreservation, biological samples go through many processes before being cryopreserved. In the case of IVF cells are subjected to a phase of procurement, fertilization, development, transfer and finally cryopreservation. This represents an approximate 6-day process in which many factors can affect the contamination of the sample at the end of the process. We can find contamination or cross-contamination in the following cases (15):

- Handling contaminated biological samples (semen, follicular fluid, tissue, etc.). Without precautions to avoid contamination outside the base plate to be used for conservation (cryotube, straw, etc.). It is very important to disinfect and clean the container before filling it with LN2 (16). In this regard to ensure an adequate level of biosafety a study is needed of infectious diseases transmissible from any patient or donor who wants to freeze any samples. According to Castilla (17) the clinic policy for a donor with infectious diseases is radically different to that of a patient with any of these diseases wanting to freeze biological material for autologous use. In the first situation, the biological material at hand will not freeze. In the second, the biological material should be frozen but with measures that we discuss later. Screenings for infectious diseases that normally must be submitted are: To analyse serological studies for syphilis, hepatitis and HIV. To analyse the clinical studies infective clinical phases: toxoplasmosis, rubella, herpes virus, cytomegalovirus (CMV), Neisseria gonorrhoeae and Chlamydia trachomatis. These tests are required for donors of semen every 6 months. As the risk of disease transmission during storage in LN2 is mainly viral. Interestingly, the American Society of Fertilisation (18), ESHRE (19), British Andrology Society (BSA) (20) and the Spanish Association of Tissue Banks (AEBT) (21) also recommend serologic screening for CMV, not just clinical. The presence of CMV in semen has been associated with active disease (anti-CMV IgM + or recent seroconversion anti-CMV IgG +). Similarly, these companies recommend performing serologic tests for HTLV-I and HTLV-II. But although it is clearly demonstrated the transmission of human papilloma virus by using LN2 cryotherapy and has been shown IUI transmission of herpes simplex virus (HSV) (22), none of the scientific associations mentioned above recommend a culture for detection or serological studies of HPV donors or patients with infection who are going to freeze biological material because the analysis to detect these deceases are not very sensitive. As rubella serologic screening of donors, its low prevalence in this population means that serological tests have a low positive predictive value, making it unadvisable. Finally, we believe a patient who needs to freeze some reproductive biological material should have at least one serology for HIV, hepatitis B and C. This proposal is consistent with the recommendations of the AEBT for cryopreservation of semen (21).

- Use of contaminated culture media. In these cases the degree of cross-contamination would reach very high levels having an impact on many patients. Although the preparation of embryo culture media and sperm extenders from specific ingredients are avoided in human clinics, it continues to be a common practice in animal ART (23). Nevertheless, many ingredients of embryo culture media and sperm extenders act as stabilizers for many micro-organisms at freezing temperatures (milk, serum or serum albumin, sucrose, sorbitol and other sugars). Unfortunately, the most common cryoprotectors (CPs) in applied oocyte cryopreservation and embryo (glycerol, DMSO, ethylene glycol, propylene glycol, methanol etc.) are toxic for cells. Also bacteria and viruses efficiently protect from cryoinjuries, eg Concentrations of DMSO as low as 5% enveloped viruses defend against the trauma of freezing (24). The Fact That micro-organisms survive in association with germplasm is not only important from the potential of disease transmission by embryo transfer to recipients, but also in approaches to the storage of samples for testing and health certification of embryos for international movement. On the other hand we must also bear in mind that all culture media containing antibiotics to prevent or limit survival of microorganisms.

- Conservation of contaminated material or straws cryotubes closed or sealed badly flawed causing the breakdown of the frozen straw, leaving the contaminated sample directly exposed to the LN2 tank risking contaminating the other samples. Closed systems can be sealed in many ways (thermal sealer, ultrasound sealer, radiofrequency sealer, polyvinyl alcohol powders, and solid caps). Given the sealing time and the temperature reached does not affect the cryopreserved sample, we have to ensure that the seal is airtight and that the device is built of resistant material to low temperatures of LN2 (Ionomeric resins, quartz glass capillary, Polyvinyl chloride, Polyethylene glycol tetralato, etc).
- Using contaminated LN2 during the freezing process. In this case we have proposed some solutions that we will see later.
- Poor source management of LN2 from our supplier contaminating commercial LN2 that comes to our lab in the process of manufacture or transportation and filling our containers.
- For transportation of contaminated material in containers. Storage containers should be emptied and cleaned periodically due to the risk of lost straws or small particles of contaminated material that falls to the bottom of a large container (25,26). Most of the companies of LN2 containers provide cleaning protocols. The main problem is the cleaning of transport cylinders called "dry" because the material that absorbs the LN2 in these bottles is difficult to sterilize. Bielanski (27) describes a method of disinfection of commercial dry shippers with two different types of a LN absorbent. Based on the results presented, it appears that solutions of sodium hypochlorite and ethylene oxide are equally useful for the disinfection of dry shippers constructed with a hydrophobic LN absorbent. In contrast, for dry shippers without a hydrophobic LN absorbent it is advisable to use gas only for decontamination in order sterilization to avoid their damage by liquid disinfectants.
- The air in the room. If the air that reaches the lab comes from another area that could be contaminated and there isn't a good filter. Some laboratories do not have filtration systems or positive pressure to prevent air contamination.
- Operators. If they are infected then that can lead to contamination by contact or peeling during processing of samples or the handling of cryogenic tanks. Staff must meet certain health and hygiene conditions: negative serology for HIV, HCV, HBV and vaccination against hepatitis B and other viral diseases for which there is a vaccine available. We must also have a detailed description of their jobs, tasks and responsibilities. In addition the centre must provide the worker training in freezing techniques for updating and improving procedures.
- Use of open devices. In recent times there is much talk of closed or open system and the possibility of contamination, so many countries have banned open systems and the trend is to ban the high risk of sample contamination. In a closed or semi-closed device the nitrogen of common container is never in contact with biological material frozen on the inside so cross-contamination cannot produced. In the open system, the biological material is in contact with the common nitrogen so contamination from the sample is very easy if the LN2 is contaminated or contamination of LN2 if the sample is contaminated. The latest study done by Criado and his group (28) showed 45% of contamination in an open device (Cryotop) Vs 0% of contamination in a semi-close device (Ultravit) equal and using a contaminated laboratory LN2.

2.2 Cooling solution contamination

The cooling solution plays a significant role in avoiding contamination of biological samples. It means that we will freeze the sample and we will deposit it for a long storage until thawed and used. Normally the LN2 cooling solution is the most widely used in cryopreservation and survival of pathogens at high temperatures (-196 ° C) has already been proven by many studies (1-3,27,28) cases also involved in seeing cross-contamination of human papillomavirus (14,29).

The need for better cooling rates to avoid formation of crystals in cryopreservation has resulted in the discovery and use of new cooling solutions (slush, slurry, etc.). So there are more components to consider when contamination is to be avoided. Using these new cooling solutions gives a lower temperature than the LN2 temperature and much faster transmission. The Slush nitrogen is obtained by a vacuum pump (Telstar TOP-3; Telstar S.A., Terrassa, Spain) that solidifies part of the LN2 in a few minutes. On return to normal atmospheric pressure, the nitrogen collapses, and the subcooled LN2 has solid particles in it commonly referred to as "slush" (30).The advantage of Slush nitrogen lies not only in the temperature difference with respect to LN2 (-196°C Vs -210°C) but also in the reduction of the Leiden frost effect, which is the formation of a layer of vapor around the sample when immersed in the cryogenic liquid from room temperature decreasing the cooling rate (31,32). It has not yet been demonstrated the survival or non survival of pathogens in this cooling solution of 15-20 ° C difference in LN2, this is obtained by vacuum pressure, which can lead to rupture of the cell wall of pathogens to balance internal and external pressure of these in the process of forming Slush. The 'Slurry' nitrogen is a mix of LN2 with different particles for example copper powder. At present investigations are being carried out as an alternative to LN2 to increase the cooling rate because with this cooling solution the thermal conduction is increased. Likewise, experiments are ongoing with various solutions to increase the thermal conduction and the cooling rate.

These cooling solutions "alternatives" are only used at the time of freezing the sample and once frozen, it passes to the general container that is filled with LN2, although these solutions where they freeze cool samples have to be sterile we have to ensure that the general LN2 container does not have contact with the frozen sample in order to not contaminate the sample and the LN2 if the sample is positive for any pathogen.

Retrospective studies in which commercial LN2 cryotanks were examined after 35 continuous years of service revealed various bacterial and fungal contaminations in the LN2 detritus (23). Many of the identified bacteria isolated in these studies were ubiquitous environmental micro-organisms and were rare opportunistic pathogens of low significance in producing disease in humans or animals (Table I). It should be acknowledged that some of the isolates may have been derived from laboratory contamination during semen and embryo processing for cryopreservation rather than genuinely being present within the sample. In agreement with Bielansky and Vajta the risk of contamination by human pathogens seems to be rather low. Components of the standard LN2 production system comprise a compressor, a cryogenerator and containers. From a practical point of view, the complete sterilization and maintenance of sterility in such a robust system might be a very demanding task, if possible at all. Accordingly, some ubiquitous bacterial agents can be expected in any commercially produced LN2. Nevertheless, it is an 'in and out' system and only air-borne contaminants are supposed to enter it (LN2 compressor) via air used for LN2 production. As they are not air-borne, it is unlikely that viral agents of human concern such as HIV, hepatitis and herpes viruses would enter the LN2 production system.

Sample tank	Identified microbial contamination			Years of storage	Total no. of stored samples
	Liquid nitrogen	Semen	Embryos		
Research					
Laboratory tanks					
1	*Staphylococcus auricularis*	Nd	Nd	20	580
2	*Bacillus licheniformis, Bacillus* spp.	*Stenotrophomonas maltophilia, Staphylococcus sciuri*	Nd	15	840
3	CDC group IVc-2, *Alcaligenes faecalis*	*Proteus vulgaris*	Nd	8	460
4	*Brevundimonas vesicularis*	*E. coli*	Nd	15	1350
5	*Stenotrophomonas maltophilia*	—	Nd	12	650
6	*Staphylococcus capitis*, unidentified Gram-negative rod, *Comamonas acidovorans*	*Morganella morganii, Gemella morbillorum, Stenotrophomonas maltophilia, Citrobacter koseri*	*Bacillus subtilis, Ochrobactrum anthropi, Staphylococcus epidermidis*	15	1200
7	*Stenotrophomonas maltophilia, Comamonas testosteroni*	*Stenotrophomonas maltophilia, Citrobacter koseri*	*Stenotrophomonas maltophilia, Bacillus* spp., *Pseudomonas fluorescens, Acinetobacter lwoffi*	18	1480
8	*Bacillus pumilus, Eikenella corrodens*	—	*Stenotrophomonas maltophilia, Bacillus* spp., unidentified Gram-negative rod	10	960
Commercial tanks					
9	Nd	Nd	—	10	58 456
10	Nd	*Aspergillus* spp.	—	30	138 430
11	*Aspergillus* spp.	*Corynebacterium xerosis, Bacillus sphaericus*	—	30	34 962
12	Nd	*Stenotrophomonas maltophilia*	—	35	150 000
13	Nd	*Stenotrophomonas maltophilia*	—	15	280 864
14	*Aspergillus* spp.	*Photobacterium damsela*	—	15	260 912
15	Nd	*Bacillus sphericus, Corynebacterium* spp., *Staphylococcus sciuri*	—	12	262 642
16	*Stenotrophomonas maltophilia*	*Ralstonia picketti*	—	12	404 955

Nd, not detected; —, not available for testing

Table 1. Microbiological contamination of embryos and semen during storage in LN2 (23)

One of the biggest discussions recently in the world of cryopreservation focuses on the importance of the sterility of LN2. As shown in Table I and in total agreement with Bielansky and Vajta and many other authors the commercial LN2 reaching our lab is not contaminated enough to cause any infection to freeze biological material. The major problem is common containers where the samples are deposited with a LN2 stored for months, years or even decades in contact with many samples, which, many clinics do not empty and do not disinfect, so it is in common containers where we can find the highest risk of contamination and cross contamination.

As a possible solution to minimize the risk of freezing biological material some systems have been proposed where we sterilize the LN2 and where we ensure that the sample is not in contact with LN2 with the use of semi-close devices or devices that are the only ones that guarantee a hermetic sealing of the device and avoid any risk of breakage of the solder thus ensuring the aseptic samples. The fact that LN2 can be quickly and safely sterilized could encourage the clinical application of human cell/tissue vitrification, both with open carriers and with closed systems. The problem is that if this device is an open device and is passed to the general container where all the other cryopreserved samples there is a huge risk of cross contamination, so it has not helped.

- LN2 Filtration: One of the solutions that have been developed is the filtration of LN2. Air Liquid has marketed CERALIN a liquid filtration system through LN2 ceramic filters. The CERALIN ON LINE consists of two elements of liquid filtration connected in series and inserted into a section of vacuum transfer line. The ceramic membrane is made from multiple layers formed into a multi-channel element. It is housed in a vacuum insulated pipe, itself installed close to the end-use point. The filter minimizes the pressure drop and avoids the vaporization of the LN2. Thus it avoids nitrogen losses. Several sizes are available, depending on the nitrogen flow. The efficiency of this equipment was investigated and proved in laboratory. The filter is located downstream of the nitrogen vessel. During operation, LN2 flows through the filter and over the ceramic membrane. The result is high-purity LN2 with a bacteria count of less than 1 CFU/L gas. Additionally, the large filtration area of the membrane and low level of contamination of LN2 means it is likely to be several decades before filter saturation.

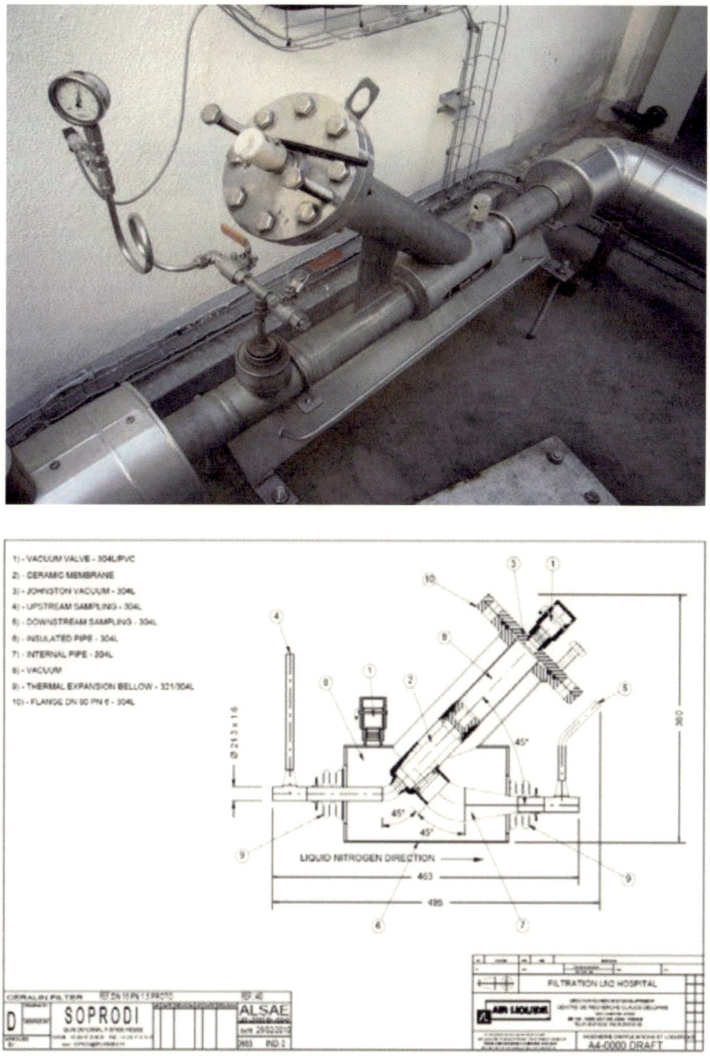

Fig. 1. CERALIN

- UV Sterilization: This method is based on emitting the minimum dose on UV radiation necessary to kill micro-organisms that can survive at the boiling point of nitrogen (-196°C) and which is irradiated in a temperature-controlled regimen, within a short time interval, before the LN2 completely evaporates. The extremely radiation-resistant bacterium Deinococcus radiodurans is inactivated (>- 4log) by administe-ring 400.000 μWs/cm² per each sterilization cycle. An adequate amount of UV radiation de-activates the growth of all kinds of micro-organisms, from viruses like Hepatitis (which require an 8.000 UV dose) to fungi like Aspergillus Niger (330.000 UV dose) (33). At CRYO 2011 Dr. Parmegiani spoke about a new dispositive of UV sterilization of the common containers with cells or tissues inside but the scientific community thinks that is too dangerous biological samples exposed to UV rays without any protection. Although his group is proposing special canisters "not transparent" I think they have to do many more tests to rule out damage to the samples because the common view is confirmed that UV light is harmful, even if used just overnight decreased embryo developmental rates.

- LN2 Steam: As an alternative to hermetical storage in LN2, cryostorage contamination might be avoided by storing the carrier containing the vitrified oocytes in LN2 vapour (34, 35). However, Grout and Morris (36) maintain that storage in the vapour phase of LN2 still carries a risk of sample contamination. Storage of semen in LN2 vapours was discarded early in the development of sperm cryopreservation techniques and it was found that long-term viability of sperm was reduced compared with LN2 storage (37,38). However, recent experiments with new materials have succeeded in developing the technique with acceptable results for both semen and embryos (39,40) and in our last experiment we demonstrated 0% of contamination in vapor nitrogen in a experimental contaminated laboratory LN2 (non published). The drawback of the generalization of this form of storage is the need for careful monitoring of temperature in different parts of the container, which makes the marketing of these containers type (40) more difficult.
- Before entering discussions regarding the sterility in LN2 used for vitrification, we should debate the use of communal containers, which is where cross-contamination can be found, as there is a possibility that the "contaminated cells" could come into contact with each other, and where a number of viruses and bacteria are found, which would never be found in the commercial LN2.

2.3 Contamination in transport

To carry out a safe transportation of biological material we should clearly distinguish a number of concepts (17).

1. Infectious substances: those that contain viable microorganisms (bacteria, virus, prions, parasite, fungus) or bacterial toxins that are known or believed to cause disease in animals or humans.
2. Diagnostic specimens: human or animal materials (body fluids, blood, tissue, tissue fluids, etc.). Obtained for diagnostic or investigational (41).

Most often transported biological reproductive materials are cryopreserved semen donor and follicular fluid when the laboratory is separated from the follicular puncture site. In both cases, we consider the recommendations to follow are those of diagnostic specimens. There are several documents related to the transport of biological material, such as the Universal Postal Union (UPU), the International Aviation Organization (ICAO) and International Air Transport Association (IATA) (42-44).

At European level, all documents related to transport are based on the recommendations of the Committee of Experts of the United Nations Dangerous Goods (UN) (45). There is also a european agreement on international transport of dangerous goods by road (ADR), approved by RD 2115/9838 (46). We will describe some aspects of the mentioned regulations on the transport of diagnostic specimens. The basic system consists of packaging:

1. Primary container, watertight, leak proof, labeled and contains the sample. This container should be wrapped in absorbent material. In terms of labeling, according to AEBT, if it is a semen sample from a donor, must contain an alphanumeric code that identifies the donor and the sample number of the donor. On the other hand, if the sample is for autologous use may be noted also the surname of the patient (21).
2. Secondary container, sealed, leak-proof and protects the primary container. You can place multiple primary containers wrapped in a secondary container. This should be sufficient absorbent material used to protect all primary containers and avoid collisions between them.
3. Outer shipping container: the secondary container is placed in a shipping package that protects the secondary container and its contents from outside elements, such as physical damage and water.

The data forms, letters and other identifying information of the sample should be placed taped outside the secondary container. The label for submitted materials consists of:

1. Basic triple packaging.
2. Does not require signs from United Nations (UN).
3. No substances require pictogram or declaration from the sender.
4. "Biological material for clinical use" must be indicated.
5. Tag address:
   - Name, address of destination, as detailed as possible, and phone number.
   - Name, address, telephone number and contact person at the semen bank.
6. The documents included with the storage conditions and special instructions for shipping. One of the special considerations that we must have in mind when transporting a sample of semen is not breaking the cold chain, so you must use a container or LN2 as well as avoiding the possible use of dry ice.
7. Permission for import / export and declaration.
8. Label orientation.
9. Date and time of departure of Semen Bank (21).

The requirements to be met for local transport are as follows:

1. Sealed and resistant containers.
2. Threaded tubes upright (rack, tray ...).
3. Use of resistant boxes and perfect closure.
4. Secured box in the transport vehicle.
5. Appropriate Labeling.
6. Have the forms with necessary details.
7. Vehicle with kit (gloves, absorbent material, disinfectant, waste container, etc.).

You must ensure perfect coordination of transport between the sender, carrier and recipient to ensure delivery. Thus, each party involved should carry out its part perfectly and appropriately. So stand out from other actions that the sender must ensure the proper identification, packaging, labeling and documentation according to established biosafety guidelines in the "Recommendations of the Committee of Experts of the United Nations Transport of Dangerous Goods" transporting must be kept in appropriate conditions (temperature, light ...) the material from which the sender receives it until it is delivered to your destination and have the appropriate licenses to perform this type of transport, and finally, the recipient must confirm with national authorities that the material can be legally imported.

According to AEBT (21), the possibility of returning a material that hasn't been used should be avoided, as a rule, the return of the semen that has been provided by the Bank, as it will only accept the return of the displayed when you meet the following 3 conditions:

1. The sample wasn't thawed.
2. You can demonstrate the integrity of the packaging (the seals are intact).
3. The temperature of the sample was maintained throughout the transport.

3. Open device

Following recent studies of cell and tissue contamination in freezing and the recent debates regarding the sterility of LN2 in vitrification processes the devices play an important role in the asepsis of the frozen sample. There are many different types of device and of various materials but from the point of view of sterility, devices can be divided into Open, Closed and Semi-closed devices.

There is a lot of controversy and confusion about the concept of Open device. For most cryobiologists Open devices are devices that allow direct contact of the biological sample to be frozen with the cooling solution but when there is contact with the interior of the device but not with the sample to be frozen it would not be considered an open device. Once inside the common cooling containers the cooling solution enters and leaves the device keeping all frozen samples in contact.

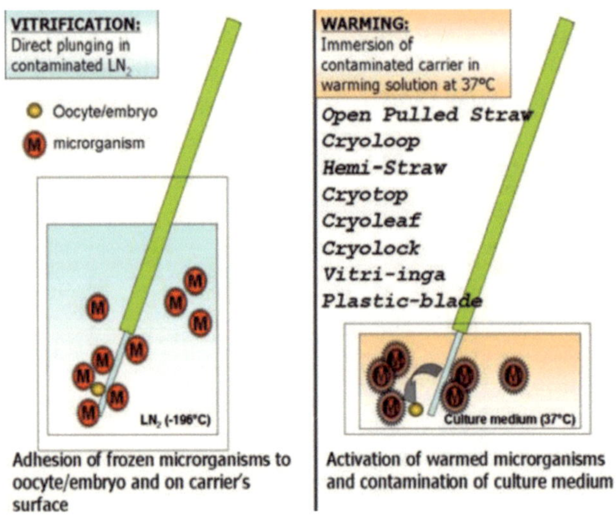

Fig. 2. Risk of Contamination with open devices (47)

Generally, using open devices the achieved cooling rates are approximately 20.000-30.000 ° C / min which favor good vitrification of the sample. The problem is that being in direct contact with the cooling solution there is a risk of pathogen transmission to the biological sample at the time of freezing and a high risk of cross contamination in the common cooling containers. They have been prohibited in many countries for this high risk of contamination and the global trend is to ban them for use with human samples. Recent microbiological studies indicate a 45% pathogen contamination (Pseudomonas and E-coli) with a simple 10-second contact with the open device (Cryotop) with contaminated cooling solution (28).

The most known and used open devices are the following:

Open pulled straw

In the OPS method, 0.25 mm standard insemination plastic mini-straws were heat-softened over a hot plate and pulled manually, as originally described by Vajta et al. (48). The inner diameter and the wall thickness of the pulled part of the straw are approximately 0.8 and 0.07 mm respectively. Cells are load into the pulled straws by placing the narrow end of the pulled straw in the third droplet of medium and aspirating oocytes within a 2–3 mm long liquid column (1–1.5 µl) using capillarity. The straws are then cooled by being plunged directly into LN2 and stored briefly. For warming, the open end of the straw is immersed vertically into 4.5 ml of the warming solution at 37°C. The solidified vitrification solution became liquid within 1–2 s. A cooling rate of 16.700°C/min is obtained with this device (49).

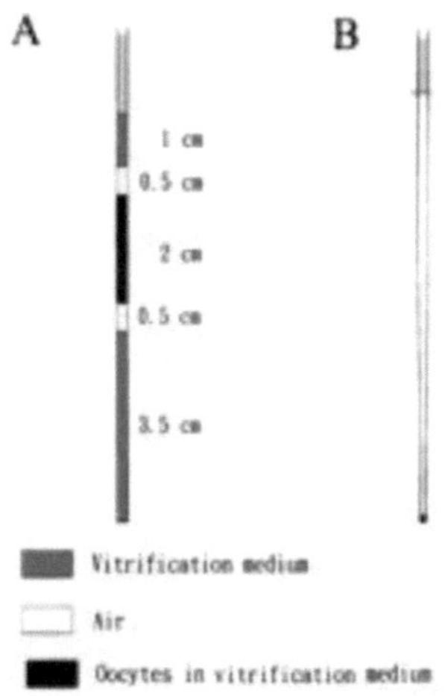

Fig. 3. A) The 0.25 ml conventional straw is loaded with 1 cm of vitrification medium, 0.5 cm of air, 2 cm of vitrification medium containing oocytes, 0.5 cm of air, and 3.5 cm of vitrification medium using a syringe. (B) The open pulled straw is loaded with vitrification solution (1–2 µl) containing oocytes by means of the capillary effect by a simple touch.

Cryoloop

The Cryoloop (Hampton Research, Aliso Viejo, CA, USA), used as a vessel in vitrification, is a thin nylon loop used to suspend a film of cryoprotectant containing the oocytes and directly immerse them in LN2. Vitrification of oocytes using the Cryoloop has advantages over conventional vitrification procedures in that the open system lacks a thermo insulating layer, together with the small volume of <1µl, results in both rapid and uniform heat exchange during cooling. A cooling rate of 20000°C/min is obtained with this device.

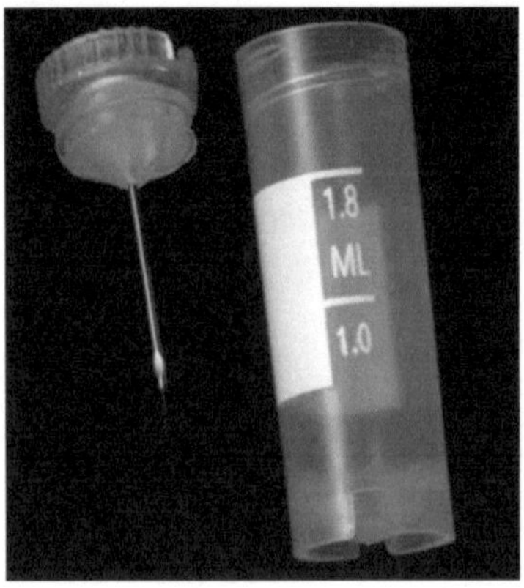

Fig. 4. Cryoloop

Hemi-straw

The Hemi-straw (Astro-Med-tec, Salzburg, Austria) is an embryo carrier that consists of a large gutter on which a small quantity of CPAs (<1µl) containing the cell is deposited. The Hemi-straw is subsequently inserted into a larger pre-cooled 0.5 ml straw (CBS, Cryo Bio System, Grenoble, France) under LN2. Prior to the commencement of the warming process the Hemi-straw is pulled out of the larger straw under LN2 and the tip of the Hemi-straw is immediately immersed into a petri dish containing a sucrose solution. A rapid cooling rate of >20.000°C/min is achieve by allowing direct contact of the biological material with LN2. (50)

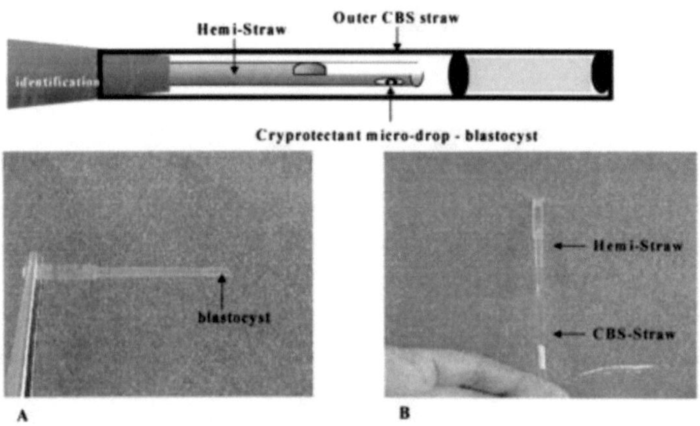

Fig. 5. Scheme of the Hemi-Straw: (A) loading the cell on the tip of the Hemi-Straw; (B) insertion of the Hemi-Straw into a larger 'CBS' straw. (51)

Cryotop

(Kitazato Supply Co, Fujinomiya, Japan) Individual oocytes were picked up in an extremely small volume (<0.1 µl) of vitrification solution and placed on top of a very fine polypropylene strip (0.4 mm wide × 20 mm long × 0.1 mm thick) attached to a hard plastic handle specially constructed according to specifications by Kitazato. The droplet volume was estimated from the length of the fluid column within the pipette tip. As soon as the oocyte was placed onto the thin polypropylene strip of the Cryotop, it was immediately submerged vertically into filtered LN2. Then, the thin strip was covered with a hard plastic cover (3 cm long) on top of the Cryotop sheet to protect it during storage in LN2 containers. For warming, the protective cover was removed from the Cryotop while it was still submerged in LN2, and the polypropylene strip of the Cryotop was immersed directly into the solution at 37°C for 1 min. A cooling rate of 23.000°C/min is obtained with this device (49).

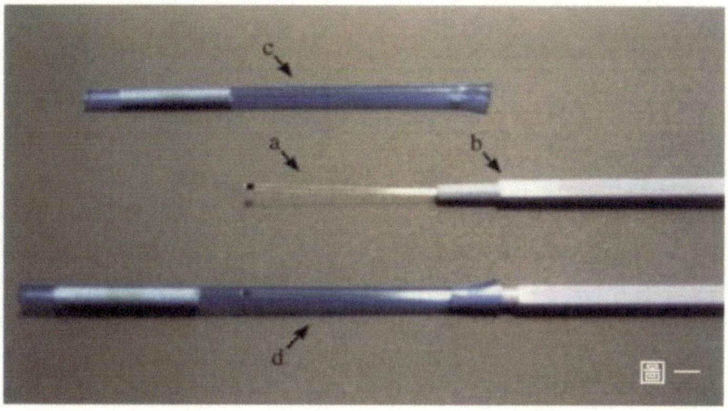

Fig. 6. Cryotop

Cryoleaf

The McGill Cryoleaf™ is very similar to Cryotop but with a number of features designed to improve the loading and storage of cells. Safety during storage has been improved, as the cells are double protected from stress and contamination through a closed cover system but not hermetically sealed leaving cells in direct contact with LN2. The McGill Cryoleaf™ and the vitrification media have been developed by Dr. Chian and Prof. Tan at McGill University, Montreal.

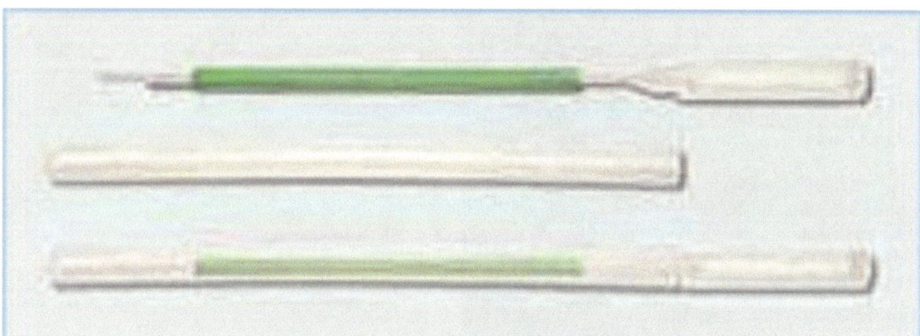

Fig. 7. Cryoleaf

Cryolock

(Biodiseño, Colombia) With this device, cells are deposited near the black mark using the minimal amount possible of vitrification solution (2 µl aprox.) The black mark eases the cover up. The Cryolock® is immediately plunged into LN2, whilst holding the Cryolock®. After this the cap is grasped with forceps and plunged into LN2 until bubbling stops, be aware to not take the Cryolock® out of the LN2 whit covered up, twist and lock gently. Finally, place the Cryolock® in the goblet with the cap downward facing and store for the desired time. For warming, remove the patients canister form the dewar and place in a styrofoam box completely cover the Cryolock® with LN2. Grasp the Cryolock® body at the indentation with forceps and remove from the goblet. Grasp the cap at the indentation with forceps, twist and pull down without taking the Cryolock® body out of the LN2, it must always remain in LN2. Finally remove the Cryolock® from the LN2 quickly and pass into thawing solution at 37°c and follow the protocol.

Fig. 8. Cryolock®

Vitri-Inga

The Vitri-Inga vitrification strip is an apparatus that consists of a fine, very thin polypropylene strip (0.7 mm thick) with a specially designed round tip, in which there is a minute hole to receive the cell; the strip is connected to a hard and thicker plastic handle. Vitri-Inga´ plastic sheaths are 0.5 ml semen straws with a cut in the middle. The total time from when the oocyte was placed into the vitrification solution till its immersion into LN2 is between 50 and 60 seconds. The plastic sheaths, which had been previously cooled for at least 2 min in LN2 vapor on the metal rack inside in the Vitri-Equip, are vertically immersed into LN2. The Vitri-Inga strip with the vitrified oocyte is then inserted into the plastic sheaths for storage, and transferred to a LN2 tank. (52)

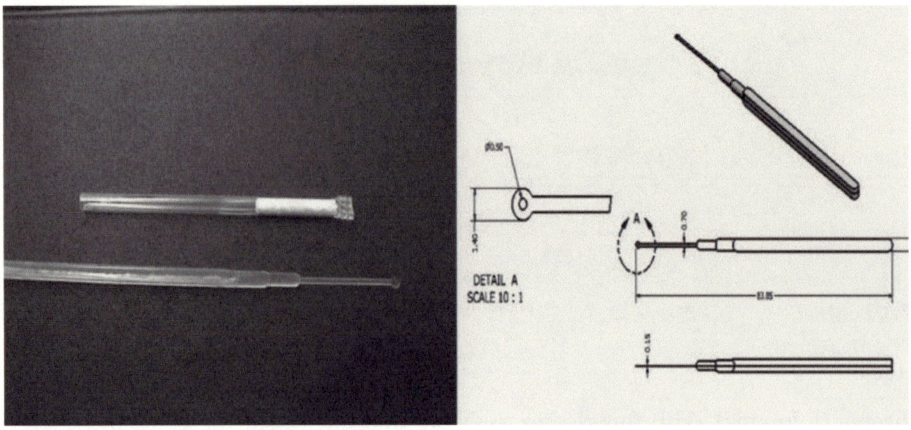

Fig. 9. Vitri-Inga

Plastic-blade

A serum Tube (Sumitomo Bakelite, Tokyo) was employed as a vessel for cryopreservation. A clear polyethylene terephthelate film (50 mm in thick) was cut into a T-shaped piece. As shown in Fig. 10, the horizontal arm of the "T" shape was rolled and fit securely to the inner wall of the cap. After equilibration with cryo-medium, the embryo for vitrification was placed at the center of a plastic blade, the vertical limb projected from the cap, and five embryos were the maximum allowed on one blade. The width of the plastic blade was significantly wider than that of the Cryotop which was commercially available tool for the storage of the vitrified human embryo. The blade was submerged directly into LN2 and inserted into the tube that was pre-cooled with LN2, then the cap was fastened on the serum tube, which accomplished preservation in the LN2 container. For warming, the serum tube that contained the plastic blade was submerged under LN2, and the serum tube was opened and the plastic blade containing vitrified cells was removed from the LN2 and placed directly into the well of the base medium at 37°C. (53)

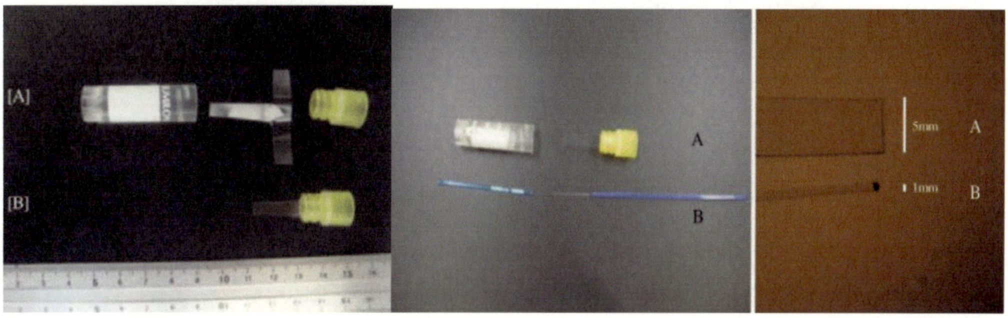

Fig. 10. Plastic-Blade

4. Closed device

Closed systems were born as a need to solve this direct contact with the open devices. The pioneer and first major proponent of close systems for freezing embryos, oocytes or sperm was F. Ostashko in 1960 as an aseptic alternative to the Cassou method. In such systems the biological sample is not in contact with the cooling solution at the time of freezing or at the time of storage in the common containers. This prevents contamination by contact and cross contamination from shared containers. The main feature is that the cooling rate is much lower with these closed devices. By lowering the cooling rate most vitrification protocols with closed systems have a high concentration of cryoprotectants to prevent crystal formation making them "dangerous" protocols for the cell due to the cytotoxicity of the cryoprotective substances. Many comparative studies of open and closed devices listing very similar results.

In the market there are many closed systems to vitrify and more appear daily due to the emphasis that cryobiologists put into to finding the perfect vitrification system that will prevent contamination of the sample and cross-contamination allowing a survival and cell viability with a protocol free or low of cryoprotective substances. The closed devices can be closed or sealed in many ways but most importantly a hermetic seal must be made, preventing entry to the inside and leakage of pathogens to the outside. Thermo seal, radiofrequency seal and ultrasound seal are some of the most used systems that ensure that the stalled sample remains suspended in time.

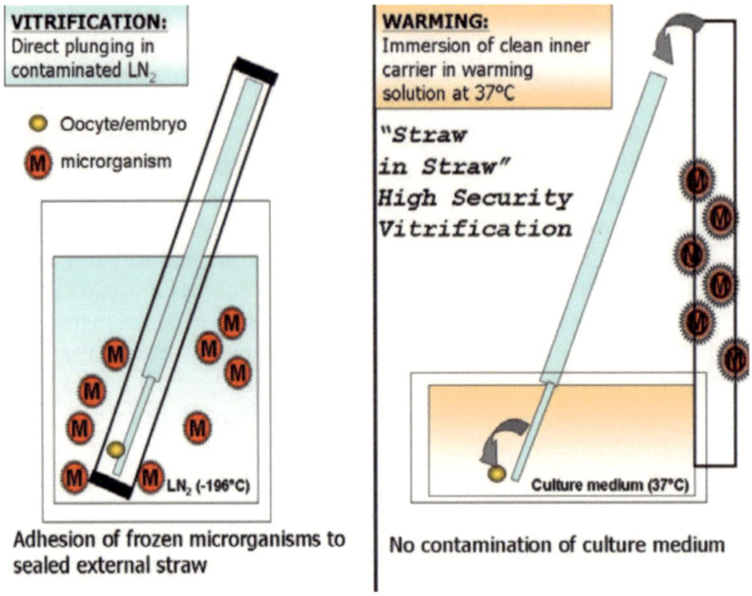

Fig. 11. Risk of Contamination with closed devices (47)

Amongst the most commonly used closed systems are the following:

25 to 0.5 ml Straw

This was one of the first devices used to freeze semen, oocytes, embryos or tissues. If the device is not hermetically sealed they are open devices, but if they are hermetically sealed they are closed systems, as the sample does not come into contact with the cooling solution. The main problem is the cooling rate, as the device is constructed from PVC or ionomeric resin and having a substantional wall thickness there is little temperature transmission (CBS). The cooling rate achieved by these straws is approximately 2.500 ° C / min (49)

Fig. 12. straw

CVM Ring Fibre Plug

CVM™ (Cryologic, Australia) involves the rapid cooling of specimens without their immersion in, or direct contact with the cooling solution. This reduces the risk of any potential contamination by pathogenic microorganism that may be present in the cooling solution. The specimens are put into a droplet which is transferred to the hook at the end of a custom designed fibre called a Fibreplug™. The Fibreplug™ is then transferred to the specially treated surface of a CVM™ Block that has been chilled to LN2 temperature. The droplet vitrifies into a glassy bead and the Fibreplug™ is placed securely into a pre-cooled CVM™ sleeve. A cooling rate of 10.000°C/min is obtained with this device. Besides the cooling rate, another main problem with this method is that to cool the CVM block the surface makes contact with LN2 thus "contaminating" the surface, which could then provoke a contamination of the sample.

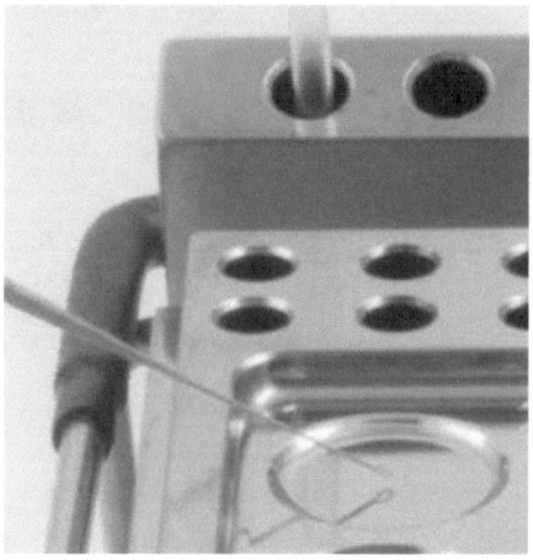

Fig. 13. CVM Ring Fibre Plug

Rapid-i™

Rapid-i (Vitrolife, Sweden) is based on the same principle as the open vitrification system of the Cryoloop meaning that the embryos are place in a minute volume of vitrification solution in a hole and held there by surface tension. The Rapid-i™ holding the embryos is in turn placed in a pre-cooled RapidStraw™ sitting in the container filled with LN2. This unique feature of the Rapid-i™ vitrification System means that vitrification actually takes place in super-cooled air reducing contamination risks. The straw is sealed after vitrification making the critical time frames of the dehydration steps easier to keep and creating an aseptic vitrification system without any contact between vitrified material and LN2. The main problem is that a cooling rate of 1.200°C/min is achieved with this device (54).

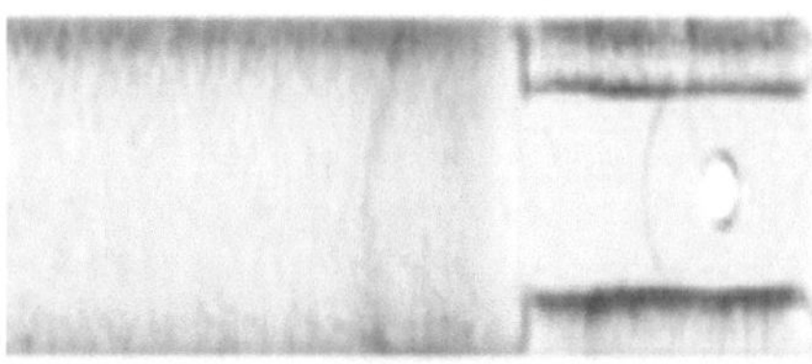

Fig. 14. Rapid-i

Vitrisafe

Is a modification of the previous Hemi-straw vitrification plug that allows a complete insertion in high security 0.5 ml straw. The Vitrisafe consists of a large gutter that is totally inserted into a larger pre-cooled 0.5 ml straw (CBS, Cryo Bio System, Grenoble, France). Only after welding both ends of the 0.5 ml straw to ensure the complete isolation of the biological sample is the complete straw plunged into LN2. For warming, the gutter is removed from the outer straw without contact with LN2 and the tip containing the biological material is directly plunged into the dilution solution in order to archive a rapid warming. A cooling rate of 1.300°C/min is archived in the vitrification process (50).

Fig. 15. Vitrisafe

High security vitrification straw

The CBS™ High Security vitrification straws are made from an ionomeric resin that is chemically inert, biocompatible and has physical characteristics resistant to ultra low temperatures and pressures created by expanding liquids and LN2. Sealed straws are resistance tested to 150 kg/cm2 (2133 lb/sq.inch), both the seals and the material should resist in order to have the batch approved. The HSV (High Security Vitrification) kit is composed of a High Security ionomeric resin straw, a capillary tube with a pre-formed gutter attached to a colored handling rod and a blue plastic insertion device. For freezing the sample is deposited into the gutter a few millimeters from the end using a micropipette. The drop holding the sample must be under 0.5 μL. immediately place the capillary rod and handler into the straw and push until the rectangular portion of the handler comes in contact with the flared end of the straw. While still holding the straw in place, seal the open end, hold the straw using tweezers in the area of the handling rod and quickly plunge the entire straw into LN2 vertically. For thawing lift the straw enough to expose the colored handling rod. Make sure the end with the sample remains immersed in the LN2. Holding the straw, use the opening device for HSV kit to section the straw and immediately (within 2 seconds), plunge the gutter into the first dilution media. A cooling rate of 2.000°C/min is archived in the vitrification process (55).

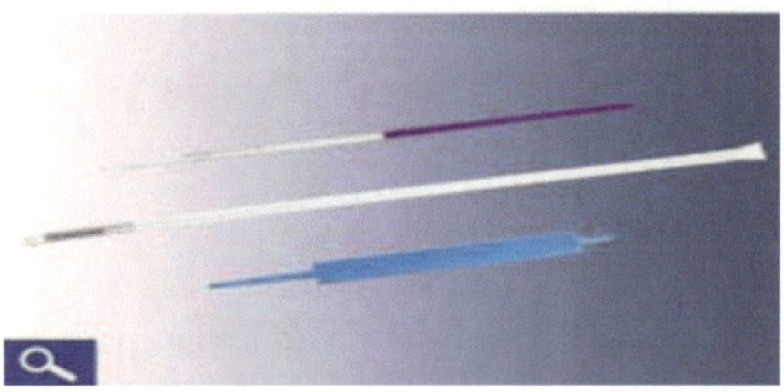

Fig. 16. High security vitrification straw

CryoTip

(Irvine Scientific) A plastic straw container which can be sealed as a closed device to hold gametes or embryos in a specialized medium during cryopreservation procedures and subsequent long term storage in a LN2 tanks. CryoTip consist of a drawn plastic straw with an ultra fine tip and a protective metal cover sleeve. This device has been optimized as a closed system for cryopreservation procedures. For freezing aseptically remove one CryoTip when ready to use. Aseptically attach the wide end of the CryoTip to an aspiration tool, such as a luer tip syringe, using the Connector. When specimens are ready to load into the CryoTip, aseptically slide the metal cover sleeve carefully along the straw to expose the fine tip end. Gently load the specimens into the CryoTip by aspiration using the plunger on the syringe to control the uptake of medium and specimens. Heat seal the fine tip below the 1st mark, then slide the metal cover sleeve down over the fine tip to protect it and plunge the sealed CryoTip into the LN2 reservoir. A cooling rate of 12.000ºC/min is obtained with this device (49)

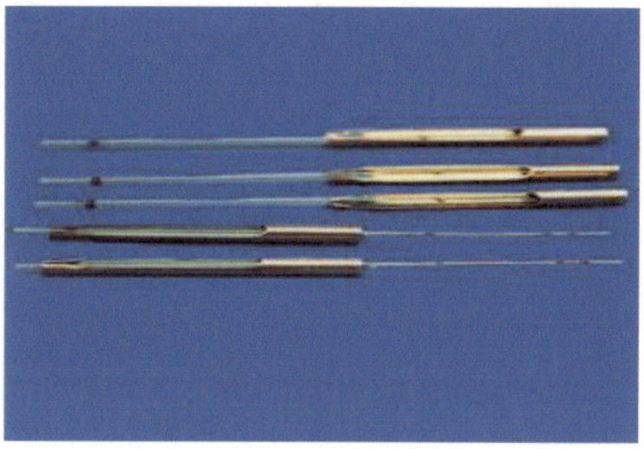

Fig. 17. Cryotip

Cryopette®

(Origio) It is derived from the original STRIPPER® family of denudation tools. It includes a sterile STRIPPER® tip with an integrated bulb to facilitate loading. This eliminates awkward external handles, rods, and pick-up tools required with other devices and guarantees simplicity, speed and ease of use. The bulb is designed to deliver the sample to the desired location every time. The maximum load volume is 1.2 µl, producing a cooling rate of approximately 23.700ºC/min.

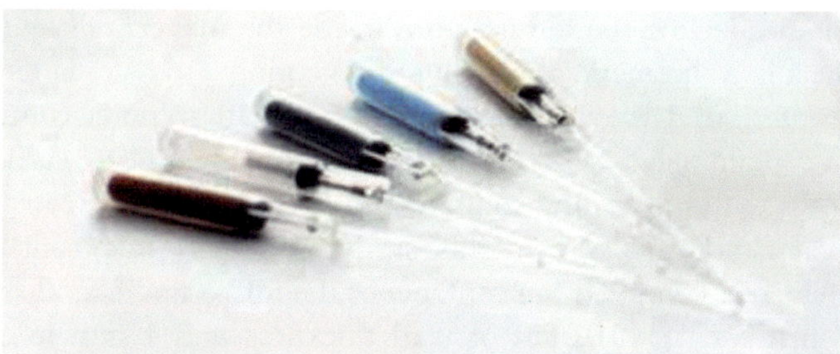

Fig. 18. Cryopette

These last two devices have been criticized by Parmegiani and his group's latest articles and in the last CRYO congress 2011, for the "potential danger" they have in his opinion: In the thawing process the external part of the device is in direct contact with the warming solution and any pathogen that could be on the exterior could pass onto the sample. In my opinion it is a very remote hypothesis and the probability that this could occur is minimal in comparison with potential contamination of an open device, a bacteriological study would be necessary that could demonstrate that with a high % of contamination, a simple contact of the end of the device with the warming solution is sufficient to contaminate the sample.

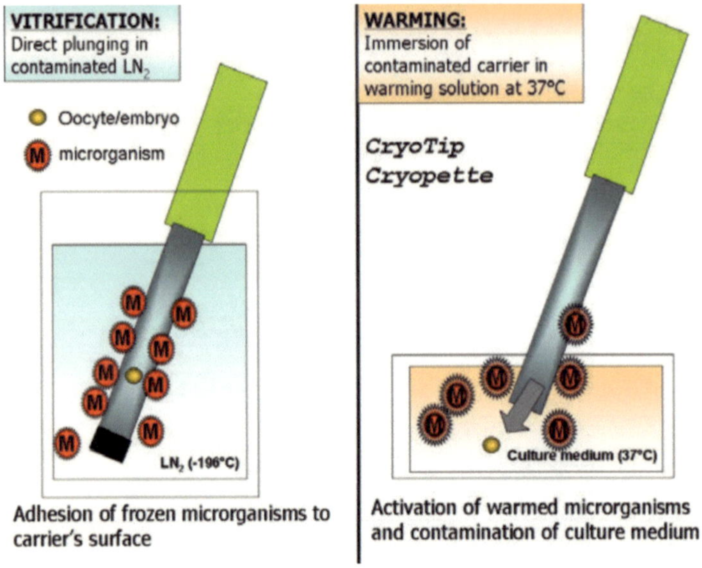

Fig. 19. Risk of Contamination with some closed devices (47)

Ultravit

Ultravit is a novel device composed of a 0.3 mm internal diameter quartz glass microcapillary tube and a flexible, transparent inert sheath that has been designed to protect and prevent it floating in the LN2. Loading the internal microcapillary tube and removing the cells from the device is very simple and easy using a syringe. Before warming, the protective sheath is cut and the internal microcapillary tube is placed in a sterile medium at 37ºC after the thawing protocol. The open end of the sheath can be sealed ultrasonically in milliseconds without affecting the temperature inside the microcapillary tube, closing the system and ensuring a hermetic seal, thus preventing cross-contamination. The last microbiological control of Ultravit showed that the 5-10 seconds contact of Ultravit's internal part with contaminated LN2 (E.Coli and Psudomonas) is not sufficient to produce direct contact of cells with the cooling solution and does not result in contamination (0% of Ultravit Vs 45% of Cryotop) (28). In this study we didn't find contamination in the microdrops into which we emptied the contents of the microcapillary, also submerging the end of this (0.2mm diameter, 0.01mm of wall thickness and 1 mm in contact with the warming drop). There is a great difference between thawing with an open device (in which the entire strip is submerged in the warming solution with a surface of 42-50 mm2) and thawing with Ultravit with 95.5% less of surface in contact with the warming solution at 37ºC (1.7-1.9 mm2). With Ultravit protocol, only the end of the microcapillary touches the base of the dish used to thaw, but at no time does the external part touch the warming drop (56). The following diagram shows the use of Ultravit, presuming that the cooling solution is contaminated:

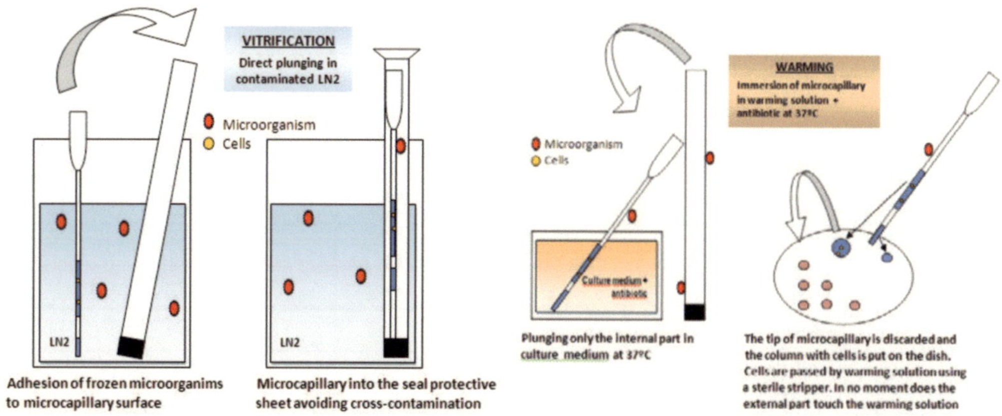

Fig. 20. Risk of Contamination with Ultravit

Our work has demonstrated that the microorganisms that may be in the cooling solution, on the outside of the microcapillary, cannot come into contact with ultra-vitrified cells inside due to the loading procedure, the contact time with the cooling solution and the diameter and surface of Ultravit making it a secure device and is enough to exclude the theoretical danger of contamination. The cooling rate obtained with Ultravit was 250.000 C°/min (57) with Slush nitrogen allowing ultra-vitrification with low concentration of CPA (1.5-2 M) and a morphological survival rate of 92% of human mature oocytes and 59.1% of blastulation rate in mouse embryos.

5. Semi-closed device

As a consequence of the necessity of a device which avoids contact of the biological sample with the cooling solution but that would achieve cooling rates high enough to ensure a high rate of vitrification, semi-closed devices were designed. Gabor Vajta was one of the first to hypothesize the enclosure of open carriers (after direct contact of cells/LN2) in pre cooled hermetical containers (48). There are systems in which there is direct contact of the biological sample with the cooling solution only at the time of vitrification. Once the sample is vitrified , the device is placed in a protective sheath which is hermetically sealed before being passed to the communal container. This ensures no cross-contamination in the tanks. As stated earlier in this chapter, microbiological studies performed (23) showed that many of the Identified bacteria isolated in the commercial tanks are ubiquitous environmental micro-organisms and are rare opportunistic pathogens of low significance in producing disease in humans or animals so these devices are an important tool for high survival of biological samples to avoid cross-contamination. Theoretically all open systems can become semi-closed systems to protect the biological sample from cross-contamination with a high cooling rate but the most important is:

OPS safe method

Vajta in 1997 devised a vitrifying system with OPS, but once submerged in LN2, the OPS straw is transferred to a 0.5ml CBS straw. Using this method a cooling rate of approximately 16.700 °C/min is achieved, but once passed into the 0.5ml CBS straw it is protected from cross-contamination in the common tanks.

A summary of the cooling rates obtained with each device is as follows:

DEVICE	VOLUME	COOLING RATE
CRYOLOOP	>1 μl	20.000 °C/min
HEMI-SATRAW	>1 μl	>20.000 °C/min
CRYOLEAF	>1 μl	23.000 °C/min
VITRI-INGA	1 μl	20.000 °C/min
CVM-RING	>1 μl	10.000 °C/min
VITRISAFE	>1 μl	1.300 °C/min
HSS	0.5 μl	2.000 °C/min
0.25 ML STRAW	25 μl	2.500 °C/min
OPS	1 μl	16.700 °C/min
CRYOTOP	0.1 μl	23.000 °C/min
CRYOTIP	1 μl	12.000 °C/min
RAPID-I	0.5 μl	1.200 °C/min
CRYOPETTE	1.2μl	23.700 °C/min
ULTRAVIT	0.2 μl	250.000 °C/min

Table 2. Different cooling rates

6. Cooling rate Vs closed systems

Today the differences between Slow freezing and Vitrification are known worldwide. We all know that slow freezing is characterized by a prolonged cooling curve and the use of low concentrations of cryo-protectors generally "non-toxic" for the cells (1–2 M) with cell injury due to ice formation (58) and that Vitrification is characterized by the rapid procedure and the use of a high concentration of cryo-protectors (4-6M) to prevent cell damage that is toxic to most mammalian cells (59-67).Thus, vitrification with a semi-close devices have a better cooling rate without cross-contamination or novel cyopreservation techniques are needed that allow rapid cooling to achieve vitrification in the absence of high concentration of CPA or if is possible without CPA.

The requirements and relationships for conditions to achieve satisfactory vitrification in the area of mammalian ART are well displayed in the equation of Yavin and Arav (68)

$$\text{Probability of vitrification} = \frac{\text{Cooling and warming rates} \times \text{Viscosity (CPA concentration)}}{\text{Volume}}$$

Fig. 21. Probability of vitrification by Yavin and Arav (68)

The main points to be gathered from this relationship are that the smaller volume of the vitrification solution in which the cellular material is placed for the vitrification process, the faster cooling and warming rate that can be achieved and the lower concentration of CPAs needed reducing the detrimental effect of the inherent toxicity of CPAs and increasing the overall success of the procedure (50).

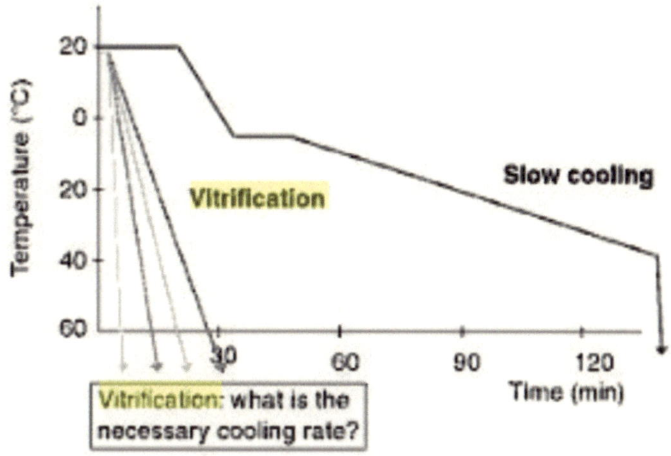

Fig. 22. necessary cooling rate to have a good probability of vitrification

What would happen if We could vitrify without CPA's or with a low concentration of CPA's? What would happen if We could combine the advantages of Slow freezing and Vitrification and vitrify with low concentrations of CPA's with a secure and free contamination device? That is Kinetic vitrification (Ultra-vitrification). Perfecting the techniques of Vitrification has been achieved a morphological survival rate comparable to normal Vitrification protocol (30) or a 59.1% of blastulation rate in mouse embryo (69) with Kinetic vitrification and concentrations of CPA's typical of Slow freezing.

Previous studies have tried to achieve high cooling rates for cell vitrification. However, none of them utilized low CPA concentrations (1.5-2 M). In 1985, Rall and Fahy successfully vitrified mouse embryos in 6.5 M cryoprotectant cocktail solution (70). In that case the method consisted in a 0.25 ml straw container plunged into LN2; the cooling rate was 2.500 °C/min. When this container was plunged into Slush nitrogen, the cooling rate increased up to 4000 °C/min (71). The use of OPS (instead of the 0.25 ml straw) in LN2 increases this cooling rate up to 5.300 °C/min (71) and to 10.000–20.000 °C/min if plunged in Slush nitrogen (71,72). Similar cooling rates were achieved in the case of a Cryoloop quenched in Slush nitrogen (73). The use of electron microscope copper grids has also been investigated, but the cooling rates were in the same order of magnitude that the afore mentioned works: 11.000–14.000 °C/min in the case of plunging the grid in LN2 (74) and 24.000–30.000 °C/min if plunged in Slush nitrogen (74,75). From Boutron's theory, none of these approaches reaches the critical cooling rate to achieve vitrification with low concentration of CPA (1.5-2M). It´s impossible to use open devices with Slush nitrogen as the cell is on the outside and there is a possibility of detaching from the device.

Adjusting to Yavin and Arav formula the Ultra-vitrification technique arose achieving a cooling rate above 250.000 °C/min and of 90.000 °C/min in thawing. This rate is one order of magnitude higher than the highest cooling rate achieved in different strategies (electron microscope copper grids in Slush nitrogen (74,75), whilst keeping all the advantages of a straw-like form for the container and being in the range of the necessary cooling rate to achieve vitrification. To have this increase in the cooling rate a few changes were made to the normal vitrification process:

Slush Nitrogen

As a cooling agent this technique uses Slush nitrogen, much colder than LN2 (-196ºc Vs - 210ºC) and with the property of avoiding the Leiderfrost Effect. When something is submerged in LN2, bubbles rise to the surface through the device, varying the thermal conductivity from the outside into the inside of the device. This does not happen with Slush nitrogen. Slush nitrogen is achieved with a vacuum pump in 5 to 10 minutes and it remains slush for a further 5 – 10 minutes before returning to liquid.

It was shown for oocytes and embryos that increasing the cooling rate would improve survival rates by up to 37% (76)

Model	Survival slush (%)	Survival LN (%)	Sig.	Publication
Bovine MII	48	28	$P<0.05$	Arav & Zeron (1997)
Ovine GV	25	5	$P<0.05$	Isachenko *et al.* (2001)
Porcine blastocysts	83	62	$P<0.05$	Beebe *et al.* (2005)
Bovine MII	48	39	$P<0.05$	Santos *et al.* (2006)
Mouse four-cell embryos with biopsied blastomere	87	50	$P<0.05$	Lee *et al.* (2007)
Rabbit embryos	92	83	NS	Papis *et al.* (2009)
Porcine blastocysts	89	93	NS	Cuello *et al.* (2004)
Mouse MII	>80	>80	NS	Seki & Mazur (2009)
Rabbit oocytes	82	83	NS	Cai *et al.* (2005)

LN, liquid nitrogen; GV, germinal vesicle; Sig., statistical significance; NS, not significant.

Table 3. Effect of cooling rate on survival; comparison between LN2 and Slush nitrogen (75)

Quartz Micro-capillary

Another determining factor to achieve a high cooling rate is the device used. To increase the thermal conductivity and minimize the volume, this technique has used a quartz micro-capillary. This has a 0.2-0.3 diameter allowing to ultra-vitrify 0.1-0.2 µl with a 0.01mm wall, a lot thinner than any other device (0.075 mm in OPS). Another important characteristic is the material it is made from: Quartz. The thermal conductivity of quartz glass is a lot higher than that of plastic of which other devices are made of. This converts it in one of the materials that best conducts the temperature (77)

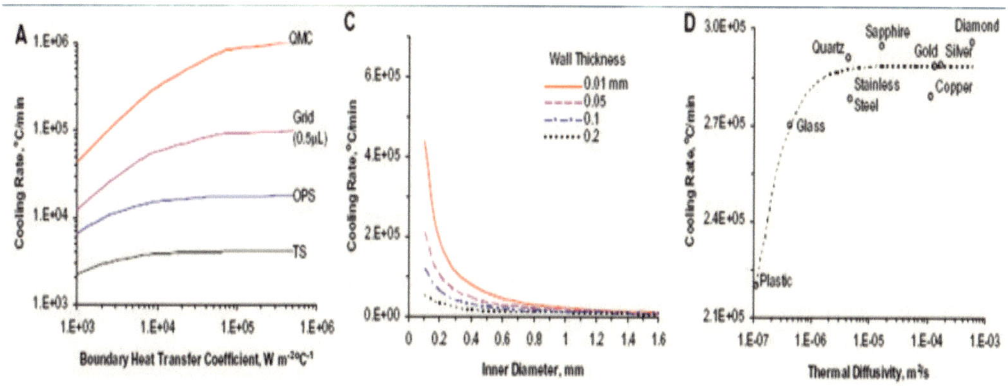

Fig. 23. Xiaoming He et al (77)

In a thermal performance of quartz capillaries for vitrification done by Risco and his group (57) a commercially available version of the OPS (MTG Medical Technological Vertriebs, GmbH) was used. The thermal conductivity of these PVC straws was 0.19 W m_1 K_1. The inner diameter is 0.800 mm and the thickness of its wall is 0.075 mm (Fig. 1a). The QC used (The Charles Supper Company, Inc.) have an inner diameter of 0.180 mm and a wall thickness of 0.010 mm. These geometrical improvements (4.44 times smaller in diameter and 7.50 times thinner) translate not only into a faster heat transfer, but also into a 20 times reduction in volume of the contained solution (for a given height). This is beneficial because the thermal conductivity of the quartz glass is 1.3 W m_1 K_1, that is almost one order of magnitude higher than that of PVC.

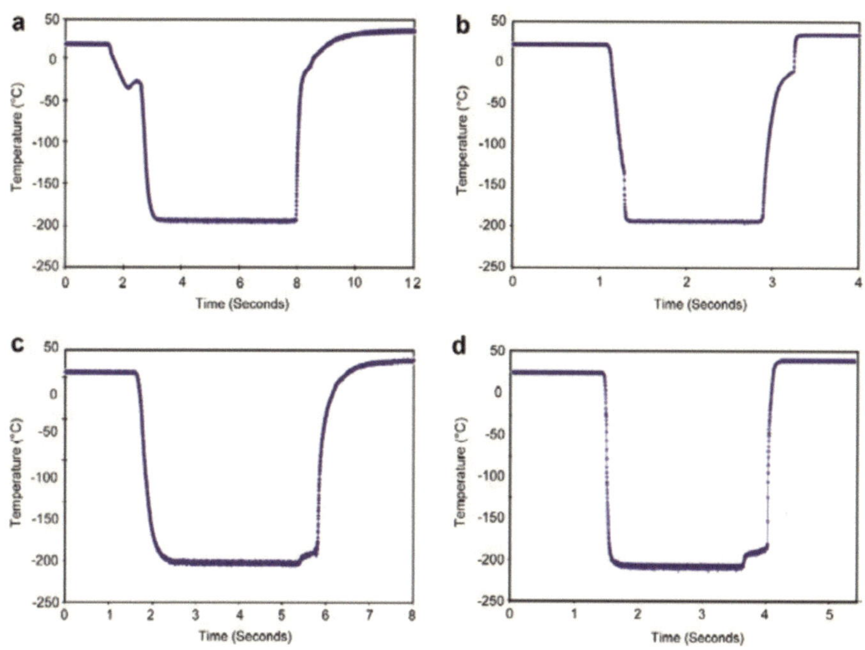

Fig. 24. (a) Thermal history for the OPS when filled with a 1.5M propane-1,2-diol and 0.3M sucrose cryoprotectant solution quenched in LN2 and then thawed in a water bath at 37°C.

A clear heat release peak is present during cooling as well as melting during rewarming. (b) Thermal history for QC when filled with a 1.5M propane-1,2-diol and 0.3M sucrose cryoprotectant solution quenched in LN2 and then thawed in a water bath at 37ºC. Crystallization of water is not obvious during cooling, but melting is shown during rewarming. (c) Thermal history for OPS when filled with a 1.5M propane-1,2-diol and 0.3 M sucrose cryoprotectant solution quenched in Slush nitrogen and then thawed in a water bath at 37ºC. In this case, crystallization during cooling and melting during rewarming was not recorded. However, visual inspection reveals the presence of ice. (d) Thermal history for QC when filled with a 1.5Mpropane-1,2-diol and 0.3 Msucrose cryoprotectant solution quenched in Slush nitrogen and then thawed in a water bath at 37ºC. The sample keeps its transparency over all the cooling-rewarming cycle, an indication of the capability of this approach to vitrify the studied solution.

All these changes have allowed us to maintain a concentration of cryoprotectors typical of slow freezing, 2 M PrOH+0.5 M sucrose, obtaining a morphological survival rate of 92 % in human oocytes (31). Dr. Ho-Joon Lee et al (69) tested this new technique on mouse ooctyes and they saw that using Ultra-vitrification with low concentrations of cryoprotectors improved the fertilization rate and above the blastulation rate. Only the use of Ultravit device in this technique ensures the non contamination of the sample or cross-contamination in communal containers.

%	Slow Freezing (78)	Vitrification (78)	Ultra-vitrification mourine oocytes (69)	Ultra-vitrification human oocytes (31)
Surv. rate	61	91,8	92.5	92
Fert. rate	61,3	67,9	75	?
Blast. rate	12	33.1	59.1	?

Table 4. Comparison between slow Freezing, Vitrification and Ultravitrification (78, 69, 31)

This comparison demonstrates the use of low concentration of cryoprotectant in the Ultra-vitrification protocol favours the morphological survival (92%) and increases the blastulation rate (59.1%). Thus confirming the hypothesis that cryoprotectants are toxic to the biological sample and if we could find a vitrification protocol that would allow us to vitrify without cryoprotectant, we would achieve a better embryo development and a greater chance of pregnancy in the case of freezing eggs or embryos. A lot more studying is needed regarding this new technique but a priori the results indicate that we can hopefully lower the concentration of the cryoprotectants decreasing the toxicity in cells.

7. Conclusion

Cryopreservation protocols have improved and resulted in a much higher efficiency in outcomes in the last years. However, it remains important to always seek for amelioration on cryopreservation protocols and devices to ensure a major benefit and patients' safety during procedures. We need to find a method that combines a high cooling and warming rate, high survival and function of cells and tissues and is made in a way that ensures patient safety.

We must be clear that survival after a vitrification process is a "morphological survival" and that this cell has to fertilize perfectly and develop normally to rule out any damage in the process of cryopreservation and to consider a real survival.

With so much variety of devices and many different protocols, laboratories have to find the protocol and the device that best suits their skills, provided they ensure the sterility of vitrified samples and prevent cross-contamination in general containers. It has created an exaggerated paranoia about the vitrification cell and tissue contamination of the cooling solution that everything can contaminate our samples. In my opinion we must think scientifically, leaving aside the commercial interests of many of us and worrying about more logical things and research data: the probability of contamination must be demonstrated with % of contamination and leave a little aside science fiction and theoretical assumptions. For many cryobiologists avoiding cross-contamination of samples in the general containers is the most important matter as it has been proven that commercial LN2 does not have a high enough risk to have to sterilize it. The use of a closed or semi-closed device that would allow us a high cooling rate and a sealing prior to being deposited in the general container should be enough to ensure sterility and good survival and development results.

We must centre all of our interest in more practical things like finding a vitrification protocol that allows us to a vitrify without cryoprotectant, discovering new cooling solutions, discovering new materials, new devices, new procedures in which we can safely freeze samples without cryoprotectants toxicity problems, thus ensuring a good development cell after thawing, all in secure systems which ensure sterility of the sample and avoid cross-contamination. Finding an "Universal Protocol" risen several times by Dr. Katkov (79), allowing us to use the same protocol worldwide, as movement of frozen specimens around the world has increased dramatically and the lack of component preparation in the laboratory that is to thaw the frozen sample with a protocol and a device different to those they know, would not give us 100% guarantee of sample survival. Today freezing has become a luxury and not all patients can afford the excessive cost of the freezing products, if we could find a protocol without the need of cryoprotectants or a universal protocol that all could use, the price of products would decrease and the patients would benefit economically.

8. Acknowledgment

The author thanks Ms Ana Yus and Ms Cristina Gonzalez for the critical and technical support.

9. References

[1] Karrow AM, Critser JK. Reproductive tissue banking: scientific principles. San Diego: Academic Press, 1997.

[2] Vajta G, Nagy ZP. Are programmable freezers still needed in the embryo laboratory? Review on vitrification. Reprod Biomed Online 2006;12:779-96.

[3] Kuwayama M. Highly efficient vitrification for cryopreservation of human oocytes and embryos: the Cryotop method. Theriogenology 2007;67:73–80.
[4] Kleegman SJ. Therapeutic donor insemination. Conn Med 1967;31:705–13.
[5] Barwin BN. Transmission of Ureaplasma urealyticum by artificial insemination by donor. Fertil Steril 1984;41:326–7.
[6] Stewart GJ, Tyler JP, Cunningham AL, Barr JA, Driscoll GL, Gold J, et al. Transmission of human Tcell lymphotropic virus type III (HTLV-III) by artificial insemination by donor. Lancet 1985;2:581–5.
[7] Berry WR, Gottesfeld RL, Alter HJ, Vierling JM. Transmission of hepatitis B virus by artificial insemination. JAMA 1987;257:1079–81.
[8] Mascola L, GuinanME. Semen donors as the source of sexually transmitted diseases in artificially inseminated women: the saga unfolds. JAMA 1987;257:1093–4.
[9] Araneta MR, Mascola L, Eller A, O'Neil L, Ginsberg MM, Bursaw M, et al. HIV transmission through donor artificial insemination. JAMA 1995; 273:854–8.
[10] Englert Y, Lesage B, Van Vooren JP, Liesnard C, Place I, Vannin AS, et al. Medically assisted reproduction in the presence of chronic viral diseases. Hum Reprod Update 2004;10:149–62.
[11] Stewart GT, Tiler GP, Cunningham AL, Barr JA, Driscoll GL, Gold J, et al. Transmision of human T cell lymphotropic virus type III (HTLV-III) by artificial ibsemination donor. Lancet a985; 2:581-5
[12] 12 Tedder RS, Zuckerman MA, Goldstone AH, Hawkins AF, Fielding A, Briggs EM, et al. Hepatitis B transmission from contaminated cryopreservation tank. Lancet 1995; 346:137-40.
[13] Bielanski A, NadinDavis S, Sapp T, Lutze-Wallace C. Viral contamination of embryos cryopreserved in liquid nitrogen. Cryobiology 2000;40:110–6.
[14] Bielanski A, Bergeron H, Lau PC, Devenish J. Microbial contamination of embryos and semen during long term banking in liquid nitrogen. Cryobiology 2003;46:146–52.
[15] Clara González, Montserrat Boada. Bio-seguridad en el laboratorio de criobiología y aspectos legales de la criopreservación. Cuadernos de medicina reproductiva 2008.
[16] Russell PH, Lyaruu VH, Millar JD, Curry MR, Watson PF. The potencial transmission of infectious agents by semen packaging during storage for artificial insemination. Anim Reprod Sci 1997;47:337-42.
[17] Castilla JA, Magan R, Martinez L, Fernandez A, Ramirez JP, Yoldi A, Vergara F. Seguridad biologica en crioconservacion y transporte de material biológico reproductivo. Aula de formación en embriología clínica N°4 2003.
[18] American Society of Reproductive Medicine. 2002 Guidelines for gamete and embryo donation. Fertil Steril 2002;77 suppl5.
[19] Barratt C, Englert Y, Gottlieb C, Jouannet P. Gamete donation guidelines. The Corsendonk consensus document for the European Union. Hum Reprod 1998;13 supplement 2:7-9.
[20] British Andrology Society. British Andrology Society guidelines for the screening of semen donors for donor insemination (1999). Hum Reprod 1999; 14:1823-1826.
[21] Miralles A, Veiga A, Bassas L, Castilla JA, Mirabet V, Villalba R. Semen. En
[22] Estándares de la Asociación Española de Bancos de Tejidos (2ª ed), 2002.

[23] Moore DE, Ashley RL, Zarutskie PW et al. Transmisión of genital herpes by donor insemination. J Am Med Assoc 1989;261:3441-3443.

[24] Bielanski A and Vajta G. Risk of contamination of germplasm during cryopreservation and cryobanking in IVF units. Human reproduction 2009;24:2457-2467.

[25] Wallis C, Melnik JL. Stabilization of enveloped viruses by dimethyl sulfoxide. J virol 1968; 2:953-954.

[26] Steyaert SR, Leroux-Roels GG, Dhont M. Infections in IVF: review and guidelines. Hum Reprod 2000;6:432-41.

[27] Hargreave TB, Ghosh C. The impact of HIV on a fertility problems clinic. J Reprod Inmunol 1998;41:261-70.

[28] Bielansky A. Experimental microbial contamination and disinfection of dry (vapour) shipper dewars designed for short-term storage and transportation of cryopreserved germplasm and other biological specimens. Theriogenology 2005;63:1946-1957.

[29] Criado E, Moalli F, Polentarutti N, Albani E, Morreale G, Menduni F, et al. Experimental contamination assessment of a novel closed ultravitrification device. Fertil Steril 2011;95(5):1777-9.

[30] Charles CR, Sire DJ. Transmision of papovavirus by cryoherapy applicator. JAMA 1971;218:1435.

[31] Criado E, Albani E, Novara PV, Smeraldi A, Cesana A, Parini V, et al. Human oocyte ultravitrification with a low concentration of cryoprotectants by ultrafast cooling: a new protocol. Fertl Steril 2011;95:1101-1103.

[32] Cowley CW, Timson WJ, Sawdye JA. Ultra rapid cooling techniques in the freezing of biological materials. Biodynamica 1961;8:317–29.

[33] Steponkus PL, Myers SP, Lynch DV, Gardner L, Bronstheyn V, Leibo SP, et al. Cryopreservation of Drosophila melanogaster embryos. Nature 1990;345:170–2.

[34] Parmegiani L., Cognigni G.E., Filicori M. Ultra-violet sterilization of liquid nitrogen prior to vitrification. Hum. Reprod 2009;24,2969.

[35] Eum J.H., Park J.K., Lee W.S., Cha K.R., Yoon T.K., Lee D.R. Long-term liquid nitrogen vapor storage of mouse embryos cryopreserved using vitrification or slow cooling. Fertil Steril 2009. 91,1928–1932.

[36] Cobo A., Romero J.L., Perez S., de Los Santos M.J., Meseguer M., Remohi J. Storage of human oocytes in the vapor phase of nitrogen. Fertil. Steril 2010;94:1903–1907. 36. Grout B.W., Morris G.J. Contaminated liquid nitrogen vapour as a risk factor in pathogen transfer. Theriogenology 2009;71:1079–1082.

[37] Sawada Y. The preservation of human semen by deep freezing. Int J Fertil 1964;9:525-32.

[38] Ackerman DR. Damage to human spermatozoa during storage at warming temperatures. Int j Fertil 1968;13:220-5.

[39] Tomlinson M, Sakkas D. Is a review of standards procedures for cryopreservation needed? Safe and effective cryopreservation-should sperm banks and fertility centres move toward storage in nitrogen vapour?. Human Reprod 2000;15:2460-3.

[40] Clarke GN. Sperm cryopreservation: is there a significant risk of cross-contamination?. Hum Reprod 1999;14:2941-3.

[41] Calero S. Transporte material biologic. Curso Trabajar con seguridad en el laboratorio de microbiología. Aula científica 2003.

[42] The Universal Postal Union (UPU): "Manual of the Universal Postal Convention". Lays down detailed regulation for the transport of biological substances by post/mail.1995.
[43] The International Civil Aviation Organization (ICAO). "ICAO technical instructions". Safe transport of dangerous goods, 1984. Last ed:1999.
[44] The International Air Transport Association (IATA). "Dangerous goods regulations (40ª ed.)". 1999.
[45] The United Nations Committee of Expert on the Transport of Dangerous Goods (UN ECOSOC). Recommendations on the transport of dangerous goods (10ª ed). 1997
[46] Loza E, Alomar P, Bernal A, Harto A, Perez JL, Picazo JJ, et al. Seguridad en el laboratorio de Microbiología Clínica. En procedimientos de Microbiología Clínica, recomendaciones de la sociedad Española de Enfermedades Infecciosas y Microbiologia Clínica. Picazzo JJ(ed) 2000;10:1-42.
[47] L Parmegiani, GE Cognigni, S Bernardi, S Cuomo, W Ciampaglia, FE Infante, C Tabarelli de Fatis, A Arnone, AM Maccarini, M Filicori. Efficiency of aseptic open vitrification and hermetical cryostorage of human oocytes. Reproductive BioMedicine Online (2011) 23, 505- 512.
[48] Vajta G., Holm P., Kuwayama M. et al. Open pulled straw (OPS) vitrification: a new way to reduce cryoinjuries of bovine ova and embryos. Mol Reprod 1998. Dev., 51, 53–58.
[49] Shee-Uan Chen, Yu-Shih Yang. Slow Freezing or Vitrification of Oocytes: Their Effects on Survival and Meiotic Spindles, and the Time Schedule for Clinical Practice. Taiwanese Journal of Obstetrics and Gynecology 2009;48:15-22.
[50] Ri-Cheng Chian and Patrick Quinn. Cryopreservation. Cambridge University Press 2010.107.
[51] Vanderzwalmen P, Bertin G, Debauche Ch, Standaert V, Bollen N, van Roosendaal E, Vandervorst M, Schoysman R, Zech N. Vitrification of human blastocysts with the Hemi-Straw carrier: application of assisted hatching after thawing. Hum Reprod. 2003;18(7):1504-11.
[52] Fachini F, Almodin CG, Minghetti-Cámara VC, Fernandes Moron A, Nakano R, Shimabukuru L. Vitri-Inga. A new vitrification protocol. SBRA 2008.
[53] Sugiyama R, Nakagawa K, Shirai A, Sugiyama R, Nishi Y, Kuribayashi Y, et al. Clinical outcomes resulting from the transfer of vitrified human embryos using a new device for cryopreservation (plastic blade). J Assist Reprod Genet 2010;27(4):161-167.
[54] Larman MG and Gardner DK. Vitrification of mouse embryos with super-cooled air. Fertil Steril 2011;95(4):1462-6.
[55] Vanderzwalmen P. Oral presentation ESHRE 2010.
[56] Criado E. Autor reply about Contamination of single-straw carrier for vitrification. Fertl Steril 2011;95(8):e69;author reply e70.
[57] Risco R, Elmoazzen H, Doughty M, He X, Toner M. Thermal performance of quartz capillaries for vitrification. Cryobiology 2007;55:222–9.
[58] Mazur P. Freezing of living cells: mechanisms and implications. Am J Physiol 1984;247:125–42.

[59] Fahy GM, Levy DI, Ali SE. Some emerging principles underlying the physical properties biological actions and utility of vitrification solutions. Cryobiology 1987;24:196–213.

[60] Fahy GM, Wowk B, Wu J, Paynter S. Improved vitrification solutions based on the predictability of vitrification solution toxicity. Cryobiology 2004;48:22–5.

[61] Fahy GM,Wowk B,Wu J, Phan J, Rasch C, Chang A, et al. Cryopreservation of organs by vitrification: perspectives and recent advances. Cryobiology 2004;48:157–78.

[62] Fowler A, Toner M. Cryo-injury and biopreservation. Ann NYAcad Sci 2005;1066:119–35.

[63] Heng BC, Kuleshova LL, Bested SM, Liu H, Cao T. The cryopreservation of human embryonic stem cells. Biotechnol Appl Biochem 2005;41:97–104.

[64] Hunt CJ, Pegg DE, Armitage SE. Optimising cryopreservation protocols for haematopoietic progenitor cells: a methodological approach for umbilical cord blood. Cryo Letters 2006;27:73–83.

[65] Jain JK, Paulson RJ. Oocyte cryopreservation. Fertil Steril 2006;86:1037–46.

[66] Pegg DE. The role of vitrification techniques of cryopreservation in reproductive medicine. Hum Fertil 2005;8:231–9.

[67] Rall WF, Fahy GM. Ice-free cryopreservation of mouse embryos at -196 degrees C by vitrification. Nature 1985;313:573–5.

[68] Yavin S and Arav A. Measurement of essential physical properties of vitrification solutions. Theriogenology 2007;67:81–89

[69] Lee H, Elmoazzen H, Wright D, Biggers J, Rueda BR, Heo YS, et al. Ultra-rapid vitrification of mouse oocytes in low cryoprotectant concentrations. Reprod Biomed Online 2010;20:201–8.

[70] Rall W.F., Fahy G.M. Ice-free cryopreservation of mouse embryos at -196 degrees C by vitrification. Nature 1985;313:573–575.

[71] Nowshari M.A., Bren G. Effect of freezing rate and exposure time to cryoprotectant on the development of mouse pronuclear stage embryos, Hum. Reprod. 15 (2001) 2368–2373.

[72] Vajta G., Holm P., Greve T., Callesen H. Vitrification of porcine embryos using the Open Pulled Straw (OPS) method, Acta Vet. Scand. 38 (1997) 349–352.

[73] Mukaida T., Nakamura S., Tomiyama T., Wada S., Kasai M., Takahashi K. Successful birth after transfer of vitrified human blastocysts with us of a cryoloop containerless technique, Fertil. Steril. 76 (2001) 618–620.

[74] A. Arav, Y. Zeron, A. Ocheretny, A new device and method for vitrification increases the cooling rate and allows successful cryopreservation of bovine oocytes, Theriogenology 53 (2000) 248.

[75] Steponkus P.L., Myers S.P., Lynch D.V., et al. Cryopreservation of Drosophila melanogaster embryos, Nature 345 (1990) 170–172.

[76] Saragusty J and Arav A. Current progress in oocyte and embryo cryopreservation by slow freezing and vitrification. Society for Reproduction and Fertility 2011;1470-1626.

[77] He X, Park EYH, Fowler A, Yarmush ML, Toner M. Vitrification by ultra-fast cooling at a low concentration of cryoprotectants in a quartz micro-capillary: a study using murine embryonic stem cells. Cryobiology 2008;56:223–32.

[78] Yun-Xia Cao, Qiong Xing, Li Li, Lin Cong, Zhi-Guo Zhang, Zhao-Lian Wei et al. Comparison of survival and embryonic development in human oocytes cryopreserved by slow-freezing and vitrification. Ferti and Steril Volume 92, Issue 4.2009.1306-1311
[79] Katkov I.I. Race for the pace: is the universal cryoprotocol a dream or reality?. Cryobiology 2011:61;374-375.

Part 3

Farm / Pet / Laboratory Animal ART

7

Cryopreservation of Boar Spermatozoa: An Important Role of Antioxidants

Kampon Kaeoket
Faculty of Veterinary Science, Semen laboratory, Department of Clinical Science and Public Health, Mahidol University Thailand

1. Introduction

Artificial insemination (AI) is one of the first reproductive biotechnologies has been established and developed in the pig production system. In most case, liquid stored semen or fresh semen is used for AI in commercial swine herds (Wagner and Thibier, 2000). The use of FT boar semen for AI is limited due to the low fertility outcomes compared to extended fresh semen (Johnson et al., 2000; Wagner and Thibier, 2000). The first success of boar semen cryopreservation was reported in 1956 (Polge, 1956) and the first pregnancy was achieved with FT boar semen using surgical insemination in 1970 (Polge, 1970). Currently, the attempt to develop the boar semen cryopreservation technique is ongoing. Nevertheless, the success of boar semen cryopreservation is relatively variable because the factors responsible for the cryosurvival of boar spermatozoa have not been entirely elucidated.

Cryopreservation of boar semen is useful for preservation of genetic resources, improve the genetic progress and enhance the transportation of genetic material across countries (Almlid and Hofmo, 1996; Johnson, 1998). In addition, the frozen-thawed (FT) boar semen is also used with other reproductive technologies, such as in vitro fertilization (IVF), embryo transfer (ET) and sex pre-selection (Gerrits et al., 2005). Unfortunately, the advancement of sperm cryopreservation in pigs is slow, partly due to the pig producer is satisfied with the liquid stored semen and low conception rate and litter size remain the major problems when using FT boar semen (Eriksson et al., 2002.). Under field conditions, low fertility is still obtained even using FT boar semen with a sufficient motility and number of spermatozoa for insemination (Johnson et al., 2000, Eriksson et al., 2002).

The use of frozen-thawed (FT) boar semen has been developed for artificial insemination (AI) in pig long time ago in Europe and USA (Larsson and Einarsson, 1976). In Thailand, few studies on boar semen cryopreservation have been established (Buranaamnuay et al., 2006 [a,b]). However, a great variation on the survival rate of post-thawed spermatozoa are obtained, due to the lack of biological background concerning the cryopreservation technique (Buranaamnuay et al., 2006 [a,b]). During the recent years, studies on FT boar semen have dramatically improved boar semen cryopreservation technique, for instance, optimum freezing protocols (Eriksson and Rodrigrez-Martinez, 2000), types of freezing package (Bwanga et al., 1991;Berger and Fisherleitner, 1992; Bwanga et al., 1991; Eriksson and Rodriguez-Martinez, 2000), semen centrifugation methods (Carvajal et al., 2004), thawing process (Eriksson and Rodrigrez-Martinez, 2000; Córdova-Izquierdo et al., 2006) and the supplement of some additives to the semen extender (Peña et al., 2003, Gadea et al., 2004; Roca et al., 2004, 2005).

Boar semen differs in several aspects from the semen of other domestic animals, for instance, the semen is produced in a large volume and highly sensitive to cold shock, the viability of the sperm cells is dramatically reduced when expose to temperatures below 15 °C (Gilmore et al., 1996). Therefore, the manipulation of boar semen requires special consideration during cryopreservation process (Johnson et al., 2000). Many factors that should be concerned for the boar semen cryopreservation included composition of diluents, type and concentration of cryoprotective agent, equilibration time, cooling rate and thawing procedure.

The relatively low fertility of FT boar semen is associated with many factors including a highly sensitive plasma membrane of boar spermatozoa against the changing in temperature during cooling, freezing and thawing process (Holt, 2000; Watson, 2000). This problem is related to the lipid composition of the sperm plasma membrane. The plasma membrane of the boar spermatozoa contains a high level of polyunsaturated fatty acids (PUFAs) i.e., docosapentaenoic acid (DPA) and docosahexaenoic acid (DHA), and had a low cholesterol to phospholipids ratio. DPA and DHA are dominant fatty acids in the plasma membrane of boar spermatozoa (Johnson et al., 1969).

During cryopreservation, PUFAs decrease dramatically due to lipid peroxidation. This is initiated when the spermatozoa is attacked by reactive oxygen species (ROS) (De Lamirande and Gagnon, 1992; Sikka et al., 1995). In mammals, the major sources of ROS formation include leucocyte, defective and dead spermatozoa (Aitken et al., 1994; Silva, 2006). The excessive ROS formation influence sperm motility, mid-piece abnormalities and sperm-oocyte fusion (Chatterjee et al., 2001; Agarwal et al., 2005).

The supplement of antioxidant compounds and some fatty acid to the semen extender, to minimize ROS formation and protect the plasma membrane function, have been used in many species (Peña et al., 2003; Gadea et al., 2004; Roca et al., 2004; 2005; Maldjian et al., 2005). It has been demonstrated that the proportion of DHA was significantly higher in the semen diluted with an extender supplemented with n-3 enriched hen egg yolks compared with the semen diluted with normal hen egg yolks (Maldjian et al., 2005). However, no study has been demonstrated clearly whether or not the supplement of DHA could improve the quality of the boar spermatozoa after cryopresevation. Rooke et al. (2001) found that DHA supplement in the boar feed increase progressive motility and normal acrosome and decrease abnormal spermatozoa. Recently, Kaeoket et al. (2008) reported that the supplement of DHA-enriched fish oil improved the FT boar semen quality. It has been shown that the supplement of cryoprotective agents (e.g., glycerol and Equex®), cholesterol analogue (Zeng and Terada, 2001) and antioxidants (e.g., Vitamin E, alpha-tocopherol, glutathione, taurine, cysteine, butylated hydroxytoluene, superoxide dismutase and catalase) in the semen extenders does improve the freezing ability of spermatozoa of many species such as stallion (Aurich et al., 1997; Ball et al., 2001), bull (Beconi et al., 1993; Bilodeau et al., 2001), ram (Uysal and Bucak, 2007; Bucak et al., 2007), avian (Donoghue and Donoghue, 1997), boar (Cerolini et al., 2000) and some wildlife (Leibo and Songsasen, 2000). Studies have demonstrated that the supplement of alpha-tocopherol (Peña et al., 2003), butylated hydroxytoluene (Roca et al., 2004), superoxide dismutase and catalase (Roca et al.,

2005) in the semen extenders reduces the ROS formation and improve post-thawed sperm motility and viability of FT boar semen. In addition, it was found that the supplement of extended boar semen with 5 mM of cysteine improved the viability and functional status of the chilled boar spermatozoa (Funahashi and Sano, 2005).

2. History of cryopreservation

Nowadays, there are 2 techniques for cryopreservation in boar semen, traditional nitrogen method and controlled rate freezing method. Verheyen (1993) reported a significantly better post-thaw sperm outcome when computer controlled rate freezing was used compared to non-controlled rate freezing. In human, it has been reported that controlled rate freezer method provided significant superior post-thaw sperm motility, viability, and cryosurvival rate, compared with traditional nitrogen method (Petyim and Choavaratana, 2006). However, the study of Thalchil (1981) did not confirm the different outcome of these 2 methods. Besides breed-specific fertility, data from field trials found that the mean motility of frozen–thawed semen between Norwegian Landrace and Duroc boars was difference. In the different breeds of boars found the differences in membrane lipid composition, can neither explain the major differences in post-thaw survival and fertility between breeds (Waterhouse et al., 2006).

Major limitation of frozen-thawed semen (FT-boar semen) have been observed, i.e. low conception rate and low litter size after AI (Johnson et al., 2000; Buranaamnuay et al., 2006). The relatively low fertility of FT-boar is associated with many factors. It has been reported that reactive oxygen species (ROS) generation, induced by the cryopreservation process, can be responsible for mammalian sperm damage (Griveau and Le Lannou, 1997) ROS production has been associated with reduction of sperm motility and decreased capacity for sperm–oocyte fusion. Spermatozoa are sensitive to lipid peroxidation due to their high content of polyunsaturated fatty acids, and are unable of resynthesizing their membrane components, although this may not be the sole mechanism by which sperm function is impaired by ROS. Many studies have shown that the supplementation of antioxidants in extenders improved the qualities of both fresh boar semen (Bamba and Cran, 1992; Funahashi and Sano, 2005) and frozen boar semen (Breininger et al., 2005; Gadea et al., 2005; Pena et al., 2003; Roca et al., 2004; Roca et al., 2005). Earlier studies showed that the supplementation of some antioxidant such as, water soluble Vitamin E 200 μM to semen extenders for freezing boar spermatozoa reduced post-thaw ROS generation and improved sperm motility and viability (Pena et al., 2003). Funahashi and Sano (2005) reported that supplementation extended boar semen with glutathione or *L*-cysteine of 5 mM improved the viability and functional status of boar spermatozoa during liquid storage at 10 °C for at least 14 day. A recent report (Gadea et al., 2005) demonstrated that supplementation with 1 mM of reduced glutathione to freezing media resulted in a protective effect on sperm function.

3. Cryopreservation of animal spermatozoa

Cryopreservation of boar semen need to be developed for AI in the pig industry due to a number of reasons including preservation of a good genetic resource, increase genetic improvement, distribution of genetic lines across countries and reduce boar transportation

(Almlid and Hofmo, 1996; Johnson, 1998). The widespread exchange of genetic material between breeding populations with liquid stored semen is difficult because of the short life span of the spermatozoa (Wagner and Thibier, 2000; Johnson et al., 2000). The first FT boar spermatozoa have been reported since 1956 (Polge, 1956). Unfortunately, the FT spermatozoa has a very low fertilizing ability. In 1970, the first pregnancy was achieved with FT boar semen using a surgical insemination technique (Polge et al., 1970). In 1971, many studies have reported the pregnancies after intra-cervical insemination using FT boar semen in pig (Crabo et al., 1971; Pursel et al., 1971).

In general, there are many important factors in the process of FT boar semen that affect the post-thawed semen quality. For instance, the semen collection technique, equilibration time, type of semen extender, type and concentration of cryoprotectant, freezing package, freezing rate and thawing procedure (Johnson et al., 2000). Many types of freezing package have been used for frozen boar semen, such as medium straws, maxi straws (Bwanga et al., 1991; Berger and Fisherleitner, 1992), plastic bags (Bwanga et al., 1991) and FlatPack®/MiniFlatPack® (Eriksson and Rodriguez-Martinez, 2000). Most containers has been developed for suitable storage, transport, post thawed semen quality and practical insemination.

The freezing and thawing procedure have a significant impact on the survival rate of sperm after cryopreservation (Johnson et al., 2000). However, optimal freezing and thawing rates vary depending on the type and concentration of the cryoprotectant (Mazur et al., 1970; Fiser et al., 1993). Currently, the optimal rates for boar sperm freezing appear to be 30°C/min with 3% glycerol as cryoprotectant when freezing in 0.5 ml straws (Fiser et al., 1990) and 16°C/min with 3.3% glycerol in 5 ml straws (Pursel et al., 1985). For both these methods the optimal thawing rate is 1200 °C/min (Westerndorf et al., 1975; Fiser et al., 1993). Eriksson and Rodriguez-Martinez, (2000) found that the optimal freezing rate was 50 °C/min in 3% glycerol with a 900 °C/min thawing rate for flattened plastic bags (FlatPack®) container. A variety of cryoprotectants are used in the freezing extender of different in species. Glycerol, egg yolk and sodium dodecyl sulphate (SDS) (Equex STM or Orvus ES paste) is commonly used as cryoprotectants for the cryopreservation of boar semen (Westerndorf et al., 1975; Pursel et al., 1978; Holt, 2000b). The optimal concentration of glycerol was approximately 3 % in pig (Holt, 2000a). Egg yolk and SDS are non-permeable cryoprotectant used in freezing extender and provide protective effect to spermatozoa and improved post thawed sperm quality (Pursel et al., 1978). It has been suggested that SDS enhances the cryoprotective properties of the egg yolk to protect the sperm membrane from cryoinjuries (Buhr et al., 1996)

4. Boar semen cryopreservation methods

4.1 Semen collection

Three ejaculates from each boar are collected using gloved-hand method. During collection the semen is filtered through gauze and only sperm rich fractions are collected. Within 30 min after collection, semen volume, pH, sperm motility, concentration, percentage of live and dead sperm and morphology are determined. Only ejaculates with motility of ≥70% and ≥80 % morphologically normal are used for cryopreservation.

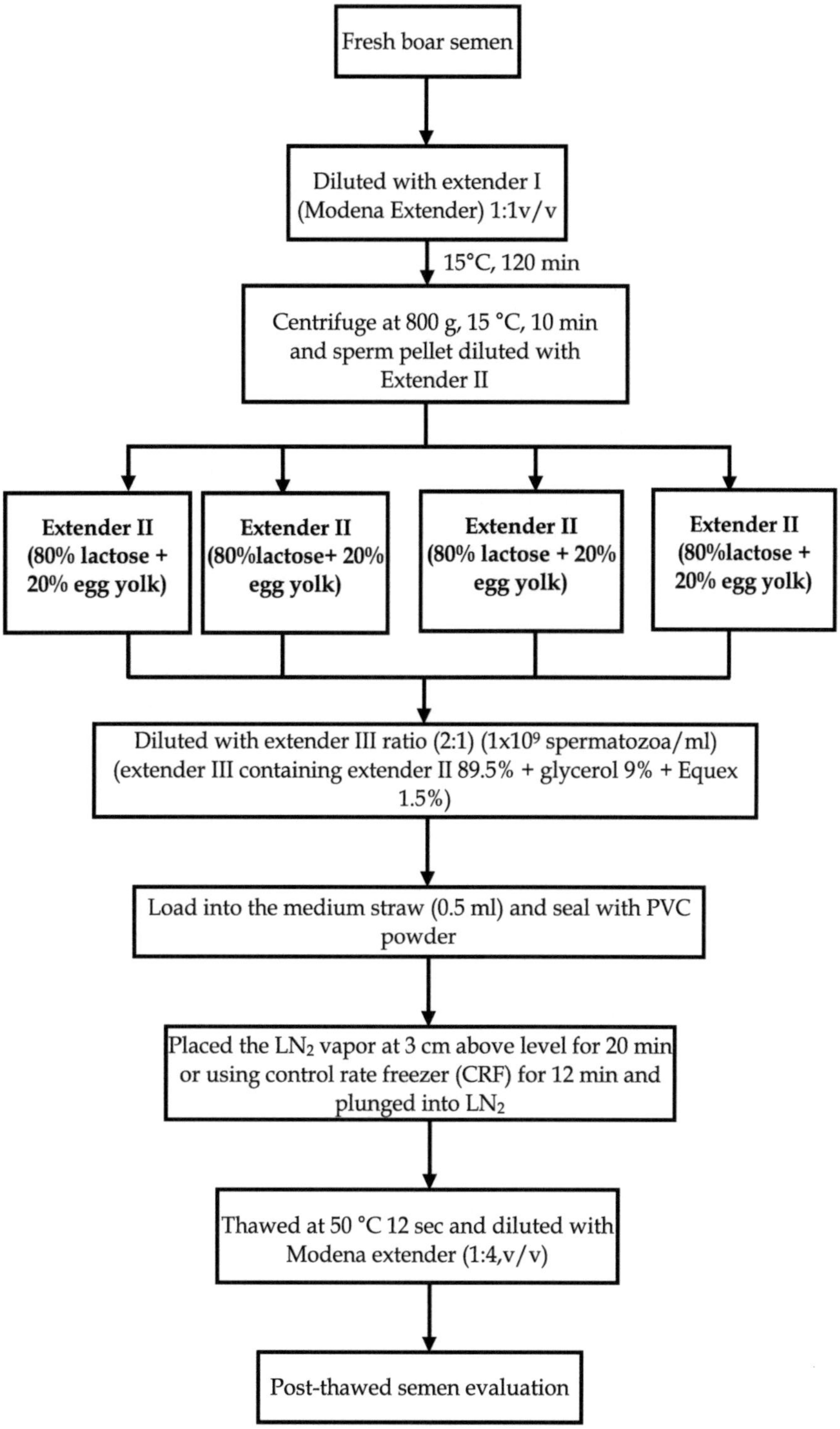

Fig. 1. Flow chart of the boar semen freezing processes, thawing and evaluation

4.2 Semen freezing and thawing procedures

Shortly after collection, the semen is diluted (1:1 v/v) with extender I (Modena™, Swine Genetics International, Ltd., Iowa, USA). The diluted semen is transferred to 50 ml centrifuge tubes, equilibrated at 15 °C for 120 min and centrifuged at 800x g for 10 min. The supernatant is discarded and the sperm pellet was re-suspended (about 1-2:1) with extender II (80 ml of 11% lactose solution and 20 ml egg yolk) to a concentration of $1.5x10^9$ spermatozoa/ml. The diluted semen is cooled to 5 °C for 90 min. Then, two parts of the semen are mixed with one part of extender III (89.5% of extender II with 9% glycerol and 1.5% Equex-STM®). The final concentration of semen is approximately $1.0x10^9$ spermatozoa/ml and contained 3% glycerol (modified after modified after Westerndorf et al., 1975 and Gadea et al., 2004). The processed semen is loaded into 0.5 ml straws (Bio-Vet, Z.I. Le Berdoulet, France). The straws are sealed with PVC powder before being placed in contact with nitrogen vapour about 3 cm above the liquid nitrogen level for 20 minutes in an expandable polystyrene box. Then the straws are plunged into liquid nitrogen (-196 °C) for storage. Thawing is achieved by immersing the straws in water at 50°C for 12 sec (Selles et al., 2003). Immediately after thawing, the semen is diluted (1:4) with a Modena™ extender. Post-thawed sperm qualities are evaluated after incubation in a 37°C water-bath for 15 min.

4.3 Semen extender

After incubation in extender I, the semen is divided into 4 groups according to the composition of extender II. Group I, extender II containing 80 ml of 11% lactose solution and 20 ml egg yolk. Group II, extender II is supplemented with 0.29 g of fish oil (Fish oil 1000; Blackmores LTD, New Southwell, Australia; containing DHA 120 mg/g fish oil) per gram of egg yolk. Normal egg yolk contains approximately 3.15 mg DHA per gram of egg yolk, as was analyzed at the Institute of Nutrition, Mahidol University (AOAC, 2007). Group III is supplemented with a combination of fish oil 0.29 g and L-cysteine 5 mM (Fluka Chemie GmbH, Sigma-Aldrich, Switzerland). Group IV is supplemented with a combination of fish oil 0.29 g and L-cysteine 10 mM.

5. Harmful effect of cryopreservation to spermatozoa

The use of FT boar semen under field conditions results in low conception rate and reduced number of total piglets born per litter (Eriksson et al, 2002). These problems occur because of the poor post-thawed semen quality and low survival rate of boar spermatozoa after cryopreservation (Hammerstedt et al., 1990; Curry et al., 2000). The detrimental effects of cooling, freezing and thawing caused subsequently impaired the membrane integrity, structure and function of the spermatozoa, eventually fertilizing ability (Hammerstedt et al., 1990; Guthrie and Welch, 2005).

It is well documented that the boar spermatozoa are highly susceptible to temperatures below 15 °C. The viability of the spermatozoa are dramatically reduced within a few hours after expose to cooling below 15 °C, so call 'cold shock' (Gilmore et al., 1996). Cold shock caused the damage of plasma membranes and alterations in the metabolism of the spermatozoa. This caused by changes in the arrangement of plasma membrane compositions especially phospholipids (reviewed by Medeiros et al., 2002).The sperm damage cause by cold shock is characterized by an irreversible loss of motility and the loss of sperm permeability. Boar spermatozoa seem to acquire a cold shock resistance when the semen is held at room temperature in seminal plasma for 1-5 hours (Pursel et al., 1972). It was found that viability and fertilizing ability of the boar spermatozoa was significantly improved when the fresh semen was held at 15 °C for over 3 hours before cryopresenvation (Almlid and Johnson, 1988; Eriksson et al., 2001)

During the process of freezing, the decrease of temperature from -15 °C to -60 °C causes sperm damage (Mazur, 1985). This causes by the intracellular ice formation and cellular dehydration (osmotic stress). The subsequent physical events depend on the cooling rates. Intracellur ice formation occur during a rapid cooling when intracellular water does not leave the cell to maintain equilibration. If cooling is slow, the spermatozoa will lose water rapidly avoid to intracellular ice formation. However, if spermatozoa are cooled too slowly, they will expose to high concentration of solutes which caused intracellular water to diffuse out of the cell, dehydration both the cell and plasma membrane (also known as solution effects) (Mazur, 1970; Parks and Graham, 1992). Gilmore et al. (1996) demonstrated that the boar spermatozoa are sensitive to osmotic stress. Similar findings were also found in dog (Songsasen et al., 2002), cat (Pukazhenthi et al., 2000), ram (Curry and Watson, 1994), stallion (Ball and Vo, 2001) and bull (Liu and Foote, 1998).

Cellular damage due to intracellular ice formation and dehydration, oxidative stress is another important cause of sperm damage leading to abnormal sperm structure and function and subfertility. Currently, several studies have been reported oxidative stress affect the damage of sperm membrane, proteins and DNA in human (Agarwal et al., 2003), stallion (Baumber et al., 2000; 2003), bull (Bilodeau et al., 2001) and boar (Roca et al., 2004; 2005)(Fig. 2).

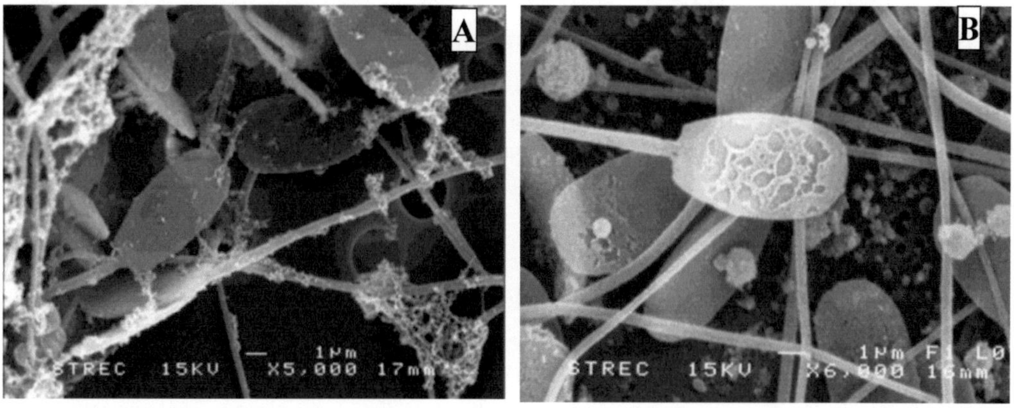

Fig. 2. Scanning electron microscopic (SEM) picture of fresh boar semen with normal plasma membrane (A) as compare with SEM picture of frozen boar semen with plasma membrane damage (B).

6. Capacitation like changes of the frozen-thawed boar semen and its influence on *in vivo* fertility

After ejaculation, the spermatozoa are not able to fertilize the oocyte. They undergo the activation process that called ' capacitation' and acrosome reaction which spermatozoa will able to reach the ampulla of the oviduct, penetrate the cumulus oophorus, bind to the zona pellucida, activate the acrosome reaction and eventually fertilize the oocyte (Yanagimachi, 1994). Normally, capacitation *in vivo* occurs in the female reproductive tract, but capacitation can also be induced in vitro by the incubation in capacitating media which the most of media contain the bicarbonate, calcium and serum albumin (Yanagimachi, 1994). Furthermore, It has been demonstrated that the cooling and freezing process can also induced the capacitation like change which affect to the low fertilizing capacity of spermatozoa in boar and other mammalian species (Maxwell and Johnson, 1997; Green and Watson, 2001; Barrios et al., 2000). During cooling, freezing and rewarming process, it is hypothesized that change in low temperature cause the modification and destabilization of the lipid content in the sperm plasma membrane, reducing the selective permeability resulted in the cholesterol efflux and intracellular calcium uptake leading to the capacitation like change (Green and Watson, 2001; Tardif et al., 2001). To improve the the FT spermatozoa, there are some studies about the addition of cholesterol-loaded cyclodextrins increased the cryosurvival of boar, ram and bovine spermatozoa because cyclodextrins used to deliver cholesterol to the sperm plasma membrane which against cold shock (Purdy and Graham, 2004; Bailey et al., 2008; Mocé et al., 2009). In addition, the supplement of antioxidants such as vitamin E or alpha-tocopherol decreased the capacitation like change of cryopreserved boar spermatozoa (Satorre et al, 2007). Furthermore, the addition of seminal plasma to boar spermatozoa has been shown to reduce the capacitated spermatozoa in chilled and FT boar semen (Kaneto et al., 2002; Suzuki et al., 2002; Vadnais et al., 2005a,b; Okazaki et al., 2009, Garcia et al., 2009).

7. Laboratory method for semen quality assessment

7.1 Sperm concentration and progressive motility

Sperm concentration will be assessed by direct cell count using a Bürker haemocytometer (Boeco, Humburg, Germany) (Beardon and Fuquay, 1997). The visual progressive motility of both fresh and FT sperm is evaluated at 38°C under a phase contrast microscope at 200x and 400x magnification. The motility is assessed by the same person throughout the experiment.

7.2 Computer-assisted sperm analysis (CASA)

The motility patterns of diluted FT semen are assessed using the CASA system (Halminton Thorne Biosciences IVOS, Version 12 TOX IVOS, Beverly, USA). Each FT thawed semen samples is diluted with pre-warmed Modena extender (37°C) to obtain a final concentration of 50×10^6 spermatozoa/ml. A 5 µl of diluted semen is pipetted into the chamber and allowed the 1 min before analysis fore sample distribution and pre-warming (Iguer-Ouada and Verstegen, 2001). After the first assessment (T0), the diluted semen is evaluated after incubation at 37 °C for 30 min (T30) and 60 min (T60). The camera will recognize the position of the sperm heads in successive frames. Spermatozoa heads are marked with a different color to enable the observer and the analyzer to differentiate between the different motility patterns. Each semen sample is measured twice, 3 fields are evaluated and counted at least 1000 cells per analysis. Motility patterns including (1) Curvilinear velocity (VCL, µm/s), the average velocity measured in the progression line along the whole track of cell path; (2) Average pathway velocity (VAP, µm/s) , the average velocity of the smoothed cell path; (3) Straight line velocity (VSL), the average velocity measured in a straight line from the beginning to the end of the track (µm/s); (4) The amplitude of the lateral head displacement (ALH), the mean width of the head oscillation as the sperm cells swim (µm); (5) The beat cross-frequency (BCF, Hz), frequency of the sperm head crossing the average path in either direction; (6) The straightness (STR, %) = average value of the ratio VSL/VAP; (7) The Linearity (LIN, %) = average value of the ratio VCL/VAP.

7.3 Sperm viability

The percentages of sperm viability will be determined by 2 methods. The first one is eosin-nigrosin staining (Dott and Foster, 1972). The semen sample (50 µl) are mixed well with a drop of eosin-nigrosin dyes (Fluka Chemie GmbH, Sigma-Aldrich, Switzerland), and the mixture (10 µl) is smeared and dried on a glass slide. Evaluation is undertaken by counting 200 spermatozoa with 1000x magnification. Spermatozoa with an unstained head are regarded as live spermatozoa. The second method is evaluated by SYBR-14/Ethidiumhomodimer-1 (EthD-1) (Fertilight®, Sperm Viability Kit, Molecular Probes Europe, Leiden, The Netherlands). This technique is modified after Axnér et al. 2004 and Garner and Johnson, 1995). Ten µl of diluted semen are mixed with 2.7 µl of the user solution of SYBR-14 and 10 µl of EthD-1. The user solution is SYBR-14 diluted (1:100) in dimethyl sulfoxide (DMSO), fractionated and frozen in eppendorfs. After incubation at 37 °C for 20 min, two hundred spermatozoa will be assessed (x1000) under fluorescent microscope. The nuclei of the spermatozoa with an intact plasma membrane are stained green with SYBR-14, while those with damaged membranes stained red with EthD-1.

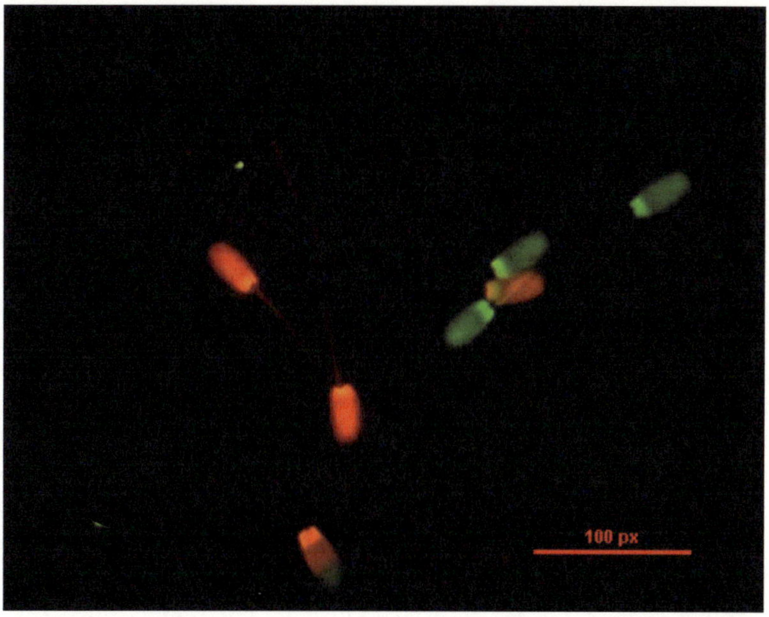

Fig. 3. Spermatozoa stained with SYBR-14/EthD-1 or PI: live spermatozoa stained green with SYBR-14 while dead spermatozoa stained red with EthD-1 or PI.

Spermatozoa are classified into three types; live spermatozoa stained green with SYBR-14, dead spermatozoa stained red with EthD-1 and moribund spermatozoa stained both green and red (Axnér et al. 2004; Garner and Johnson, 1995).The results are expressed as the percentage of live spermatozoa with intact plasma membranes.

7.4 Acrosome integrity

Acrosome integrity will be assessed using fluorescein isothiocyanate-labeled peanut (*Arachis hypogaea*) agglutinin (FITC-PNA) staining. Ten µl of the diluted semen is mixed with 10 µl of Ethidiumhomodimer-1 and incubated at 37 °C for 15 min. Five µl of the mixture is smeared on a glass slide and fixed with 95 % ethanol for 30 second. Fifty µl Fit C-PNA (dilute Fit C-PNA with PBS 1:10 v/v) is spread over the slide and incubated in a moist chamber at 4 °C for 30 min. After incubation, it is rinsed with cold PBS and air dried. Two hundred spermatozoa are assessed under fluorescent microscope at 1000x magnification and classified as intact acrosome, damaged acrosome and missing acrosome (Cheng et al., 1996; Axner et al., 2004). The results are scored as the percentage of intact acrosome spermatozoa.

7.5 The functional integrity of the sperm plasma membrane

The functional integrity of the sperm plasma membrane will be assessed using a short hypo-osmotic swelling test (sHOST) (Perez-Llano et al., 2001). Spermatozoa are incubated, at 38 °C for 30 min, with 75 mOsm/kg a hypo-osmotic solution that consist of 0.368 % (w/v) Na-citrate and 0.675 % (w/v) fructose (Merck, Germany) in distilled water. Following this incubation time, 200 µl of the semen-hypo-osmotic solution is fixed in 1000 µl of a hypo-osmotic solution plus 5 % formaldehyde (Merck, Germany), for later evaluation. Two hundred spermatozoa are assessed under a phase contrast microscope at 400x magnification. The coiled tail (sHOST positive) spermatozoa found following incubation are functional intact plasma membrane.

7.6 DNA damage

DNA damage can be evaluated by Acridine orange (AO) staining or Halomax staining method (Fig. 4). The technique is modified after Thuwanut et al. (2008). Briefly, two smears from each sample were prepared on glass slide and air-dried. Each smear is fixed overnight in Carnoy's solution, freshly prepared with methanol and glacial acetic acid (3:1 v/v). The slide is removed from the fixative solution, air-dried, and then stained with 1% (100 mg/ml) AO (Sigma) in distilled water for 10 min. The AO staining solution is prepared by adding 10 ml of 1% AO in distilled water to 40 mL of 0.1 M citric acid (Merck, Darmstadt, Germany) and 2.5 ml of 0.3 M Na2HPO4.7H2O (Merck, Darmstadt, Germany) pH 2.5. The AO staining solution will be prepared daily and stored in the dark at room temperature until use. After staining, the slide is gently washed by distilled water and covered with the cover slip. One thousand spermatozoa are evaluated under the fluorescence microscope. The heads of the sperm cells with normal DNA (double-stranded) have green fluorescence, while those with damaged or single stranded DNA showed orange or red fluorescence. The results are expressed as the proportion of the damage/single stranded DNA per 1,000 counted spermatozoa.

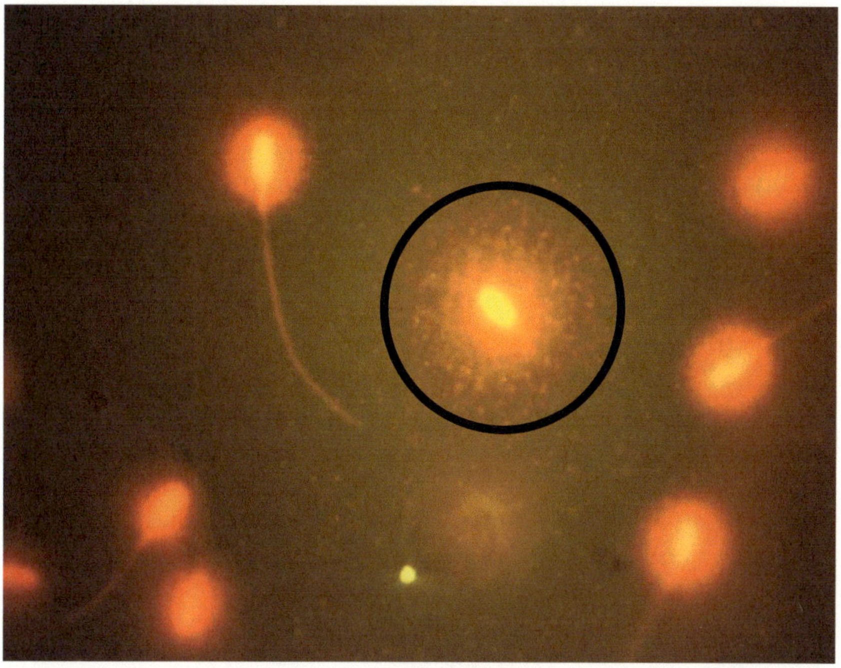

Fig. 4. Sperm DNA damage (black circle) stained with a commercial kit (Halomax staining)

7.7 Chlortetracycline (CTC) assay

The CTC assay is slightly modified from as described previously (Harayama et al., 2000). The CTC staining solution containing 750 µm CTC, 5 mM DL-cysteine, 130 mM NaCl and 20mM Tris (hydroxymethyl aminomethane) (pH 7.8) is prepared immediately before use. This solution is protected from light until analysis. Briefly, 50 µl of sperm suspension will be mixed with 50 µl of CTC staining solution for 30 sec, followed by the addition of 10µl 12.5% paraformaldehyde in 0.5 M Tris-HCl (pH 7.4) as a fixative. Then, 10 µl of the sperm suspension is mixed well with equal volume of antifade solution (0.22 M of 1,4-diazabicyclo[2,2,2]octane ;DABCO) in glycerol:PBS (9:1) on the microscopic slide and gently compressed with coverslip. Two slides are prepared from each sample and stored in the dark at at 4 °C until evaluation. Two hundred spermatozoa per slide will be under a Nikon fluorescence microscope at 400x under blue-violet illumination (excitation at 400–440 nm and emission at 470 nm). The spermatozoa are classified in to three staining patterns described by Fraser et al. (1995). F-pattern, fluorescence over the whole region of the sperm head are considered to be "non-capacitated spermatozoa". B-pattern, fluorescence in the acrosomal region except post-acrosomal region are considered to be "capacitated spermatozoa". AR-pattern, low or no fluorescence over the whole head except thin bright ban in the equatorial segment are considered to be "acrosome-reacted spermatozoa".

7.8 Annexin-V/PI assay

Apoptosis will be evaluated by apoptosis detection kit ApopNexin™ (Chemicon Int., USA) using a fluorescent microscope. This assay will detect the phosphatidylserine translocation from inner to outer leaflet of cell plasma membrane which is the hallmark of apoptosis during the degradation phase. Following manufacturer instructions, sperm cells are washed twice with PBS (pH 7.4) by centrifugation at 400 g for 5 min. Sperm pellet are resuspended with HEPES buffer (10mM HEPES/NaOH, pH7.4, 150mM NaCl, 5mM KCl, 1mM MgCl2, 1.8mM CaCl2, 2x106 sperm/ml). One hundred µl of sperm suspension are mixed well with 5 µl annexinV-FITC conjugate and 3 µl of Propidium iodide (PI;20µg/ml) and incubated for 15 min at room temperature in the dark. Two hundred spermatozoa are assessed under fluorescent microscope at 400x magnification. The apoptotic sperm cells will fluorescence green while necrotic sperm cells fluorescence red. Alternatively, flow cytometry analysis can also be used instead of fluorescent microscope (Fig. 5).

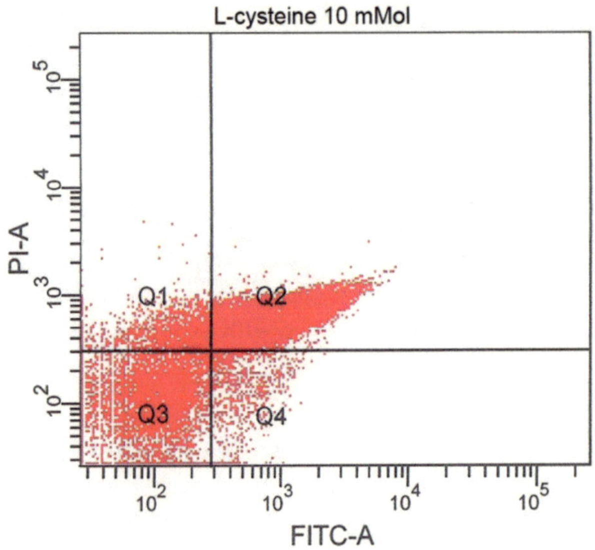

Fig. 5. FCS/SCC two-dimensional histogram, flow cytometry analysis of frozen-thawed boar spermatozoa (supplemented with L-cysteine) stained with Annexin-V/PI: Q3 represent the viable spermatozoa with intact plasma membrane, while Q2 represent dead spermatozoa with non-intact plasma membrane

8. Lipid composition of sperm plasma membrane

The lipid compositions of the plasma membrane of the mammalian spermatozoa are markedly different from those of somatic cells. In general, the sperm plasma membrane contains approximately 70% phospholipids, 25% neutral lipids, and 5 % glycolipids (Flesch and Gadella, 2000). All lipid components located in the sperm membranes responsible for the fluidity of membrane lipid bilayers, regulation of sperm maturation, spermatogenesis, capacitation, acrosome reaction and membrane fusion (Parks and Hammerstedt, 1985; Martinez and Morros, 1996; Sanocka and Kurpisz, 2004).

Sperm plasma membrane are made up of a phospholipids bilayer, with the major phospholipids were choline phosphoglycerides (CP), ethanolamine phosphoglycerides (EP) and sphingomyelin which their proportions differed between species. These phospholipids contain a high proportion of long chain, polyunsaturated docosapentanoyl (22:5) and docosahexanoyl (22:6) groups which both lipids represent approximately 50 to 60 % of total phospholipids in boar and bull spermatozoa (Pursel and Graham, 1967; Johnson et al., 1969; Parks and Lynch, 1992). Cholesterol was the major sterol in sperm lipids of all species. Cholesterol to phospholipid molar ratios were 0.26, 0.30, 0.36, and 0.45 for sperm plasma membrane of the boar, rooster, stallion, and bull, respectively (Parks and Lynch, 1992).Glycolipids represented less than 10% of total polar lipids for all species.

The susceptibility of spermatozoa to cold shock differ among species because of the differences of lipid composition of the sperm plasma membrane among species (Flesch and Gadella, 2000).The resistance of the mammalian spermatozoa to cold shock was high in species in which the cholesterol to phospholipids molar ratio and the phospholipids saturation is high (Darin-Bennett and White, 1977). The avian spermatozoa have a high level of cold shock resistant and have a higher level of saturated phospholipids compared to mammalian sperm (Parks and Lynch, 1992). The plasma membrane of the boar spermatozoa is characterized by a high protein, low cholesterol and high proportion of EP compared to other species (Parks and Lynch, 1992; Nikolopoulou et al., 1985). In contrast, the protein content and EP proportion of rooster sperm plasma membrane is low while the cholesterol content is intermediate (Parks and Lynch, 1992).

As mentioned above the sperm plasma membrane has a very high amounts of polyunsaturated fatty acids (PUFAs) especially docosapentaenoic acid (DPA) and docosahexaenoic acid (DHA) (Johnson et al., 1969; Parks and Lynch, 1992). It has been suggested that the proportion of unsaturated fatty acid influence the properties of sperm plasma membrane (Miller et al., 2005). High levels of long chain PUFAs, DPA and DHA, are associated with an increased membrane fluidity (Quinn, 1985). During cryopreservation, the fluidity of the plasma membrane from boar spermatozoa is significantly decreased when compared to fresh spermatozoa which tend to restrict the post-thawed sperm quality (Buhr et al., 1994). In human, sperm with a high level of membrane fluidity had a higher post-thawed motility compared to sperm with a low level of membrane fluidity after cryopreservation (Giraud et al., 2000).

9. Docosahexaenoic acid (DHA)

Docosahexaenoic acid (commonly known as DHA; 22:6 (n-3)) is an omega-3 essential polyunsaturated fatty acid. DHA is most often found in cold water fatty fish (salmon fish, tuna fish) and in fish oil supplements, along with eicosapentaenoic acid (EPA). DHA is the main fatty acid composition of the spermatozoa as well as the brain and the retina (Neuringer et al., 1988). For the sperm plasma membrane, DHA play a major role in regulating membrane fluidity in sperm and in the regulation of spermatogenesis (Haidl and Opper, 1997; Ollero et al., 2000). DHA content is significantly higher in immature spermatozoa than mature spermatozoa.

Studies have demonstrated that the supplement of PUFAs in the feed of the boar improve the quality of the boar spermatozoa (Paulenz et al., 1999; Rooke et al., 2001; Strezezek et al., 2004 ;Maldjian et al., 2005). In addition, Rooke et al. (2001) found that tuna oil supplemented in the boar diet increase viability, progressive motility and normal morphology. The supplementation of PUFAs also enhanced the survival rate of post-thawed boar spermatozoa (Strezezek et al., 2004). DHA improved the reproductive performance of the male turkey (Blesbois et al., 2004). Maldjian et al. (2005) found that the use of DHA-enriched hen egg yolk for the semen extender and the supplement of 3% fish oil in the boar feed increased the DHA content of the boar spermatozoa post-thawed. However, the authors could not demonstrate the improvement of the quality of post-thawed spermatozoa.

10. Oxidative stress and sperm function

Oxidative stress is a condition associated with an increasing rate of cellular damage, induced by oxygen and oxygen-derived oxidants, commonly known as ROS (Sikka et al., 1995). ROS are highly reactive oxidizing agents belonging to the class of free-radicals, which contains one or more unpaired electrons. Normally ROS included superoxide anion (O_2^-), hydrogen peroxide (H_2O_2), peroxyl radical (ROO^-) and the very reactive hydroxyl radicals (OH^-). The nitrogen-derived free radical nitric oxide (NO) and peroxynitrite anion ($ONOO^-$) play an important role to the fertilization. Two main resources of ROS in semen include leukocytes and immature or defective spermatozoa (Aitken et al., 1992; Silva, 2006).

It is known longtime ago that ROS harm the spermatozoa (Macleod, 1943). Nowadays, studies have shown that the cryopreservation of spermatozoa induces the ROS formation and led to lipid peroxidation as well as DNA oxidation. These events attribute to the decrease of sperm function and infertility (Aitken et al., 1991; Alvarez and Storey, 1992; Agarwal, 2003). Nevertheless, spermatozoa normally produce a small amounts of ROS, needed for capacitation and acrosome reaction (Agarwal et al., 2005; De Lamirande and Gagnon, 1993).

Oxidative stress is the result of an imbalance between ROS generation and scavenging activities (Sikka et al., 1995; Sharma and Agarwal, 1996). Spermatozoa are sensitive to oxidative stress because of low concentrations of scavenging enzymes in the cytoplasm (de Lamirande and Gagnon, 1995; Saleh and Agarwal, 2002) and the plasma membranes contain high amounts of PUFAs (Alvarez and Storey, 1995). ROS act as triggers a chain of reaction. Lipid peroxidation (LPO) (De Lamirande and Gagnon, 1992; Sikka et al., 1995). LPO of sperm plasma membrane is the key mechanism of ROS-induced sperm damage (Alvarez et al., 1987)..??.

LPO of sperm membranes is an autocatalytic self-reaction composed of 3 steps. Firstly, initiation step, this is the abstraction of a hydrogen atom from an unsaturated fatty acid. Secondly, propagation step, this is the formation of alkyl radical which followed by its rapid reaction with oxygen to form a peroxyl radical is capable of abstracting a hydrogen atom from an unsaturated fatty acid with the concomitant formation of a lipid radical and lipid peroxide such as hydrogen peroxide(H_2O_2). Since the peroxyl and alkyl radicals are regenerated, the cycle of propagation could continue indefinitely. Finally, the termination step, the substrates is consumed or stopped by the radical-radical reaction which produce a non-radical species (Sanocka and Kurpisz, 2004). LPO has been reported to affect the sperm dysfunction associated with decreased membrane fluidity, loss of membrane integrity and function of spermatozoa (Sanocka and Kurpisz, 2004). Furthermore, LPO also damage DNA and proteins resulted in an increased the susceptibility to be attacked by the macrophage (Aitken et al., 1994).

10.1 Effect of antioxidants on oxidative stress and sperm function

Antioxidants are compounds that suppress the formation of ROS and protect spermatozoa against ROS (Sikka, 1995). Studies have demonstrated that seminal plasma contains a number of enzymatic antioxidants such as superoxide dismutase (SOD; Alvarez et al., 1987), glutathione peroxidase/glutathione reductase (GPX/GRD) and catalase. These antioxidants protect the spermatozoa against LPO (Lenzi et al., 1996; Sikka et al., 1996; Saleh and Agarwal, 2002). SOD spontaneously dismutates (O_2^-) anion to form O_2 and H_2O_2. Catalase converts H_2O_2 to O_2 and H_2O. In addition, glutathione peroxidase, a selenium-containing antioxidant enzyme with glutathione, is an electron donor removes peroxyl (ROO^-) radicals from various peroxides including H_2O_2 (Sikka et al., 1996). In addition, seminal plasma contains a variety of non-enzymatic antioxidants such as ascorbic acid (vitamin C), alpha-tocopherol (vitamin E), and reduced glutathione (Lenzi et al., 1994;Saleh and Agarwal, 2002; Silva, 2006).

Vitamin C is a major chain–breaking antioxidant present in the extracellular fluid (Saleh and Agarwal, 2002). It neutralized hydroxyl, superoxide and hydrogen peroxide radicals and prevent sperm agglutination (Agarwal et al., 2004). Vitamin E is a chain-breaking antioxidant in the cell membrane, inhibits LPO by scavenging peroxyl and alkoxyl radicals. Glutathione is the most abundant antioxidant , plays a role in protecting lipids, proteins and nucleic acids against oxidative stress.

Studies have shown that the supplementation of antioxidants in extenders both chilled and frozen-thawed semen such as alpha-tocopherol, butylated hydroxytoluene, superoxide dismutase and catalase, cysteine or glutathione have been reported to improve the semen quality in boar (Pursel, 1979; Bamba and Cran, 1992; Brezezinska-Slebodzinska E, 1995; Cerolini et al., 2000; Penã et al., 2003; Gadea et al., 2004, Roca et al., 2004, 2005; Funahashi and Sano, 2005; Breininger et al., 2005; Satorre et al., 2007), bull (Bilodeau et al., 2001), turkey (Donoghue and Donoghue, 1997), stallion (Aurich et al., 1997; Ball et al., 2001) and ram (Uysal and Bucak, 2007).

10.2 Effect of L-Cysteine on frozen boar semen

L-cysteine, an amino acid containing a sulphydryl group, is a precursor of intracellular glutathione biosynthesis. L-cysteine plays a role in the intracellular protective mechanism against oxidative stress, membrane stabiliser and capacitation inhibitor (Johnson et al., 2000). Glutathione is the most common non-thiol protein in mammalian cells which protects plasma membrane from LPO, scavenges superoxide and minimized O_2^- formation. It has been demonstrated that the supplementation of *L-cysteine* in the semen extender prevents the loss of sperm motility by minimizing hydrogen peroxide of FT semen in the bull (Bilodeau et al., 2001). Funahashi and Sano (2005) found that the supplement of L-Cysteine for 5 mM improved the viability and functional status of the boar spermatozoa during chilled storage.

During the past few years, many studies have been carried out by supplementation of various antioxidants (e.g. Vitamin E, Glutathione, Taurine) in the freezing extenders of frozen boar semen in order to minimize the detrimental effect of ROS which occurred during the freezing process (Pena et al., 2003; Roca et al., 2004; Breininger et al. 2005; Gadea et al., 2005). Funahashi and Sano (2005) demonstrated that supplement of L-cysteine (5 mM) could improve the viability and progressive motility in fresh boar semen, and also the same case found in frozen bovine semen (Bilodeau et al., 2001). This L-cysteine is also improve survival time of semen and sperm chromatin structure in fresh chilled boar semen at 15°c

(Szczesniak-Fabianczyk et al. 2003). In frozen dog semen, Micheal et al. (2007) reported that supplement of L-cysteine resulted in increased viability and rapid steady forward movement (RSF movement). Recently, Kaeoket et al. (2008b) also found that addition of 5 mM L-cysteine (the same concentration used for fresh boar semen preservation by Funahashi and Sano, 2005) has a tendency (not significant difference) to improve post-thawed semen quality when compare with the addition of glutathione and water-soluble vitamin E.

10.3 Effect of L-Cysteine x DHA on frozen boar semen

The characteristics of the sperm cryoinjury included the destabilization of lipid bilayer of the sperm plasma membrane, change in permeability of plasma membrane and a reduction of the viability of sperm. It is well documented that the boar sperm are highly susceptible to temperature below 15 °C mainly due to a relatively poor lipid composition and structure of plasma membrane compared to other domestic species. The boar sperm plasma membrane consisted of a high levels of polyunsaturated fatty acids (PUFAs) especially docosapentaenoic acid (DPA) and docosahexaenoic acid (DHA) and low level of cholesterol: phospholipids ratio. It has been shown that the level of PUFA content play an important role in the sperm membrane fluidity and cause sperm susceptible to lipid peroxidation (LPO). During frozen-thawed (FT) process, sperm are attacked by reactive oxygen species (ROS) owing to LPO and leading to significantly decrease in the PUFAs content of their plasma membrane. ROS were mainly produced by the defective or dead spermatozoa and result in a reduction of sperm motility, sperm viability and eventually fertilizing ability. In order to minimize the sperm cryoinjury, the supplement of antioxidant compounds and some fatty acid to the semen extender have been reported to minimize ROS formation and enhance the plasma membrane function in many species. *L*-cysteine, a precursor of intracellular glutathione, plays an important role in the protecting sperm from oxidative stress and act as capacitation inhibitor. Earlier studies have demonstrated that *L*-cysteine supplement in the semen extender improve the motility of FT bull semen, prolonged sperm survival time and reduced chromatin damage in FT boar sperm. In addition, the use of n-3-enriched hen egg yolk in the semen extender increased the proportion of DHA content in the boar sperm. Our previous study found that the addition of *L*-cysteine directly into lactose egg yolk (DHA-enriched) base extender significantly improves the sperm motility and intact acrosome of FT boar sperm. In addition, Kaeoket et al. (2010) found that the supplement of DHA (fish oil) improves the sperm motility, viability and acrosome integrity of the FT boar sperm.

10.4 Effect of seminal plasma on post-thawing semen quality and reproductive performance after artificial insemination

Seminal plasma is the liquid constituent of an ejaculate, comprising a combination of fluids secreted by the male accessory glands (i.e., mainly from the seminal vesicle in boars) during an ejaculation. There is evidence that seminal plasma is able to arrest or reverse cryoinjury and perhaps extend the longevity of the sperm by inhibiting or reversing capacitation and acrosome reactions, and also by its antioxidant activity (Brzezińska-Ślebodzińska et al. 1995; Strzezek et al. 1999; Suzuki et al. 2002; Vadnais and Roberts 2007; Bailey et al. 2008). During the cryopreservation process of boar semen, seminal plasma is normally not required, and discarded by the centrifugation at the beginning of semen preparation, which may result in a lack of a significant contribution (i.e., antioxidant property, inhibiting or reversing capacitation, inhibiting and acrosome reaction and binding ability of its protein to the sperm plasma membrane) of seminal plasma in protecting sperm from cryoinjury. It has been demonstrated in rams that seminal plasma protein is able to revert the cold shock damage on the sperm membrane (Barrios et al. 2000). This effect has also been reported in boars, in that holding boar spermatozoa in its seminal plasma before cooling and freezing defends against the cold shock (Pursel et al. 1973). In addition, it has been shown that adding seminal plasma to the post-thawing solution increased the percentage of stallion sperm motility (Alghamdi et al. 2005). In dogs, it has been shown that frozen semen diluted with their prostatic fluid yielded a higher post-thawing motility (Rota et al. 2007). As a result, it seems likely that seminal plasma or prostatic fluid constituents have a positive effect on post-thawing sperm motility. In addition, semen with a high percentage of progressive motility illustrates their plasma membrane integrity and superb metabolism (Johnson et al., 2000) and also reflects their ability for fertilization (Vyt et al. 2004; Estienne et al. 2007). It can be hypothesized that the presence of supernatant (semen plasma plus semen extender) during the thawing process may improve frozen-thawed boar sperm motility.

Seminal plasma is the liquid constituent of an ejaculate, comprising a combination of fluids secreted by the male accessory glands (i.e. seminal vesicle) during an ejaculation. It is evidence that seminal plasma able to arrest or reverse cryoinjury and perhaps extend the longevity of the sperm by inhibiting or reversing capacitation and acrosome reaction (Suzuki et al., 2002). Besides, seminal plasma appears to play an important role in the female reproductive tract after insemination, e.g. attenuate the post insemination inflammatory response in the uterus of the sow which may influence the chances of conception (Rozeboom et al.,1999) and its component such as hormone estrogen may also resulted in a release of prostaglandins from the pigs endometrium to the utero-ovarian veins and lymphatic vessels which in turn decrease duration time from standing oestrus to ovulation in gilts (Clause et al.,1987,1990; Weitze et al.,1990), Therefore, it seems likely that seminal plasma constituents (both the oestrogen and the protein fraction) have an effect on ovulation time in sows (Waberski et al.,1995; Kaeoket and Tummaruk, 2002b).

Generally, limitations to achieve a high reproductive performance in swine arise from a failure of sows to express estrus, failure to accurately determine onset of estrus for artificial inseminations and failure to determine the ovulation time after standing oestrus. The best predictor for time of ovulation is frequent detection of oestrus, because time of ovulation occurs approximately 38 to 48 h after onset of estrus (Anderson et al., 1993; Weitze et al., 1994; Soede et al., 1995). The effects of the timing of insemination relative to ovulation on fertilization rate has been study by Soede et al.(1995).Targeting insemination within 24 h before ovulation seems optimal to achieve a high farrowing rate and large little size (Nissen et al., 1997). Insemination between 0 and 24 h before ovulation results in high fertilization rates and consequently, a low number of re-breeders and a slightly higher litter size (Kemp et al., 1996; Kaeoket et al., 2002a, 2005).

10.5 Effect of long term versus short term extenders as freezing extender I on quality of frozen boar semen

In pig industry, the boar semen used for artificial insemination is extended with semen extender and kept in cold storage at 18-20°C for few days before artificial insemination. It has been recently reported that using of long term extenders (i.e. Androstar®Plus,

Modena™, Vitasem LD) to preserve fresh semen for 7 days yield a superior fresh boar semen qualities compare with those using of short term extender (Kaeoket et al., 2010d). In addition, the different in extended fresh semen qualities were also found depending on each type of long term extender used. This indicated that some constituents in each long term extender may assist sperm to overcome cold shock during cold storage. Generally, the difference between the short term and long term extenders are the ingredients contained in the extenders. Long term extenders contain complex buffering agent (i.e. HEPES, Tris, TES and MOPs) and antioxidants (i.e., bovine serum albumin (BSA), beta-carotene, cysteine, taurine, vitamin E and ascorbic acid) (Alvarez and Storey, 1995; Gadea, 2003; Funahashi and Sano, 2005), which can maintain semen qualities during cold storage for a longer period than short term extender.

10.6 Effect of differents sugars in LEY freezing extender on frozen boar semen quality

During cryopreservation, both physical and chemical factors including the rapid change in temperature or thermal stress, the intracellular ice formation, oxidative stress and osmotic stress led to the sperm plasma membrane damage (Meideros et al., 2000). Generally, the freezing extender consists of cryoprotectant, sugars, buffer, and some antibiotics (Johnson et al., 2000). Glycerol is the most common permeable cryoprotectant used for cryopreservation of boar semen (Holt, 2000). Egg yolk is a common non-permeable cryoprotectant. Different types of sugars, such as, trehalose, lactose, fructose, have been used in the freezing extender of boar semen (Purdy et al., 2006). Sugar is not only a source of energy but also protects the spermatozoa from dehydration and intracellular ice formation during the cryopreservation process (Watson, 2000). In general, lactose is the most common sugar used for the cryopreservation of boar semen (Johnson, 1985; Buranaamnuay et al., 2009; Chanapiwat et al., 2009; Chanapiwat et al., 2010; Kaeoket et al., 2010a; Kaeoket et al., 2010b; Kasettrut and Kaeoket, 2010). The effect of either type or concentration of sugar supplement in the freezing extender on the post-thawed semen qualities has been reported in dog (Yildiz et al., 2000; Yamashiro et al., 2007), ram (Aisen et al., 2002), bovine (Woelders, et al., 1997; Hu et al., 2010) and boar (Roca et al., 2008; Gutiérrez-Pérez et al., 2009; Malo et al., 2010; Mercado et al., 2010). For instance, the supplement of 55 mM glucose improved the motility pattern of the FT boar spermatozoa compared to 0 and 180 mM (Roca et al., 2008). Hu et al. (2009) found that the addition of 100 mM trehalose in the extender improved post-thawed boar sperm motility, viability and acrosome integrity compared to 0, 25, 50 and 200 mM. In addition, Malo et al. (2010) found that the trehalose-based freezing extender enhances the sperm survival rate and the fertilization rate by *in vitro* fertilization (IVF) compared to lactose and glucose based freezing extender. In order to improve the post-thawed boar sperm quality, it is important to investigate the influence of different sugars on FT boar sperm.

11. Phytosterol on frozen boar semen quality

During the past decade, several studies have focused on supplementation with a variety of antioxidants (e.g. vitamin E, vitamin C, L-cysteine, glutathione, taurine, pyruvate, SOD, catalase) in the freezing extenders of frozen boar semen with an attempt to minimize the detrimental effects of ROS, which occur during the freezing process.

Gamma-oryzanol, a phytosteryl ferulate mixture extracted from rice bran oil, has received a great deal of attention because of its significant various health-promoting functions such as antioxidant activity, inhibition of lipoperoxidation by its scavenging activity, reduction in LDL cholesterol and induction of HDL cholesterol, inhibition of platelet aggregation [20], its potential implications as a UV-A filter in sunscreen cosmetics, treatment of type 2 diabetes mellitus and allergic reactions. These data suggest that gamma-oryzanol, especially, with its antioxidant and scavenging activities can be useful as an antioxidant and lipid peroxidation inhibitor (i.e., membrane stabilizing) during cryopreservation. Rice bran oil is widely used in salad dressing and cooking oil in Asian countries including China, India, Japan and Thailand. At present, it is becoming to gain acceptance in Western countries as well. However, no scientific information is available on its antioxidant and scavenging activities in minimizing the detrimental effects of ROS during the cryopreservation of semen.

12. Artificial insemination with frozen boar semen

In pigs, it is well-documented that the optimal insemination time for fresh semen to maximize the good fertilisation rate is within 24 h before ovulation. It has been shown that the fertile life span of the pig oocyte is limited to between 8-12 h after ovulation. At suboptimal times for artificial insemination (AI) leads to inferior FR and litter sizes results. In addition, Kaeoket et al. (2002; 2005) demonstrated that when sows were inseminated after ovulation, fertilised oocytes and developed embryos were observed up to Day 11 but no embryos were found at Day 19. Subsequently, these sows returned to oestrus with a prolonged interval.

It has been demonstrated that the duration of oestrus is related to the WOI, i.e., sows with a short WOI (3-4 days) on average have a long oestrus duration, which is associated with a longer time from onset of oestrus to ovulation. On the contrary, sows with a WOI of 5-6 days or longer, have a shorter time from onset of oestrus to ovulation, and therefore should be inseminated (with fresh semen) sooner after the onset of oestrus to ensure that the first insemination occurs before ovulation. This recommendation is in accordance with the observation that the average timing of ovulation varies between 64 and to 72% of the duration of oestrus.

For frozen boar semen, it is predictable that insemination with frozen-thawed semen will result in lower PR, low FR and litter sizes. During the last decade, most of the experiments with fertility tests (field trial) of frozen-thawed boar semen have been carried out by using deep intrauterine insemination (DIUI, dose ranged from 150 million to 1 billion spermatozoa). Nevertheless, for fresh semen, an intrauterine insemination (doses ranging from 1-3 billion) has been performed with a high fertility results (i.e., high PR, FR and litter sizes). Recently, it has been shown that a satisfactory fertility outcome was accomplished by performing IUI (doses raning from 1.5-3 billion) together with fixed-time insemination (using a correlation of WOI-Oestrus duration-Ovulation time). This strategy may improve fertility of frozen boar semen when one performs insemination in a commercial pig farms (a field trial).

13. Conclusion

Based on above review, the conclusion can be drawn as follows: (I) some antioxidants, such as, Oryzanol, L-cysteine and its combination with DHA from fish oil, Vitamin E, Vitamin C, can be used in order to improve the quality of frozen boar semen" (II) the artificial insemination (i.e. timing, dose and AI techniques) by using frozen boar semen on pig farm need further investigations.

14. References

Agarwal, A., Saleh, R.A. and Bedaiwy, M.A. 2003. Role of reactive oxygen species in the pathophysiology of human reproduction. Fertil. Steril. 79:829-843.

Almlid, T. and Hofmo, P.O. 1996. A brief review of frozen semen application under Norwegian AI service conditions. Reprod. Dom. Anim. 31:169-173.

Almlid, T. and Johnson, L.A. 1988. Effects of glycerol concentration, equilibration time and temperature of glycerol addition on post-thaw viability of boar spermatozoa frozen in straws. J. Ani. Sci. 66:2899-2905.

Alvarez, J.G. and Storey, B.T. 1992. Evidence for increased lipid peroxidative damage and loss of superoxide dismutase activity as a mode of sublethal cryodamage to human sperm during cryopreservation. J. Androl. 13:232-241.

Axnér, E., Hermansson, U. and Linde-Forsberg, C. 2004. The effect of Equex STM paste and sperm morphology on frozen-thawed survival of cat epididymal sperm. Anim. Reprod. Sci. 84: 179–91.

Bailey, J.L., Lessard, C., Jacques, J., Breque, C., Dobrinski, I., Zeng, W. and Galantino-Homer, H.L. 2008. Cryopreservation of boar semen and its future importance to the industry. Theriogenology. 70:1251-1259.

Ball, B.A. and Vo, A. 2001. Osmotic tolerance of equine spermatozoa and the effects of soluble cryoprotectants on equine sperm motility, viability, and mitochondrial membrane potential. J. Androl. 22:1061-1069.

Barrios, B., Perez-Pe, R., Gallego, M., Tato, A., Osada, J., Muino-Blanco, T. and Cebrian-Perez, J.A. 2000. Seminal plasma proteins revert the cold shock damage on ram sperm membrane Biol. Reprod. 63:1531-1537.

Baumber, J., Ball, B.A., Gravance, C.G., Medina, V. and Davies-Morel, M.C. 2000. The effect of reactive oxygen species on equine sperm motility, viability, acrosomal integrity, mitochondrial membrane potential, and membrane lipid peroxidation. J. Androl. 21:621-628.

Baumber, J., Ball, B.A., Linfor, J.J. and Meyers, S.A. 2003. Reactive oxygen species and cryopreservation promote DNA fragmentation in equine spermatozoa. J. Androl. 24:895-902.

Beardon, H.J. and Fuquay, J.W. 1997. Semen evaluation. Appl Anim Rep 4th ed. 158-170.

Beconi, M.T., Francia, C.R., Mora, N.G. and Affranchino, M.A. 1993. Effect of natural antioxidants on frozen bovine semen preservation. Theriogenology. 40:841–851.

Berger, B. and Fischerleitner, F. 1992. On deep freezing of boar semen: investigations on the effects of different straw volumes, methods of freezing and thawing extenders. Reprod. Dom. Anim. 27:266-270.

Bilodeau, J. F., Blanchette, S., Gagnon, C. and Sirard, M.A. 2001. Thiols prevent H_2O_2-mediated loss of sperm motility in cryopreserved bull semen. Theriogenology. 56:275-286.

Blesbois, E., Douard, V., Germain, M., Boniface, P. and Pellet, F. 2004. Effects of n-3 polyunsaturated dietary supplementation on the reproductive capacity of male turkeys. Theriogenology. 61:537-549.

Breininger, E., Beorlegui, N.B., O'Flaherty, C.M. and Beconi, M.T. 2005. Alpha-tocopherol improves biochemical and dynamic parameters in cryopreserved boar semen. Theriogenology. 63:2126-2135.

Brzezińska-Ślebodzińska, E., Slebodziński, A.B., Pietras, B. and Wieczorek, G. 1995. Antioxidant effect of vitamin E and glutathione on lipid peroxidation in boar semen plasma. Biol. Trace. Elem. Res. 47:69-74.

Buranaamnuay, K., Tummaruk, P. and Techakumphu, M. 2010. Intra-uterine insemination using frozen-thawed boar semen in spontaneous and induced ovulating sows under field conditions. The 36th Annual Conference of the IETS/the 23rd Annual Meeting SBTE. 9-12 January 2010. Cordoba, Argentina

Buranaamnuay, K., Tummaruk, P., Singlor, J., Rodriguez-Martinez, H. and Techakumphu, M. 2009. Effects of straw volume and Equex-STM® on boar sperm quality after cryopreservation. Reprod. Dom. Anim. 44:69-73.

Bwanga, C.O., Einarsson, S. and Rodriguez-Martinez, H. 1991. Deep freezing of boar semen packaged in plastic bags and straws. Reprod. Dom. Anim. 26:117-125.

Cerolini, S., Maldjian, A., Surai, P. and Noble, R. 2000. Viability, susceptibility to peroxidation and fatty acid composition of boar semen during liquid storage. Anim. Reprod Sci. 58:99-111.

Chanapiwat, P., Kaeoket, K. and Tummaruk, P. 2009. Effects of DHA-enriched hen egg yolk and L-cysteine supplementation on quality of cryopreserved boar semen. Asian J Androl. 11: 600-608.

Chanapiwat, P., Kaeoket, K. and Tummaruk, P., 2010. The sperm DNA damage aftercryopreservation of boar semen in relation to post-thawed semen qualities, antioxidant supplementation and boars effects. Thai. J. Vet. Med. 40, 187-193.

Chanapiwat, P., Kaeoket, K. and Tummaruk, P., 2011. Cryopreservation of boar semen by egg-yolk based extenders containing lactose or fructose is better than sorbitol. J. Vet. Med. Sci. In press.

Cheng, F.P, Fazeli, A., Voorhout, W.F., Marks, A., Bevers, M.M. and Colenbrander, B. 1996. Use of peanut agglutinin to assess the acrosomal status and the zona pellucida-induced acrosome reaction in stallion spermatozoa. J Androl 1996; 17: 674–82.

Crabo, B. and Einarsson, S. 1971. Fertility of deep frozen boar spermatozoa. Acta. Vet. Scand. 12:125-127.

Curry, M.R. and Watson, P.F. 1994. Osmotic effects on ram and human sperm membranes in relation to thawing injury. Cryobiology. 31: 39-46.

Donoghue, A.M. and Donoghue, D.J. 1997. Effects of water and lipid-soluble antioxidants on turkey sperm viability, membrane integrity, and motility during liquid storage. Poult. Sci. 76:1440–1455.

Dott, H.M. and Foster, G.C.1972. A technique for studying the morphology of mammalian spermatozoa which are eosinophilic in a differential live/dead stain. J. Reprod. Fert. 29:443-445.

Eriksson, B.M. and Rodriguez-Martinez, H. 2000. Effect of freezing and thawing rate on the post-thaw viability of boar spermatozoa frozen in FlatPack and Maxi-straws. Anim. Reprod. Sci. 63: 205-220.

Eriksson, B.M., Petersson, H. and Rodriguez-Martinez, H. 2002. Field fertility with exported boar semen frozen in the new flatpack container. Theriogenology. 58: 1065-1079.

Eriksson, B.M., Vazquez, J.M. and Martinez, E.A. 2001. Effects of holding time during cooling and of type of package on plasma membrane integrity, motility and in vitro oocyte penetration ability of frozen-thawed boar spermatozoa. Theriogenology. 55:1593-1605.

Fiser, P., Fairfull, R.W., Hansen, C., Panich, P.L., Shrestha, J.N.B. and Underhill, L. 1993. The effect of warming velocity on motility and acrosomal integrity of boar sperm as influenced by the rate of freezing and glycerol level. Mol. Reprod. Dev. 34:190-195.

Fiser, P.S. and Fairfull, R.W. 1990. Combined effect of glycerol concentration and cooling velocity on motility and acrosomal integrity of boar spermatozoa frozen in 0.5 ml straws. Mol. Reprod. Dev. 25: 123-129.

Funahashi, H. and Sano, T. 2005. Select antioxidants improve the function of extended boar semen stored at 10 degrees C. Theriogenology. 63:1605-1616.

Gadea, J., Selles, E., Marco, M.A., Coy, P., Matas, C., Romar, R. and Ruiz, S. 2004. Decrease in gluthathione content in boar sperm after cryopreservation, effect of the addition of reduced glutathione to the freezing and thawing extender. Theriogenology. 62:690-701.

Garcia, J.C., Dominguez, J.C., Pena, F.J., Alegre, B., Gonzalez., Castro, M.J., Habing, G.G. and Kirkwood, R.N. 2009. Thawing boar semen in the presence of seminal plasm: effects on sperm quality and fertility. Anim. Reprod. Sci. (In press)

Garner, D.L. and Johnson, L.A. 1995. Viability assessment of mammalian sperm using SYBR-14 and propidium iodide. Biol. Reprod. 53:276-284.

Gerrits, R.J., Lunney, J.K., Johnson, L.A., Pursel, V.G., Kraeling, R.R., Rohrer, G.A. and Dobrinsky, J.R. 2005. Perspectives for artificial insemination and genomics to improve global swine populations. Theriogenology. 63:283-299.

Gilmore, J.A., Junying, D., Jun, T., Peter, A.T. and Crister, J.K. 1996. Osmotic properties of boar spermatozoa and their relevance to cryopreservation. J. Reprod. Fertil. 107:87-95.

Green, C.E. and Watson, P.F. 2001. Comparison of the capacitation-like state of cooled boar spermatozoa with true capacitation. Reproduction. 122:889-898.

Hernández, M., Roca, J., Calvete, J.J., Sanz, L., Muiño-Blanco, T., Cebrián-Pérez, J.A., Vázquez, J.M. and Martínez, E.A. 2007. Cryosurvival and in vitro fertilizing capacity post thaw is improved when boar spermatozoa are frozen in the presence of seminal plasma from good freezer boars. J. Androl. 28:689-697.

Holt, W.V. 2000a. Fundamental aspects of sperm cryobiology: the importance of species and individual differences. Theriogenology. 53:47-58.

Holt, W.V. 2000b. Basic aspects of frozen storage of semen. Anim. Reprod. Sci. 62:3–22.

Hu, J.H., Li, Q.W., Jiang, Z.L., Yang, H., Zhang, S.S. and Zhao, H.W. 2009. The cryoprotective effect of trehalose supplementation on boar spermatozoa quality. Anim. Reprod. Sci. (In press).

Iguer-Ouada, M. and Verstegen, J.P. 2001. Evaluation of the Hamilton thorne computer-based automated system for dog semen analysis. Theriogenology. 55: 733-749.

Johnson, L. A. 1985. Fertility results using frozen boar spermatozoa: 1970 to 1985. In L. A. Johnson, K. Larsson (Ed.), Deep freezing of boar semen. Uppsala, Swedish Univ Agric Sci: 199-222.

Johnson, L.A. 1998. Current developments in swine semen: preservation, artificial insemination and sperm sexing. Proceedings 15th International Pig Vet Society, Birmingham, UK, 1:225-229.

Johnson, L.A., Weitze, K.F., Fiser, P. and Maxwell, W.M.C. 2000. Storage of boar semen. Anim. Reprod. Sci. 62:142-172.

Kaeoket, K. and Tummaruk, P., 2002. Seminal plasma: its function and importance for pig artificial insemination. Thai J. Vet. Med. 32, 15-25.

Kaeoket, K., Chanapai, P., Chanchaiyaphoom, P. and Chanapiwat, P. 2011. Effect of using long term and short term extenders during cooling process on the quality of frozen boar semen. Thai. J. Vet. Med. 41, 283-288.

Kaeoket, K., Chanapiwat, P., Tummaruk, P. and Techakumphu, M. 2010. Supplementaleffect of varying L-cysteine concentrations on quality of cryopreserved boar semen. Asian. J. Androl. 12, 260-265.

Kaeoket, K., Chanapiwat, P., Tummaruk, P., Kunavongkrit, A. and Techakumphu, M., 2011. A preliminary study on using autologous and heterologous boar sperm supernatant from freezing processes as post-thawing solution: its effect on sperm motility. Trop. Anim. Health. Prod. 43, 1049-1055.

Kaeoket, K., Chanapiwat, P., Wongtawan, T. and Kunavongkrit, A. 2010. Successful intrauterine insemination (IUI) with frozen boar semen: effect of dose, volume and fixed time–AI on fertility. Thai. J. Agri. Sci. 43, 31-37.

Kaeoket, K., Persson, E. and Dalin, A.M., 2002. The influence of pre- and post-ovulatory insemination on sperm distribution in the oviduct, accessory sperm to the zona pellucida, fertilisation rate and embryo development in sows. *Anim. Reprod. Sci.* 71, 239-248.

Kaeoket, K., Sang-urai, P., Thamniyom, A., Chanapiwat, P. and Techakumphu, M. 2010. Effect of Decosahexaenoic acid (DHA) on quality of cryopreserved boar semen in different breeds. Reprod. Dom. Anim. 45, 458-463.

Kaeoket, K., Tantasuparuk, W. and Kunavongkrit, A., 2005. The effect of post-ovulatory insemination on the subsequent oestrous cycle length, embryonic loss and vaginal discharge in sows. Reprod. Domest. Anim. 40, 492-494.

Kaeoket, K., Tantiparinyakul, K., Kladkaew, W., Chanapiwat, P. and Techakumphu M. 2008. Effect of different antioxidants on quality of cryo-preserved boar semen in different breeds. Thai. J. Agri. Sci. 41, 1-9.

Kaneto, M., Harayama, H., Miyake, M. and Kato, S. 2002. Capacitation-like alterations in cooled boar spermatozoa: assessment by the chlortetracycline staining assay and immunodetection of tyrosine-phosphorylated sperm proteins Anim. Reprod. Sci. 73:197-209.

Kasetrtut, C. and Kaeoket, K. 2010. Effect of using supernatant for post-thawing solution and semen extender prior to insemination on sow reproductive performance. Thai. J. Vet. Med. 40, 171-178.

Kunavongkrit, A., Sang-Gasanee, K., Phumratanaprapin, C., Tantasuparuk, W. and Einarsson S. 2003. A study on the number of recovered spermatozoa in the uterine horns and oviducts of gilts, after fractionated or non-fractionated insemination. J. Vet. Med. Sci. 65:63–67.

Larsson, K. and Einarsson, S. 1976. Fertility of deep frozen boar spermatozoa. Acta vet. Scand. 17:43-62.

Leibo, S.P. and Songsasen, N. 2002. Cryopreservation of gametes and embryos of non-domestic species. Theriogenology. 57:303-326.

Liu, Z. and Foote, R.H. 1998. Bull sperm motility and membrane integrity in media varying in osmolality. J. Dairy. Sci. 81:1868-1873.

Maldjian, A., Pizzi, F., Gliozzi, T., Cerolini, S., Penny, P. and Noble, R. 2005. Changes in sperm quality and lipid composition during cryopreservation of boar semen. Theriogenology. 63:411-421.

Martin, M.J., Edgerton, S. and Wiseman, B. 2000. Frozen semen: a breeding protocol that results in high fecundity. Swine. Health. Prod. 8:275-277.

Martinez, E.A., Vazquez, J.M., Roca, J., Lucas, X., Gil, M.A., Parrilla, I., Vazquez, J.L. and Day, B.N. 2001. Successful non-surgical deep intrauterine insemination with small number of spermatozoa in sows. Reproduction 122, 289–296.

Maxwell, W.M.C. and Johnson, L.A. 1999. Physiology of spermatozoa at high dilution rates: the influence of seminal plasma. Theriogenology. 52:1353-1362.

Maxwell, W.M.C., Welch, G.R. and Johnson, L.A. 1997.Viability and membrane integrity of spermatozoa after dilution and flow cytometric sorting in the presence or absence of seminal plasma. Reprod. Fertil. Dev. 8:1165-1178.

Mazur, P. 1970. Cryobiology: the freezing of biological systems. Science. 168:939-949

Medeiros, C.M.O., Forell, F., Oliveira, A.T.D. and Rodrigues, J.L. 2002. Current status of sperm cryopreservation: why isn't it better? Theriogenology. 57:327-44

Moce, E., Purdy, P.H. and Graham, J.K. 2009. Treating ram sperm with cholesterol-loaded cyclodextrins improves cryosurvival. Anim. Reprod.Sci. (In press)

Nissen, A. K., Soede, N. M., Hyttel, P., Schmidt, M. and D Hoore, L. 1997. The influence of time of insemination relative to time of ovulation and farrowing frequency and litter size in sows, as investigated by ultrasonography. Theriogenology. 47: 1571-1582.

Okazaki, T., Abe, S., Yoshida, S. and Shimada, M. 2009. Seminal plasma damages sperm during cryopreservation, but its presence during thawing improves semen quality and conception rates in boars with poor post thaw semen quality. Theriogenology. 71:491–498.

Parks, J.E. and Graham, J.K. 1992. Effects of cryopreservation procedures on sperm membranes. Theriogenology. 38:209-222.

Penã, F. J., Johannisson, A., Wallgren, M. and Rodriguez Martinez, H. 2003. Antioxidant supplementation in vitro improves boar sperm motility and mitochondrial membrane potential after cryopreservation of different fractions of the ejaculate. Anim. Reprod. Sci. 78:85-98.

Perez-Llano, B., Lorenzo, J.L., Yenes, P., Trejo, A. and Garcia-Casado, P. 2001. A short hypoosmotic swelling test for the prediction of boar sperm fertility. Theriogenology. 56:387-398.

Polge, C. 1956. Artificial insemination in pigs. Vet. Rec. 68: 62-76.

Polge, C., Salamon, S. and Wilmut, I.1970 Fertilizing capacity of frozen boar semen following surgical insemination. Vet. Rec. 87:424-429.

Pukazhenthi, B., Noiles, E., Pelican, K., Donoghue, A., Wildt, D.E. and Howard, J.G. 2000. Osmotic effects on feline spermatozoa from normospermic versus teratospermic donors. Cryobiology. 40:139-150.

Purdy, P.H. and Graham, J.K. 2004 Effect of adding cholesterol to bull sperm membranes on sperm capacitation, the acrosome reaction and fertility. Biol. Reprod. 71:522-527.

Pursel, V. G., Johnson, L. A. and Schulman, L. L.1973. Effect of dilution, seminal plasma and incubation period on cold shock susceptibility of boar spermatozoa. J. Anim. Sci. 37: 528-531

Pursel, V.G., Johnson, L.A. and Schulman, L.L. 1972. Interaction of extender composition and incubation period on cold shock susceptibility of boar spermatozoa. J. Anim. Sci. 35:580-584.

Pursel, V.G., Schulman, L.L. and Johnson, L.A. 1978a. Effect of glycerol concentration on frozen boar sperm Theriogenology. 9:305-312

Pursel, V.G., Schulman, L.L. and Johnson, L.A. 1978b. Effect of Orvus ES Paste on acrosome morphology, motility and fertilizing capacity of frozen-thawed boar sperm J. Anim. Sci. 47:198-202.

Roca, J., Gil, M.A., Hernandez, M., Parrilla, I., Vazquez, J.M. and Martinez, E.A. 2004. Survival and fertility of boar spermatozoa after freeze-thawing in extender supplemented with butylated hydroxytoluene. J. Androl. 25:397-405.

Roca, J., Rodriguez, J.M., Gil, M.A., Carvajal, G., Garcia, E.M., Cuello, C., Vazquez, J.M. and Martinez, E.A. 2005. Survival and in vitro fertility of boar spermatozoa frozen in the presence of superoxide dismutase and/or catalase. J. Androl. 26:15-24.

Rodriguez-Martinez, H. 2009. Freezing of boar semen:state of the art. Proceedings 4th Asian Pig Vet Society, Tsukuba, Japan. 55-58.

Rooke, J.A., Shao, C.C. and Speake, B.K. 2001. Effects of feeding tuna oil on the lipid composition of pig spermatozoa and in vitro characteristics of semen. Reproduction. 121:315-322.

Rozeboom, K.J., Troedsson, M.H.T., Hodson, H.H., Shurson, G.C. and Crabo, B. 2000. The importance of seminal plasma on the fertility of subsequent artificial inseminations in swine J. Anim. Sci. 78:443-448

Saleh, R.A. and Agarwal, A. 2002. Oxidative stress and male infertility: From research bench to clinical practice. J. Androl. 23:737-749

Satorre, M.M., Breininger, E., Beconi, M.T. and Beorlegui, N.B. 2007. Alpha-tocopherol modifies tyrosine phosphorylation and capacitation-like state of cryopreserved porcine sperm. Theriogenolgy. 68:958-965.

Selles, E., Gadea, J., Romar, R., Matas, C. and Ruiz, S. 2003. Analysis of in vitro fertilizing capacity to evaluate the freezing procedures of boar semen and to predict the subsequent fertility. Reprod. Dom. Anim. 38:66-72.

Sharma R.K. and Agarwal, A. 1996. Role of reactive oxygen species in male infertility (Review). Urology. 48:835–850.

Sikka, S.C., Rajasekaran, M. and Hellstrom, W.J. 1995. Role of oxidative stress and antioxidants in male infertility. J. Androl. 16:464-468.

Silva, P.F.N. 2006. Physiology of peroxidation process in mammalian sperm. PhD Thesis. Utrecht University Ridderprint, Ridderkerk, 4-45.

Soede, N. M, and Kemp, B. 1997. Expression of oestrus and timing of ovulation in pigs. J Reprod Fertil (Suppl.) 52: 91-103.

Soede, N. M., Wetzels, C. C., Zondag, W., de Koning, M. A. and Kemp, B. 1995. Effects of time of insemination relative to ovulation, as determined by ultrasonography, on fertilization rate and accessory sperm count in sows. J. Reprod. Fertil. 104: 99-106.

Songsasen, N., Yu, I., Murton, S., Paccamonti, D.L., Eilts, B.E., Godke, R.A. and Leibo, S.P. 2002. Osmotic sensitivity of canine spermatozoa. Cryobiology. 44:79-90.

Strzezek, J., Lapkiewicz, S. and Lecewicz, M. 1999. A note on antioxidant capacity of boar seminal plasma. Anim. Sci. P. 17:181-188

Suzuki, K., Asano, A., Eriksson, B.M., Niwa, K., Nagai, T. and Rodriguez-Martinez, H. 2002. Capacitation status and in vitro fertility of boar spermatozoa: effects of seminal plasma, cumulus-oocyte-complexes-conditioned medium and hyaluronan Int. J. Androl. 25:84-93

Szczesniak-Fabianczyk, B., Bochenek, M., Smorag, Z. and Silvestre, M.A. 2003. Effects of antioxidants added to boar semen extender on the semen survival and sperm chromatin structure. Reprod. Biol. 3:81-87.

Tardif, S., Dube, C., Chevalier, S. and Bailey, J.L. 2001. Capacitation is associated with tyrosine phosphorylation and tyrosine kinase-like activity of pig sperm proteins. Biol. Reprod. 65: 784–792.

Thurston, L.M., Siggins, K., Mileham, A.J., Watson, P.F. and Holt, W. 2002 Identification of amplified restriction fragment length polymorphism markers linked to genes controlling boar sperm viability following cryopreservation. Biol Reprod. 2002 66:545-54.

Thuwanut, P., Chatdarong, K., Techakumphu, M. and Axnér, E. 2008. The effect of antioxidants on motility, viability, acrosome integrity and DNA integrity of frozen-thawed epididymal cat spermatozoa. Theriogenology. 70:233–240.

Tummaruk, P., Sumransap, P., Techakumphu, M. and Kunavongkrit, A. 2007. Distribution of spermatozoa and embryos in the female reproductive tract after unilateral deep intra uterine insemination in the pig. Reprod. Dom. Anim. 42:603-609.

Uysal, O. and Bucak, M.N. 2007. Effects of Oxidized Glutathione, Bovine Serum Albumin, Cysteine and Lycopene on the Quality of Frozen-Thawed Ram Semen. Acta. Vet. Brno. 76:383-390.

Vadnais, M.L. and Roberts, K.P. 2007. Effects of seminal plasma on cooling induced capacitative changes in boar sperm. J. Androl. 28. 416–422.

Vadnais, M.L., Kirkwood, R.N., Sprecher, D.J. and Chou, K. 2005b. Effect of extender, incubation temperature, and added seminal plasma on capacitation of cryopreserved, thawed boar sperm as determined by chlortetracycline staining. Anim. Reprod. Sci. 90. 347–354.

Vadnais, M.L., Kirkwood, R.N., Tempelman, R.J., Sprecher, D.J. and Chou, K. 2005a. Effect of cooling and seminal plasma on the capacitation status of fresh boar sperm as determined using chlortetracycline assay. Anim. Reprod. Sci. 87. 121–132.

Waberski, D., Sudhoff, H., Hahn, T., Jungblut, P.W., Kallweit, E., Calvete, J.J., Ensslin, M., Hoppen, H.-O., Wintergalen, N., Weitze, K.F. and Topfer-Petersen, E. 1995. Advanced ovulation in gilts by the intrauterine application of a low molecular mass pronase-sensitive fraction of boar seminal plasma. J. Reprod. Fertil. 105:247-225.

Waberski, D., Weitze, K. F., Gleumes, T., Schwarz, M., Willmen, T. and Petzoldt, R. 1994. Effect of time of insemination relative to ovulation on fertility with liquid and frozen boar semen. Theriogenology. 42: 831-840.

Wagner, H.G. and Thibier, M. 2000. World statistics for artificial insemination in small ruminants and swine. Proc 14th ICAR, Stockholm, Sweden. vol. 2, 15:3

Watson, P. F. 2000. The causes of reduced fertility with cryopreserved semen. Anim Reprod. Sci. 60-61:481-492.

Watson, P. F. and Behan, J. R. 2002. Intra-uterine insemination of sows with reduced sperm numbers: results of a commercially based field trial. Theriogenology 57: 1683-1693.

Weitze, K. F., Wagner-Rietschel, H., Waberski, D., Richter, L. and Krieter, J. 1994. The onset of heat after weaning, heat duration, and ovulation as major factors in AI timing in sows. Reprod. Dom. Anim. 29: 433-443.

Westendorf, P., Richter, L. and Treu, H. 1975. Zur Tiefgefrierung von Ebersperma. Labor- und Besamungsergebnisse mit dem Hulsenberger Pailletten-verfahren. Dtsch. Tierarztl. Wschr, 82:261-267.

Yanagimachi R. Mammalian fertilisation. In: The Physiology of Reproduction Knobil E, 2nd ed. New York: Raven Press. 189–317.

8

Cryopreservation of Rat Sperm

Hideaki Yamashiro[1] and Eimei Sato[2]
[1] *Laboratory of Animal Genetics and Reproduction, Faculty of Agriculture, Niigata University*
[2] *Laboratory of Animal Reproduction, Graduate School of Agricultural Science, Tohoku University*
Japan

1. Introduction

The laboratory rat, *Rattus norvegicus*, was the first mammalian species domesticated for scientific research, which work dating back to before 1850. From this auspicious beginning, the rat has become the most widely studied experimental animal model for biomedical research (Jacob, 1999). Since the development of the first inbred rat strain by King 1909, over 500 inbred rat strains have been developed for a wide range of biochemical and physiological phenotypes and different disease models (Aitman et al., 2008, Canzian. 1997). In the last decade, there has been an extraordinary increase in rat genomic resources (Gibbs et al., 2004, Pennisi. 2004), and the advent of knock-out technology allow the insertion or deletion of individual genes into the rat by advances in stem-cell technology (Geurts et al., 2009, Izsvak et al., 2010, Tong et al., 2010). Thus, a wide array of research opportunities now open up, especially in studies involving the laboratory rat (Hamra. 2010). However, protocol for sperm cryopreservation and oocytes fertilized *in vitro* by using cryopreserved sperm are still under development for preservation of most rat strains. Therefore, greater use of the cryopreservation of rat sperm may provide an essential resource to preserve and increase the number of valuable genetic strains for research and application.

In this chapter, we will introduce several of these approaches to cryopreserving the rat sperm. It will be valuable for developing new freezing extender for cryopreservation of rat sperm, and might be applied to other reproductive technologies in this species for preservation of valuable rat strains.

2. Reporting studies of rat sperm cryopreservation

The first live-born rat derived from frozen-thawed sperm were successfully reported artificial insemination (AI) by using frozen-thawed rat sperm, and also reported cryopreservation of several strains of rat sperm, including those from mutant and transgenic rats (Nakatsukasa et al., 2001, Nakatsukasa et al., 2003). More recent publication by same group confirmed that cryopreserved rat sperm can be revitalized and result in the birth of live offspring through embryo transfers after in vitro fertilization (*IVF*) (Seita et al., 2009a). Although the authors mentioned by another publication that intracytoplasmic sperm injection (ICSI) is the only way to routinely obtain offspring routinely derived from oocytes fertilized in vitro using fresh and cryopreserved sperm (Seita et al., 2009b). Hagiwara et al, (2009) also pointed out that only one group has reported successful cryopreservation of rat sperm, subsequently used for AI and yielding live offspring. This result has not yet been repeated by other investigators and labs, and further investigation of the ability to yield viable rat sperm after cryopreservation is urgently needed. Further, each of cryopreservation procedures has not been completed for several mammalian species to date, the basic science and technology required to do so is rapidly becoming available and this should be completed for a number of species, in the future (Agca and Critser, 2002). Those of the information will also allow further improvements in cryopreservation of rat sperm from various mammalian species.

3. Cryodiluent for rat sperm

The characteristics of rodent sperm *in vitro* differ from that of other mammalian sperm largely due to differences in sperm membrane lipid content or composition (Parks and Lynch, 1992). The morphology of rodent sperm shows a longer tail when compared to that of sperm from domestic animals (Cardullo and Baltz, 1991). Rat sperm is extremely sensitive to a number of environmental changes, such as centrifugation, pH, viscosity, osmotic stress (Varisli et al., 2009a: Nakatsukasa et al., 2003: Chularatnatol, 1982: Si et al., 2006). Rat sperm have therefore proven to be more difficult to cryopreserve than other mammalian sperm, including that of the mouse, and current survival rates for sperm cryopreservation are still inadequate for AI, *IVF* and safe preservation of most rat strains. Important factors affecting sperm cryopreservation are cooling, freezing, thawing, and the composition of cryoprotectant in the freezing extender. In the rat, Nakatsukasa et al. (2001) employed a freezing medium that contained lactose monohydrate, Equex STM, and egg yolk solution. Based on the experimental conditions and extender components described in the current study, we offer the following suggestions to those attempting to cryopreserve epididymal rat sperm. Moreover, Varisli et al., (2009b) investigated that effect of chilling on the motility and acrosomal integrity of rat sperm in the presence of various extenders. They found that the addition of glycerol or propylene glycol to either Tris–citrate or TEST extender or of DMSO into lactose monohydrate, Tris–citrate, or TEST extender resulted in optimal motility rates.

4. Identified optimal energy substrates and other components of rat sperm cryodiluent

4.1 Freezability of rat sperm induced by raffinose in modified Krebs–Ringer bicarbonate (mKRB) based extender solution.

We first studied to develop an ideal freezing extender and method for rat sperm cryopreservation (Yamashiro et al., 2007). Experiments were conducted to study its post-thaw characteristics when freezing with raffinose-free buffer or various concentrations of raffinose and egg yolk dissolved in distilled and deionised water, PBS, or mKRB based extender. Different concentrations of glycerol, or Equex STM dissolved in either PBS or mKRB containing egg yolk were also tested. Based on the data from these experiments, further experiments tested how different sugars such as raffinose, trehalose, lactose, fructose, and glucose dissolved in mKRB with Equex STM and egg yolk supplementation affected the post-thaw characteristics of cryopreserved sperm. Beneficial effects on the post-thaw survival of sperm were obtained when raffinose in mKRB was used with Equex STM, and egg yolk (Fig.1).

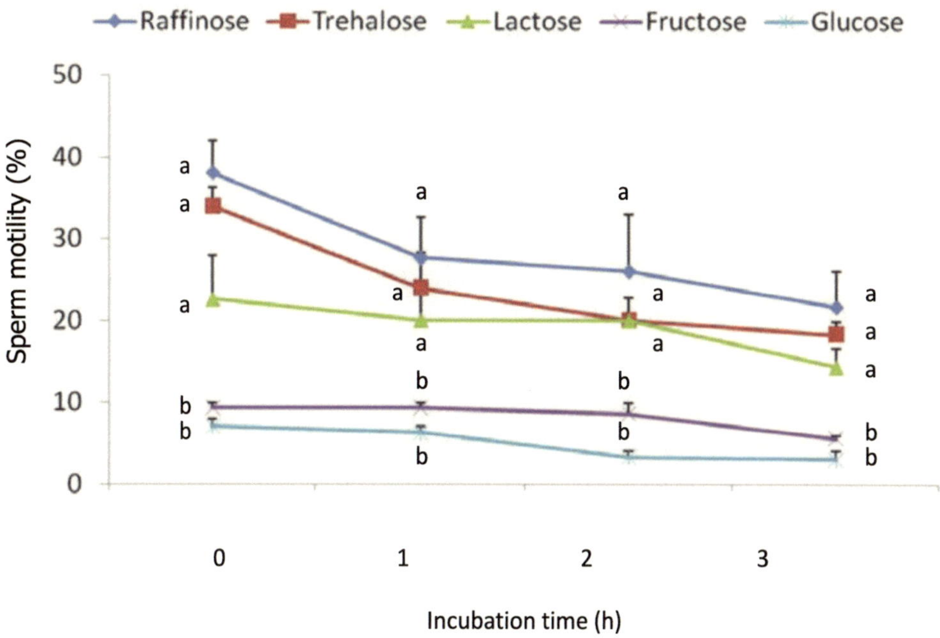

Fig. 1. Motility of frozen-thawed rat sperm which were frozen in different sugars dissolved in an mKRB egg yolk, and then subjected to a thermal resistance test at 37°C. Values are the mean ±SEM (n=3). [a-b] Different superscripts within the same column denote significant differences ($P < 0.05$).

Mammalian sperm provide energy for metabolic requirements by both mitochondrial oxidative respiration and glycolysis, and sperm motility is driven by the flagellum and is dependent on the availability of an adequate and continued supply of ATP (Cummins and Woodall, 1985). ATP is used by the dynein ATPases that function as the flagellar motors and in protein kinase A-mediated signal transduction pathways to regulate motility throughout the tail (Cao et al., 2006). In the mouse sperm, Mukai and Okuno (2004) reported that glycolysis has an important role in providing the ATP required for mouse sperm motility than mitochondrial respiration throughout the length of the flagellum. Bunch et al. (1998) also suggested that mouse sperm utilize glycolysis to generate ATP in the principal piece of the tail. While, Odet et al., (2011) demonstrated that lactate dehydrogenase (LDH) is responsible for the maintenance of energy metabolism in progressive and hyperactivated in mouse sperm. In rat sperm, Gallina et al. (1994) was studied the operation of shuttle functions for the ATP reconstituted systems in the mitochondria present in the middle place of rat, mouse and rabbit. They showed that the redox couple lactate/pyruvate and lactate dehydrogenase are active with rat and rabbit mitochondria, and it does not work with mouse. From these examining the energy metabolism of sperm, the glycolytic activity is exclusively responsible for the generation of mouse sperm metabolism, and mitochondrial metabolism seems to interacted with motility activity of rat sperm. These findings in conjunction with the present study indicate that successful cryopreservation of rat sperm in the presence of glucose, lactate and pyruvate in mKRB egg yolk extender solutions may be achieved through the ability to synthesize ATP, which could have profound effects on sperm metabolism and thereby impart a greater endurance against freeze–thawing damage.

4.2 Lactate and adenosine triphosphate in the extender enhance the cryosurvival of rat sperm

On the basis of the results of the previous experiments, we hypothesized that the metabolic state of sperm before cryopreservation would influence their survival during this stressful process (Yamashiro et al., 2010a). We evaluated the cryosurvival of rat sperm preserved in raffinose–mKRB egg yolk extender supplemented with various energy-yielding substrates (glucose, pyruvate, lactate, and ATP) and assessed the effect on sperm oxygen consumption. The incubation of sperm in lactate-free extender decreased sperm motility and oxygen consumption before and after thawing compared with those of sperm in glucose- and pyruvate-free mediums. We then focused on the effect of supplementing the extender with lactate and found that sperm frozen and thawed in extender supplemented with lactate exhibited the highest motility (Fig. 2). When we supplemented extender containing lactate with ATP, sperm frozen and thawed in the extender supplemented with ATP exhibited considerably higher motility and viability than those of sperm frozen and thawed in ATP-free extender (Fig. 3). Especially, exogenous ATP was observed that it dramatically induced the cryosurvival of rat sperm (Fig.4). Moreover, this may involve a lactate-transport system

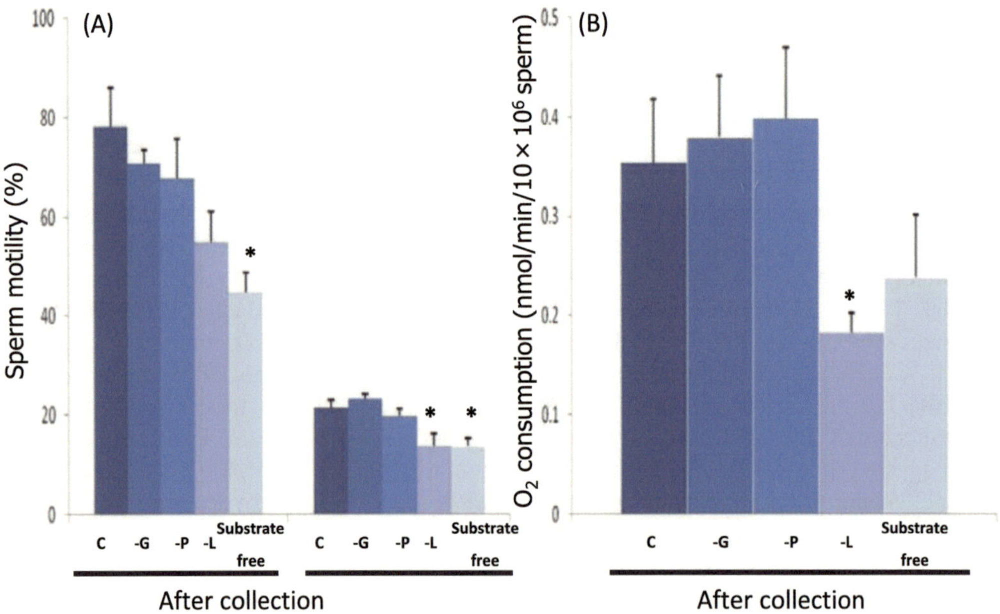

Fig. 2. Effect of the substrates in raffinose-mKRB egg yolk extender on the characteristics of after collected and frozen-thawed rat sperm (A) and oxygen consumption of after collected sperm during incubation at 37°C for 10 min (B). Indicates C; control, G; glucose, P; pyruvate, L; lactate. Values are mean ± SEM (n=5), respectively. Statistical difference (P<0.05) in comparison to the control is indicated by asterisk (*).

for regenerating cytoplasmic ATP throughout the principal piece of rat sperm in Fig.5 (Yamashiro et al., 2009). This thought is also in concert with the presence of a unique pathway that utilizes lactate and extracellular ATP in the rat sperm; this suggests new possibilities for energy production and translocation mechanisms related to motility, fertility, and freezability of rat sperm. These results provide the first evidence that supplementation of the raffinose-mKRB egg yolk extender with lactate and ATP increases of number of motile sperm before freezing and enhances the cryosurvival of rat sperm.

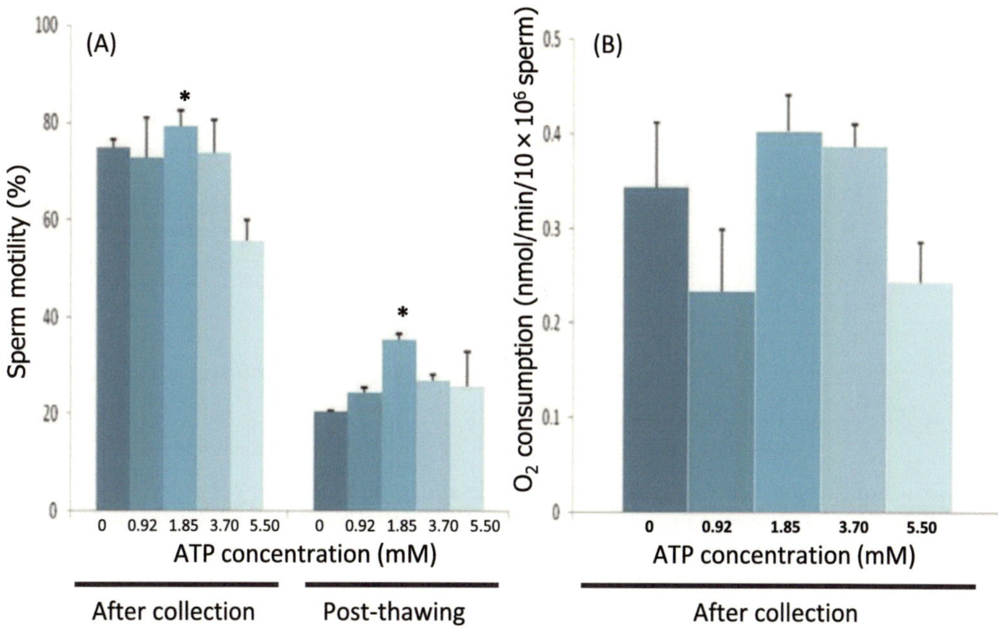

Fig. 3. Effect of different concentrations of ATP in raffinose-mKRB- gg yolk extender containing lactate on the characteristics of after collected and frozen-thawed rat sperm (A) and oxygen consumption of after collected sperm during incubation at 37°C for 10 min (B). Values are mean±SEM (n=5), respectivaly. Statistical difference ($P < 0.05$) in comparison to the control is indicated by asterisk (*).

The sperm-specific enzyme LDH isozyme C4 is located in the cytosol and the matrix of the mitochondria in the midpiece of rat sperm. Further, a study (Gallina et al., 1994) has revealed that both a shuttle involving the redox couple lactate-pyruvate and LDH C4 are active in rat sperm mitochondria. In another study (Harris et al., 2005), the lactate concentration in oviductal fluids was 10-fold higher than the glucose concentration, and the lactate concentration in the uterine fluids was 15-fold higher than the glucose concentration during the murine estrous cycle. Therefore, it is very likely that lactate is used by rat sperm as an essential substrate to maintain highly regulated ATP production and dissipation: lactate in the cytosol and mitochondrial matrix is oxidized to pyruvate by mitochondrial LDH isozyme C4, and pyruvate is oxidized through the Krebs cycle and electron transport chain (Brooks et al., 1994, Brooks. 2002, Montamat et al., 1988, Poole and Halestrap, 1993). To our knowledge, our findings are the first evidence showing that rat sperm can use exogenous lactate in the cryodiluent as an essential substrate to maintain highly regulated metabolic capacity and that this lactate acts as an energy substrate for mitochondria to the mobilization of fresh and frozen-thawed sperm.

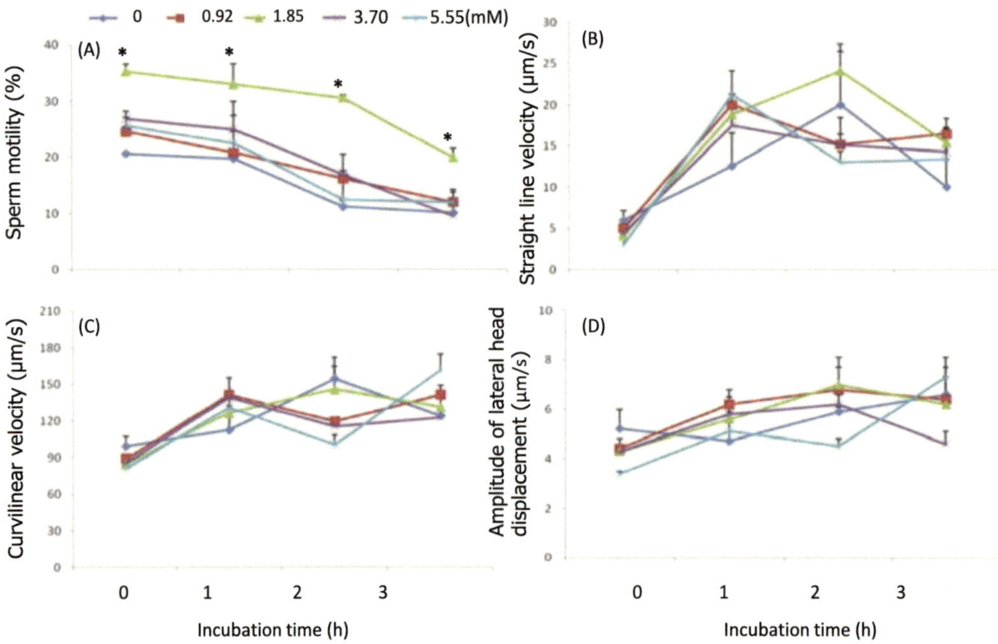

Fig. 4. Effect of different concentrations of ATP in raffinose-mKRB egg yolk medium containing lactate on the motility (A), straight line velocity (B), curvilinear velocity (C), and amplitude of lateral head displacement (D) of frozen-thawed rat sperm during incubation at 37 °C for 3 h. Values are represented as mean ± SEM (n = 3). Statistical difference ($P<0.05$) in comparison to the control is indicated by asterisk (*).

Mitochondria, the site of ATP generation due to oxidative phosphorylation, are localized solely in the midpiece of sperm (Millette et al., 1973). The oxidative production of ATP through the Krebs cycle is an essential function of the midpiece mitochondria for motility (Suarez et al., 2007). The mitochondrial biochemical pathways of oxidative phosphorylation are 15 times more efficient than is anaerobic glycolysis for ATP production (Cardullo and Baltz. 1991, Ruiz-Pesini et al., 2007). These findings also support our arguments that the energy production and dissipation in rat sperm are highly dependent on the mitochondria.

The present study showed that supplementation of raffinose-mKRB egg yolk extender with lactate and exogenous ATP considerably increases sperm motility before freezing, thus improving the survivability of sperm after cryopreservation. Exogenous ATP in the freezing medium may be responsible for the generation of multiple metabolic signals that appear to be related to the sperm motility through a rise in calcium levels (Gibbons.1963, Kinukawa et al., 2006, Litvin et al., 2003, Luria et al., 2002, Ren et al., 2001, Rodriguez-Miranda et al., 2007); this reaction increases de novo ATP synthesis before freezing and may contribute to the remobilization of sperm after freezing-thawing. The motility of ram sperm was restored by exogenous ATP that crossed plasma membrane when the membrane was damaged by cryopreservation (Holt et al., 1992). In light of that finding, we cannot discount that our result is caused by the facultative transport of ATP across plasma membrane because of damage during freezing, thereby allow ing substrates to directly access ATP and allowing adenosine triphosphatase to use ATP directly to generate energy for the mobilization of rat sperm.

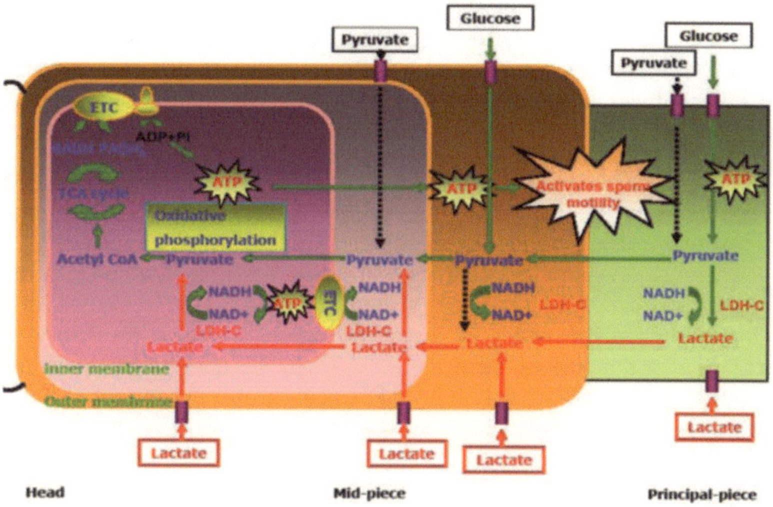

Fig. 5. Hypothesis of lactate-transport system for energy production and translocation in the rat sperm.

4.3 Extracellular ATP and dibutyryl cAMP enhance the freezability of rat sperm

The basic mechanochemical event underlying sperm motility is ATP-induced microtubule sliding (Brokaw. 1972). ATP associated with the dynein arms on outer-doublet microtubules provides the energy required for this process (Warner and Mitchell, 1980). ATP, calcium, and cAMP have received considerable attention as potential primary regulators of sperm motility in several species of animals (Aoki et al., 1996, Lindemann. 1978, Lindemann and Gibbons. 1975). Extracellular ATP acts on sperm by triggering a purinergic receptor-mediated increase in the intracellular calcium level; this increase may produce several downstream effects that enhance sperm motility (Luria et al., 2002, Rodriguez-Miranda et al., 2008). Increased calcium levels presumably activate soluble adenylyl cyclase, thereby increasing the cAMP concentration in sperm (Cook and Babcock. 1983, Garbers. 2001, Ren et al., 2001). cAMP induces protein phosphorylation by activating protein kinase A (Fujinoki et al., 2004, Kinukawa et al., 2006) and mediates calcium influx into sperm via the CatSper calcium ion channels (Cook and Babcock, 1983, Garbers. 2001, Ren et al., 2001). In addition, cAMP may elevate mitochondrial calcium levels (Degasperi et al., 2006), thereby activating the calcium-dependent dehydrogenases involved in the Krebs cycle and providing ATP required for sperm motility.

Previously, we showed that rat sperm become freezable when diluted in ATP-containing raffinose-mKRB egg yolk extender (Yamashiro et al., 2010a). This finding indicates the existence of a unique pathway that utilizes extracellular ATP in rat sperm and suggests that extracellular ATP produces several downstream effects that improve sperm motility by increasing calcium levels or by activating cAMP signal transduction pathways. Further elucidation of the role of extracellular ATP in the energy-synthetic processes and motility-regulation system of rat sperm could lead to improved motility, freezability, and fertilizing ability of the sperm. We, therefore, evaluated the freezability of rat sperm preserved in raffinose-mKRB egg yolk extender with ATP, ionomycin (a calcium ionophore), and dibutyryl cAMP (dbcAMP; a membrane-permeable cAMP analog) under various conditions. We also determined the effects of these agents on oxygen consumption by sperm. Sperm cryopreservation was considered successful if frozen–thawed sperm fertilized oocytes. To improve the effectiveness of in vitro fertilization (*IVF*), we determined whether ATP- and dbcAMP-supplemented *IVF* media improve the fertilizing ability of sperm. We also attempted artificial insemination with frozen–thawed rat sperm.

Results showed that rat sperm become freezable when diluted in ATP, and dbcAMP-containing raffinose-mKRB egg yolk extender (Yamashiro et al., 2010b). This finding indicates the existence of a unique pathway that utilizes extracellular ATP in rat sperm and suggests that extracellular ATP produces several downstream effects that improve sperm motility by increasing calcium levels or by activating cAMP signal transduction pathways. Further elucidation of the role of extracellular ATP in the energy-synthetic processes and motility-regulation system of rat sperm could lead to improved motility, freezability, and fertilizing ability of the sperm (Fig. 6). The results showed that the cryopreservation of rat

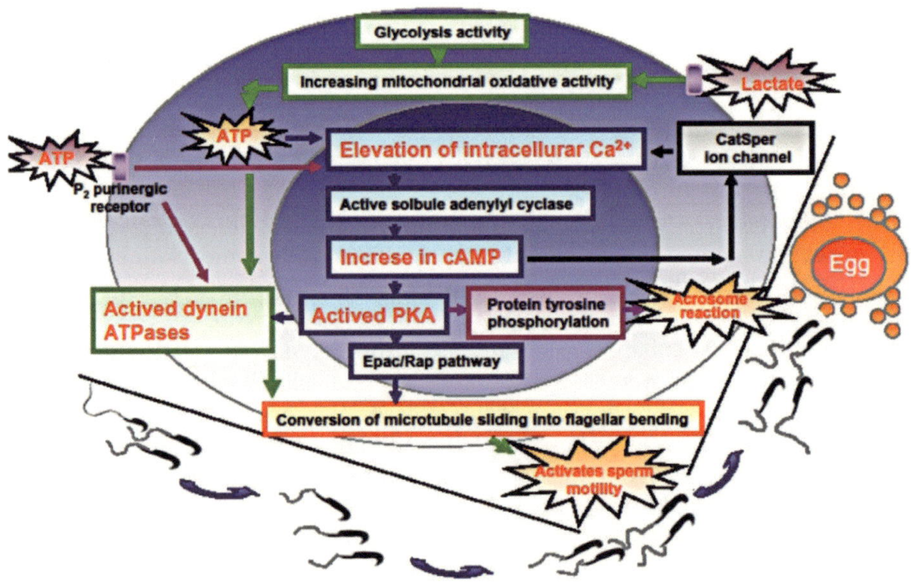

Fig. 6. Hypothesis of utilization of exogenous ATP pathway in the rat sperm.

sperm in raffinose–mKRB egg yolk extender supplemented with ATP and dbcAMP rendered sperm from rats freezable. This finding indicates that ATP- and dbcAMP-containing extenders improved the postthaw motility and fertilizing ability of cryopreserved rat sperm. Moreover, the *IVF* medium developed in the current study may be effective for the *in vitro* production of embryos from cryopreserved rat sperm. More recently, Vasudevan et al., (2011) reported that treatment of mouse sperm with extracellular ATP enhanced *IVF* rates in outbred and hybrid mice. Thus, this chapter was introduced how we identified optimal energy substrates and of rat sperm cryodiluent and *IVF* medium.

5. Concluding remarks

In this chapter described in detail the effects of the various components of cryodiluent that are used for rat sperm cryopreservation. We found that lactate, ATP and dbcAMP conferred freezability on rat sperm and enhanced oocyte fertilization by frozen-thawed sperm. In addition, if the rat sperm can be controlled flagellar movement in a way that changed by energy-yielding substrates such as lactate, it would be possible to enhance their fertilizing ability of rat sperm for *in vitro* fertilization by using a both of the fresh and frozen-thawed sperm (Fig. 7). However, there is still a lack of information for physiological trigger

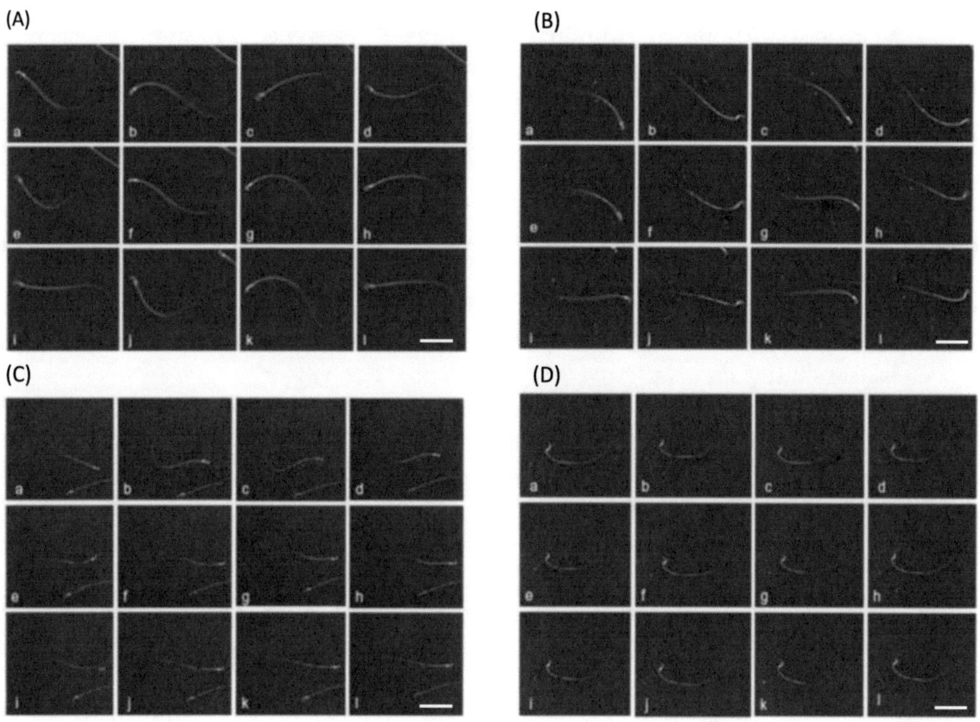

Fig. 7. Representative patterns of rat sperm movement which were extended in the presence (A) or absence (C and D) of lactate in raffinose-mKRB egg yolk solution. In B, sperm was incubated at 37°C for 3 h in the lactate containing solution. Panels a to l indicate the turn. Scale bars = 50μm

regarding how rat sperm switches on their flagellar movement at the real-time of fertilization, such as "hyper-activation" or "ultra-activation" (Yamashiro et al., 2009). Further, we believe that the *IVF* medium developed in our study is effective for the *in vitro* production of embryos from cryopreserved rat sperm. These results not only indicate that the cryopreservation of rat sperm with the present method can be applied to reproductive technologies but also indicate that exogenous lactate, rather than glucose and pyruvate, exerts a mediating effect on energy-dependent synthetic processes.

In conclusion, this is one of the very few information in the field that actually test the components of a cryodiluent in a logical manner for rat sperm cryopreservation. From this point of view, the chapter is likely to be important not only for freezing of rat sperm but also for freezing of sperm of other species in general.

6. References

Agca, Y., & Critser, J.K. (2002). Cryopreservation of spermatozoa in assisted reproduction. Semin Reprod Med. 20, pp.15–23.

Aitman, T.J., Critser, J.K., Cuppen, E., Dominiczak, A., Fernandez-Suarez, X.M., Flint, J., Gauguier, D., Geurts, A.M., Gould, M., Harris, P.C., Holmdahl, R., Hubner, N., Izsvák, Z., Jacob, H.J., Kuramoto, T., Kwitek, A.E., Marrone, A., Mashimo, T., Moreno, C., Mullins, J., Mullins, L., Olsson, T., Pravenec, M., Riley, L., Saar, K., Serikawa, T., Shull, J.D., Szpirer, C., Twigger, S.N., Voigt, B., & Worley, K. (2008). Progress and prospects in rat genetics: a community view. *Nat Genet.* 40, pp. 516–522.

Aoki, F., Sakai, S., & Kohmoto, K. (1999). Regulation of flagellar bending by cAMP and Ca^{2+} in hamster sperm. *Mol Reprod Dev.* 53, pp. 77–83.

Brokaw, C.J. (1972). Flagellar movement: a sliding filament model. *Science* 178, pp. 455–462.

Brooks, G.A., Dubouchaud, H., Brown, M., Sicurello, J.P., & Butz, C.E. (1999). Role of mitochondrial lactate dehydrogenase and lactate oxidation in the intracellular lactate shuttle. *Proc Natl Acad Sci USA.* 96, pp. 1129–1134.

Brooks. G,A. (2002). Lactate shuttles in nature. *Biochem Soc Trans.* 30, pp. 258–264.

Bunch, D.O., Welch, J.E., Magyar, P.L., Eddy, E.M., & O'Brien, D.A. (1998). Glyceraldehyde 3-phosphate dehydrogenase-S protein distribution during mouse spermatogenesis. *Biol Reprod.* 58, pp. 834–841.

Canzian, C. (1997). Phylogenetics of the laboratory rat rattus norvegicus. *Genome Res.* 7, pp. 262–267.

Cao, W., Haig-Ladewig, L., Gerton, G.L. & Moss, S.B. (2006). Adenylate kinases 1 and 2 are part of the accessory structures in the mouse sperm flagellum. *Biol Reprod.* 75, pp. 492-500.

Cardullo, R.A. & Baltz, J.M. (1991). Metabolic regulation in mammalian sperm: Mitochondrial volume determines sperm length and flagellar beat frequency. *Cell Motil Cytoskeleton.* 19, pp. 180–188.

Chularatnatol, M. (1982). Motility initiation of quiescent spermatozoa from rat caudal epididymis: effects of pH, viscosity, osmolality and inhibitors, *Int J Androl.* 5, pp. 425–436.

Cook, S.P., & Babcock, D.F. (1983). Activation of Ca^{2+} permeability by cAMP is coordinated through the pHi increase induced by speract. *J Biol Chem.* 268, pp. 22408–22413.

Cummins, J.M., & Woodall, P.F. (1986). On mammalian sperm dimensions. *J Reprod Fertil.* 75, pp. 153–175.

Degasperi, G.R., Velho, J.A., Zecchin, K.G., Souza, C.T., Velloso, L.A., Borecký, J., Castilho, R.F., & Vercesi, A.E. (2006). Role of mitochondria in the immune response to cancer: a central role for Ca^{2+}. *J Bioenerg Biomembr.* 38, pp. 1–10.

Fujinoki, M., Kawamura, T., Toda, T., Ohtake, H., Ishimoda-Takag,i T., Shimizu, N., Yamaoka, S., & Okuno, M. (2004). Identification of 36 kDa phosphoprotein in fibrous sheath of hamster spermatozoa. *Comp Biochem Physiol B Biochem Mol Biol.* 137, pp. 509–520.

Gallina, F.G., Deburgos, N.M.G., Burgos, C., Coronel, C.E., & Blanco, A. (1994). The lactate-pyruvate shuttle in spermatozoa: operation in vitro. *Arch Biochem Biophys.* 308, pp. 515–519.

Garbers, D.L. (2001). Ion channels. Swimming with sperm. *Nature* 413, pp. 581–582.

Geurts, A.M., Cost, G.J., Freyvert, Y., Zeitler, B., Miller, J.C., Choi, V.M., Jenkins, S.S., Wood, A., Cui, X., Meng, X., Vincent, A., Lam, S., Michalkiewicz, M., Schilling, R., Foeckler, J., Kalloway, S., Weiler, H., Ménoret, S., Anegon, I., Davis, G.D., Zhang, L., Rebar, E.J., Gregory, P.D., Urnov, F.D., Jacob, H.J., & Buelow, R. (2009). Knockout rats via embryo microinjection of zinc-finger nucleases. *Science* 325, pp. 433.

Gibbons, I.R. (1963). Studies on the protein components of cilia from Tetrahymena pyriformis. *Proc Natl Acad Sci USA.* 50, pp. 1002–1010.

Gibbs, R.A., Weinstock, G.M., Metzker, M.L. et al., (2004). Genome sequence of the Brown Norway rat yields insights into mammalian evolution. *Nature* 428, pp. 493–521.

Hagiwara, M., Choi, J.H., Devireddy, R.V., Roberts, K.P., Wolkers, W.F., Makhlouf, A., & Bischof, J.C. (2009). Cellular biophysics during freezing of rat and mouse sperm predicts post-thaw motility. *Biol Reprod.*1, pp. 700–706

Hamra, F.K. (2010). Gene targeting: Enter the rat. *Nature* 467, pp. 211–213.

Harris, S.E., Gopichandran, N., Picton, H.M., Leese, H.J., & Orsi, N.M. (2005). Nutrient concentrations in murine follicular fluid and the female reproductive tract. *Theriogenology* 64, pp. 992–1006.

Holt, W.V., Head, M.F., & North, R.D. (1992). Freeze-induced membrane damage in ram spermatozoa is manifested after thawing: observations with experimental cryomicroscopy. *Biol Reprod.* 46, pp. 1086–1094.

Izsvak, Z., Fröhlich, J., Grabundzija, I., Shirley, J.R., Powell, H.M., Chapman, K.M., Ivics, Z., & Hamra, F.K. (2010). Generating knockout rats by transposon mutagenesis in spermatogonial stem cells. *Nat Methods.* 7, pp. 443–445.

Jacob, H.J. (1999). Functional genomic and rat model. *Genome Res.* 9, pp. 1013–1016.

Kinukawa, M., Oda, S., Shirakura, Y., Okabe, M., Ohmuro, J., Baba, S.A., Nagata, M., & Aoki, F. (2006). Roles of cAMP in regulating microtubule sliding and flagellar bending in demembranated hamster spermatozoa. *FEBS Lett.* 580, pp. 1515–1520.

Lindemann, C.B. (1978). A cAMP-induced increase in the motility of demembranated bull sperm models. *Cell* 13, pp. 9–18.

Lindemann, C.B., & Gibbons, I.R. (1975). Adenosine triphosphate-induced motility and sliding of filaments in mammalian sperm extracted with Triton X-100. *J Cell Biol.* 65, pp. 147–162.

Litvin, T.N., Kamenetsky, M., Zarifyan, A., Buck, J., & Levin, L.R. (2003). Kinetic properties of 'soluble' adenylyl cyclase: synergism between calcium and bicarbonate. *J Biol Chem.* 278, pp. 15922–15926.

Luria, A., Rubinstein, S., Lax, Y., & Breitbart, H. (2002). Extracellular adenosine triphosphate stimulates acrosomal exocytosis in bovine spermatozoa via P2 purinoceptor. *Biol Reprod.* 66, pp. 429–437.

Millette, C.F., Spear, P.G., Gall, W.E., & Edelman, G.M. (1973). Chemical dissection of mammalian spermatozoa. *J Cell Biol.* 58, pp. 662–675.

Montamat, E.E., Vermouth, N.T., & Blanco, A. (1988). Subcellular localization of branched-chain amino acid aminotransferase and lactate dehydrogenase C4 in rat and mouse spermatozoa. *Biochem J.* 255, pp. 1053–1056.

Mukai, C. & Okuno, M. (2004). Glycolysis plays a major role for adenosine triphosphate supplementation in mouse sperm flagellar movement. *Biol Reprod.* 71, pp. 540–547.

Nakatsukasa, E., Inomata, T., Ikeda, T., Shino, M., & Kashiwazaki, N. (2001). Generation of live rat offspring by intrauterine insemination with epididymal spermatozoa cryopreserved at -196 degrees C, *Reproduction* 122, pp. 463–467.

Nakatsukasa, E., Kashiwazaki, N., Takizawa, A., Shino, M., Kitada, K., Serikawa, T., Halamata, Y., Kobayashi, E., Takahashi, R., Ueda, M., Nalashima, K., & Nakagata, N. (2003). Cryopreservation of spermatozoa from closed colonies, and inbred, spontaneous mutant, and transgenic strains of rats, *Comp. Med.* 53, pp. 639–641.

Odet, F., Gabel, S.A., Williams, J., London, .RE., Goldberg, E., & Eddy, E.M. (2011). Lactate Dehydrogenase C (LDHC) and Energy Metabolism in Mouse Sperm. *Biol Reprod.* 85, pp. 556-564

Parks, J.E., & Lynch, D.V. (1992). Lipid composition and thermotropic phase behavior of boar, bull, stallion, and rooster sperm membranes, *Cryobiology* 29, pp. 255–266.

Pennisi, E. (2004). Genetics DNA reveals diatom's complexity. *Science* 306, pp. 31.

Poole, R.C., & Halestrap, A.P. (1993). Transport of lactate and other monocarboxylates across mammalian plasma membranes. *Am J Physiol.* 264(4Pt 1), pp. C761–C782.

Ren, D., Navarro, B., Perez, G., Jackson, A.C., Hsu, S.Q., Shi, Q., Tilly, J.L., & Clapham, D.E. (2001). A sperm ion channel required for sperm motility and male fertility. *Nature* 413, pp. 603–609.

Rodriguez-Miranda, E., Buffone, M.G., Edwards, S.E., Ord, T.S., Lin, K., Sammel, M.D., Gerton, G.L., Moss, S.B., & William, C.J. (2008). Extracellular adenosine 5'-triphosphate alters motility and improves the fertilizing capability of mouse sperm. *Biol Reprod.* 79, pp. 164–171.

Ruiz-Pesini, E., Díez-Sánchez, C., López-Pérez, M.J., & Enríquez, J.A. (2007). The role of the mitochondrion in sperm function: is there a place for oxidative phosphorylation or is this a purely glycolytic process? *Curr Top Dev Biol.* 77, pp. 3-19.

Seita, Y., Sugio, S., Ito, J., & Kashiwazaki, N. (2009a). Generation of live rats produced by in vitro fertilization using cryopreserved spermatozoa. *Biol Reprod.* 80, pp. 503-510.

Seita, Y., Ito, J., & Kashiwazaki, N. (2009b). Removal of acrosomal membrane from sperm head improves development of rat zygotes derived from intracytoplasmic sperm injection. *J Reprod Dev*, 2009, 55, pp. 475–479.

Si, W., Benson, J.D., Men, H., & Critser, J.K. (2006). Osmotic tolerance limits and effects of cryoprotectants on the motility, plasma membrane integrity and acrosomal integrity of rat sperm. *Cryobiology* 53, pp. 336–348.

Suarez, S.S., Marquez, B., Harris, T.P., & Schimenti, J.C. (2007). Different regulatory systems operate in the midpiece and principal piece of the mammalian sperm flagellum. *Soc Reprod Fertil Suppl.* 65, pp. 331-334.

Tong, C., Li, P., Wu, N.L., Yan, Y., & Ying, Q.L. (2010). Production of p53 gene knockout rats by homologous recombination in embryonic stem cells. *Nature* 467, pp. 211-213

Varisli, O., Uguz, C., Agca, C., & Agca, Y. (2009a). Various physical stress factors on rat sperm motility and integrity of acrosome and plasma membrane. J Androl. 30, pp. 75–86.

Varisli, O., Uguz, C., Agca, C., & Agca, Y. (2009b). Effect of chilling on the motility and acrosomal integrity of rat sperm in the presence of various extenders. J Am Assoc Lab Anim Sci. 48, pp. 499–505.

Vasudevan, K., Sztein, J.M. Treatment of sperm with extracellular adenosine 5'-triphosphate improves the in vitro fertility rate of inbred and genetically modified mice with low fertility. (2011). *Theriogenology* 76, pp. 729-736.

Warner, F.D., & Mitchell, D.R. (1980). Dynein: the mechanochemical coupling adenosine triphosphatase of microtubule-based sliding filament mechanisms. *Int Rev Cytol.* 66, pp. 1–43.

Yamashiro, H., Han, Y.J., Sugawara, A., Tomioka, I., Hoshino, Y., & Sato, E. (2007). Freezability of rat epididymal sperm induced by raffinose in modified Krebs-Ringer bicarbonate (mKRB) based extender solution. Cryobiology 55, pp. 285-294.

Yamashiro, H., Toyomizu, M., Kadowaki, A., Takeda, Z., Nakazato, F., Toyama, N., Kobayashi, J., & Sato, E. (2009).Oxidation of exogenous lactate by lactate dehydrogenase C in the midpiece of rat epididymal sperm is essential for motility and oxidative activity. *Am J Appl Sci.* 6, pp. 1854–1859.

Yamashiro, H., Toyomizu, M., Kikusato, M., Toyama, N., Sugimura, S., Hoshino, Y., Abe, H., Moisyadi, S., & Sato, E. (2010a). Lactate and adenosine triphosphate in the extender enhance the cryosurvival of rat epididymal sperm. *J Am Assoc Lab Anim Sci.* 49, pp. 155–161.

Yamashiro, H., Toyomizu, M., Toyama, N., Aono, N., Sakurai, M., Hiradate, Y., Yokoo, M., Moisyadi, S., & Sato, E. (2010b). Extracellular ATP and dibutyryl cAMP enhance the freezability rat epididymal sperm. *J Am Assoc Lab Anim Sci*. 49, pp. 167-172.

9

Cryopreservation of Genetic Diversity in Rabbit Species (*Oryctolagus cuniculus*)

Thierry Joly, Vanessa Neto and Pascal Salvetti
Université de Lyon, France, VetAgro Sup – Isaralyon,
UPSP ICE 'Interactions between Cells and their Environment', Team Cryobio
France

1. Introduction

After the ratification of the international convention on the biodiversity in Rio, a National Cryobank was created in France in 1999 to preserve the genetic resources of domestic animals (www.cryobanque.org). Particular attention was carried on Oryctolagus cuniculus species with the extension of the national cryobanking to the rabbit (Joly et al., 1998). Nowadays, this tool is very useful for the management of animal diversity in France.

Cryopreservation corresponds of all the steps of collection and long term storage of animal populations, preserved as live cells and able to generate live animals.

Cryopreservation is not a museology action to freeze the products of the past. Contrary to that, it corresponds to the practical implementation of new biotechnologies of reproduction. It includes also all the technical means to maintain the evolutionary potential of population. So, the French National Cryobank was created to secure the biological material stored at -196°C in liquid nitrogen. It constitutes a real tool to serve all stakeholders to manage the animal diversity as part of the National Charter supported by the Genetic Resources Office (www.fondationbiodiversite.fr).

2. Genetic diversity in rabbit species

The archaeological origins of rabbit are located in Spain (Bolet et al., 2000). Actually, rabbit is widely spread all around the world and can be considered according situations as wild animal, domestic animal, pets and laboratory models. The rabbit populations can be classified into 3 categories according to their genetic originality, their specific uses and the motivations of breeders involved in the *in situ* management of populations (figure 1).

2.1 "Type I" material

Type I regroups the breeds identified according to an official standard, as well as for large breeds (Butterfly, Champagne Argente, Fauve de Bourgogne...) and endangered breeds with less than 100 females (Brun Marron de Lorraine ...). These breeds are reared by fancy breeders and animals are presented regularly in local or regional meetings supported by FFC the federation of French fancy rabbit breeders (www.ffc.asso.fr). These breeds must be preserved for their patrimonial values and socio cultural interests.

2.2 "Type II" material

Type II concerns animals with one or more specific character:

- Animals carrying an identified gene (The recessive gene Sam in "jumper Alfort rabbit" evidenced by a modified walking on anterior legs; new transgenic lines recently created)
- Animals showing a specific combination of genes (histocompatible lines; allotypic strains from Basel having used as model for histocompatibility major complex studies)
- Animals with a particular genotype from a high selected population but not from the actual breeding schemes (black Orylag rabbit and Rex albinos presenting colours differing from the selected population, the divergent strain Inra 1029 made of two foetal mortality rate-diverging lines).

Most of Type II populations mainly have a scientific interest with high experimental values. Some populations could be promoted in biomedical research and by the pharmaceutical industry.

2.3 "Type III" material

Type III regroups all the populations of rabbits selected for meat production (female lines and male lines) or fur production with high economic value.

Some strains are today completely extinct and subsist only under embryo-frozen form waiting for a possible perspective of reuse or selection scheme reorientation (Dutch hymalayan, Orylag, INRA 1077 female strains…).

Other populations are still selected and the frozen biologic material represents a selection control to measure genetic progress precisely. Moreover, for security, several populations commercially spread are subject to regular cryoconservation in order to save the selection core from a sanitary risk (Inra 2066, commercial lines of Hypharm and Hycole societies…)

Rabbit « Fauve de Bourgogne » (type I) adapted to a traditionnal breeding system

Rabbit « Brun Marron de Lorraine » (type I), endangered breed

Rabbit « Orylag® castor (type III) bred for its "Rex" fur

Rabbit "Sauteurs d'Alfort" (type II). They move on forlegs when they are stressed

Fig. 1. Examples of different categories of rabbits

3. Principal ways of cryoconservation of rabbit genetic resources

To keep rabbit genetic resources, we can distinguish three principal ways according to the nature of frozen biological material (Figure 2).

3.1 Germinal cells

Based on the conservation of germinal cells, this way helps to conserve the gene pool of particular male or female individuals.

Mature spermatozoids (sperm collected in artificial vaginas) or immature (epididymal spermatozoids, spermatogonies in gonadic tissue) helps to save the male line. The freezing of an individual's semen (n generation) helps to obtain progeny (n+1 generation) after thawing and artificial insemination (AI) of females.

Semen samples of bad quality can be promoted by new technologies of *in vitro* fertilization (FIV) or intracytoplasmic sperm injection (ICSI), which then permits *in vitro* embryo production and progeny delivering after embryo transfer in synchronized recipient female (Daniel *et al*, 2007). Nowadays, semen freezing is not a reliable or repeatable method yet, and results after AI with rabbit frozen semen are still too inconstant to plan a routine utilization of this technique (Vicente *et al*, 1996; Moce *et al*, 2003). Only half of the sampled males can produce semen with freezable quality and approximately 50% of females give birth after insemination of thawed semen with a large variability [15%~80%]. However, this is the only available method to preserve precious males semen, mainly for type I and II.

Mature oocytes (picked up in oviduct 15 hours after ovulation) or immature oocytes (present in follicles of ovarian tissue) permits to save genetic resources by female way and to preserve cytoplasmic heredity. Mature oocytes freezing of a female individu (n generation) would permit to obtain progenies (n+1 generation) after FIV or ICSI, but no young rabbit has been obtained from thawed oocytes yet (Salvetti *et al*, 2010). But, recently, young rabbits obtained from frozen ovarian tissue are born from females transplanted by orthotopic autograft (Almodin *et al.*, 2004; Neto *et al.*, 2007). In emergency situations (sanitary problems, injured animals), ovarian cortex freezing, even if this method is not yet completely under control (Neto *et al.*, 2008), can be proposed to save the heredity pool of an important female of type II.

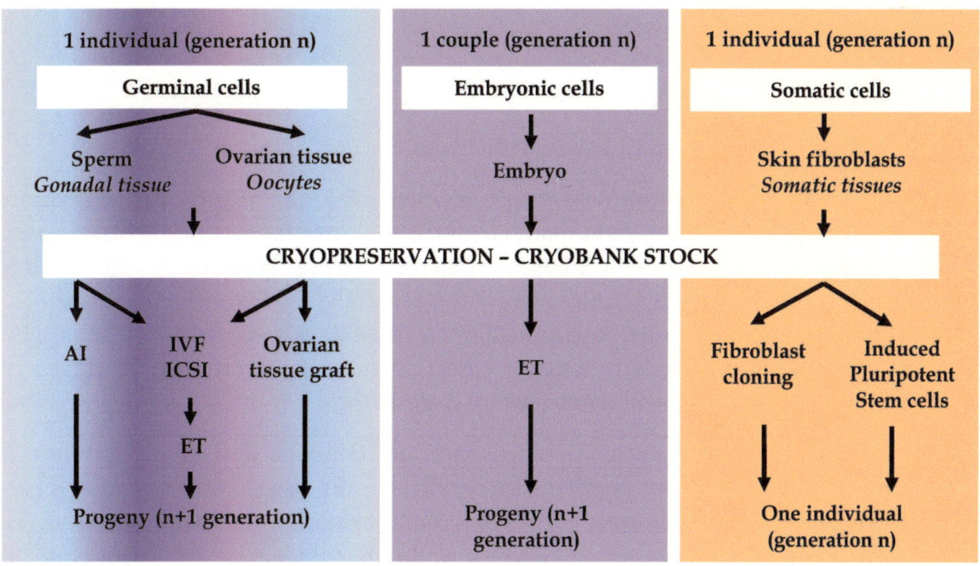

Fig. 2. Biotechnologies of reproduction applied to the cryopreservation of the rabbit genetic resources

3.2 Embryonic cells

Embryo freezing, allowing the preservation of both male and female ways, is the main way to recreate quickly a population from thawed embryo transfer. Most of the time, embryos are frozen at compacted morula stage (65h~72h *post coïtum*) even if freezing at earlier stage (4 cells stage) is possible. They can be produced *in vivo* from superovulated females or not, or *in vitro* after *In Vitro* Fertilization and *in vitro* culture. Embryos freezing from a planned mating (at n generation) permits to obtain progenies (at n+1 generation) after thawing and transfer in recipient females.

3.3 Somatic cells

This way has the main advantage of being easy to do, which permits to sample tissues quickly and simply on a large number of individuals. But the main difficulty is the recreation of animals after somatic cloning, still not well controlled. The cells can be isolated from different tissues: skin, cartilage, bones, blood… These cells, present in a large number in each organism, would allow the reconstitution of an individual of the same generation than the tissues giver after somatic cloning (Chesne *et al.,* 2002).

More recently, a new opportunity came with induced pluripotent stem cells, which are able to differentiate in many tissues. These cells are potential vectors for genetic traits transmission and could be used after thawing as an important core source to regenerate many individuals after nuclear transfer (Honda *et al,* 2010).

4. Frozen embryo, the main way for rabbit cryoconservation

The embryo is the favored biological material to conserve most of rabbit populations, by both male and female ways, and then preserve cytoplasmic heredity. This method has been widely proved for 15 years with more than 32 000 produced embryos but can only be applied to living and fertile animals.

4.1 Embryos production

As rabbit is naturally prolific species, two methods can be used to produce embryos:

- Without ovarian stimulation: one single injection of a GnRH analog (gonadotropic hormon) induces females ovulation at the same time as artificial insemination. Sometimes, an injection of 20-30 UI of eCG (equine chorionic gonadotrophin) three days before mating increases the females receptivity, especially the sexual resting ones;
- With ovarian stimulation by superovulation: a three days treatment of FSH injection stimulates the ovarian activity and increases the number of produced embryos per donor female (Kauffman *et al,* 1998; Salvetti *et al.,* 2007).

Overall, 73% of the treated and collected females are embryos donors and produce an average of 9.2 embryos without ovarian stimulation or with a slight eCG stimulation, and two to three times more (21.2 embryos) after a superovulation treatment (**table 1**). The embryo recovering rates (number of collected embryos/number of corpus luteum) are between 70% and 80% and those of freezable embryos (number of frozen embryos/number of collected embryos) are from 75% to 85%.

Treatment	Nb of treated females	% donor	Nb of frozen embryos	Nb of embryos per donor
No ovarian stimulation	1226	77%	8633	9.2
Superovulation (FSH)	1589	71%	23833	21.2
Total	**2815**	**73%**	**32466**	

Table 1. Method of embryo production (activity from 1998 to 2011)

4.2 Embryos recovering and freezing

Slaughter the females with uterine tract washing is the most effective method to collect rabbit embryos. This method is simple and easy to implement directly on the breeding place of the animals. Cryoconservation of a population, about 40 to 50 females is feasible in one single day. The method is to keep for large population of type I and III because of systematic slaughter of females.

In addition, a new method of embryos collection by endoscopy allows renewing about four times the operation on the same female, preserving the female's integrity (Garcia *et al.*, 1991; Besenfelder *et al.*, 1998). A lot heavier to implement, it is reserved to rare and precious animals of type II or to small effective populations of type I.

Embryos were frozen in the same cryoprotective solution containing 1.5 M DMSO and by the same slow freezing process, even more vitrification could provide good results but this method is not totally under control (Mehaisen *et al.*, 2006).

4.3 Embryo transfer and births

After thawing of a part of these embryos, the embryo transfer results vary according to the environmental conditions defined by the recipient female genotype and breeding conditions (**table 2**).

Transfer conditions	Nb of recipients	% of delivering female	Nb of thawed embryos	Nb of borned pups	Embryo developmental rate
Standard controlled	277	82%	2746	1118	41%
On Field transfer	87	56%	846	176	21%
Total	**364**	**76%**	**3592**	**1294**	**36%**

Table 2. Pups production after transfer of thawed embryos (activity from 1998 to 2011)

The optimal environmental conditions are defined by the recipient from a mother female line placed in a control environment in an aboveground breeding (16 hours of light per day), while uncontrolled conditions are defined by a recipient placed in an uncontrolled environment (sanitary, light, temperature) in a traditional farm's breeding conditions.

Globally, 76% of the recipients give birth to young rabbits after thawed embryo transfer with an embryo development rate of 36%. It is particularly important to control the recipient's genotype and the breeding conditions in order to guarantee the population recreation after cryoconservation.

The efficiency of the cryoconservation method has been concretely proved after transfer of embryos stored for more than 15 years in liquid nitrogen. In November 2006, 69 Brun Marron de Lorraine young rabbits got born after thawing and transfer of 134 embryos in recipients. The young rabbits born from frozen embryos of 1992 were presented to public at the international agriculture show in Paris in March 2007, in collaboration with fancy breeders and the FFC (Salvetti *et al.*, 2007).

5. The national cryobank: A saving tool of rabbit genetic resources

In March 2008, the rabbit collections presented a large genetic diversity, which could be ranged according to three types of materials previously defined. Nearly sixty of rabbit populations have been cryopreserved. More than 19.000 embryos from about 1.300 rabbit doe donor have been frozen and are currently stored in liquid nitrogen at -196°C. For all these populations, three methods of embryos production have been applied according to the situation and the physiological state of rabbit does (superovulation, eCG, simple induction of ovulation). Then, all these embryos were treated according to the same freezing protocol (Joly *et al*, 1998). All the actors working for the rabbit populations management are henceforth convinced of the interest of this tool. They have actively participated to the building of the rabbit French cryobank. Public and private selectors, associations of fancy breeders and different groups of independent breeders are regular depositor.

6. A reality

The patrimonial cryobank for rabbit species is henceforth a reality. It allows to satisfy the expectations of breeders, selectors and the actors of research and biomedical industries. This last actor takes a more and more important rules in the rabbit production.

A standard method of rabbit embryos cryopreservation has been routinely applied for 15 years. Its implementation is performed on the field with a portative controlled rate freezer. This method requires a simple stimulation of the females before the collection, and an appropriate choice of the males for the breeding combinations. However, the complete conditions are not always encountered and this method is not efficient in emergency situation (for example, when an animal is rugged or during sanitary crisis).

So, another complementary methods are studied. These new way of research includes the cryopreservation of testicular tissue and epididymal spermatozoa in the male which are dead for less than two days. It includes also the cryopreservation of the rabbit ovarian tissue. Nowadays, the freezing of ear's fibroblasts is studying to produce Induced Pluripotent Stem Cells in order to use it for chimaeric animal production as tool of regeneration of initial population without genetic drift.

7. References

Almodin, C.G., Minguetti-Camara, V.C., Meister, H., Ferreira, J.O., Franco, R.L., Cavalcante, A.A., Radaelli, M.R., Bahls, A.S., Moron, A.F., Murta, C.G. (2004). Recovery of fertility after grafting of cryopreserved germinative tissue in female rabbits following radiotherapy. *Hum Reprod,* Vol.19, No.6, pp:1287-1293,

Besenfelder, U., Strouhal, C., Brem, G. (1998). A method for endoscopic embryo collection and transfer in the rabbit. *Zentralbl Veterinarmed A,* Vol.45, No.9, pp: 577-579,

Bolet, G., Monnerot, M., Arnal, C., Arnold, J., Belle, D., Bergoglio, G., Besenfelder, U., Bosze, S., Boucher, S., Brun, J.M., Chanteloup, N., Ducourouble, M.C., Durand-Tardif, M., Esteves, P.J., Ferrand, N., Hewitt, G., Joly, T., Koehl, P.F., Laube, M., Lechevestrier, S., Lopez, M., Masoero, G., Piccinin, R., Queney, G., Saleil, G., Surridge, A., Van Der Loo, W., Vanhommerig, J., Vicente, J.S., Virag, G., Zimmermann, J.M. (2000). Evaluation and conservation of european rabbit genetic resources : first results and inferences. *World rabbit Science,* Vol. 8, pp:281-315,

Chesne, P., Adenot, P.G., Viglietta, C., Baratte, M., Boulanger, L., Renard, J.P. (2002). Cloned rabbits produced by nuclear transfer from adult somatic cells. *Nat Biotechnol,* Vol.20, pp:366-369,

Daniel, N., Chesne, P., Baratte, M., Renard, J.P. (2007). Lapins produits par injection intra-cytoplasmique de spermatozoïdes (ICSI) stockés à température ambiante. *12èmes Journées de la Recherche Cunicole,* Le Mans, France, pp.33-36,

Garcia-Ximenez, F., Vicente, J.S., Santacreu, M.A. (1991). Embryo transfer in lactating rabbit does by laparoscopy. *Anim Reprod Sci,* Vol.24, pp:343-346,

Honda, A., Hirose, M., Hatori, M., Matoba, S., Miyoshi, H., Inoue, K., Ogura, A. (2010). Generation of Induced Pluripotent Stem Cells in Rabbits : Potential experimental models for human regenerative medicine. *J. Biol. Chem.* Vol.285, pp :31362-31369,

Joly, T., Rochambeau, H. de., Renard, J.P. (1998). Etablissement d'une cryobanque d'embryons pour la conservation ex situ de la diversité génétique chez le lapin : aspects pratiques. *Genet.Sel.Evol.* Vol.30, No.1, pp:259-269,

Kauffman, R.D., Schmidt, P.M., Rall, W.F., Hoeg, J.M. (1998). Superovulation of rabbits with FSH alters *in vivo* development of vitrified morulae. *Theriogenology,* Vol. 50, pp: 1081-1092,

Mehaisen, G.M.K., Viudes-de-Castro, M.P., Vicente, J.S., Lavara, R. (2006). *In vitro* and *in vivo* viability of vitrified and non-vitrified embryos derived from eCG and FSH treatment in rabbit does. *Theriogenology,* Vol.65, pp:1279-1291,

Mocé, E., Vicente, J.S., Lavara, R. (2003). Effect of freezing-thawing protocols on the performance of semen from three rabbit lines after artificial insemination. *Theriogenology,* Vol.60, pp:115-123,

Neto, V., Joly, T., Salvetti, P., Lefranc, A.C., Corrao, N., Buff, S., Guérin, P. (2007). Ovarian tissue cryopreservation in the doe rabbit: from freezing to birth. *Cryobiology,* Vol.55, pp:344,

Neto, V., Buff, S., Lornage, J., Bottollier, B., Guerin, P., Joly, T. (2008). Effects of different freezing parameters on the morphology and viability of preantral follicles after cryopreservation of doe rabbit ovarian tissue. *Fertil. Steril.* Vol.89, No.3, pp:1348-1356,

Salvetti, P., Joly, T., Boucher, S., Hurtaud, J., Renard, JP. (2007). Viability of rabbit embryos after 15 years storage in liquid nitrogen. *44th Annual Meeting of the Society for Cryobiology CRYO 2007,* Lake Louise, Canada,

Salvetti, P., Theau-Clement, M., Beckers, J.F., Hurtaud, J., Guerin, P., Neto, V., Falieres, J., Joly, T. (2007). Effect of the luteinizing hormone on embryo production in superovulated rabbit does. *Theriogenology,* Vol.67, pp:1185–1193,

Salvetti, P., Buff, S., Afanassieff, M., Daniel, N., Guérin, P., Joly, T. (2010). Structural, metabolic and developmental evaluation of ovulated rabbit oocytes before and after cryopreservation by vitrification and slow freezing. *Theriogenology,* Vol.74, pp:847 - 855,

Vicente, J.S., Viudes-de-Castro, M.P. (1996). A sucrose-DMSO extender for freezing rabbit semen. *Reprod Nutr Dev.* Vol.36, No.5, pp:485-492.

10

Review on Ovarian Cryopreservation in Large Animals and Non-Human Primates

Milan Milenkovic, Cesar Díaz-Garcia and Mats Brännström
Department of Obstetrics and Gynecology, Sahlgrenska Academy
University of Gothenburg
Sweden

1. Introduction

The ultimate aim of ovarian cryopreservation research is naturally to increase the effectiveness of this fertility preservation procedure in female cancer victims and much of this research on whole ovary cryopreservation, ovarian cortex cryopreservation and transplantation has been performed in animal models. However, ovarian cryopreservation could also be used in the future in programs with the purpose to rescue endangered species (Santos et al., 2010) and certain specific strains of animals (Dorsch et al., 2004). Due to the ethical barriers in the research on human tissue and the shortage of human premenopausal ovarian tissue for research purposes, there is a need to find animal models that are reasonably analogous to the human. As a general rule, animal models have to be comparable in biochemical, physiological and anatomical characteristics to the human so that the results can be applicable to human conditions (VandeBerg, 2004). Regarding research of ovarian cryopreservation for human fertility preservation, a similar tissue architecture and size of the ovary (Table 1) as well as being a mono-ovulatory species with the primordial follicles located superficially in the cortex would be of advantage (Gerritse et al., 2008). The previous research on ovarian cryopreservation using bovine, porcine, sheep and non-human primate models will be presented in this chapter.

Species	Ovarian volume	Tissue architecture	Ovulation pattern	Cycle length (days)
Cow	14.3 (+/- 5.7) cm^3 (Gerritse et al. 2008)	Similar to the human	Mono, di-ovulatory	21
Pig	7.3 (+/- 2.2) cm^3 (Gerritse et al. 2008)	Less fibrous than the human ovary	Multi ovulatory	18-24
Sheep	1.0 (+/- 0.4) cm^3 (Gerritse et al. 2008)	Similar to the human	Mono, tri-ovulatory	16-17 (Seasonal)
Non-human primate (cynomolgus macaque)	0.258 (+/- 0.159) cm^3 (Jones, 2011)	Similar to the human	Mono-ovulatory	28-32

Table 1. Ovarian characteristics of different large animal models

2. Cow

Cow ovaries may be used in research on ovarian cryopreservation. They are much larger (14.3+/- 5.7 cm^3) (Gerritse et al., 2008) as compared to the human, but on the other hand the tissue architecture is similar. The function (Yang&Fortune, 2006) and the structure (Rodgers&Irving-Rodgers, 2010) of the bovine ovary has been extensively studied. Also, the granulosa cells and the extracellular matrix of bovine follicles of various developmental stages are well described (Lavranos et al., 1994; Irving-Rodgers et al., 2006).

There exist one report of avascular ovarian transplantation in the cow after cryopreservation by vitrification using a solution consisting of 20% ethylene glycol (EG) and 20% dimethylsulphoxide (DMSO) in TCM-199 medium (Kagawa et al., 2009). In this study, vitrification was performed using the Cryotissue method (Kagawa et al., 2007), where the ovarian cortex was positioned on a thin metal strip that was plunged into liquid nitrogen (LN) and inserted into a protective container for storage in LN. Rapid post-thaw warming was done with immersion of the metal strip into TCM-199, supplemented with sucrose at 40°C followed by washing in the identical solution but with decreasing sucrose concentration. After warming, the tissue was grafted subcutaneously to the neck or orthotopically to oophorectomized cows, and resumed cyclicity was seen in both groups within two months. Histological analysis of grafted tissue showed normal morphological appearance and about 80% viability among preantral follicles, as demonstrated by fluorescent staining. These results may be regarded as encouraging towards clinical application of the vitrification procedure for ovarian cortex cryopreservation.

In another study, aimed towards in vitro follicle maturation of bovine follicles, slow-frozen bovine ovarian cortical pieces were incubated after thawing for durations between 1 and 48 h (Paynter et al., 1999). The major finding was that the thawed tissue had a capacity to recover from damage during the subsequent incubation period. This idea was further utilized in later research, where bovine ovarian cortical pieces were cultured for six days followed by isolation of secondary follicles and culture in the presence of inhibin (McLaughlin&Telfer, 2010) showing significant estradiol (E2) secretion and oocytes growth up to a diameter > 100 µm.

The notion that antioxidants may enhance survival of frozen-thawed tissue was studied using bovine ovarian cortex (Kim et al., 2004). After cryopreservation by slow freezing with 1.5 M DMSO, bovine ovarian cortex was in vitro cultured for periods up to 48 h in minimal essential medium (MEM) with or without ascorbic acid. Interestingly, there was no difference between the two groups in apoptosis rate evaluated by terminal deoxynucleotidyl transferase dUTP nick end labelling test (TUNEL) or deoxyribonucleic acid laddering. Nevertheless, protective effects by ascorbic acid were seen in stromal cells that were cultured for 24 h. In addition, this study also demonstrated that stromal cells are more susceptible to damage mechanisms than primordial follicles, which is a finding also observed in frozen-thawed human ovarian tissue (Gook et al., 1999; Hreinsson et al., 2003). The model of supplementation of antioxidant agents to the cryoprotectant (CPA) should be investigated further.

The toxic effect of various concentrations and types of CPAs that are frequently used for slow freezing was compared using bovine ovarian cortical strips (Lucci et al., 2004; Celestino et al., 2008). Among other CPAs, DMSO at 1.5 M and 3 M were evaluated in both studies.

While the study by Celestino and colleagues (Celestino et al., 2008) showed increased toxicity with rising concentration of DMSO, the other study (Lucci et al., 2004) showed slightly higher proportion of normal follicles in the 3 M DMSO group when assessed by conventional histology. However, ultrastructural analysis by transmission electron microscopy (TEM) revealed some irregularities in the cytoplasm of granulosa cells when 3 M DMSO was applied (Lucci et al., 2004).

There is one study on bovine ovarian cortical strips comparing slow freezing with vitrification and this study demonstrated higher efficiency of slow-freezing (Gandolfi et al., 2006). Furthermore, another study also demonstrated advantages of the slow-freezing method as compared to vitrification, when bovine ovarian cortical pieces were used, but on the other hand an advantage of the vitrification method was seen when whole ovaries with vasculature were used as the model system (Zhang et al., 2011). In the latter study, the effectiveness of the different cryo techniques was evaluated by Trypan blue test, histology as well as E2 and progesterone levels obtained from supernatant after in vitro culture of the tissue. The same research group (Zhang et al., 2011) performed controlled-rate slow freezing with DMSO of whole bovine ovary and compared different cooling rates and ice seeding temperatures. The cooling rate of 0.2°C/min and ice seeding temperature of -5°C showed superiority in comparison to different combinations of the cooling rates of 2°C/min and 0.1°C/min and the ice-seeding temperatures of -2°C, -5°C and -8°C.

Additionally, one study on bovine ovarian tissue was designed to evaluate the effect of the thickness of the ovarian cortex strip on follicular morphology after incubation for 20 min and slow freezing with 1.5 M propylene glycol (PROH) (Ferreira et al., 2010). Ovarian cortex pieces of 10 x 3 mm, with a thickness of either 2 or 4 mm, were compared and considerably higher proportion of normal follicles were found in the 2 mm group compared to the 4 mm in both fresh and cryopreserved tissue. This result may be explained by superior tissue impregnation with CPA in the 2 mm group, but the exact mechanisms remain to be clarified.

3. Pig

The pig is a species that has been used in biomedical research, particularly regarding development and training of surgical techniques for later use in the human. The reproductive cycle of the pig lasts for 18-24 days and generally 8-15 oocytes are released from each ovary at ovulation (Soede et al., 2011). The size of the pig ovary is about 7.3 (+/- 2.2) cm^3 (Gerritse et al., 2008), which is comparable to the human ovarian size of 6.5 (+/- 2.9) cm^3 (Munn et al., 1986). The equal ovarian size of the sow and human female, as well as the possibility to get fresh pig ovaries from slaughterhouses, renders the pig as a good model for ovarian cryopreservation research. Nevertheless, a fairly low number of studies in this area with the pig ovary as an experimental model have been performed, as shown below.

One recent study on pig ovarian tissue evaluated whether the size of ovarian cortical pieces is important for the cryopreservation outcome (Jeremias et al., 2003). Cortical strips were all of 1 mm thickness but either 1x1 mm or 5x1 mm in surface area. The pieces were cryopreserved by slow freezing in 1.5 M DMSO and after rapid thawing the size of the surviving primordial follicle pool, was compared to fresh tissue (1x1 mm) (Jeremias et al., 2003). The freezing method was uncontrolled-rate freezing with the cryovials containing ovarian tissue were placed in a freezer at -20ºC for 30 min followed by plunging in LN vapor for 30 min and then stored in LN. The results of the experiment showed similar density of primordial follicles of the 5x1x1 mm group as compared to the fresh tissue, while lower number of primordial follicles was observed in the small (1x1 mm) frozen-thawed pieces. It was not further discussed why the larger pieces were more resistant to cryoinjury.

In one study, using porcine ovarian cortex samples, programmed slow-freezing was performed with four different CPAs (glycerol (GLY)-10%; DMSO-1.5 M, EG-1.5 M, PROH-1.5 M) (Borges et al., 2009). The ovarian cortex pieces were incubated after thawing, and histological analysis, including light microscopy and TEM, demonstrated that the follicular viability was decreased after freezing with better results obtained by DMSO and EG as compared to PROH and GLY. This result correlates to that pregnancies in the human so far only have been demonstrated after ovarian cryopreservation in either DMSO or EG as CPAs (Donnez et al., 2004; Andersen et al., 2008). However, it should be emphasized that species differences exist in this regard, as demonstrated in a comparative study between ovarian cortex of human, bovine and porcine ovarian tissue (Gandolfi et al., 2006). In that study, Gandolfi and colleagues study showed that DMSO and PROH were equally effective to protect primordial and primary follicles of the pig ovary to cryoinjury and the pig ovarian tissue was also more resistant to cryoinjury as compared to the bovine and the human ovarian tissue.

The pig model was also used to study vitrification procedures (Gandolfi et al., 2006; Moniruzzaman et al., 2009). In one study, ovarian strips from 15-day old pigs, were vitrified using 15% EG, 15% DMSO and 20% fetal calf serum with addition of either 0 M, 0.25 M or 0.5 M sucrose (Moniruzzaman et al., 2009). Histological evaluation after warming showed higher percentage of healthy primordial follicles when CPA solution with 0.25 M sucrose was applied as compared to 0 M and 0.5 M solution. Moreover, the higher oocyte shrinkage was observed with the sucrose supplementation of 0.5 M and these results can be explained by unsatisfactory cell dehydration without sucrose as well as to excessive dehydration at 0.5 M sucrose. In one study, xenografting of vitrified ovarian tissues under the kidney capsule of nude mice was performed (Moniruzzaman et al., 2009). Histological evaluation, two months after grafting, revealed decrease of the primordial follicle density by around 20% in both fresh and vitrified grafts, but in the vitrified grafts, the follicles did not developed beyond the secondary stage. Hence, it may be that follicular development is disturbed after vitrification or that a follicular stage dependant developmental blockage is incurred by cryoinjury. Another study using pig ovarian cortex demonstrated low survival rate of primordial follicles with either EG or a combination of EG and DMSO as CPA (Gandolfi et al., 2006)

The pig ovary has been used in one study of whole ovary cryopreservation (Imhof et al., 2004). The ovary was perfused through the ovarian artery with 1.5 M DMSO and cryopreserved with a slow freezing protocol. Light microscopy evaluation after thawing at 25ºC demonstrated a lower proportion of viable primordial follicles in the frozen-thawed ovaries (84%) as compared to fresh controls (98%). Furthermore, about 20% of healthy primordial follicles were seen when the ovaries were positioned directly into LN without previously perfusion with CPA. Noteworthy is that TEM did not demonstrate any major cellular difference between the fresh and frozen-thawed tissue.

4. Sheep

A variety of in vivo and in vitro sheep models have been used in research on both ovarian cortex cryopreservation and whole ovary cryopreservation, most probably due to the large knowledge about the physiology of the sheep ovary. In addition, the sheep ovary has, similarly to the human ovary, a collagen-dense outer stroma containing the pool of primordial follicles (Arav et al., 2005). Nevertheless, the size of the sheep ovary is only around 20% of the human premenopausal ovary (Munn et al., 1986; Gerritse et al., 2008) and this fact has to be taken into consideration, particularly in comparative studies concerning whole ovary cryopreservation.

The pioneering study in ovarian cryopreservation was published in 1994 by the Edinburgh group (Gosden et al., 1994), where live births (Table 2), after ovarian cortex cryopreservation and transplantation in a large experimental animal, were reported for the first time. In fact, this study opened up the field for future clinical fertility preservation and this is also accentuated in its conclusion by stating "that frozen storage and replacement of a patient´s own ovarian tissue might be practicable when fertility potential is threatened by chemotherapy/radiotherapy". The cryopreservation protocol applied in that study has subsequently been widely used in research and clinical practice, often with minor modifications. It should be emphasized that this slow freezing-thawing protocol was adapted from a study on cryopreservation of mouse primordial follicles (Carroll&Gosden, 1993) and the authors also wrote that they "have no evidence whether it is optimal". In the Gosden study, ovarian cortex pieces were positioned in cryovials with Leibowitz L-15 solution containing donor serum and 1.5 M DMSO and the cryovials were held on ice for 15 min. The ovarian cortex strips were then considered equilibrated in the CPA and cooled in a programmable freezer at a rate of 2°C/min to -7°C and maintained at -7°C for 10 min before seeding. Freezing was performed further by reducing the temperature at a rate by 0.3°C/min to -40°C following by 10°C/min until -140°C before plunging into LN. Thawing was performed by exposure of frozen tissue to air temperature for 2 min and then placed in a water bath at room temperature for an unspecified time. The frozen-thawed slices were attached adjacent to the left ovarian pedicle at the second laparoscopy of the sheep. High progesterone level four months later indicated reestablished normal cyclicity. Eight months after grafting, one sheep gave birth and one more lamb was delivered by Cesarean section at gestational age of about 144 days.

In a follow up study the authors reported that the longevity of the transplants was at least 22 months, but at that time very few primordial follicles were found in the graft (Baird et al., 1999). The levels of follicle-stimulating hormone (FSH) and luteinizing hormone (LH) in animals which had received grafts were higher as compared to controls. However, all grafted animals demonstrated normal cyclical pattern. These findings were corroborated by the lower levels of inhibin, indicating reduced number of inhibin-producing small antral follicles. In the second part of the study Baird and colleagues grafted both fresh and frozen-thawed ovarian pieces under the kidney capsule of nude mice to assess whether the freezing-thawing procedure per se or if the warm ischemic time post transplantation was the main reason for loss of the follicle pool already within two years. A follicular depletion of about 65% was seen after grafting of fresh tissue as compared to fresh non-grafted ovarian tissue. Additionally 7% was lost after transplantation of frozen-thawed pieces. Thus, this study demonstrated that ischemia rather than cryopreservation is the main cause of follicle depletion.

Reference	Ovarian tissue	Cryopreservation	Cryoprotectant	Transplantation	Results
Gosden et al. 1994.	Cortical pieces	Slow freezing	1.5 M DMSO in Leibovitz L-15 with 10% donor calf serum	Avascular, orthotopic	2 live births out of 6 transplants
Salle et al. 2002, Salle et al. 2003.	Hemi-ovarian cortex	Slow freezing	2 M DMSO in BM1 medium with 10% fetal cord serum	Avascular, orthotopic	11 live births (of which 3 sheep delivered twins) out of 6 transplants; 2 sheep delivered for the second time
Bordes et al. 2005.	Hemi-ovarian cortex	Vitrification	2.62 M DMSO +2.6 M acetamide+1.3M propylene glycol+0.0075 M polyethylene glycol in BM1 medium	Avascular, orthotopic	4 live births (1 twins) out of 6 transplants
Imhof et al. 2006.	Whole ovary	Slow freezing	1.5 M DMSO in RPMI 1640 with 10% autologous sheep blood serum	Vascular, orthotopic	1 live birth out of 8 transplants

DMSO- dimethylsulphoxide; RPMI-Roswell Park Memorial Institute medium.

Table 2. Live-births after transplantation of cryopreserved ovarian tissue in large animal models

Depletion of the major follicular pool during the ischemic post grafting period was also demonstrated in a recent study in the sheep, where fresh and frozen ovarian cortical tissue were grafted either subcutaneously in the anterior abdominal wall or the uterine horn (Aubard et al., 1999). The cryopreservation procedure was a slight modification of the original Gosden method (Gosden et al., 1994). About 5% of primordial follicles survived grafting at evaluation after seven months. No pregnancy was demonstrated, but mature oocytes were retrieved after gonadotropin stimulation. In addition, poor fertilization rate and cleavage arrest at 4 cells were recorded in oocytes from both fresh and frozen ovarian tissue. These results open up the issue of about cytoplasmatic maturation as well as quality of the oocytes from grafted and frozen-thawed tissue.

In other studies, a larger part of sheep ovaries, containing also parts of the medulla, were evaluated regarding viability after cryopreservation and avascular transplantation (Salle et al., 1998). In that study, the ovarian tissue was cryopreserved using the slow freezing technique published by Gosden (Gosden et al., 1994). Transplantation was performed to the ovarian hilus. At evaluation after six months, well preserved morphology with follicles of all stages was demonstrated (Salle et al., 1998) and progesterone secretion was reestablished (Salle et al., 1999). The same research group reported live births (Table 2) after cryopreservation and avascular transplantation of hemi-ovarian cortex (Salle et al., 2002). However, the cryoprotocol had been slightly modified using higher concentration (2 M) of DMSO and the temperature was decreased by a rate of 2°C/min to -35°C followed by 25°C/min to -140°C. Semiautomatic seeding was initiated at -11°C. Orthotopic transplantation was performed. Four out of six ewes achieved pregnancy of which two delivered twins. The remaining four ewes were observed further for two years (Salle et al., 2003). All animals became pregnant (Table 2) and delivered lambs of which two sheep gave birth for a second time. Noteworthy is that considerable loss of follicles was seen in all grafts. The results of this study also emphasizes that the avascular transplantation per se induces major damage to the tissue.

In one study, vitrification of sheep hemi-ovaries was performed with the VS1 CPA solution, containing 2.62 M DMSO, 2.60 M acetamide, 1.31 M PROH and 0.0075 M polyethylene glycol in BM1 medium (Bordes et al., 2005). The ovarian tissue was equilibrated in increasing concentration of VS1 (12.5%, 25%, 50% and 100%) and then placed in cryovials containing 100% of VS1 followed by direct plunging into LN. The vitrified tissue was warmed in water at 37°C for 10 min. Notably, regained cyclicity was shown four months to one year after grafting and 3/6 sheep delivered offspring (Table 2). However, histological evaluation of the graft after delivery revealed very few follicles (6-58 follicles per graft).

A group from the Cleveland Clinic performed microvascular fresh ovary transplantation in the sheep (Jeremias et al., 2002) and soon after published their results after autotransplantation of the whole cryopreserved ovary applying the identical surgical technique (Bedaiwy et al., 2003) with comparison of the outcome to avascular grafting of frozen-thawed cortical pieces (frozen with Gosdens protocol). The whole ovaries were flushed by a solution consisting of Leibowitz L-15 medium, 10% fetal calf serum and 1.5 M DMSO and then allocated to controlled-rate freezing. The temperature was reduced at 2°C/min until seeding temperature of -7°C. Further reduction of temperature was done at 2°C /min until -35°C followed at by 25°C /min until -140°C when cryovials were positioned into LN. At thawing, the cryovials were placed in a water bath at 37°C. After that, the ovaries were flushed with Leibovitz L15 medium containing 10% fetal calf serum for 20 min followed by end-to-end anastomosis between the ovarian and the inferior epigastric vessels. Immediate vascular patency was showed in all grafts, but only 27% of the whole ovary transplants showed patent blood vessels eight to ten days after transplantation. Histological evaluation demonstrated large necrotic areas in the grafts of the non-patent anastomosis group and the primordial follicle pool was severely reduced in that group as compared to the patent group. Also, severe injuries with focal transmural necrosis of intraovarian vessels may be regarded as a cryoinjury of the vasculature. The follicular viability, evaluated by Trypan blue test, was around 80%, in the cortical avascular transplantation and the patent whole ovary transplantation group, with a slightly higher rate of apoptotic cells in the whole ovary group. It should be emphasized that this study evaluated only short-term (8-10 days) results.

The long-term outcome, using the same microsurgical technique as described above, was assessed in a study of ovarian viability approximately five months after transplantation of frozen-thawed ovaries (Grazul-Bilska et al., 2008). A similar cryoprotocol was applied as in the earlier experiments (Bedaiwy et al., 2003) and the only difference was that the temperature from the seeding point was decreased by the rate 0.3°C/min until -40°C in contrast to 2°C/min until -35°C in the previous study. Histological evaluation revealed normal follicular development in only 25% of the transplanted ovaries. Furthermore, oocytes (n=3) from the larger follicles could be matured, but fertilization was not achieved. The vessels of the patent grafts appeared normal, with expression of marker proteins such as factor VIII, vascular endothelial growth factor (VEGF) and smooth muscle cell actin (SMCA). In view of the fact that the normal follicular development was seen in only 25% of the transplanted ovaries, and that mature oocytes did not fertilize, the study points to that major improvements in the fields of whole ovary cryopreservation and retransplantation are needed.

Main advances in the whole ovary cryopreservation field may be to optimize the freezing technique and also to improve the anastomosis technique as evaluated by an Israeli research group (Revel et al., 2004). After perfusion via the ovarian artery with 1.4 M DMSO in University of Wisconsin (UW) solution for 3 min, the sheep ovary was cryopreserved by the directional freezing technology. This technique provides identical cooling rate through the whole organ and allows constant cooling rate. The temperature was decreased by a rate of 0.6°C/min until seeding and 0.3°C/min to the temperature of -30°C before placement in LN. Rapid thawing was accomplished by placement of the cryovials into a water bath at 68°C for 20 s and 37°C for 2 min. The ovary was transplanted by microvascular anastomosis to an orthotopic site by end-to-end anastomosis to the remaining parts of the ovarian vessels. Cyclicity was demonstrated in three out of eight animals at around six months after the procedure. At laparotomy eight weeks after surgery, adhesions were seen in only one animal. In a follow up study, presenting long-term outcome of this technique (Arav et al., 2005), three animals demonstrated cyclicity two to three years after transplantation. Oocytes obtained from these animals could be parthenogenically activated with divisions until the 8-cell stage. In an extensive long-term follow up study (6 years) of the three whole ovary transplanted sheep (Arav et al., 2010) two of the ovaries responded to gonadotropin stimulation and these ovaries were normal at post mortem histological evaluation. The third sheep did not respond to FSH stimulation and histology revealed a fibrotic ovary and absence of follicles. The importance of this study is that cryopreserved whole ovary can survive for a long time and indicates beneficial effect of directional slow-freezing method. Nevertheless, it should be pointed out that the ultimate end point of healthy offspring has not been demonstrated by the use of this cryopreservation technique.

There is only one report on live-birth after whole ovary cryopreservation and vascular transplantation in sheep (Imhof et al., 2006). In that study, the ovaries were cryopreserved using the protocol of Gosden and coworkers (Gosden et al., 1994), but naturally the ovaries were cannulated and perfused with CPA before controlled-rate slow-freezing. Thawing was done by exposure of the frozen ovaries in air for two minutes followed by placement in a water bath at 25°C for seven minutes and perfusion by Roswell Park Memorial Institute medium (RPMI) to remove CPA. Transplantation was performed after removal of the contralateral ovary and the frozen-thawed ovary was orthotopicaly transplanted using microvascular (9-0 sutures) anastomosis to the ovarian vessels which is a comparable technique to that described by the Israeli group (Arav et al., 2005). Cyclicity was reported in four out of nine transplanted sheep. Importantly, one pregnancy occurred in this resulted in delivery of a healthy lamb around 1.5 years after grafting. At histological examination of ovaries 18-19 months after transplantation, the size of the primordial follicle pool was only 2-8% of that in non transplanted control ovaries. The authors discussed that the major follicular loss occurred during the freezing-thawing procedure. However, this suggestion relied on results of histological assessment, which is probably an unreliable method to evaluate viability and should be combined with other methods. Nevertheless, this single large animal species live-birth after whole ovary cryopreservation is a proof of the concept, which should encourage further research in this area.

It is of considerable importance to recognize the mechanisms behind the low success rate of whole ovary cryopreservation and also to understand what cell compartments are affected by the cryopreservation and thawing procedures. It seems that the follicular survival and the ovarian function are directly correlated to the vascular patency, as demonstrated in one elegant study of heterotopic autotransplantation of the frozen-thawed whole ovary (Onions et al., 2009). The CPA and freezing protocols were similar to the traditional Gosden protocol (Gosden et al., 1994) and the microvascular anastomosis was by aortic patch and utero-ovarian vein to carotid artery and jugular vein, as developed more than 40 years ago (Goding et al., 1967). The control group was the animals which received non-frozen heterotopic transplants. Eight months after transplantation, 7/8 cryopreserved transplants and 3/4 fresh ovarian transplants demonstrated patency. However, regardless of vascular patency, 5/7 frozen-thawed ovaries with vascular patency did not regain cyclicity and eight months after transplantation, a follicular loss of about 90% was seen in both the fresh and frozen group. A possible damage of endothelial cells during cannulation was discussed as one detrimental factor.

In vitro studies of vitrification of the whole sheep ovary preceded the trials in vivo. It was demonstrated in an elaborate study that the whole sheep ovary could be vitrified and that the VS4 cryoprotectant solution (mixture of 2.75 M DMSO, 2.76 M formamide and 1.97 M PROH) was superior to the VS1 solution (mixture of 2.62 M DMSO, 2.60 M acetamide, 1.31 M PROH and 0.0075 M polyethylene-glycol) (Courbiere et al., 2005). After thawing, the higher primordial follicular density (50% vs 23%) as well as proportion of histologically normal primordial follicles (53% vs 25%) was demonstrated in theVS4 group. However, it should be underlined that endothelial damage of the vascular pedicle was more evident in the VS4 group.

In a subsequent study by the same research group, the thermodynamic properties of VS4 in RPS-1 medium were studied (Courbiere et al., 2006). Evaluation of the cooling rate was done by differential scanning calorimeter by connection of the thermocouples to the ovarian cortex, the medulla and the CPA solution. The rate of cooling was above 300°C/min, with the measured cooling rate of the cortex being slightly higher than that of the medulla. Furthermore, the cooling rate of the CPA solution was higher in comparison to the ovarian medulla and cortex. This finding may be explained by differences in tissue architecture and vascularity of the ovary that leads to uneven distribution of the CPA. Results of this study also pointed out that it is not likely that the ovarian tissue can be completely vitrified at the end of the procedure. Ice crystallization during warming was also observed, indicating that the warming rate did not exceed the critical warming rate. Nevertheless, in contrast to the former study (Courbiere et al., 2005), injury of the endothelial layer of the ovarian vasculature was not demonstrated which possibly may be the result of the two-step-warming procedure to avoid ice crystallization (Pegg et al., 1997) as used in this study (Courbiere et al., 2006).

The same research group demonstrated in a subsequent study that the warming rate of the cortex was slightly higher than the medulla (Baudot et al., 2007). This corroborates the irregular distribution of CPA as well as complexity of the ovarian tissue. As in the previous study (Courbiere et al., 2005), a primordial follicle survival rate of about 50% was demonstrated after warming. In addition, ice crystals were observed during the cooling and the authors discussed that maybe limited ice crystallization can be acceptable for clinical application of this procedure.

Subsequent to the in vitro research on whole ovary vitrification, as described above (Courbiere et al., 2005; Courbiere et al., 2006), the efficiency of this method in vivo was assessed. In one study only one out of five vitrified-warmed ovaries resumed endocrine function (Courbiere et al., 2009) after orthotopic vascular transplantation. Vitrification was obtained using VS4 solution (2.75 M DMSO, 2.76 M formamide and 1.97 M PROH) in RPS-1 medium and the ovaries were perfused by gradually increasing concentrations of CPA. At warming, the ovaries were kept in LN vapor following by placement in a water bath at 45°C. The rationale behind this two-step warming was to attempt to avoid fractures of the vessels during the glassy state (Pegg et al., 1997). A total follicular loss in the vitrified group was demonstrated one year after transplantation. Vascular thrombosis occurred in three out of four vitrified ovaries, with patent vessels seen in the fresh group. One possible explanation may be that the warm ischemic time was longer in the vitrified group (median 287 min) in comparison to the control group (median 129 min) although it is more likely that major injuries occurred during the cryopreservation procedures.

It is obvious from the results presented above that there is a need for more systematic studies on the different stages of cryopreservation/transplantation procedures to get better results. As it relates to whole ovary cryopreservation, the viability and ovarian function should be evaluated after each stage of the procedure.

The effect of diverse cryoprotocols with accent of CPA toxicity was studied using the sheep hemiovary model (Demirci et al., 2001). The hemiovaries were incubated for 10 minutes in various concentrations (2-10 M) of DMSO and PROH and evaluated before and after slow-freezing-thawing (1, 1.5 and 2 M CPAs) concerning primordial follicle survival (Trypan blue test and histology). The follicular survival after incubation was higher at decreasing concentration of the CPA regardless of the type of CPA.

In vitro ovarian perfusion methodology, which was initially developed for evaluation of ovarian physiology (Brannstrom et al., 1987), may be used for evaluation of frozen-thawed ovaries (Fig. 1). The in vitro perfusion system highly mimics the physiological in vivo situation, and has shown that a complex processes, such as ovulation, occur during in vitro perfusion (Lofman et al., 1989). Recently, in vitro ovarian perfusion was used together with live-dead assay, histology and cell culture to assess the viability of frozen-thawed whole sheep ovaries (Wallin et al., 2009). The ovaries were frozen by slow uncontrolled-rate freezing (Martinez-Madrid et al., 2004) using PROH, stored in LN and then thawed in a water bath at 37°C. The in vitro perfusion results demonstrated compromised ovarian function of the ovary after cryopreservation in PROH, when compared to fresh controls.

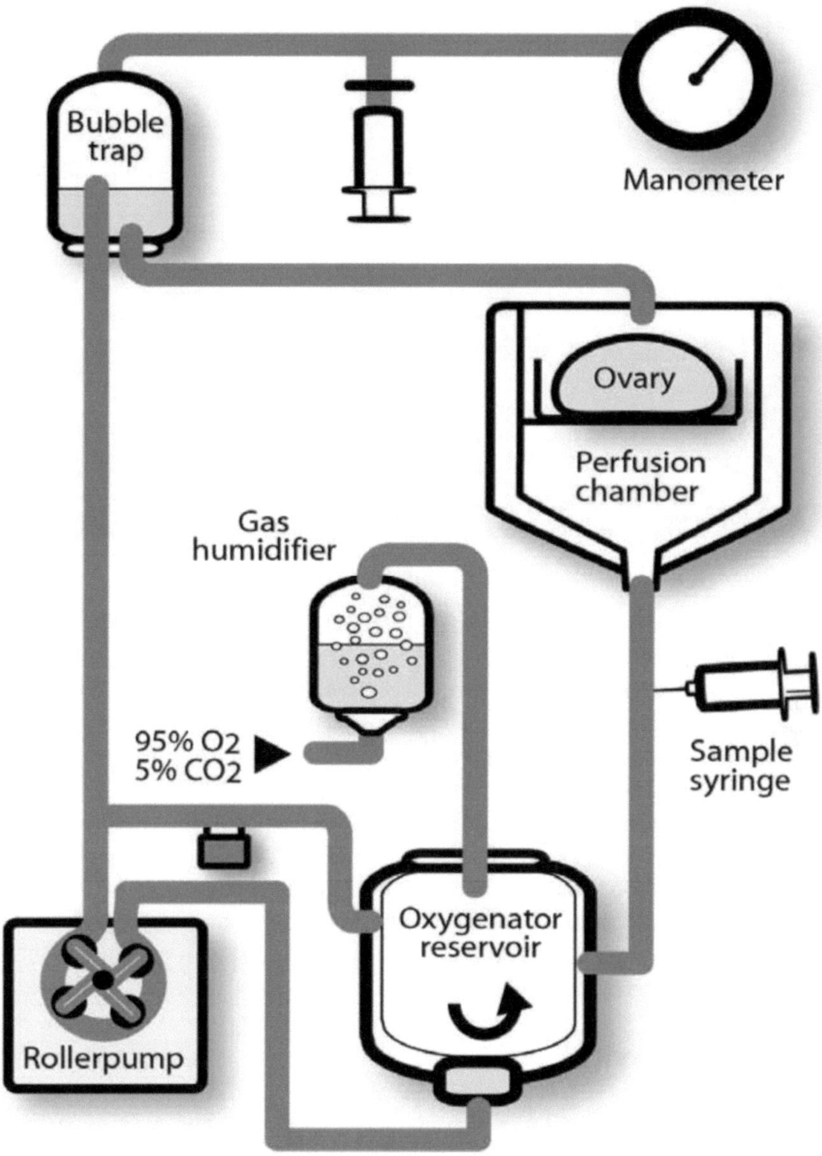

Fig. 1. Schematic drawing of the in vitro perfusion system

In another study of the same research group, the sheep ovaries were cryopreserved using the same uncontrolled-rate slow freezing with DMSO (Milenkovic et al., 2011). The ovaries in control group were frozen without CPA. Interestingly, steroid production after in vitro perfusion (Fig. 2) and cell culture was also demonstrated in control group which can be explained by adequate equilibration between cell dehydration and extracellular ice formation. However, a clear benefit of DMSO presence was seen.

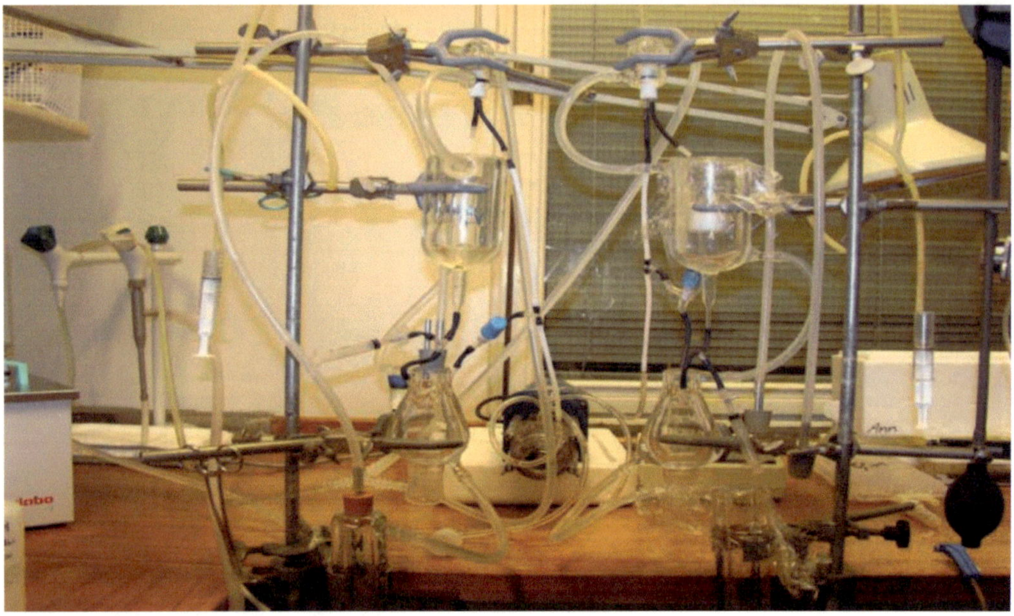

Fig. 2. Photograph of the in vitro perfusion apparatus

As discussed above, cryopreservation injury may not only occur within the follicles but also in the vascular bed of the ovary or in the large vessels. It was hypothesized that perfusion with the anti-apoptopic agent sphingosine-1-phosphate (S-1-P) before cryopreservation-perfusion by 1.5 M DMSO (Onions et al., 2008) may protect the tissue and diminish the cryoinjury. Histological examination after thawing showed arterial endothelial disruption of the vascular pedicle tissue in the cryopreserved ovaries, with the most extensive injury in the area where the cannula had been placed, followed by the hilus region and with less extensive injury on the venous side. No protective effect of the addition of sphingosine-1-phosphate could be demonstrated. In summary, this study was able to demonstrate the vulnerable state of the vasculature in whole ovary cryopreservation especially on the arterial side. It was also shown by proliferation and apoptosis markers that granulosa cells of antral follicles remain viable after cryopreservation.

5. Non-human primates

A number of non-human primate species have been used in research involving reproductive physiology and development of methods later used clinically in reproductive medicine. The benefit of these experimental animals is that the physiology and anatomy are similar to the human (Stevens, 1997; Weinbauer et al., 2008), although the ovarian size of the most studied cynomolgus monkey is considerably smaller (0.258+/- 0.159 cm^3) (Jones et al., 2010) as compared to the human. Even if a lot of procedures are introduced in the human without appropriate tests in non-human primate species, it is advisable to include these animal models in preclinical research.

There is one study on non-human primate that has assessed the function of cryopreserved ovarian cortex tissue in vivo. The ovarian pieces were cryopreserved by controlled-rate freezing with 1.5 M DMSO and both fresh and frozen-thawed strips were grafted subcutaneously to the upper arm of cynomolgus macaque monkeys (Schnorr et al., 2002). Regained menstrual cycle and ovarian steroidogenesis were demonstrated in 80% of the animals after fresh ovarian cortex transplantation and in 50% of the animals after cryopreserved transplants, demonstrating that the cryopreservation technique may have damaging effect on the tissue. Additionally, only one mature oocyte was aspirated from subcutaneous ovarian tissue after gonadotropin stimulation performed in four animals. In the other part of this study (Schnorr et al., 2002) an attempt to enhance angiogenesis was performed by local administration of VEGF for two weeks after grafting, but no beneficial effect could be demonstrated.

There is one study on ovarian cortex from macaques when vitrification (3.4 M GLY, 4.5 M EG) and slow freezing (1.5 M EG) were compared and evaluated whether post-thawing co-culture on feeder cells (mouse fetal fibroblast monolayer) with addition of follicle stimulating hormone, insulin, transferrin and selenium, would increase viability (Yeoman et al., 2005). The post-thaw viability, as assessed by live-dead fluorescent staining, was comparable (around 70%) in the two groups, which was only marginally lower than the follicular viability of the fresh tissue (76%). Interestingly, follicular viability was increased after post-thawing co-culture, indicating rescue of partly damaged follicles and possible beneficial effect of co-culture.

In another study on ovarian cortex tissue of rhesus macaques, controlled-rate slow-freezing with 1.5 M EG was done and expressions of activin subunits as well as the phosphorylated form of the signalling protein were investigated (Jin et al., 2010). Activin subunits and the phosphorylated form of the signalling protein are markers for early follicular development. Immunohistochemistry revealed that these proteins were regularly distributed in primordial and primary follicles of both cryopreserved and fresh cortex. One interesting finding was that a higher rate of post thawing apoptosis was found in the stromal cells as compared to oocytes and granulosa cells. Another recently published study was able to demonstrate follicular development after in vitro three dimensional culture of cryopreserved secondary follicles from rhesus macaque (Ting et al., 2011). The ovarian cortex from rhesus macaque was cryopreserved by slow freezing (1.5 M EG) and vitrification (3 M GLY and 4.5 M EG). After thawing, the secondary follicles were mechanically isolated, encapsulated in 0.25% alginate and cultured for five weeks in MEM at 37°C. The development of antral follicles, although delayed and functionally compromised as compared to fresh tissue was demonstrated after both cryopreservation methods with slightly better results obtained after vitrification in comparison to slow freezing protocol. This observation again supports the need for further improvement of cryopreservation techniques.

6. Conclusions

Regardless pioneering success and reporting pregnancies after cryopreservation and avascular transplantation of ovarian tissue in human (Donnez et al., 2011) the future course regarding cryopreservation of ovarian tissue should be focused on further research using animal models. The size of ovarian cortex pieces, whole ovary freezing or not, choice and exposure time to CPA, stepwise or direct addition of CPA are some of the unresolved questions. The difference between species, complexity of ovarian tissue and uneven distribution of primordial follicles should be taken into consideration when creating an optimal study design. Vitrification seems promising, but differences between the protocols used make the interpretation of the data difficult. However, further research on animal models should lead to better understanding and improvement of cryopreservation techniques and then to higher efficiency, when used as a clinical procedure.

7. References

Andersen, C. Y., M. Rosendahl, A. G. Byskov, A. Loft, C. Ottosen, M. Dueholm, K. L. Schmidt, A. N. Andersen&E. Ernst. (2008). Two successful pregnancies following autotransplantation of frozen/thawed ovarian tissue. *Hum Reprod*, 23, 2266-2272, 1460-2350 (Electronic)

Arav, A., Z. Gavish, A. Elami, Y. Natan, A. Revel, S. Silber, R. G. Gosden&P. Patrizio. (2010). Ovarian function 6 years after cryopreservation and transplantation of whole sheep ovaries. *Reprod Biomed Online*, 20, 48-52, 1472-6491 (Electronic) 1472-6483 (Linking)

Arav, A., A. Revel, Y. Nathan, A. Bor, H. Gacitua, S. Yavin, Z. Gavish, M. Uri&A. Elami. (2005). Oocyte recovery, embryo development and ovarian function after cryopreservation and transplantation of whole sheep ovary. *Hum Reprod*, 20, 3554-3559, 0268-1161 (Print)

Aubard, Y., P. Piver, Y. Cogni, V. Fermeaux, N. Poulin&M. A. Driancourt. (1999). Orthotopic and heterotopic autografts of frozen-thawed ovarian cortex in sheep. *Hum Reprod*, 14, 2149-2154, 0268-1161 (Print) 0268-1161 (Linking)

Baird, D. T., R. Webb, B. K. Campbell, L. M. Harkness&R. G. Gosden. (1999). Long-term ovarian function in sheep after ovariectomy and transplantation of autografts stored at -196 C. *Endocrinology*, 140, 462-471, 0013-7227 (Print)

Baudot, A., B. Courbiere, V. Odagescu, B. Salle, C. Mazoyer, J. Massardier&J. Lornage. (2007). Towards whole sheep ovary cryopreservation. *Cryobiology*, 55, 236-248, 1090-2392 (Electronic)

Bedaiwy, M. A., E. Jeremias, R. Gurunluoglu, M. R. Hussein, M. Siemianow, C. Biscotti&T. Falcone. (2003). Restoration of ovarian function after autotransplantation of intact frozen-thawed sheep ovaries with microvascular anastomosis. *Fertil Steril*, 79, 594-602, 0015-0282 (Print)

Bordes, A., J. Lornage, B. Demirci, M. Franck, B. Courbiere, J. F. Guerin&B. Salle. (2005). Normal gestations and live births after orthotopic autograft of vitrified-warmed hemi-ovaries into ewes. *Hum Reprod*, 20, 2745-2748, 0268-1161 (Print)

Borges, E. N., R. C. Silva, D. O. Futino, C. M. Rocha-Junior, C. A. Amorim, S. N. Bao&C. M. Lucci. (2009). Cryopreservation of swine ovarian tissue: effect of different cryoprotectants on the structural preservation of preantral follicle oocytes. *Cryobiology*, 59, 195-200, 1090-2392 (Electronic) 0011-2240 (Linking)

Brannstrom, M., B. M. Johansson, J. Sogn&P. O. Janson. (1987). Characterization of an in vitro perfused rat ovary model: ovulation rate, oocyte maturation, steroidogenesis and influence of PMSG priming. *Acta Physiol Scand*, 130, 107-114, 0001-6772 (Print)

Carroll, J.&R. G. Gosden. (1993). Transplantation of frozen-thawed mouse primordial follicles. *Hum Reprod*, 8, 1163-1167, 0268-1161 (Print) 0268-1161 (Linking)

Celestino, J. J., R. R. Dos Santos, C. A. Lopes, F. S. Martins, M. H. Matos, M. A. Melo, S. N. Bao, A. P. Rodrigues, J. R. Silva&J. R. De Figueiredo. (2008). Preservation of bovine preantral follicle viability and ultra-structure after cooling and freezing of ovarian tissue. *Anim Reprod Sci*, 108, 309-318, 1873-2232 (Electronic) 0378-4320 (Linking)

Courbiere, B., L. Caquant, C. Mazoyer, M. Franck, J. Lornage&B. Salle. (2009). Difficulties improving ovarian functional recovery by microvascular transplantation and whole ovary vitrification. *Fertil Steril*, 91, 2697-2706, 1556-5653 (Electronic) 0015-0282 (Linking)

Courbiere, B., J. Massardier, B. Salle, C. Mazoyer, J. F. Guerin&J. Lornage. (2005). Follicular viability and histological assessment after cryopreservation of whole sheep ovaries with vascular pedicle by vitrification. *Fertil Steril*, 84 Suppl 2, 1065-1071, 1556-5653 (Electronic)

Courbiere, B., V. Odagescu, A. Baudot, J. Massardier, C. Mazoyer, B. Salle&J. Lornage. (2006). Cryopreservation of the ovary by vitrification as an alternative to slow-cooling protocols. *Fertil Steril*, 86, 1243-1251, 1556-5653 (Electronic)

Demirci, B., J. Lornage, B. Salle, L. Frappart, M. Franck&J. F. Guerin. (2001). Follicular viability and morphology of sheep ovaries after exposure to cryoprotectant and cryopreservation with different freezing protocols. *Fertil Steril*, 75, 754-762, 0015-0282 (Print) 0015-0282 (Linking)

Donnez, J., M. M. Dolmans, D. Demylle, P. Jadoul, C. Pirard, J. Squifflet, B. Martinez-Madrid&A. Van Langendonckt. (2004). Livebirth after orthotopic transplantation of cryopreserved ovarian tissue. *Lancet*, 364, 1405-1410, 1474-547X (Electronic)

Donnez, J., S. Silber, C. Y. Andersen, I. Demeestere, P. Piver, D. Meirow, A. Pellicer&M. M. Dolmans. (2011). Children born after autotransplantation of cryopreserved ovarian tissue. a review of 13 live births. *Ann Med*, 43, 437-450, 1365-2060 (Electronic) 0785-3890 (Linking)

Dorsch, M., D. Wedekind, K. Kamino&H. J. Hedrich. (2004). Orthotopic transplantation of rat ovaries as a tool for strain rescue. *Lab Anim*, 38, 307-312, 0023-6772 (Print) 0023-6772 (Linking)

Ferreira, M., A. Bos-Mikich, N. Frantz, J. L. Rodrigues, A. L. Brunetto&G. Schwartsmann. (2010). The effects of sample size on the outcome of ovarian tissue cryopreservation. *Reprod Domest Anim*, 45, 99-102, 1439-0531 (Electronic) 0936-6768 (Linking)

Gandolfi, F., A. Paffoni, E. Papasso Brambilla, S. Bonetti, T. A. Brevini&G. Ragni. (2006). Efficiency of equilibrium cooling and vitrification procedures for the cryopreservation of ovarian tissue: comparative analysis between human and animal models. *Fertil Steril*, 85 Suppl 1, 1150-1156, 1556-5653 (Electronic)

Gerritse, R., C. C. Beerendonk, M. S. Tijink, A. Heetkamp, J. A. Kremer, D. D. Braat&J. R. Westphal. (2008). Optimal perfusion of an intact ovary as a prerequisite for successful ovarian cryopreservation. *Hum Reprod*, 23, 329-335, 1460-2350 (Electronic)

Goding, J. R., J. A. Mccracken&D. T. Baird. (1967). The study of ovarian function in the ewe by means of a vascular autotransplantation technique. *J Endocrinol*, 39, 37-52, 0022-0795 (Print)

Gook, D. A., D. H. Edgar&C. Stern. (1999). Effect of cooling rate and dehydration regimen on the histological appearance of human ovarian cortex following cryopreservation in 1, 2-propanediol. *Hum Reprod*, 14, 2061-2068, 0268-1161 (Print)

Gosden, R. G., D. T. Baird, J. C. Wade&R. Webb. (1994). Restoration of fertility to oophorectomized sheep by ovarian autografts stored at -196 degrees C. *Hum Reprod*, 9, 597-603, 0268-1161 (Print)

Grazul-Bilska, A. T., J. Banerjee, I. Yazici, E. Borowczyk, J. J. Bilski, R. K. Sharma, M. Siemionov&T. Falcone. (2008). Morphology and function of cryopreserved whole ovine ovaries after heterotopic autotransplantation. *Reprod Biol Endocrinol*, 6, 16, 1477-7827 (Electronic)

Hreinsson, J., P. Zhang, M. L. Swahn, K. Hultenby&O. Hovatta. (2003). Cryopreservation of follicles in human ovarian cortical tissue. Comparison of serum and human serum albumin in the cryoprotectant solutions. *Hum Reprod*, 18, 2420-2428, 0268-1161 (Print)

Imhof, M., H. Bergmeister, M. Lipovac, M. Rudas, G. Hofstetter&J. Huber. (2006). Orthotopic microvascular reanastomosis of whole cryopreserved ovine ovaries resulting in pregnancy and live birth. *Fertil Steril*, 85 Suppl 1, 1208-1215, 1556-5653 (Electronic)

Imhof, M., G. Hofstetter, H. Bergmeister, M. Rudas, R. Kain, M. Lipovac&J. Huber. (2004). Cryopreservation of a whole ovary as a strategy for restoring ovarian function. *J Assist Reprod Genet*, 21, 459-465, 1058-0468 (Print)

Irving-Rodgers, H. F., K. D. Catanzariti, W. J. Aspden, M. J. D'occhio&R. J. Rodgers. (2006). Remodeling of extracellular matrix at ovulation of the bovine ovarian follicle. *Mol Reprod Dev*, 73, 1292-1302, 1040-452X (Print) 1040-452X (Linking)

Jeremias, E., M. A. Bedaiwy, R. Gurunluoglu, C. V. Biscotti, M. Siemionow&T. Falcone. (2002). Heterotopic autotransplantation of the ovary with microvascular anastomosis: a novel surgical technique. *Fertil Steril*, 77, 1278-1282, 0015-0282 (Print)

Jeremias, E., M. A. Bedaiwy, D. Nelson, C. V. Biscotti&T. Falcone. (2003). Assessment of tissue injury in cryopreserved ovarian tissue. *Fertil Steril*, 79, 651-653, 0015-0282 (Print)

Jin, S., L. Lei, L. D. Shea, M. B. Zelinski, R. L. Stouffer&T. K. Woodruff. (2010). Markers of growth and development in primate primordial follicles are preserved after slow cryopreservation. *Fertil Steril*, 93, 2627-2632, 1556-5653 (Electronic) 0015-0282 (Linking)

Jones, J. C., S. E. Appt, S. R. Werre, J. C. Tan&J. R. Kaplan. (2010). Validation of multi-detector computed tomography as a non-invasive method for measuring ovarian volume in macaques (Macaca fascicularis). *Am J Primatol*, 72, 530-538, 1098-2345 (Electronic) 0275-2565 (Linking)

Kagawa, N., M. Kuwayama, K. Nakata, G. Vajta, S. Silber, N. Manabe&O. Kato. (2007). Production of the first offspring from oocytes derived from fresh and cryopreserved pre-antral follicles of adult mice. *Reprod Biomed Online*, 14, 693-699, 1472-6483 (Print) 1472-6483 (Linking)

Kagawa, N., S. Silber&M. Kuwayama. (2009). Successful vitrification of bovine and human ovarian tissue. *Reprod Biomed Online*, 18, 568-577, 1472-6491 (Electronic)

Kim, S. S., H. W. Yang, H. G. Kang, H. H. Lee, H. C. Lee, D. S. Ko&R. G. Gosden. (2004). Quantitative assessment of ischemic tissue damage in ovarian cortical tissue with or without antioxidant (ascorbic acid) treatment. *Fertil Steril*, 82, 679-685, 0015-0282 (Print)

Lavranos, T. C., H. F. Rodgers, I. Bertoncello&R. J. Rodgers. (1994). Anchorage-independent culture of bovine granulosa cells: the effects of basic fibroblast growth factor and dibutyryl cAMP on cell division and differentiation. *Exp Cell Res*, 211, 245-251, 0014-4827 (Print) 0014-4827 (Linking)

Lofman, C. O., M. Brannstrom, P. V. Holmes&P. O. Janson. (1989). Ovulation in the isolated perfused rat ovary as documented by intravital microscopy. *Steroids*, 54, 481-490, 0039-128X (Print) 0039-128X (Linking)

Lucci, C. M., M. A. Kacinskis, L. H. Lopes, R. Rumpf&S. N. Bao. (2004). Effect of different cryoprotectants on the structural preservation of follicles in frozen zebu bovine (Bos indicus) ovarian tissue. *Theriogenology*, 61, 1101-1114, 0093-691X (Print) 0093-691X (Linking)

Martinez-Madrid, B., M. M. Dolmans, A. Van Langendonckt, S. Defrere&J. Donnez. (2004). Freeze-thawing intact human ovary with its vascular pedicle with a passive cooling device. *Fertil Steril*, 82, 1390-1394, 0015-0282 (Print)

Mclaughlin, M.&E. E. Telfer. (2010). Oocyte development in bovine primordial follicles is promoted by activin and FSH within a two-step serum-free culture system. *Reproduction*, 139, 971-978, 1741-7899 (Electronic) 1470-1626 (Linking)

Milenkovic, M., A. Wallin, M. Ghahremani&M. Brannstrom. (2011). Whole sheep ovary cryopreservation: evaluation of a slow freezing protocol with dimethylsulphoxide. *J Assist Reprod Genet*, 28, 7-14, 1573-7330 (Electronic) 1058-0468 (Linking)

Moniruzzaman, M., R. M. Bao, H. Taketsuru&T. Miyano. (2009). Development of vitrified porcine primordial follicles in xenografts. *Theriogenology*, 72, 280-288, 1879-3231 (Electronic) 0093-691X (Linking)

Munn, C. S., L. C. Kiser, S. M. Wetzner&J. E. Baer. (1986). Ovary volume in young and premenopausal adults: US determination. Work in progress. *Radiology*, 159, 731-732, 0033-8419 (Print)

Onions, V. J., M. R. Mitchell, B. K. Campbell&R. Webb. (2008). Ovarian tissue viability following whole ovine ovary cryopreservation: assessing the effects of sphingosine-1-phosphate inclusion. *Hum Reprod*, 23, 606-618, 1460-2350 (Electronic)

Onions, V. J., R. Webb, A. S. Mcneilly&B. K. Campbell. (2009). Ovarian endocrine profile and long-term vascular patency following heterotopic autotransplantation of cryopreserved whole ovine ovaries. *Hum Reprod*, 24, 2845-2855, 1460-2350 (Electronic)

Paynter, S. J., A. Cooper, B. J. Fuller&R. W. Shaw. (1999). Cryopreservation of bovine ovarian tissue: structural normality of follicles after thawing and culture in vitro. *Cryobiology*, 38, 301-309, 0011-2240 (Print) 0011-2240 (Linking)

Pegg, D. E., M. C. Wusteman&S. Boylan. (1997). Fractures in cryopreserved elastic arteries. *Cryobiology*, 34, 183-192, 0011-2240 (Print) 0011-2240 (Linking)

Revel, A., A. Elami, A. Bor, S. Yavin, Y. Natan&A. Arav. (2004). Whole sheep ovary cryopreservation and transplantation. *Fertil Steril*, 82, 1714-1715, 0015-0282 (Print)

Rodgers, R. J.&H. F. Irving-Rodgers. (2010). Morphological classification of bovine ovarian follicles. *Reproduction*, 139, 309-318, 1741-7899 (Electronic) 1470-1626 (Linking)

Salle, B., B. Demirci, M. Franck, C. Berthollet&J. Lornage. (2003). Long-term follow-up of cryopreserved hemi-ovary autografts in ewes: pregnancies, births, and histologic assessment. *Fertil Steril*, 80, 172-177, 0015-0282 (Print) 0015-0282 (Linking)

Salle, B., B. Demirci, M. Franck, R. C. Rudigoz, J. F. Guerin&J. Lornage. (2002). Normal pregnancies and live births after autograft of frozen-thawed hemi-ovaries into ewes. *Fertil Steril*, 77, 403-408, 0015-0282 (Print)

Salle, B., J. Lornage, B. Demirci, F. Vaudoyer, M. T. Poirel, M. Franck, R. C. Rudigoz&J. F. Guerin. (1999). Restoration of ovarian steroid secretion and histologic assessment after freezing, thawing, and autograft of a hemi-ovary in sheep. *Fertil Steril*, 72, 366-370, 0015-0282 (Print)

Salle, B., J. Lornage, M. Franck, L. Isoard, R. C. Rudigoz&J. F. Guerin. (1998). Freezing, thawing, and autograft of ovarian fragments in sheep: preliminary experiments and histologic assessment. *Fertil Steril*, 70, 124-128, 0015-0282 (Print) 0015-0282 (Linking)

Santos, R. R., C. Amorim, S. Cecconi, M. Fassbender, M. Imhof, J. Lornage, M. Paris, V. Schoenfeldt&B. Martinez-Madrid. (2010). Cryopreservation of ovarian tissue: an emerging technology for female germline preservation of endangered species and breeds. *Anim Reprod Sci*, 122, 151-163, 1873-2232 (Electronic) 0378-4320 (Linking)

Schnorr, J., S. Oehninger, J. Toner, J. Hsiu, S. Lanzendorf, R. Williams&G. Hodgen. (2002). Functional studies of subcutaneous ovarian transplants in non-human primates: steroidogenesis, endometrial development, ovulation, menstrual patterns and gamete morphology. *Hum Reprod*, 17, 612-619, 0268-1161 (Print)

Soede, N. M., P. Langendijk&B. Kemp. (2011). Reproductive cycles in pigs. *Anim Reprod Sci*, 124, 251-258, 1873-2232 (Electronic) 0378-4320 (Linking)

Stevens, V. C. (1997). Some reproductive studies in the baboon. *Hum Reprod Update*, 3, 533-540, 1355-4786 (Print) 1355-4786 (Linking) Ting, A. Y., R. R. Yeoman, M. S. Lawson&M. B. Zelinski. (2011). In vitro development of secondary follicles from cryopreserved rhesus macaque ovarian tissue after slow-rate freeze or vitrification. *Hum Reprod*, 26, 2461-2472, 1460-2350 (Electronic) 0268-1161 (Linking) Vandeberg, J. L. (2004). Need for primate models in biomedical research. *Gynecol Obstet Invest*, 57, 6-8, 0378-7346 (Print) 0378-7346 (Linking) Wallin, A., M. Ghahremani, P. Dahm-Kahler&M. Brannstrom. (2009). Viability and function of the cryopreserved whole ovary: in vitro studies in the sheep. *Hum Reprod*, 24, 1684-1694, 1460-2350 (Electronic) 0268-1161 (Linking)

Weinbauer, G. F., M. Niehoff, M. Niehaus, S. Srivastav, A. Fuchs, E. Van Esch&J. M. Cline. (2008). Physiology and Endocrinology of the Ovarian Cycle in Macaques. *Toxicol Pathol*, 36, 7S-23S, 1533-1601 (Electronic) 0192-6233 (Linking)

Yang, M. Y.&J. E. Fortune. (2006). Testosterone stimulates the primary to secondary follicle transition in bovine follicles in vitro. *Biol Reprod*, 75, 924-932, 0006-3363 (Print) 0006-3363 (Linking)

Yeoman, R. R., D. P. Wolf&D. M. Lee. (2005). Coculture of monkey ovarian tissue increases survival after vitrification and slow-rate freezing. *Fertil Steril*, 83 Suppl 1, 1248-1254, 0015-0282 (Print)

Zhang, J. M., Y. Sheng, Y. Z. Cao, H. Y. Wang&Z. J. Chen. (2011). Cryopreservation of whole ovaries with vascular pedicles: vitrification or conventional freezing? *J Assist Reprod Genet*, 28, 445-452, 1573-7330 (Electronic) 1058-0468 (Linking)

Zhang, J. M., Y. Sheng, Y. Z. Cao, H. Y. Wang&Z. J. Chen. (2011). Effects of cooling rates and ice-seeding temperatures on the cryopreservation of whole ovaries. *J Assist Reprod Genet*, 28, 627-633, 1573-7330 (Electronic) 1058-0468 (Linking)

11

New Approaches of Ovarian Tissue Cryopreservation from Domestic Animal Species

Vanessa Neto[1] et al.*
Université de Lyon, VetAgro Sup – Veterinary Campus of Lyon, UPSP ICE 'Interactions between Cells and their Environment', Team Cryobio France

1. Introduction

Cryopreservation has been attempted for most of the developmental stages of both male and female reproductive cells, ranging from the immature gametes residing in ovarian or testicular tissues through the mature oocytes and spermatozoa.

However, each variant of the reproductive cells has introduced their own problems, and it has been realized that many aspects of the particular physiology of the cells will dictate how they respond to cryopreservation. Both male and female gametes have acquired highly specialized structural components (essential to fertilization and development) that may respond to the freezing process in ways different from that of basic cell structures.

1.1 Cryopreservation of spermatozoa

Semen is one of the most practical means of storing germplasm due to its abundant availability and ease of application (Holt and Pickard, 1999; FAO, 2007). Stored spermatozoa could be introduced back into existing populations either immediately or decades or centuries afterwards. In this way, cryopreservation of spermatozoa associated with artificial insemination (AI) or *in vitro* fertilization (IVF) facilitates the management of domestic animals herds, especially cow dairy herds where it is now used since 60 years. Cryopreservation better allows the use of semen from genetically superior males of threatened livestock breeds and has the potential to protect existing diversity and maintain heterozygosity while minimizing the movement of living animals and the transmission of venereal diseases (Johnston and Lacy, 1995; Andrabi and Maxwell, 2007).

Spermatozoon is a very small cell containing low amounts of cytoplasm and consequently low quantity of water. Furthermore sperm nuclear material is compacted and protected

*Thierry Joly[1], Loris Commin[1], Pierre Bruyère[1], Anne Baudot[2], Gérard Louis[2], Pierre Guérin[1] and Samuel Buff[1]
[1]*Université de Lyon, VetAgro Sup – Veterinary Campus of Lyon, UPSP ICE 'Interactions between Cells and their Environment', Team Cryobio, France*
[2]*Université Paris Descartes, Sorbonne Paris Cité, UFR Biomedical, France*

against physical injuries. For these reasons, cryopreservation of spermatozoa gives excellent results in term of viability and fertility, and is today widely used in human and animal assisted reproduction.

Semen from most mammalian and a few avian species has been successfully frozen in the past several years (FAO, 2007). Indeed, much better results have been obtained with sperm cryopreservation than with oocytes (and embryos) in term of viability. For these reasons, sperm cryobanking is used since the middle of the last century in domestic animal species and lately in human. Numerous storage facilities such as the French National Cryobank were also created for the preservation of valuable or endangered species. However, protocols currently used to conserve semen are still suboptimal and cannot be easily applied across species (Woods et al., 2004). First-service conception rates vary drastically between different breeding programs, but on average conception rates are fairly high in cattle, pigs, goats, and sheep. In wild species cryopreservation of gametes is currently used to preserve endangered species or breeds, and to overcome some fertility troubles. Breed reconstruction solely from semen is possible through a series of back-cross generations; however, the entire genetics of the original breed will not be recovered (Boettcher et al., 2005).

In human, gametes cryopreservation has also been developed to overcome fertility troubles (eg. genital duct problems, sexual dysfunction …). Sperm cryopreservation that has produced live birth has been available for over 50 years (Sherman and Bunge, 1953). Anonymous donor sperm banking has been a fundamental concept of reproductive medicine for several decades, and artificial insemination and donor sperm cryobanking are widely used in reproductive medicine centers (Critser, 1998). The availability of frozen donor sperm has become a mainstay for the treatment of serious male infertility worldwide. More recently, gametes cryopreservation has been used to preserve fertility of men (or women) submitted to gonadotoxic treatments and elective sperm cryopreservation programs have been provided from cancer patients all over the world. Using vitrification of sperm obtained from testicular biopsy, epididymal fluid, or a semen sample after electroejaculation could create new hope for infertile men (Edge et al., 2006).

1.2 Cryopreservation of testicular tissue

Cryopreservation of testicular tissue has been studied since about ten years in animals (Jezek et al., 2001; Jahnukainen et al., 2007; Milazzo et al., 2008; Zeng et al., 2009; Abrishami et al., 2010; Milazzo et al., 2010; Curaba et al., 2011) and human (Bahadur and Ralph, 1999; Bahadur et al., 2000; Guerin, 2005; Revel and Mejia, 2010). It is the only available solution for pre-pubertal boys who must receive a gonadotoxic treatment (eg. cancer therapy;(Keros et al., 2005; Keros et al., 2007; Wyns et al., 2007; Wyns et al., 2008; Wyns et al., 2010).

In contrast to the situation in the ovary, it is well established that spermatogonial stem cells exist in the testis and are responsible for maintaining spermatogenesis from puberty for the lifetime of the male. Human testicular cells might be harvested, cryopreserved before a gonadotoxic treatment and re-introduced into the testis upon its completion. Possibilities include transplantation back into the inactive testes (ipsigenic germ cell transplantation), maturation *in vivo* in another host (xenogenic germ cell transplantation), or *in vitro* spermatogenesis. Mature sperm could then be used for fertilization by ICSI.

Sperm obtained after stem cell transplantation were shown to be able to fertilize mouse oocytes. Fertile offspring were obtained through artificial reproductive technologies following the establishment of complete spermatogenesis by grafting testis tissue from newborn mice, pigs, or goats into mouse host (Honaramooz et al., 2002; Schlatt et al., 2002). Different freezing protocols have been developed in several species but without a clearly identified procedure (Travers et al., 2011). Testicular tissue from prepubertal boys facing gonadotoxic treatment could be banked for several years for spermatogonial stem cell transplantation. Pregnancies have been achieved with ICSI using immature spermatids and secondary spermatocytes extracted from testicular tissue in men with spermatogenic arrest (Fishel et al., 1997; Vanderzwalmen et al., 1997; Sofikitis et al., 1998).

1.3 Cryopreservation of oocytes

Oocyte cryopreservation offers many advantages. It permits to preserve endangered species and low effective breeds, and to preserve fertility of high genetic value females. In human it permits to overcome some fertility problems. Oocyte banks would also enlarge the gene pool, facilitate several assisted reproductive procedures, salvage female genetics after unexpected death, and avoid controversy surrounding the preservation of embryos (Ledda et al., 2001; Checura and Seidel, 2007). Like semen, oocyte cryopreservation is beneficial for international exchange of germplasm, as it avoids injury and sanitary risks involved in live animal transportation (Pereira and Marques, 2008). But oocyte cryopreservation gives much lower results when compared with spermatozoa.

This is due essentially to the much important size and complexity of oocytes. For example, nuclear material is much more exposed in prophase I or metaphase II oocyte than in compacted chromatin of sperm cells. Oocytes collected from slaughterhouse derived ovaries are at the germinal vesicle (GV) stage in which the genetic material is contained within the nucleus. Since this stage has no spindle present, GVs are assumed to be less prone to chromosomal and microtubular damage during cryopreservation. However, oocytes can also be cryopreserved at the metaphase II stage of maturation. During the metaphase II stage, the cumulus cells surrounding the oocyte are expanded, microfilaments of actin are involved in cell shape and movements, and microtubules form the spindle apparatus (Massip, 2003). Moreover, cryopreservation of oocytes necessitates the success of the following steps: *in vitro* maturation, *in vitro* fertilization and embryo culture. Progress in female gametes cryopreservation has gone hand in hand with that for *in vitro* maturation and embryo culture. In livestock animals, oocytes collected by *in vivo* pickup or at slaughter can be frozen for extended periods of time for subsequent IVF to produce embryos. However, in some species such as *canidae* the collection of oocytes is difficult and *in vitro* maturation (IVM), *in vitro* fertilization (IVF) and embryo culture are not yet under control (Luvoni et al., 2006). For these reasons precise cooling and thawing rates and the use of a programmable cell freezer are necessary for oocyte (and also embryo) cryoconservation and very few studies have been conducted in animals.

Oocytes are extremely sensitive to chilling, and the technique is not as established as in semen or embryos, due to the fact that oocytes typically have a low permeability to cryoprotectants (Woods et al., 2004). The major differences between oocytes and embryos are the plasma membrane, presence of cortical granules, and spindle formation at metaphase II stage of meiosis (Chen et al., 2006; Salvetti et al., 2010). To date, there has been no consistent oocyte cryopreservation method established in any species, although, there has been significant progress and offspring have been born from frozen-thawed oocytes in cattle, sheep, and horses (Otoi et al., 1996; Maclellan et al., 2002; Woods et al., 2004). During the process of cryopreservation, oocytes suffer considerable morphological and functional damage, although, the extent of cryoinjuries depends on the species and the origin (*in vivo* or *in vitro* produced). The mechanism for cryoinjuries is yet to be fully understood, and until more insight is gained, improvement of oocyte cryopreservation will be difficult. Also, it is noticeable that immature oocyte present in primordial follicles seems more resistant to cryopreservation when compared to mature oocyte. Consequently, cryopreservation of ovarian cortex may constitute an interesting alternative to isolated mature oocyte cryopreservation.

1.4 Cryopreservation of ovarian tissue

At birth, the ovaries contain the lifetime complement of primary oocytes which are arrested in the prophase stage of meiosis 1 and are surrounded by a single-layered epithelium to form the primordial follicles. Ovarian cortex presents several advantages when compared with isolated oocytes: *(i)* it contains the important pool of growing follicles; *(ii)* it does not necessitate the *in vitro* maturation /*in vitro* fertilization /embryo culture steps if it is associated with grafting; *(iii)* no previous ovarian stimulation is necessary. Consequently cryopreservation of ovarian cortex is an alternative to cryopreservation of isolated oocytes or embryos. It could be used as an emergency preservation and as infertility therapy method for valuable animals. Ovarian cortex cryopreservation has been developed in human in order to preserve fertility in young women submitted to gonadotoxic therapy (Stahler et al., 1976; Gook et al., 2004). In human newborns were obtained after orthotopic autograft of frozen-thawed ovarian cortices (Donnez et al., 2004).

It is obvious that, to achieve successful cryopreservation of ovarian tissue, it is essential to maintain the functional status of the whole mixture of different cell types: oocytes, granulosa cells, epithelial cells, fibroblasts… This represents a major difficulty, because the optimum kinetic of cooling is different for each cell type. Oocytes are large cells, with a low surface to volume ratio, surrounded by zona pellucida. Immediately adjacent to the oocyte are corona radiata cells that have long cytoplasmic extensions which penetrate the zona pellucida, ending in oocyte membrane. These processes and gap junctions are important in the metabolic cooperation between the oocyte and surrounding layers of granulosa cells, which form the cumulus-oocyte complex during the growth phase. Consequently, at the opposite to cryopreservation of isolated cells, a cryopreservation protocol for a tissue represents a compromise between the requirements of the different constitutive cells.

The early work on ovarian tissue cryopreservation was performed in animal studies: rabbit (Smith, 1952) and rat (Parkes and Smith, 1953; Deanesly, 1954). The earliest positive results were obtained when glycerol (15%) plus serum were used as cryoprotective agents (CPAs) for cryopreservation of rabbit granulosa cells, via a slow cooling protocol (Smith, 1952). An equilibration period was necessary to achieve CPA penetration into the tissue. For this reason small samples were recommended. A rapid rewarming by plunging the samples into a water bath at 40°C was the most effective procedure (Parkes and Smith, 1953; 1954). Normal offspring were obtained from mice with orthotopic ovarian grafts of tissue that had been frozen and stored at -79°C (Parrott, 1960). Vitrification of ovarian tissue was also investigated. Nevertheless, Isachenko et al suggested that in human, low freezing protocols were more promising than vitrification protocols (Isachenko et al., 2009).

This technique has also been developed in rabbit (Neto et al., 2007a), mouse (Candy et al., 1997), rat (Aubard et al., 1998), ewe (Gosden et al., 1994; Demirci et al., 2001), cow (Paynter et al., 1999). We have obtained newborn rabbits after autografting of cryopreserved ovarian cortex (Neto et al., 2007b). Also, our team developed this technique in cat (Neto et al., 2006) and dog (Commin et al., 2011).

Several techniques have been applied to ovarian cortex cryopreservation: slow freezing, vitrification. Simultaneously to ovarian tissue cryopreservation, numerous researches have been conducted about ovarian tissue grafting: orthotopic, heterotopic, auto-, allo- and hetero-grafting (Pullium et al., 2008).

2. Development of optimized methods for the cryopreservation of the ovarian tissue in domestic mammalian species

The most common cryopreservation method is the slow freezing procedure, consisting of an initial slow, controlled-rate cooling to subzero temperatures followed by rapid cooling as the sample is plunged into liquid nitrogen for storage (−196°C). At such a low temperature, biological activity is effectively stopped, and the cells functional status may be preserved for centuries. However, several physical stresses damage the cells at these low temperatures. Intracellular ice formation is one the largest contributors to cell death; therefore, freezing protocols use a combination of dehydration, freezing point depression, supercooling, and intracellular vitrification in an attempt to avoid cell damage.

Currently used ovarian cortex cryopreservation protocols have been direct, or slight modifications of the methods developed for isolated oocytes and embryos. There were primarily developed by trial and error adjustments of cooling and warming rates, and choice of CPA and CPA concentrations. However, because there are a large number of protocol variables potentially affecting cell viability, an exhaustive experimental search for the optimal combination of these parameters has long been considered to be prohibitively expensive in terms of time and resources.

2.1 Chemical and physical parameters affecting equilibration and freezing processes of ovarian tissue in mammalian species

The result of a cryopreservation process is influenced by several chemophysical parameters affecting directly or not the functions and the integrity of the ovarian cells along the freezing process, from the equilibration to the thawing. Among these parameters, the method of equilibration, the freezing rate, the composition of the freezing solution and notably the nature of the permeating CPAs and the non-permeating CPAs, the concentration of each CPA, the use of serum, or the rate of thawing may be investigated to know the relative influence of each of them and the induced cell injuries.

In general, we can expect coupled flows of water and CPAs when CPAs are added, during freezing, thawing and when CPAs are removed from the cells, resulting in a series of anisosmotic conditions. During freezing, the cells dehydrate and shrink and remain shrunken during storage, but return to their isosmotic volume upon thawing. Finally, the cells are subjected to potentially lethal swelling upon CPA dilution and removal. During the controlled slow cooling extracellular ice formation is induced (seeding) at a temperature just below the solutions' freezing point, and then the cooling continues at a given rate in the presence of a growing extracellular ice phase, which raises the extracellular solute concentration in the unfrozen fraction and results in water being removed from the cell via exosmosis.

Permeating CPAs, such as glycerol, dimethyl sulfoxide, ethylene glycol or propylene glycol are typically included in the cryoprotective medium, to protect the cells against injury from the high concentrations of electrolytes that develop as water is removed from the solution as ice. During the equilibration step the inner cell water is partly replaced by the permeating CPAs. However, the CPAs can be damaging to the cells, especially when it is used at high concentrations. The toxicity can be reduced by decreasing the time or the temperature of the equilibration step (Karlsson and Toner, 1996). But equilibration at low temperatures requires increasing the exposition time to freezing solution. Furthermore, the CPAs may have dramatic osmotic effects upon the cells during their addition and their removal. Consequently, the use of several steps of increasing concentrations of CPAs during the equilibration allows reducing the osmotic gradient. The cells exposed to such permeating CPAs undergo initial dehydration, followed by rehydration, and potential gross swelling upon removal. This osmotic shock may generate membrane damages by mechanical means and predisposition of the cell to injuries during the other steps of cryopreservation, or even cell death (Mazur and Schneider, 1986). These kinds of damages could be reduced by using cells surfactant such as serum. During the freezing step, the follicular preservation depends on the nature and the concentration of the CPAs.

Control of the cooling and warming rates is also crucial, as the freezing/thawing rates and the temperature of seeding also influence the ice properties. If cells are cooled too rapidly during the controlled slow cooling process, water does not exit the cells fast enough to maintain equilibrium and, therefore, the oocytes and other ovarian cells freeze intracellularly, resulting in death in most cases. If cooling is too slow, the long duration can cause 'solution effects' injury resulting from the high concentration of extra- and intracellular solutes, probably due to the effects of the solutes on the cellular membrane or through osmotic dehydration. During warming the small intracellular ice crystals might subsequently undergo recrystallization, forming bigger ice crystals that rupture the cell membrane, thus leading to fatal damage. Finally, the thawing and the removal of the CPA depend on the temperature and on the presence of non-permeating CPA limiting the osmotic swelling during rinsing.

2.2 Use of fractional experimental design

The influence of the multiple chemical and physical parameters cannot be exhaustively performed as it would require too much time and resources. Even if the number of factors, k, in a design is small, the 2^k runs specified for a full factorial can quickly become very large. For example, $2^5 = 32$ runs is for a two-level, full factorial design with five factors. To this design we would need to add a good number of centerpoint runs and we could thus quickly run up a very large resource requirement for runs with only a modest number of factors.

Moreover, while the approach is sequential in nature, it is potentially increasing in complexity as the knowledge and understanding of the application and domain evolves. Design of experiments techniques provides a systematic, effective and efficient approach to the investigation of a phenomenon. The main advantages of this strategy were the saving in times and resources expended compared to other approaches and the resulting mathematical models that helps us to better understand the phenomena under investigation more fully. To analyze the response of ovarian cortices from different species to the freeze/thaw process, the authors decided to use fractional (2^{n-p}) experimental designs (Mechakra et al., 1999).

Using this statistical tool, the authors used only a fraction of the runs specified by the full factorial design; which runs to make and which to leave out was one of our subjects of picked. The authors used various strategies that ensure an appropriate choice of the runs. As for an example, fractional experimental designs $2^{(5-2)}$ presented in table 1 aim to evaluate the combined effect of five different factors according to two modalities for each of them. For each experimental design, eight combinations of factors were performed. For each of them, the ratio of morphological preservation and the ratio of viability of isolated preantral follicles were recorded. While the designs were similar for each of the species that were evaluate, the parameters were chosen according to each species (Table 2 to Table 5).

	Variable						
Run	I	X_1	X_2	X_3	X_4	X_5	$X_1.X_2$
1	+1	-1	-1	-1	+1	-1	+1
2	+1	+1	-1	-1	+1	+1	-1
3	+1	-1	+1	-1	-1	+1	-1
4	+1	+1	+1	-1	-1	-1	+1
5	+1	-1	-1	+1	-1	+1	+1
6	+1	+1	-1	+1	-1	-1	-1
7	+1	-1	+1	+1	+1	-1	-1
8	+1	+1	+1	+1	+1	+1	+1

Table 1. Fractional experimental design $2^{(5-2)}$ used to evaluate the cryopreservation protocols in the doe rabbit, in the queen and in the cow

Variables	Level -1	Level +1
X_1: Permeating CPA	DMSO	PROH
X_2: Concentration of permeating CPA	1.5M	2M
X_3: Non permeating CPA	trehalose	sucrose
X_4: Freezing rate	0.3°C/min	2°C/min
X_5: Equilibration	1 step	3 steps

Table 2. Dependent variable list evaluated in the rabbit doe

These $2^{(n-2)}$ experimental designs allowed discriminating between five factors influencing the cryopreservation process (variables X_1 to X_5) and the simultaneous interactions between two of them. The linearity (structural and estimated model) of the experimental model was evaluated by an ANOVA test. One randomly chosen assay was replicated three times to estimate the experimental error (*E*).

So, eleven experiments were randomly performed. Multi-linear regression was performed using all the variables in order to evaluate experimental results according to this model:

$$\hat{y} = \beta_0 + \beta_1.X_1 + \beta_2.X_2 + \beta_3.X_3 + \beta_4.X_4 + \beta_5.X_5 + \beta_{1,2}.X_1.X_2 + E$$

Variables	Level -1	Level +1
X1: Permeating CPA	PROH	DMSO
X2: add of sucrose	no	yes
X3: Freezing rate	0.5°C/min	2°C/min
X4: manual seeding	no	Yes
X5: Freezing rate after -40°C	In the freezing chamber	Direct immersion in LN_2

Table 3. Dependent variable list evaluated in the queen

Variables	Level -1	Level +1
X1: Permeating CPA	PROH	DMSO
X2: Concentration of permeating CPA	1.5M	2.5M
X3: Non permeating CPA	sucrose	trehalose
X4: Medium	Euroflush®	Medium with choline
X5: Cell surfactant	Fetal calf serum	Albumax®

Table 4. Dependent variable list evaluated in the cow

Variables	Level -1	Level +1
X1: Permeating CPA	DMSO	PROH
X2: Non permeating CPA	trehalose	sucrose
X3: Freezing rate	0.3°C/min	2°C/min
X4: Equilibration	1 step	3 steps

Table 5. Dependent variable list evaluated in the bitch

Results of experimental designs for each species were completed by at least one additional biological evaluation and one quantitative evaluation of normal and viable follicle rates after freezing, according to the best combination of factors chosen with experimental designs.

2.3 Criteria to assess the quality of frozen-thawed cortices

A survey of nearly all quality assays available to the preservation scientist reveals that they can be grouped into different categories. The following assay tier is not specific to cryopreservation, but is presented below as a support that can guide those who work with preserved tissues. The authors decided to assess the quality of the protocols developed in different species using all or part of alternate test presented below. The morphology and the viability of the ovarian follicles were systematically assessed, in combination with the investigation of the ultrastructure of the follicles, and their capacity to resume folliculogenesis after graft.

2.3.1 Morphology and ultrastructure

Assessment of the morphology of ovarian cells required the ovarian fragments were fixed in a preliminary defined fixative agent adapted to the species, before being processed for classical light microscopy. Primordial to primary follicles - from oocytes surrounded by flattened granulosa cells until oocytes surrounded by one layer of cuboidal granulosa cells (Gougeon and Chainy, 1987) - were usually classified into four types of morphological defects: *(Type I)* follicle without any morphological defect - follicle is regular, with joined follicular cells; cytoplasm of the oocyte is homogeneous and chromatin is diffused and regular; *(Type II)* follicle with cytoplasmic defect - cytoplasm of the oocyte is vacuolated or eosinophil; *(Type III)* follicles with nuclear defect - nucleus of the oocyte is picnotic, without apparent nuclear membrane or with an irregular nuclear membrane; *(Type IV)* degenerated follicle - oocyte with combined cytoplasmic and nuclear defects or follicle with irregular shape or with disjoined follicular cells or with swelled follicular cells.

The ultrastructure of ovarian follicles was also examined for the presence of apoptotic and para-apoptotic cell death. Ultrastructurally apoptosis is characterized by margination of condensed chromatin, nuclear fragmentation, and the formation of apoptotic bodies. Para-apoptosis, nonclassical apoptosis, is a specific morphologic type of non-necrotic cell death and is characterized by cytoplasmic vacuolization, condensed chromatin (but not early margination of the chromatin), and swollen mitochondria.

2.3.2 Cytolysis live/dead assay

The cytolysis assays have both a very positive and a negative attribute to them. On the positive side, there are a variety of assays that can reveal cell membrane leakage that occurs as a final stage in most forms of cell death. Yet given that cytolysis is the last stage of preservation-induced cell death, these assays do not reveal early-stage mechanisms underlying preservation-induced cell death and thus have limited use in the future as a diagnostic means to develop improved preservation formulations and protocols. The LDH assay continues today to be useful for measuring preservation-induced cytotoxicity. The concept behind this cytolysis assay is simply that if the cell membrane is compromised, then LDH will leak into the extracellular milieu where it can be measured. The trypan blue assay is also one of the most commonly used cytolysis assays. A number of investigators has used the trypan blue exclusion assay in studying preservation efficacy. It does however share the same handicap as the LDH assay given that neither can be analyzed using fluorescence and/or bioluminescence. Currently, the best cytolysis live/dead assays are those that employ fluorescent indicator dyes. Available probes can be subdivided into two different subsets, one of which is trapped by the cell and leaks out only is a membrane rupture occurs. The other subtype, exemplified by ethidium homodimer or propidium iodide, is membrane insoluble and only stains the cell if it gains access through a compromised plasma membrane. In this way, the authors did evaluate the viability of the frozen-thawed follicles using Calcein-AM and Ethidium homodimer I stains (Live/Dead® Viability/Cytotoxicity Kit, Molecular Probes) on enzymatically isolated follicles (Fig. 2).

2.3.3 In vivo functionality (autografting model) and in vivo growth potential (xenografting model)

One of the key components of any preservation protocol must be a functional assay that matches the type of cells or tissue being analyzed. In some cases this determination is quite easy. As for example, the sperm motility is a well-accepted functional assay for this system, whereas the capacity to resume folliculogenesis sounds reasonable for the ovarian cortex.

Fresh and cryopreserved rabbit doe ovarian tissues were autografted into young females. Fresh ovarian tissue was grafted on the ipsilateral ovarian pedicle immediately after ovary resection (2 grafts per female), contrary to cryopreserved ovarian tissue grafted on the controlateral ovarian pedicle 24 hours after freezing (one graft per female). Control females were ovariectomised, according to the same surgery resection that used before graft. From height week after graft, grafted rabbit doe were inseminated every 3 weeks in case of negative pregnant diagnosis. Eleven months later, ovarian grafts were removed at necropsy to observe follicular structures by histology.

In the bitch, the growth ability of frozen-thawed ovarian tissue has also been assessed by implantation of small pieces of ovarian tissue into adult female SCID mice. After removing mice ovaries, the canine frozen thawed ovarian tissue was placed intramuscularly in the gluteal superficial muscle. To study the setting up of graft resumption, the graft was harvested at one, eight or 16 weeks and the resumption of the ovarian activity was controlled by vaginal cytology assessment. After harvesting, the grafts were processed for histological assessment.

2.3.4 Vascularization

In the bitch, the alpha smooth muscle actin (alpha-SMA), a marker of the mature blood vessel, was used to assess the vessel density within the ovarian tissue, using a primary antibody directed against alpha-SMA. For each slice, image analysis was performed under direct light microscope to determine the tissue areas. The stained vessels were counted in several fields of the same slice and in several slices per animal to deduce a blood vessel density.

2.4 Experimental results

To illustrate the interest of the use of fractional experimental design, the authors decided to present some of their results obtained in the doe rabbit, in the queen, in the bitch and in the cow during the last 6 years. The doe rabbit was used as a model for the human and the animal applications of the ovarian tissue cryopreservation, because of its biological and breeding characteristics. The cat was considered as a model for the ovarian tissue cryopreservation studies of endangered wild felids; from all the felids, the domestic cat is the only non endangered feline species. The cow was used as model for ruminants, with a special interest for preserving high valuable individuals in combination to embryos and to semen cryopreservation. Finally, the bitch was studied for preserving the reproductive potential of future guide dogs submitted to neutering surgery before training.

2.4.1 In the doe rabbit

The experimental variability expressed as the repetition of one single combination, showed that both the structural and the estimated models of the experimental design were valid when considering the morphological preservation ratio of the follicles. The concentration of the permeating CPA (P = 0.67) and the number of equilibration steps (P = 0.19) seemed to have no significant effect on the morphological preservation ratio of ovarian follicles. The nature of the permeating and non-permeating CPA seemed to influence the morphological preservation ratio of the follicles (P = 0.08 and P = 0.07 respectively) although the non-significant difference. DMSO tended to reduce the morphological preservation ratio, as compared with PROH. Morphological preservation ratio was increased in the presence of trehalose compared with sucrose. The freezing rate seemed to be the factor that had the greatest impact on the morphological preservation ratio of the doe rabbit follicles. At a freezing rate of 0.3°C/min we observed a significant increasing of the follicular morphological preservation ratio, as compared with 2°C/min (P<0.01). No significant interaction was observed between the nature of the permeating CPA and its concentration.

According to the results of the experimental design, the precise evaluation of the best combination of factors influencing positively the morphological preservation ratio (3 steps equilibration protocol, 1.5M DMSO or 1.5M PROH, medium supplemented with either sucrose or trehalose) was performed. Ovarian pieces were treated according to the results obtained with experimental design. Ovarian cortices were equilibrated (3 steps) in the freezing media based on TCM 199 and 10% FCS, at room temperature. The freezing media was supplemented with 1.5 M DMSO or 1.5M PROH and 0.2 M sucrose or 0.2M trehalose. Freezing of ovarian fragments was slowly performed at 0.3°C/min from the temperature of seeding (-7°C/min) up to -35°C. Thawing, histology, viability tests and electron microscopy evaluation process were performed before and after cryopreservation as described previously.

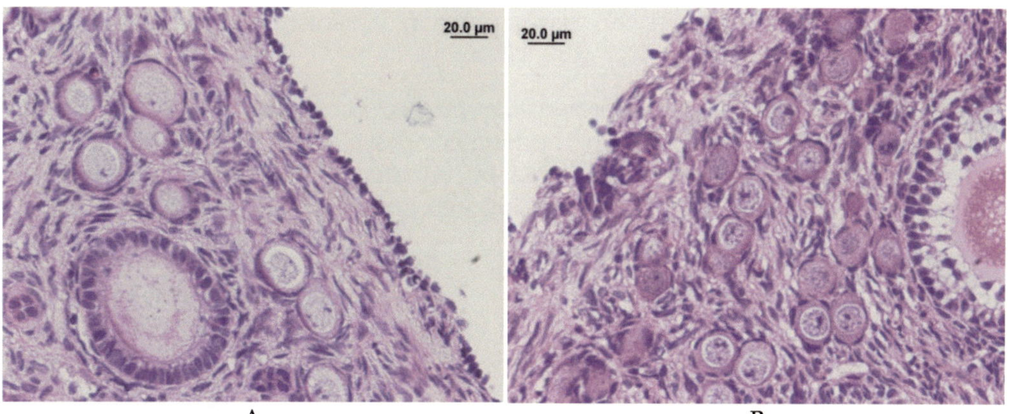

A B

Fig. 1. Rabbit follicular morphology before (A) and after (B) cryopreservation with PROH and trehalose, with a post-seeding freezing rate at 0.3°C/min

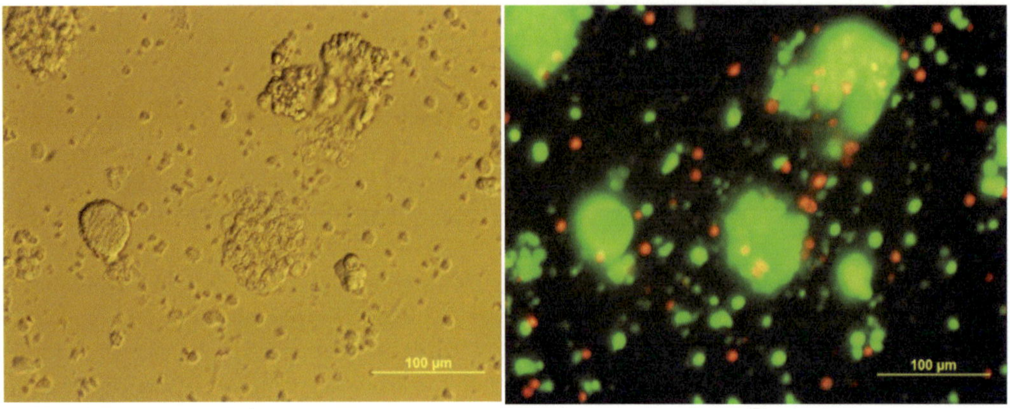

A B

Fig. 2. View of rabbit isolated follicles under direct light for selection (A) and under fluorescent light (B) after calcein AM/ethidium homodimer I stains to evaluate viability after cryopreservation with PROH, with a post-seeding freezing rate at 0.3°C/min

In control fragments, we observed 72.6 ± 2.8% and 77.7 ± 3.9% of type I follicles (no significant difference) for sucrose and trehalose control groups respectively. After cryopreservation, no statistical difference of the proportions of type I follicles was found between sucrose and trehalose (50.2 ± 4.1% *vs.* 51.1 ± 1.8% respectively) when using DMSO for cryopreservation. When using PROH as permeating CPA, the proportion of type I follicles was lower after cryopreservation with sucrose as compared to trehalose (55.0 ± 3.8% *vs.* 65.0 ± 3.3% respectively; $P<0.05$). When freezing with trehalose the proportion of type I follicles was higher with PROH as compared to DMSO (65.0 ± 3.3% *vs.* 51.1 ± 1.8% respectively; $P<0.01$). Nevertheless, the proportions of type I follicles were significantly reduced after cryopreservation (from $P\leq0.01$ to $P<0.001$), whatever the permeating and the non-permeating CPA. No significant difference was observed between the different groups of frozen ovarian cortices, when considering the morphological preservation ratio.

According to these results, the cryopreservation protocol based on a post-seeding freezing rate at 0.3°C/min and using a freezing medium composed of 1.5M PROH, supplemented with 0.2M trehalose was finally evaluated by orthotopic autografting to observe the potential of the cryopreserved follicles to resume follicular growth and to be fertilized.

Before freezing, type II follicles represented the most important part of follicles with morphological defect (19.1 ± 2.9% and 16.1 ± 3.2% in sucrose and trehalose groups respectively). After cryopreservation, follicular defect of type IV (degenerated follicles) was the most important type of morphological defect: 32.5 ± 4.8% and 24.0 ± 1.9% after freezing using DMSO, with sucrose and trehalose respectively; versus 27.2 ± 5.6% and 18.1 ± 3.0% after freezing using PROH, with sucrose and trehalose respectively. The general aspect of ovarian tissue before and after cryopreservation showed a good preservation of structural architecture (follicular structure and connective tissue). Spaces were observed in some case, in the ovarian stroma and the albuginea. Epithelial cells were often absent as compared to the fresh ovarian tissue.

Ultrastructural analysis of the preantral follicles was performed without preliminary selection on semi-thin sections. TEM analysis showed that most follicles of control ovarian tissue had normal ultrastructure, according to mitochondria, nucleus and nuclear membrane, Golgi apparatus and endoplasmic reticulum cisternae observation. They often had vacuoles in cytoplasm and vesicles. Nevertheless, vacuoles were not characteristic of apoptosis. Cellular membranes of the oocyte and follicular cells were in close connection. In general, ovarian stroma was well organised. Fibroblasts and collagen fibres were distinguishable (Fig. 3).

After cryopreservation, oocyte ultrastructure appeared to be similar to the control especially for mitochondria, Golgi apparatus, endoplasmic reticulum, interdigital structure between oocyte and follicular cells (Fig.3). Vesicles and vacuoles were rarely observed. Chromatin of the oocyte was diffused and well preserved. Nevertheless, dark follicular cells or follicular cells without any content were most frequently observed, whereas some follicles showed partial or total disruption of their nuclear membrane whatever the evaluated cryprotective solution. The most important damage observed after cryopreservation was the disorganisation of the ovarian stroma (Fig.3). Fibroblasts showed lack of cytoplasm or important vacuolisation. In general, these damages were less frequently observed after cryopreservation using PROH and trehalose.

2.4.2 Investigations in the queen

The experimental variability showed that neither the structural, nor the estimated models of the experimental design were valid when considering the morphological preservation ratio of the follicles or the viability preservation ratio. So, global discrimination of the chemo-physical parameters was not possible. Nevertheless, the influence of the freezing rate after seeding and after -40°C, and the influence of the addition of sucrose in the freezing medium composed of 1.5M CPA were evaluated and analyzed by classical ANOVA test.

Before freezing, ovarian tissue presented 72.2 ± 3.6% and 83.8 ± 2.9% of normal follicles (type I) for group 2°C/min and 0.5°C/min post-seeding freezing rate respectively. When freezing with PROH, and whatever the post-seeding freezing rate, proportions of morphologically normal follicles were not significantly reduced after freezing compared to before freezing (69.2 ± 9.1% for 2°C/min group vs. 67.4 ± 2.9°C/min for 0.5°C/min group). After freezing with DMSO, and whatever the post-seeding freezing rate, proportions of type I follicles were significantly reduced (40.8 ± 6.6% after freezing at 2°C/min and 51.6 ± 5.1% after freezing at 0.5°C/min; $P<0.05$). Whatever the post-seeding freezing rate, type III defects were the most frequently observed after freezing. General observation of the ovarian tissue showed a good preservation of the ovarian stroma cells and structure after cryopreservation.

Before freezing, ovarian tissue submitted to a free fall into the freezing chamber during the third phase of the freezing process presented 72.2 ± 3.6% of type I follicles without any difference with samples directly immersed into the liquid nitrogen (86.8 ± 2.5%). Proportion of normal follicles decreased significantly after cryopreservation except after freezing with PROH according to a free fall into the freezing chamber (68.2 ± 9.1% of type I). After freezing using a direct immersion into liquid nitrogen after -40°C, and whatever the CPA, proportions of type I follicles were decreased compared to fresh control.

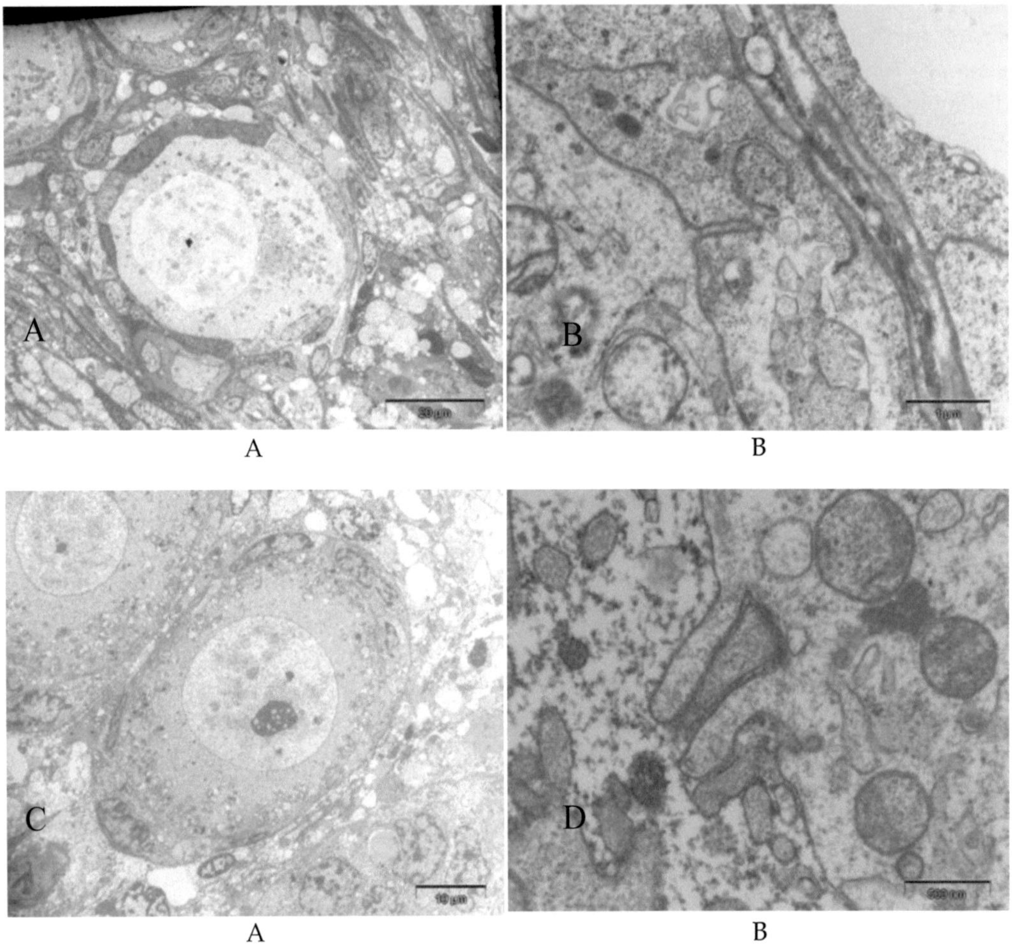

Fig. 3. Rabbit follicular ultrastructure before (A & B) and after (C & D) cryopreservation with PROH and trehalose, with a post-seeding freezing rate at 0.3°C/min

Before freezing, queen ovarian tissue showed 72.2 ± 3.6% and 74.8 ± 6.3% of type I follicles respectively for group without and with sucrose without any difference between the two control groups. After freezing, addition of sucrose allowed preserving 63.2 ± 13.6% of normal follicles versus 68.2 ± 9.1% without sucrose when associated with PROH, without any significant difference. Contrary to the results after freezing with DMSO, proportion of type I follicles was not significantly different after freezing with PROH compared to fresh control. Morphological defect of type III was the most frequently observed after freezing with PROH. Queen ovarian stroma was well preserved.

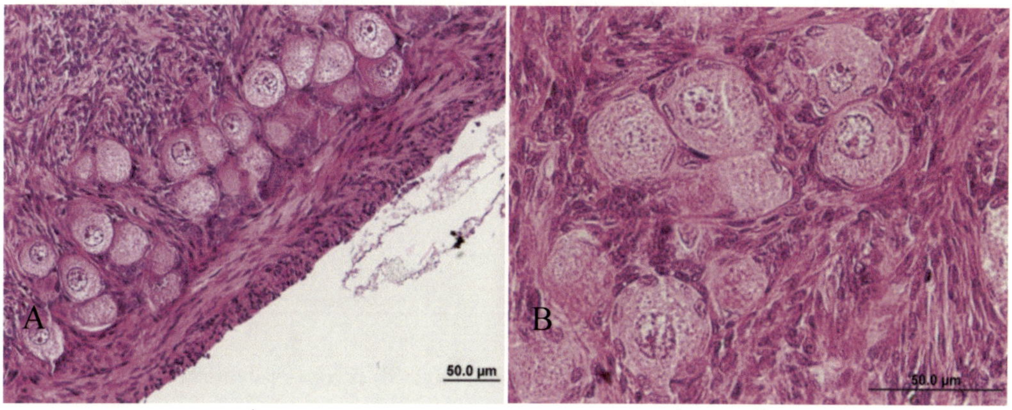

A B

Fig. 4. Queen follicular morphology before (A) and after (B) cryopreservation with 1.5M PROH and 0.2M sucrose. Post-seeding freezing rate at 2°C/min.

In conclusion, queen ovarian tissue seems to be well preserved, without any difference compared to the fresh control when freezing with PROH (according to a free fall in the freezing chamber, without influence of neither sucrose nor post-seeding freezing rate) (Fig. 4).

2.4.3 Investigation in the bitch

In the bitch, the estimated model was validated when the viability preservation ratio was considered. The nature of the non-permeating CPA ($P = 0.37$) did not influence the post thawing viability rate of the ovarian follicles. However all the other factors investigated in this experimental design presented a significant effect on the viability rate. The permeating CPA nature ($P<0.0001$) was the factor that influenced more the viability rate of the follicles. Thus, contrarily to the observations in the other species, the DMSO better preserved the evaluated parameter than PROH. The freezing rate had also a major effect on the viability rate ($P<0.0001$) and a slow freezing rate (0.3°C/min) was less deleterious than the rapid freezing rate for the follicles viability. Moreover the equilibration step also significantly affected the follicles viability, with a beneficial effect of the one step equilibration compared to the 3 steps equilibration. However, no interaction was observed in this model. As a result, the fractional experimental design developed in the bitch, suggested that ovarian tissue should be cryopreserved in a solution containing 2 M DMSO as permeating CPA supplemented with 0.2 M sucrose or trehalose in a one-step equilibration and frozen at a 0.3°C/min freezing rate (Fig. 5).

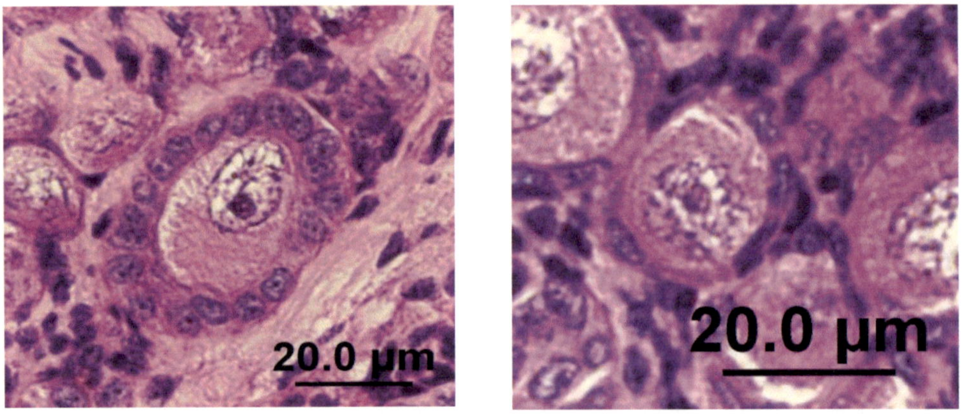

A B

Fig. 5. Bitch follicular morphology before (A) and after (B) cryopreservation

Consequently, theses parameters validated by the fractional experimental design for viability assessment were used and applied for morphological assessment of frozen-thawed bitch ovarian tissue and comparison with fresh tissue. So, the morphology of preantral follicles was compared between fresh and cryopreserved tissue. The histological analysis revealed that no significant difference was observed between fresh (89.1 ± 6.1 % type I follicles) and cryopreserved ovarian tissue (82.4 ± 4.4 % type I follicles) when type I follicles were observed. The main abnormality observed on preantral follicle after cryopreservation in the bitch was the oocyte nucleus defect (~8%). In this case, the nucleus often appeared pycnotic, with a reduced size and a densely packed chromatin. Sometimes the nuclear membrane was ruptured. However, ooplasm defect were rarely observed alone, but combined with nuclear defects. In some cases few granulosa cells were absent in the primordial or primary follicles. It is probable that ice crystal formation occurring during cooling be responsible of this partially denuded pattern, by destroying or dislodging some granulosa cell during ice expansion.

2.4.4 Investigations in the cow

As for the rabbit doe, the experimental design was valid when considering the morphological preservation ratio of the follicles, but not when considering the viability ratio of the follicles. The concentration of the permeating CPA (P = 0.59) and the medium (P = 0.76) seem to have no significant effect on the morphological preservation ratio of ovarian follicles. The nature of the permeating CPA seemed to influence the morphological preservation ratio of the follicles (P = 0.07) although the non-significant difference. PROH tended to improve the morphological preservation ratio, as compared with DMSO. The nature of the non-permeating CPA (P = 0.002) and the cells surfactant (P = 0.04) had significant influence. Trehalose and Albumax® improved morphological preservation ratio compared respectively with sucrose and FCS. No significant interaction was observed between the nature of the permeating CPA and its concentration.

In order to discriminate permeating CPAs, ovarian fragments from 5 cows were frozen using 1.5M DMSO or 1.5M PROH with 4 g/L Albumax® and 0.2M trehalose according to a post-seeding freezing rate at 0.3°C/min. Before freezing, ovarian tissue showed 40.6 ± 12.6% of type I preantral follicles. Proportion of type I follicles was reduced to 20.2 ± 3.9% after cryopreservation using DMSO and to 23.8 ± 3.4% when using PROH. No statistical difference was found between DMSO and PROH. Morphological defects of type II were the most important kind of defect (47.0 ± 11.7%). Proportion of type IV follicular defects was significantly improved compared to control for the both CPAs (47.2 ± 7.8% and 44.8 ± 4.4% for DMSO and PROH respectively). Proportion of type III follicles was constant before and after freezing. Ovarian stroma seems to be affected by cryopreservation and shows spaces and disjoined cells.

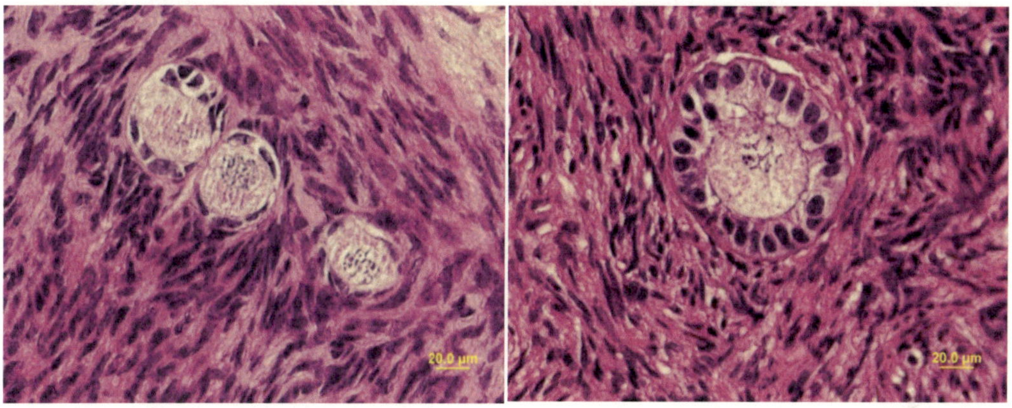

A B

Fig. 6. Cow follicular morphology before (A) and after (B) cryopreservation using PROH and slow freezing rate

As for the other species, the influence of the post-seeding freezing rate was evaluated when freezing with 2M PROH. Proportions of morphologically normal follicles were significantly reduced after freezing compared with fresh tissue, whatever the post-seeding freezing rate 17.6 ± 6.2% after freezing at 2°C/min vs. 57.8 ± 13.0% before freezing and 17.8 ± 6.5% after freezing at 0.3°C/min vs. 60.0 ± 4.9% before freezing).

2.4.5 In vivo follicle resumption from cryopreserved ovarian tissue

In the doe rabbit, nine pups were born from cryopreserved grafted group, suggesting the efficiency of the cryopreservation protocols based on PROH and trehalose and using very slow freezing rate. At necropsy, follicular structures were observed in most of females.

In the bitch, the cryopreservation method optimized with the fractional experimental design and validated by in vitro assessment (morphology, viability, and toxicity) was then evaluated by heterotopic xenografting to determine whether the ovarian tissue integrity and the follicular growth potential were maintained. The frozen-thawed tissues were grafted to female SCID mice as previously described. The histological assessments of the follicle population revealed a significant increase in the density and distribution of secondary follicles from eight weeks post grafting compared to the follicle population at 1 week ($P<0.05$). Consequently, the shift from primordial-primary follicles to secondary follicles occurred in a time laps of eight weeks. Moreover morphologically normal follicles were observed until 16 weeks post transplantation and intact secondary follicles with more than 3 layers of granulosa cells and a normal oocyte surrounded by a well-defined zona pellucida were present at this time. Despite a massive follicular loss touching particularly the early follicles and occurring just after grafting, the graft survived long term xenografting. Similarly, after an important loss just after grafting (one week grafts) the stromal cell number increased during the graft period, to reach a density comparable to fresh ovarian tissue at 16 weeks. Otherwise, the vascularization setting-up was assessed by immunohistochemistry as developed previously. The vessel density analysis revealed that already at one week post grafting the vessel density within the graft was comparable to the fresh control ovarian tissue. Moreover, the vessel density tended to increase at 16 weeks post grafting compared to the other groups (fresh control, one and eight weeks, $P<0.05$) even if no significant difference was registered. No antral follicles was present at the end of the graft time, however, persisting vaginal cornification was noticed on the recipients from 5-9 days post grafting, and estrus behavior was observed several times during the graft period in the recipients cages indicating an hormonal activity resulting from the graft. Taken together, these in vivo results confirm the good preservation of the bitch ovarian tissue by applying our cryopreservation method.

2.5 Discussion

The greater challenge of studies related to the cryopreservation of the ovarian tissue is to define a freezing protocol adapted to the different cell types such as oocytes, follicular cells, stroma cells, etc. One of the objectives of our team was to compare the effects of different freezing parameters based on the morphology and the viability of the follicles, the evaluation of the ultrastructure of the ovarian tissue, the DNA fragmentation of the oocytes or the graft of the ovarian tissue. When the mathematical model was validated, the use of experimental fractional designs allowed us to know simultaneously the individual and the relative effects of different chemo-physical freezing parameters for each species. This statistical method firstly allows a global evaluation of cryopreservation protocols and discriminates the most valuable factors. Finally, the factors which seemed to have a discriminating effect on follicular morphological preservation were evaluated in a wider population. (Neto et al., 2008).

The results of the experimental design in the doe rabbit and in the bitch show that the post-seeding freezing rate is one of the most important chemophysical factors influencing the morphology or the viability of ovarian follicles. A slow freezing rate (0.3°C/min) seems to be more appropriate for the cryopreservation of the doe rabbit and bitch ovarian tissue. Nevertheless, no influence of the freezing rate was observed in the queen and in the cow. Most of the authors use a very slow freezing rate, which is derived from embryo freezing protocols. However, few studies show the importance of this freezing parameter. In contrast to our results in the bitch and in the doe rabbit, but in accordance with our result in the queen and in the cow, Demirci *et al.* observed a high (but not significant) proportion of follicles without any morphological defect after a post-seeding freezing rate of 2°C/min in the ewe (Demirci et al., 2001). Nevertheless, Gook *et al.* also observed a better follicular preservation when using a slow freezing rate (0.3°C/min) with human ovarian tissue (Gook et al., 1999). Whereas Cleary *et al.* observed no difference in terms of follicular growth after grafting, between a conventional embryo freezing protocol (0.3°C/min) and a passive cooling at 1°C/min from 0°C to -84°C on the mouse ovarian tissue (Cleary et al., 2001). Although these two cooling rates (0.3°C/min and 2°C/min) could be considered as slow, these results may be explained by a difference in cell dehydration during the post-seeding step. With rapid cooling rates, we can hypothesise that time required for the exosmose of the cell water is insufficient and consequently promotes the formation of lethal intracellular ice. While at very slow cooling rates, high level of dehydration occurs with concomitant increasing in solute concentration (salting out). Investigations on the freezing rate were extended in the queen with the evaluation of the third freezing phase. When associated with PROH, slow cooling in solid phase seems to be more appropriate. Nevertheless, several authors use cryopreservation protocols with a direct immersion into liquid nitrogen after -40°C, such as Rodrigues et al. (2004) in the goat, Lucci et al. (2004) in the zebu cow, Santos et al. (2006) in the ewe or Lima et al. (2006) in the cat. Births had been obtained after graft of ovarian fragments frozen with such a protocol in the rabbit doe (Almodin et al., 2004b) and in the ewe (Almodin et al., 2004a), but not in the cat where in vivo follicular growth were observed when using a freezing protocol with controlled third phase (Bosch et al., 2004).

The experimental designs revealed a crucial role of the permeating CPA in the doe rabbit, in the cow and in the bitch, added to a crucial role of the non permeating CPA in the doe rabbit and in the cow. Among the various freezing protocols described in the literature, those using DMSO or PROH as permeating CPA seems to be more efficient, whatever the species. Our results suggest that PROH improve the follicular quality after freezing in the doe rabbit and in the queen. Results of experimental design obtained in the cow suggested a better morphological preservation rate when using PROH. However, in this species, standard comparison between DMSO and PROH doesn't confirm these results. Contrary to that, bitch ovarian tissue seems to be better preserved in freezing medium composed of DMSO. These results were confirmed by the follicular growth observed after xenograft. It can be hypothesized that DMSO penetrate better within the tissue than PROH. Indeed, the bitch ovarian tissue as the goat or the ewe is rich in collagenous fibbers and more fibrous than the doe rabbit ovarian tissue for example. Therefore, a good ability to penetrate within the ovarian tissue is an important characteristic for the chosen CPA Nevertheless, both CPA have sensibly the same molecular weight (PROH: 76.10 g/mol, DMSO: 78.14 g/mol) with a lower weight for PROH. Thus a better penetration of the DMSO cannot be explained by this physical parameter. The explanation may come from the toxicity of both CPAs. Our team also investigated the toxicity of the both CPAs on bitch ovarian tissue after equilibration steps at room temperature without freezing and registered a deleterious effect of PROH compared to DMSO on preantral follicle viability in this species.

Except in canine and feline models, addition of non-permeating CPA in the freezing medium seems to improve the protective effect of CPAs. The protective effect of PROH seems to be improved when it is associated with trehalose, in the doe rabbit and in the cow. This observation was confirmed by electron microscopy evaluation of the doe rabbit ovarian tissue subjected to cryopreservation. Ultrastructure of doe rabbit follicles after cryopreservation was well preserved, but stromal cells and fibroblasts were damaged. Such alterations have been observed in human tissues after cryopreservation (Navarro-Costa et al., 2005; Santos et al., 2010). However, fibroblasts can easily be reproduced by cell division, indicating that damage to the stroma can be repaired. Collagen fibers did not seem to be damaged by cryopreservation in this study, but they were sparse in the doe rabbit ovary. This observation may explain the fragility of doe rabbit ovarian tissue during the equilibration, freezing and thawing steps (Neto et al., 2005).

Sugar are not systematically associated with permeating CPA, but Marsella et al. (2008), showed the advantageous effect of sucrose. Trehalose has been frequently used in embryo cryopreservation, but not in ovarian tissue cryopreservation. Sucrose and trehalose share the property to stabilise cellular membranes and proteins via the formation of hydrogen bonds with polar residues of phospholipidic membrane. This property allows preserving the membrane integrity under anhydrous conditions. Moreover, it modifies the temperature at which the separation of lipid phase occurs during cooling (Crowe et al., 1984; Crowe et al., 1985; Crowe et al., 2001). As compared to other sugars, trehalose seems to have a higher capacity to preserve biomolecules, cellular membrane and cells in a drying or in a freezing state (Crowe et al., 1996; Storey et al., 1998; Sano et al., 1999; Welsh and Herbert, 1999).

Few comparable studies have been reported in the cryopreservation of different mammalian species. Despite encouraging results in the different studied species, and except in the queen, none of evaluated protocols allows preserving the same proportion of normal follicles than before freezing. Most of authors observed a reduction of normal follicles in frozen/thawed ovarian tissue compared with fresh control when using similar freezing protocols in the mouse (Candy et al., 1997), the goat (Rodrigues et al., 2004), the cow (Lucci et al., 2004), and the ewe (Demirci et al., 2002). As for the queen, no morphological difference was observed in human follicles before and after cryopreservation (Hovatta et al., 1996; Fabbri et al., 2006) Newton observed similar proportions of "viable" follicles after freezing when using DMSO, ethylene glycol or PROH and xenografting (Newton et al., 1996).

Live births in the rabbit doe and follicular growths observed in the bitch after grafting of cryopreserved ovarian tissue shows the efficacy of evaluated freezing protocols. Almodin et al. obtained live offspring after grafting of small fragments of cryopreserved rabbit ovarian tissue using 1.5 M DMSO and a very slow post seeding freezing rate (Almodin et al., 2004b). In the bitch, results about in vivo growth obtained after cryopreservation are relatively scarce compared to other species. Ishijima et al. (2006) tried to transplant vitrified ovarian tissue (2M DMSO, 3M PROH, 1M acetamide) into immunodeficient mice during 4 weeks and observed signs of growth in the early follicles (primordial-primary follicles). However, they noticed an important follicular loss occurring just after grafting which is in accordance with our results obtained with slow-frozen tissue. The time necessary for setting up of the neovascularization within the grafted tissue seems to be more deleterious for the cells than the cryopreservation technique itself. Except our results obtained on the bitch ovarian tissue cryopreservation no other results using slow freezing of female germ cells was obtained in this species. Furthermore, live birth has not yet been obtained after ovarian tissue cryopreservation, but the difficulties to mastered *in vitro* maturation and fertilization steps in canines do not contribute to the development of this technique.

For the first time a complete evaluation process of important factors influencing the morphology and the viability of preantral follicles has been performed after equilibration process and freezing in different species. These results suggest that cryopreservation of ovarian tissue is a promising and suitable technique that could be used as complementary tool to embryo cryopreservation, to preserve the animals' genetic resources by the female pathway. Doe rabbit could also be used as a biomedical model to investigate the long term consequences of cryopreservation on ovarian follicles and the birth of future progenies.

3. Perspectives

In definitive, the use of factorial fractional experimental design approach allowed us to develop suitable cryopreservation protocol in different species, while reducing the number of experiments and increasing the number of parameters evaluated. However it can be noticed in our model species, but also in the literature, that results can be radically different according to the species. Moreover, among the numerous articles published on ovarian tissue cryopreservation heterogeneous results can be observed in the same species after applying roughly or widely different slow-freezing protocols. One of the candidates to explain such disparity in the obtained results is the amount of ice crystals formed during slow freezing. Indeed, the strategy of slow cooling is to decrease cell temperature enough slowly to allow removal of most of the freezable intracellular water before reaching the ice nucleation temperature. The main objective of this method is to avoid intracellular ice crystal formation which is known to be lethal. However ice crystallization still occurs extracellularly with the risk of tissue shrinkage or disorganization of the tissue components.

As ice formation and melting are exothermic and endothermic phenomena respectively, they can be objectified and studied by thermodynamical measures. Among the various physical methods of analysis, Differential Scanning Calorimetry (DSC) is an interesting tool. In fact, DSC gives the opportunity to measure important parameters of a cryopreservation solution under dynamic conditions. A cryopreservation solution can thus be characterized by its thermal properties such as temperatures of phase transitions and quantity of ice crystallized and melted. Two types of DSC are commonly used: power-compensation DSC and heat-flux DSC. Our team chose the first one in order to study cryopreservation solutions with a more fundamental approach than with biological methods.

The power compensation DSC is based on the "zero balance principle" as explained as follow. The sample and a reference are placed in two microfurnaces which are continuously cooled by liquid nitrogen. The temperature of each microfurnace can be, on the one hand, precisely measured by a temperature sensor and, on the other hand, precisely adjusted by a heating resistor. Each microfurnace contains one sensor and one resistor. The principle of the power compensation DSC is to maintain the two microfurnaces under the same temperature regardless of phase transitions or reactions occurring in the sample. Thus, when a phase transition occurs in the sample, the heat released or absorbed by the sample has to be compensated by the heating resistor which is below the sample. Consequently, the calorimeter measures a difference between the heating powers provided by the two resistors. This difference reveals the phase transition. When this phase transition is crystallization, this difference allows also us to measure the quantity of ice formed.

Since recent years, several strategies are developed to avoid deleterious effects of the ice crystal formation during cooling, and thermodynamical approaches are more and more associated to these strategies.

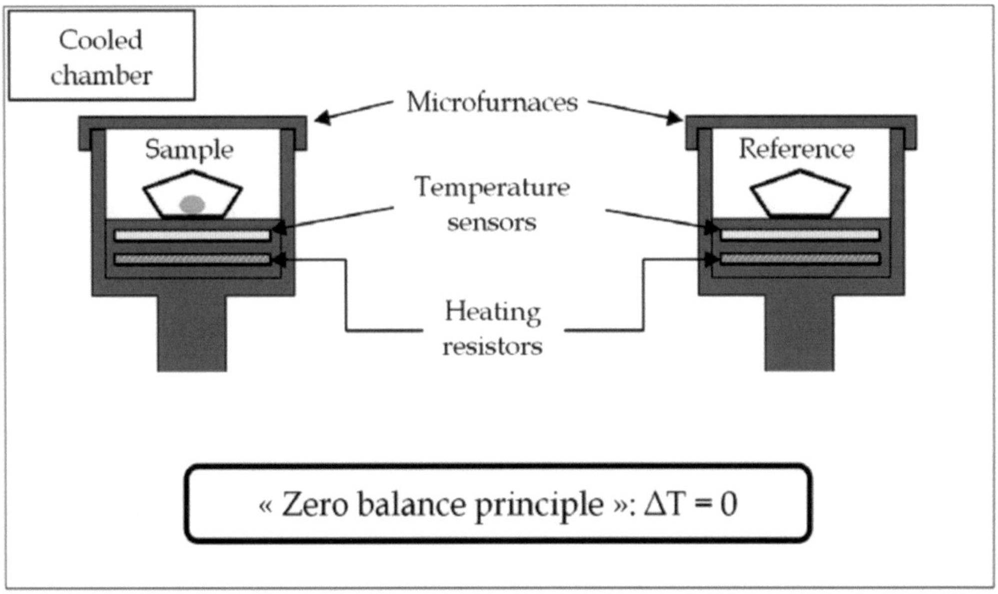

Fig. 7. Scheme of power-compensation DSC

The specimen and reference temperatures are controlled independently using separate (identical) ovens. The temperature difference between the sample and reference is maintained to zero by varying the power input to the two furnaces. This energy is then a measure of the enthalpy or hat capacity changes in the test specimen (relative to the reference).

3.1 Prevent the ice formation

The physical definition of vitrification is the glass-like solidification of solutions at low temperatures, without the formation of intracellular ice crystals. The vitrification technique is the solidification of a liquid without ice crystal formation. This phase is obtained by increasing the solute concentration of the vitrification media (increased viscosity that makes water solidify without expansion) or by using very fast cooling rate to avoid molecular rearrangement into ice crystals but into an amorphous glass. (Vajta, 2000). Consequently, the risk of injuries due to intra- or extracellular ice crystallization is avoided, which constitutes the main advantage of this technique.

Despite the fact that slow freezing is the most widely used cryopreservation technique, vitrification is a viable and promising alternative that is increasingly becoming more attractive to the commercial sector. Vitrification has been used for oocyte and tissue cryopreservation since the 1980s (Rall and Fahy, 1985). Since the first use of this approach in ovarian tissue cryopreservation, it has been well optimized, particularly by decreasing the concentration of CPAs used to reduce their toxicity, but also by increasing the cooling rate applied (Chen et al., 2006; Keros et al., 2009). These improvements of vitrification in ovarian tissue but also in embryo and oocyte vitrification place the vitrification as a gold standard method for germ cells cryopreservation (Saragusty and Arav, 2011). It has been suggested that with time, conventional slow freezing will be replaced entirely by vitrification techniques (Vajta and Kuwayama, 2006).

As mentioned above, DSC can be a precious aid for the study of vitrification. In fact, DSC allows the detection of phase transitions, both crystallization or glass transition, and can thus determine three thermal properties of vitrification solutions: the vitreous transition temperature, the critical cooling rate and the critical warming rate (Baudot et al., 2007). These three thermal properties are specific to each vitrification solution and could be used for a better utilization of the solutions. The critical cooling rate and the critical warming rate are the cooling and warming speeds above which the crystallization cannot occur between the vitreous transition temperature and the crystallization temperature. The calculation of these critical cooling and warming rates is based on a semi-empiric model developed by Boutron in 1986 according to the "classic" theory of crystallization (Boutron, 1986). This model reproduces well analytically the experimental results, but some approximations are introduced.

3.2 Limit the ice formation

As described in the introductive part, extracellular crystallization occurs during slow freezing. The penetrating CPAs like DMSO, PROH or ethylene glycol, are mainly used to reduce the volume of freezable ice within the cells, but their respective toxicity for the cells limit the potential concentration usable. However, since the last decades, non-penetrating CPAs like disaccharides have been widely used in cryoprotective solution to improve cell preservation and reduce the penetrating CPA amount necessary. To our knowledge, the first use of sucrose as cryoprotectant of ovarian tissue was in 1996 with human ovaries (Hovatta et al., 1996). Sugars have been proved to play a role on ice crystal formation and stability (Kuleshova et al., 1999). Previous studies on mouse embryos revealed that sugars increase the homogeny ice transition temperature in freezing solutions. By this way, dehydration of the cells occurs when they are still permeable to water without reaching ice nucleation temperature. Moreover, the trehalose was reported to reduce the size of ice crystals and shorten their elongation during freezing (Sei et al., 2002).

On the basis of the contribution of DSC in other fields of cryopreservation [vitrification of ovary (Baudot et al., 2007), slow-freezing or vitrification of plants (Volk and Walters, 2006; Skyba et al., 2011), slow-freezing of aquatic crustaceans (Issartel et al., 2006)], our team chose a thermodynamic approach to study slow-freezing solutions for ovarian tissue. In fact, DSC allows the measure of the maximal quantity of ice formed in a solution (Q_{max}). Q_{max} corresponds to the definition of Boutron : "The heat of solidification are represented by the numbers q of grams of ice whose solidification at 0°C would liberate the same amount of heat as that from 100g of solution on crossing the corresponding peaks. They are close to the real quantities of ice crystallized in % (w/w) of the solution when it is ice which crystallizes. One obtains the heat in calories per 100g of solution by multiplying q by 79.78" (Boutron, 1984). Our team has thus quantified the quantity of ice formed in solutions used for the cryopreservation of ovarian tissue of different species (table 6).

Species	Q_{max} in percentage (w/w) of solution
Bitch	36.81
Doe rabbit	45.79
Cow	37.64

Table 6. Maximal quantity of ice crystallized (Q_{max}) in solutions used for the cryopreservation of ovarian tissue of bitch, doe rabbit and cow.

Then, our team decided to explore two research areas. On the one hand, DSC can compare the quantity of ice formed in two different cryopreservation solutions. Thus, it seems possible to select the most suitable cryopreservation solutions for slow freezing methods. The first results obtained in doe rabbit seem to confirm this hypothesis. In fact between two cryopreservation solutions tested, those for which the quantity of ice formed was the lowest, was also the one with the best biological results. On the other hand, for a given solution, DSC allows the measure of the quantity of ice formed for different freezing kinetics. Consequently, it seems also possible to select the most suitable kinetics according to the cryopreservation solutions.

3.3 Promote the formation of a non-vulnerable extracellular ice

Another approach to optimize cryopreservation process should be to control the ice crystal growth and shape in order to promote the less deleterious crystallization. It is already assumed that intracellular ice formation is lethal. Nonetheless, the recent observations of rabbit ovarian tissue by cryoscanning electronic microscopy and cryofracture reveal that depending on the cooling rate, the extracellular ice shape is modified. Moreover, according to these results the seeding temperature influences the shape and regularity of the ice crystals resulting in large uniform crystals when seeding was induced close to the solution melting point (Gosden et al., 2010).

Otherwise, a better understanding of intracellular ice formation can also be advantageous to improve cryopreservation processes. Indeed, Han *et al* (2009) investigated the size of intracellular ice crystals in mouse oocytes by cryomicroscopy. They conclude that increasing the concentration of macromolecules in the cells by increasing the extracellular non permeating solute concentration significantly reduced the required permeating CPA concentration for intracellular vitrification. Moreover they also observed that intracellular ice melting point was always lower than extracellular ice. Taken together, this information can be helpful to optimize the cryopreservation protocols.

Regarding DSC, a recent study showed that it is possible to differentiate the crystallization of intra and extracellular ice depending on freezing kinetics (Seki et al., 2009). Consequently, in addition to the measure of the quantity of crystallized ice, DSC can provide a better control of ice formation.

4. Acknowledgment

The results presented in this chapter were partially supported financially by Région Rhône-Alpes and the breeding center (CESECAH) of the French Guide Dog Federation (FFAC). The authors also thank the LLC Hycole (Marcoing, France) for their technical support.

The authors want to thank equally regards the Electronic Microscopy Center of the University of Lyon, the Laboratory of Pathological Anatomy of the Veterinary Campus of

Lyon and the Institute Claude Bourgelat for access to their facilities. The authors are also grateful to all veterinarians which allowed the ovary collection.

5. References

Abrishami, M.; Anzar, M.; Yang, Y.&Honaramooz, A. (2010). Cryopreservation of immature porcine testis tissue to maintain its developmental potential after xenografting into recipient mice. *Theriogenology*, Vol.73, No.1, pp: 86-96,

Almodin, C. G.; Minguetti-Camara, V. C.; Meister, H.; Ceschin, A. P.; Kriger, E.&Ferreira, J. O. (2004a). Recovery of natural fertility after grafting of cryopreserved germinative tissue in ewes subjected to radiotherapy. *Fertil Steril*, Vol.81, No.1, pp: 160-164,

Almodin, C. G.; Minguetti-Camara, V. C.; Meister, H.; Ferreira, J. O.; Franco, R. L.; Cavalcante, A. A.; Radaelli, M. R.; Bahls, A. S.; Moron, A. F.&Murta, C. G. (2004b). Recovery of fertility after grafting of cryopreserved germinative tissue in female rabbits following radiotherapy. *Hum Reprod*, Vol.19, No.6, pp: 1287-1293,

Andrabi, S. M.&Maxwell, W. M. (2007). A review on reproductive biotechnologies for conservation of endangered mammalian species. *Anim Reprod Sci*, Vol.99, No.3-4, pp: 223-243,

Aubard, Y.; Newton, H.; Scheffer, G.&Gosden, R. (1998). Conservation of the follicular population in irradiated rats by the cryopreservation and orthotopic autografting of ovarian tissue. *Eur J Obstet Gynecol Reprod Biol*, Vol.79, No.1, pp: 83-87,

Bahadur, G.&Ralph, D. (1999). Gonadal tissue cryopreservation in boys with paediatric cancers. *Hum Reprod*, Vol.14, No.1, pp: 11-17,

Bahadur, G.; Chatterjee, R.&Ralph, D. (2000). Testicular tissue cryopreservation in boys. Ethical and legal issues: case report. *Hum Reprod*, Vol.15, No.6, pp: 1416-1420,

Baudot, A.; Courbiere, B.; Odagescu, V.; Salle, B.; Mazoyer, C.; Massardier, J.&Lornage, J. (2007). Towards whole sheep ovary cryopreservation. *Cryobiology*, Vol.55, No.3, pp: 236-248,

Boettcher, P. J.; Stella, A.; Pizzi, F.&Gandini, G. (2005). The combined use of embryos and semen for cryogenic conservation of mammalian livestock genetic resources. *Genet Sel Evol*, Vol.37, No.6, pp: 657-675,

Bosch, P.; Hernandez-Fonseca, H. J.; Miller, D. M.; Wininger, J. D.; Massey, J. B.; Lamb, S. V.&Brackett, B. G. (2004). Development of antral follicles in cryopreserved cat ovarian tissue transplanted to immunodeficient mice. *Theriogenology*, Vol.61, No.2-3, pp: 581-594,

Boutron, P. (1984). More accurate determination of the quantity of ice crystallised at low cooling rates in glycerol and 1,2-propanediol aqueous solutions : comparison with equilibrium. *Cryobiology*, Vol.21, 183-191,

Boutron, P. (1986). Comparison with the theory of the kinetics and extent of ice crystallization and of glass-forming tendency in aqueous cryoprotective solutions. *Cryobiology*, Vol.23, 88-102,

Candy, C. J.; Wood, M. J.&Whittingham, D. G. (1997). Effect of cryoprotectants on the survival of follicles in frozen mouse ovaries. *J Reprod Fertil*, Vol.110, No.1, pp: 11-19,

Checura, C. M.&Seidel, G. E., Jr. (2007). Effect of macromolecules in solutions for vitrification of mature bovine oocytes. *Theriogenology*, Vol.67, No.5, pp: 919-930,

Chen, S. U.; Chien, C. L.; Wu, M. Y.; Chen, T. H.; Lai, S. M.; Lin, C. W.&Yang, Y. S. (2006). Novel direct cover vitrification for cryopreservation of ovarian tissues increases follicle viability and pregnancy capability in mice. *Hum Reprod*, Vol.21, No.11, pp: 2794-2800,

Cleary, M.; Snow, M.; Paris, M.; Shaw, J.; Cox, S. L.&Jenkin, G. (2001). Cryopreservation of mouse ovarian tissue following prolonged exposure to an Ischemic environment. *Cryobiology*, Vol.42, No.2, pp: 121-133,

Commin, L.; Rosset, E.; Allard, A.; Joly, T.; Guérin, P.; Neto, V.&Buff, S. (2011). Assesment of tissue functionality after xenotransplantation of canine ovarian cortex to severe combined immunodeficient mice (SCID mice), *Theriogenology Conference and Symposia*, pp: Milwaukee, Wisconsin

Critser, J. K. (1998). Current status of semen banking in the USA. *Hum Reprod*, Vol.13 Suppl 2, 55-67; discussion 68-59,

Crowe, J. H.; Crowe, L. M.&Chapman, D. (1984). Preservation of membranes in anhydrobiotic organisms: the role of trehalose. *Science*, Vol.223, No.4637, pp: 701-703,

Crowe, J. H.; Crowe, L. M.; Oliver, A. E.; Tsvetkova, N.; Wolkers, W.&Tablin, F. (2001). The trehalose myth revisited: introduction to a symposium on stabilization of cells in the dry state. *Cryobiology*, Vol.43, No.2, pp: 89-105,

Crowe, L. M.; Reid, D. S.&Crowe, J. H. (1996). Is trehalose special for preserving dry biomaterials? *Biophys J*, Vol.71, No.4, pp: 2087-2093,

Crowe, L. M.; Crowe, J. H.; Rudolph, A.; Womersley, C.&Appel, L. (1985). Preservation of freeze-dried liposomes by trehalose. *Arch Biochem Biophys*, Vol.242, No.1, pp: 240-247,

Curaba, M.; Verleysen, M.; Amorim, C. A.; Dolmans, M. M.; Van Langendonckt, A.; Hovatta, O.; Wyns, C.&Donnez, J. (2011). Cryopreservation of prepubertal mouse testicular tissue by vitrification. *Fertil Steril*, Vol.95, No.4, pp: 1229-1234 e1221,

Deanesly, R. (1954). Immature rat ovaries grafted after freezing and thawing. *J Endocrinol*, Vol.11, No.2, pp: 197-200,

Demirci, B.; Lornage, J.; Salle, B.; Frappart, L.; Franck, M.&Guerin, J. F. (2001). Follicular viability and morphology of sheep ovaries after exposure to cryoprotectant and cryopreservation with different freezing protocols. *Fertil Steril*, Vol.75, No.4, pp: 754-762,

Demirci, B.; Salle, B.; Frappart, L.; Franck, M.; Guerin, J. F.&Lornage, J. (2002). Morphological alterations and DNA fragmentation in oocytes from primordial and primary follicles after freezing-thawing of ovarian cortex in sheep. *Fertil Steril*, Vol.77, No.3, pp: 595-600,

Donnez, J.; Dolmans, M. M.; Demylle, D.; Jadoul, P.; Pirard, C.; Squifflet, J.; Martinez-Madrid, B.&van Langendonckt, A. (2004). Livebirth after orthotopic transplantation of cryopreserved ovarian tissue. *Lancet*, Vol.364, No.9443, pp: 1405-1410,

Edge, B.; Holmes, D.&Makin, G. (2006). Sperm banking in adolescent cancer patients. *Arch Dis Child*, Vol.91, No.2, pp: 149-152,

Fabbri, R.; Pasquinelli, G.; Bracone, G.; Orrico, C.; Di Tommaso, B.&Venturoli, S. (2006). Cryopreservation of human ovarian tissue. *Cell Tissue Bank*, Vol.7, No.2, pp: 123-133,

FAO (2007). The state of the world's animal genetic resources for food and agriculture, *Food and Agriculture Organization of the United Nations*, Commission on Genetic Resources for Food and Agriculture, Rome, Italy

Fishel, S.; Green, S.; Hunter, A.; Lisi, F.; Rinaldi, L.; Lisi, R.&McDermott, H. (1997). Human fertilization with round and elongated spermatids. *Hum Reprod*, Vol.12, No.2, pp: 336-340,

Gook, D. A.; Edgar, D. H.&Stern, C. (1999). Effect of cooling rate and dehydration regimen on the histological appearance of human ovarian cortex following cryopreservation in 1, 2-propanediol. *Hum Reprod*, Vol.14, No.8, pp: 2061-2068,

Gook, D. A.; Edgar, D. H.&Stern, C. (2004). Cryopreservation of human ovarian tissue. *Eur J Obstet Gynecol Reprod Biol*, Vol.113 Suppl 1, S41-44,

Gosden, R. G.; Baird, D. T.; Wade, J. C.&Webb, R. (1994). Restoration of fertility to oophorectomized sheep by ovarian autografts stored at -196 degrees C. *Hum Reprod*, Vol.9, No.4, pp: 597-603,

Gosden, R. G.; Yin, H.; Bodine, R. J.&Morris, G. J. (2010). Character, distribution and biological implications of ice crystallization in cryopreserved rabbit ovarian tissue revealed by cryo-scanning electron microscopy. *Hum Reprod*, Vol.25, No.2, pp: 470-478,

Gougeon, A.&Chainy, G.B.N. (1987). Morphometric studies of small follicles in ovaries of women at different ages. *J Reprod Fertil*, Vol.81, 433-442,

Guerin, J. F. (2005). [Testicular tissue cryoconservation for prepubertal boy: indications and feasibility]. *Gynecol Obstet Fertil*, Vol.33, No.10, pp: 804-808,

Han, X.&Critser, J. K. (2009). Measurement of the size of intracellular ice crystals in mouse oocytes using a melting point depression method and the influence of intracellular solute concentrations. *Cryobiology*, Vol.59, No.3, pp: 302-307,

Holt, W. V.&Pickard, A. R. (1999). Role of reproductive technologies and genetic resource banks in animal conservation. *Rev Reprod*, Vol.4, No.3, pp: 143-150,

Honaramooz, A.; Snedaker, A.; Boiani, M.; Scholer, H.; Dobrinski, I.&Schlatt, S. (2002). Sperm from neonatal mammalian testes grafted in mice. *Nature*, Vol.418, No.6899, pp: 778-781,

Hovatta, O.; Silye, R.; Krausz, T.; Abir, R.; Margara, R.; Trew, G.; Lass, A.&Winston, R. M. (1996). Cryopreservation of human ovarian tissue using dimethylsulphoxide and propanediol-sucrose as cryoprotectants. *Hum Reprod*, Vol.11, No.6, pp: 1268-1272,

Isachenko, V.; Isachenko, E.; Weiss, J. M.; Todorov, P.&Kreienberg, R. (2009). Cryobanking of human ovarian tissue for anti-cancer treatment: comparison of vitrification and conventional freezing. *Cryo Letters*, Vol.30, No.6, pp: 449-454,

Ishijima, T.; Kobayashi, Y.; Lee, D. S.; Ueta, Y. Y.; Matsui, M.; Lee, J. Y.; Suwa, Y.; Miyahara, K.&Suzuki, H. (2006). Cryopreservation of canine ovaries by vitrification. *J Reprod Dev*, Vol.52, No.2, pp: 293-299,

Issartel, J.; Voituron, Y.; Odagescu, V.; Baudot, A.; Guillot, G.; Ruaud, J. P.; Renault, D.; Vernon, P.&Hervant, F. (2006). Freezing or supercooling: how does an aquatic subterranean crustacean survive exposures at subzero temperatures? *J Exp Biol*, Vol.209, No.Pt 17, pp: 3469-3475,

Jahnukainen, K.; Ehmcke, J.; Hergenrother, S. D.&Schlatt, S. (2007). Effect of cold storage and cryopreservation of immature non-human primate testicular tissue on spermatogonial stem cell potential in xenografts. *Hum Reprod*, Vol.22, No.4, pp: 1060-1067,

Jezek, D.; Schulze, W.; Kalanj-Bognar, S.; Vukelic, Z.; Milavec-Puretic, V.&Krhen, I. (2001). Effects of various cryopreservation media and freezing-thawing on the morphology of rat testicular biopsies. *Andrologia*, Vol.33, No.6, pp: 368-378,

Johnston, L. A.&Lacy, R. C. (1995). Genome resource banking for species conservation: selection of sperm donors. *Cryobiology*, Vol.32, No.1, pp: 68-77,

Karlsson, J.O.M.&Toner, M. (1996). Long-term storage of tissues by cryopreservation : critical issues. *Biomaterials*, Vol.17, 243-256,

Keros, V.; Rosenlund, B.; Hultenby, K.; Aghajanova, L.; Levkov, L.&Hovatta, O. (2005). Optimizing cryopreservation of human testicular tissue: comparison of protocols with glycerol, propanediol and dimethylsulphoxide as cryoprotectants. *Hum Reprod*, Vol.20, No.6, pp: 1676-1687,

Keros, V.; Hultenby, K.; Borgstrom, B.; Fridstrom, M.; Jahnukainen, K.&Hovatta, O. (2007). Methods of cryopreservation of testicular tissue with viable spermatogonia in pre-pubertal boys undergoing gonadotoxic cancer treatment. *Hum Reprod*, Vol.22, No.5, pp: 1384-1395,

Keros, V.; Xella, S.; Hultenby, K.; Pettersson, K.; Sheikhi, M.; Volpe, A.; Hreinsson, J.&Hovatta, O. (2009). Vitrification versus controlled-rate freezing in cryopreservation of human ovarian tissue. *Hum Reprod*, Vol.24, No.7, pp: 1670-1683,

Kuleshova, L. L.; MacFarlane, D. R.; Trounson, A. O.&Shaw, J. M. (1999). Sugars exert a major influence on the vitrification properties of ethylene glycol-based solutions and have low toxicity to embryos and oocytes. *Cryobiology*, Vol.38, No.2, pp: 119-130,

Ledda, S.; Leoni, G.; Bogliolo, L.&Naitana, S. (2001). Oocyte cryopreservation and ovarian tissue banking. *Theriogenology*, Vol.55, No.6, pp: 1359-1371,

Lima, A. K.; Silva, A. R.; Santos, R. R.; Sales, D. M.; Evangelista, A. F.; Figueiredo, J. R.&Silva, L. D. (2006). Cryopreservation of preantral ovarian follicles in situ from domestic cats (Felis catus) using different cryoprotective agents. *Theriogenology*, Vol.66, No.6-7, pp: 1664-1666,

Lucci, C. M.; Kacinskis, M. A.; Lopes, L. H.; Rumpf, R.&Bao, S. N. (2004). Effect of different cryoprotectants on the structural preservation of follicles in frozen zebu bovine (Bos indicus) ovarian tissue. *Theriogenology*, Vol.61, No.6, pp: 1101-1114,

Luvoni, G. C.; Chigioni, S.&Beccaglia, M. (2006). Embryo production in dogs: from in vitro fertilization to cloning. *Reprod Domest Anim*, Vol.41, No.4, pp: 286-290,

Maclellan, L. J.; Carnevale, E. M.; Coutinho da Silva, M. A.; Scoggin, C. F.; Bruemmer, J. E.&Squires, E. L. (2002). Pregnancies from vitrified equine oocytes collected from super-stimulated and non-stimulated mares. *Theriogenology*, Vol.58, No.5, pp: 911-919,

Marsella, T.; Sena, P.; Xella, S.; La Marca, A.; Giulini, S.; De Pol, A.; Volpe, A.&Marzona, L. (2008). Human ovarian tissue cryopreservation: effect of sucrose concentration on morphological features after thawing. *Reprod Biomed Online*, Vol.16, No.2, pp: 257-267,

Massip, A. (2003). Cryopreservation of bovine oocytes: current status and recent developments. *Reprod Nutr Dev*, Vol.43, No.4, pp: 325-330,

Mazur, P.&Schneider, U. (1986). Osmotic responses of preimplantation mouse and bovine embryos and their cryobiological implications. *Cell Biophys*, Vol.8, No.4, pp: 259-285,

Mechakra, A.; Auberger, B.; Remeuf, F.&Lenoir, J. (1999). Optimisation d'un milieu de culture pour la production d'enzymes protéolytiques acides pas *Penicilium camemberti*. *Sciences des aliments*, Vol.19, 663-675,

Milazzo, J. P.; Vaudreuil, L.; Cauliez, B.; Gruel, E.; Masse, L.; Mousset-Simeon, N.; Mace, B.&Rives, N. (2008). Comparison of conditions for cryopreservation of testicular tissue from immature mice. *Hum Reprod*, Vol.23, No.1, pp: 17-28,

Milazzo, J. P.; Travers, A.; Bironneau, A.; Safsaf, A.; Gruel, E.; Arnoult, C.; Mace, B.; Boyer, O.&Rives, N. (2010). Rapid screening of cryopreservation protocols for murine prepubertal testicular tissue by histology and PCNA immunostaining. *J Androl*, Vol.31, No.6, pp: 617-630,

Navarro-Costa, P.; Correia, S. C.; Gouveia-Oliveira, A.; Negreiro, F.; Jorge, S.; Cidadao, A. J.; Carvalho, M. J.&Plancha, C. E. (2005). Effects of mouse ovarian tissue cryopreservation on granulosa cell-oocyte interaction. *Hum Reprod*, Vol.20, No.6, pp: 1607-1614,

Neto, V.; Joly, T.&Salvetti, P. (2007a). Ovarian tissue cryopreservation in the doe rabbit: from freezing to birth *Cryobiology*, Vol.55, No.3, pp: 344,

Neto, V.; Guerin, P.; Lornage, J.; Corrao, N.; Buff, S.&Joly, T. (2005). [Follicular morphology after ovarian cortex cryopreservation in the rabbit doe(Oryctolagus cuniculus)]. *Gynecol Obstet Fertil*, Vol.33, No.10, pp: 793-798,

Neto, V.; Joly, T.; Sailley, F.; Corrao, N.; Guérin, P.&Buff, S. (2006). Assessment of different chemo-physical factors influencing the cryopreservation of queen's ovarian cortex, *5th Biannual EVSSAR Congress, 7-9 April*, pp: 283-284, Budapest, Hungary

Neto, V.; Buff, S.; Lornage, J.; Bottollier, B.; Guerin, P.&Joly, T. (2008). Effects of different freezing parameters on the morphology and viability of preantral follicles after cryopreservation of doe rabbit ovarian tissue. *Fertil Steril*, Vol.29, No.5, pp: 1348-1356,

Neto, V.; Joly, T.; Salvetti, P.; Lefranc, A.C.; Corrao, N.; Guerin, P.&Buff, S. (2007b). Ovarian tissue cryopreservation in the Doe rabbit: from freezing to birth, *44th annual Congress of the Society of Cryobiology*, pp: Lake Louise, Canada

Newton, H.; Aubard, Y.; Rutherford, A.; Sharma, V.&Gosden, R. (1996). Low temperature storage and grafting of human ovarian tissue. *Hum Reprod*, Vol.11, No.7, pp: 1487-1491,

Otoi, T.; Yamamoto, K.; Koyama, N.; Tachikawa, S.&Suzuki, T. (1996). A frozen-thawed in vitro-matured bovine oocyte derived calf with normal growth and fertility. *J Vet Med Sci*, Vol.58, No.8, pp: 811-813,

Parkes, A. S.&Smith, A. U. (1953). Regeneration of rat ovarian tissue grafted after exposure to low temperatures. *Proc R Soc Lond B Biol Sci*, Vol.140, No.901, pp: 455-470,

Parkes, A. S.&Smith, A. U. (1954). Preservation of ovarian tissue at -79 degrees C. for transplantation. *Acta Endocrinol (Copenh)*, Vol.17, No.1-4, pp: 313-320,

Parrott, D. M. (1960). The fertility of mice with orthotopic ovarian grafts from frozen tissue. *J. Reprod. Fertil*, Vol.1, 230-241,

Paynter, S. J.; Cooper, A.; Fuller, B. J.&Shaw, R. W. (1999). Cryopreservation of bovine ovarian tissue: structural normality of follicles after thawing and culture in vitro. *Cryobiology*, Vol.38, No.4, pp: 301-309,

Pereira, R. M.&Marques, C. C. (2008). Animal oocyte and embryo cryopreservation. *Cell Tissue Bank*, Vol.9, No.4, pp: 267-277,

Pullium, J. K.; Milner, R.&Tuma, G. A. (2008). Pregnancy following homologous prepubertal ovarian transplantation in the dog. *J Exp Clin Assist Reprod*, Vol.5, No.1, pp: 1,

Rall, W. F.&Fahy, G. M. (1985). Ice-free cryopreservation of mouse embryos at -196 degrees C by vitrification. *Nature*, Vol.313, No.6003, pp: 573-575,

Revel, A.&Mejia, J. (2010). Testicular tissue cryopreservation, In: Testicular tissue cryopreservation, In: Fertility cryopreservation, (Ed.)^(Eds), Cambridge University Press, 978-0-521-51778-2, United Kingdom

Rodrigues, A. P.; Amorim, C. A.; Costa, S. H.; Matos, M. H.; Santos, R. R.; Lucci, C. M.; Bao, S. N.; Ohashi, O. M.&Figueiredo, J. R. (2004). Cryopreservation of caprine ovarian tissue using dimethylsulphoxide and propanediol. *Anim Reprod Sci*, Vol.84, No.1-2, pp: 211-227,

Salvetti, P.; Buff, S.; Afanassieff, M.; Daniel, N.; Guerin, P.&Joly, T. (2010). Structural, metabolic and developmental evaluation of ovulated rabbit oocytes before and after cryopreservation by vitrification and slow freezing. *Theriogenology*, Vol.74, No.5, pp: 847-855,

Sano, F.; Asakawa, N.; Inoue, Y.&Sakurai, M. (1999). A dual role for intracellular trehalose in the resistance of yeast cells to water stress. *Cryobiology*, Vol.39, No.1, pp: 80-87,

Santos, R. R.; Rodrigues, A. P.; Costa, S. H.; Silva, J. R.; Matos, M. H.; Lucci, C. M.; Bao, S. N.; van den Hurk, R.&Figueiredo, J. R. (2006). Histological and ultrastructural analysis of cryopreserved sheep preantral follicles. *Anim Reprod Sci*, Vol.91, No.3-4, pp: 249-263,

Santos, R. R.; Amorim, C.; Cecconi, S.; Fassbender, M.; Imhof, M.; Lornage, J.; Paris, M.; Schoenfeldt, V.&Martinez-Madrid, B. (2010). Cryopreservation of ovarian tissue: an emerging technology for female germline preservation of endangered species and breeds. *Anim Reprod Sci*, Vol.122, No.3-4, pp: 151-163,

Saragusty, J.&Arav, A. (2011). Current progress in oocyte and embryo cryopreservation by slow freezing and vitrification. *Reproduction*, Vol.141, No.1, pp: 1-19,

Schlatt, S.; Kim, S. S.&Gosden, R. (2002). Spermatogenesis and steroidogenesis in mouse, hamster and monkey testicular tissue after cryopreservation and heterotopic grafting to castrated hosts. *Reproduction*, Vol.124, No.3, pp: 339-346,

Sei, T.; Gonda, T.&Arima, Y. (2002). Growth rate and morphology of ice crystals growing in a solution of trehalose and water. *Journal of Crystal Growth*, Vol.240, 218-229,

Seki, S.; Kleinhans, F. W.&Mazur, P. (2009). Intracellular ice formation in yeast cells vs. cooling rate: predictions from modeling vs. experimental observations by differential scanning calorimetry. *Cryobiology*, Vol.58, No.2, pp: 157-165,

Sherman, J. K.&Bunge, R. G. (1953). Effect of glycerol and freezing on some staining reactions of human spermatozoa. *Proc Soc Exp Biol Med*, Vol.84, No.1, pp: 179-180,

Skyba, M.; Faltus, M.; Zàmecnik, J.&Cellàrovà, E. (2011). Thermal analysis of cryopreserved Hypericum perforatum L. shoot tips: Cooling regime dependent dehydration and ice growth. *Thermochimica acta*, Vol.514, 22-27,

Smith, A.U. (1952). Culture of ovarian granulosa cells after cooling to very low temperatures. *Exp. Cell. Res.*, Vol.3, 574-583,

Sofikitis, N.; Mantzavinos, T.; Loutradis, D.; Yamamoto, Y.; Tarlatzis, V.&Miyagawa, I. (1998). Ooplasmic injections of secondary spermatocytes for non-obstructive azoospermia. *Lancet*, Vol.351, No.9110, pp: 1177-1178,

Stahler, E.; Sturm, G.; Spatling, L.; Daume, E.&Buchholz, R. (1976). Investigations into the preservation of human ovaries by means of a cryoprotectivum. *Arch Gynakol*, Vol.221, No.4, pp: 339-344,

Storey, B. T.; Noiles, E. E.&Thompson, K. A. (1998). Comparison of glycerol, other polyols, trehalose, and raffinose to provide a defined cryoprotectant medium for mouse sperm cryopreservation. *Cryobiology*, Vol.37, No.1, pp: 46-58,

Travers, A.; Milazzo, J. P.; Perdrix, A.; Metton, C.; Bironneau, A.; Mace, B.&Rives, N. (2011). Assessment of freezing procedures for rat immature testicular tissue. *Theriogenology*, Vol.76, No.6, pp: 981-990,

Vajta, G. (2000). Vitrification of the oocytes and embryos of domestic animals. *Anim Reprod Sci*, Vol.60-61, 357-364,

Vajta, G.&Kuwayama, M. (2006). Improving cryopreservation systems. *Theriogenology*, Vol.65, No.1, pp: 236–244,

Vanderzwalmen, P.; Zech, H.; Birkenfeld, A.; Yemini, M.; Bertin, G.; Lejeune, B.; Nijs, M.; Segal, L.; Stecher, A.; Vandamme, B.; van Roosendaal, E.&Schoysman, R. (1997). Intracytoplasmic injection of spermatids retrieved from testicular tissue: influence of testicular pathology, type of selected spermatids and oocyte activation. *Hum Reprod*, Vol.12, No.6, pp: 1203-1213,

Volk, G. M.&Walters, C. (2006). Plant vitrification solution 2 lowers water content and alters freezing behavior in shoot tips during cryoprotection. *Cryobiology*, Vol.52, No.1, pp: 48-61,

Welsh, D. T.&Herbert, R. A. (1999). Osmotically induced intracellular trehalose, but not glycine betaine accumulation promotes desiccation tolerance in Escherichia coli. *FEMS Microbiol Lett*, Vol.174, No.1, pp: 57-63,

Woods, E. J.; Benson, J. D.; Agca, Y.&Critser, J. K. (2004). Fundamental cryobiology of reproductive cells and tissues. *Cryobiology*, Vol.48, No.2, pp: 146-156,

Wyns, C.; Van Langendonckt, A.; Wese, F. X.; Donnez, J.&Curaba, M. (2008). Long-term spermatogonial survival in cryopreserved and xenografted immature human testicular tissue. *Hum Reprod*, Vol.23, No.11, pp: 2402-2414,

Wyns, C.; Curaba, M.; Vanabelle, B.; Van Langendonckt, A.&Donnez, J. (2010). Options for fertility preservation in prepubertal boys. *Hum Reprod Update*, Vol.16, No.3, pp: 312-328,

Wyns, C.; Curaba, M.; Martinez-Madrid, B.; Van Langendonckt, A.; Francois-Xavier, W.&Donnez, J. (2007). Spermatogonial survival after cryopreservation and short-term orthotopic immature human cryptorchid testicular tissue grafting to immunodeficient mice. *Hum Reprod*, Vol.22, No.6, pp: 1603-1611,

Zeng, W.; Snedaker, A. K.; Megee, S.; Rathi, R.; Chen, F.; Honaramooz, A.&Dobrinski, I. (2009). Preservation and transplantation of porcine testis tissue. *Reprod Fertil Dev*, Vol.21, No.3, pp: 489-497,

Part 4

Cryopreservation of Aquatic Species

12

Marine Fish Sperm Cryopreservation and Quality Evaluation in Sperm Structure and Function

Qing Hua Liu, Zhi Zhong Xiao, Shi Hong Xu,
Dao Yuan Ma, Yong Shuang Xiao and Jun Li
*Center of Biotechnology R&D, Institute of Oceanology,
Chinese Academy of Sciences, Qingdao
PR China*

1. Introduction

Long-term storage of sperm in liquid nitrogen is a valuable technique for genetic resources preservation (Kopeika et al. 2007). The research on fish sperm cryopreservation has achieved great advances since the first successful sperm cryopreservation in herring (Blaxter 1953). It provides many benefits such as ease of global germplasm shipping and supply (Tiersch et al. 2004), selective breeding and hybridization with desirable characteristics (Henderson-Arzapalo et al. 1994), and conservation of genetic diversity (Van der Walt et al. 1993; Tiersch et al. 2000; Ohta et al. 2001). Furthermore, a frozen sperm bank could maintain the continuous and stable supply of gametes for hatchery seed production or laboratory experimentation. Because of the advantages of this technique, fish sperm of over 200 freshwater and 40 marine species have been cryopreserved successfully (Gwo 2000).

Most of fish sperm cryopreservation researches have focused on freshwater species such as cyprinids (Babiak et al. 1997; Lahnsteiner et al. 2000), salmonoids (Conget et al. 1996; Cabrita et al. 2001), catfishes (Christensen and Tiersch 1997; Viveiros et al. 2000) and loach (Kopeika 2003a, b; Dzuba & Kopeika 2002). In recent years, with the rapid development of marine fish aquaculture, some experiments on germplasm cryopreservation have also been conducted in marine fish species, especially the great commercial value ones such as red seabream (Liu, et al. 2006；Liu, et al. 2007a，b；Liu, et al. 2010 a，b) turbot (Dréanno et al. 1997; Chen et al. 2004), flounder (Richardson et al. 1999; Zhang et al. 2003), and halibut (Billard et al. 1993).

Damage to sperm morphology and function usually occurs during the process of freezing and thawing. Cellular damage may greatly decrease motility, impair velocity, and reduce fertilizing capacity, even lead to DNA strand breakage or mutation (Dréanno et al, 1997; Lahnsteiner et al, 1996a; Warnecke & Pluta 2003; Kopeika et al, 2004). Although motility and fertilizing capacity are usually assessed in frozen-thawed sperm, these methods have limitations. Many factors affect the validity of these assessments, including subjectivity, microscope performance, the quality of eggs, and fertilization protocols. Some new technologies have been used in fish sperm quality analysis, such as computer-assisted sperm analysis (CASA), being used to objectively evaluate sperm motility (Lahnsteiner et al., 1996b; Lahnsteiner et al., 1998; Kime et al., 1996) ; Electron microscopy, being used to detect cryodamage (ultrastructural changes) in frozen-thawed sperm (Zhang et al, 2003, He &Woods 2004); In addition, flow cytometry of fluorescent-stained sperm have been used in mammals (Graham et al, 1990; Gravance et al, 2001) and turkeys (Donoghue et al ,1995), providing rapid, precise information regarding the viability of thousands of individual sperm. In recent years, flow cytometry has also been successfully used to assess both fresh and cryopreserved fish sperm (Ogier de Baulny et al, 1999; Segovia et al, 2000).

Red seabream is one of the most commercially important marine fish species for aquaculture in China. However, the decline of wild red seabream population has occurred due to over fishing and marine pollution in recent years. The use of cryopreserved sperm can provide an efficient method to increase its genetic population size and to help maintain genetic diversity. The aims of this study were to establish efficient methods for cryopreservation of red seabream sperm with 2-mL cryovials and to objectively measure the post-thaw sperm motility characteristics by means of CASA, to evaluate the post-thaw sperm fertilization capacity, and the cryodamage by electron microscopy and flow cytometry.

2. Sperm cryopreservation and quality evaluation

2.1 Materials and methods

2.1.1 Gametes collection

Naturally matured fishes were obtained from Qingdao hatchery during the spawning season (From the middle of March to the end of May). Twenty males and 10 females (3 kg to 4 kg individually, 10 years old) were cultivated in a 20-m3 concrete rearing pond with flow-through seawater and fed daily with cooked meat of bay mussel, *Mytilus edulis*. Prior to handling, males were firstly anesthetized in a 0.003% eugenol bath. Sperm was collected into petri dishes by gently hand-stripping the abdomen of the ripe males. Extreme care was taken to avoid the contamination of sperm with seawater, blood, urine and feces. The percentage of motile spermatozoa was checked with a Nikon-YS-100 light microscope (Nikon Corporation, Tokyo, Japan) at 250 × magnification. Sperm with motility > 85% was kept on crushed ice and transported to the laboratory for further use. Eggs were collected by abdominal pressure of the females at the time of ovulation. Good eggs were slightly yellowish, translucent and round-shaped. Eggs for fertilization trials were collected only from one female.

2.1.2 General procedure for sperm freezing and thawing

Sperm were diluted in Cortland extenders (Liu et al, 2006) containing DMSO with different concentrations (6–24% DMSO). After mixing thoroughly, 1.6 ml sperm was placed into 2-ml cryovials. The cryovials were transferred into a Kryo-360-1.7 programmable freezer (Planer Plc. Middlesex, UK), equilibrated for five minutes at 0°C, and frozen from 0 to −150°C at a cooling rate of 20°C min^{-1}, then plunged into liquid nitrogen for storage. The frozen sperm were thawed in 40°C water bath after being preserved in liquid nitrogen for one month. After that, the thawed sperm was evaluated for motility and fertilizing capacity.

2.1.3 Sperm motion characteristics analysis by using CASA

Sperm motion characteristics were assessed by using a computer-assisted sperm motion analysis system (CASAS-QH-III, Tsinghua Tongfang Inc., Beijing, China) at room temperature (18°C to 20°C). The method for computer-assisted sperm motion analysis was describied in Liu et al (2007b). The designation of the motility status was based on the level of the average path velocity (VAP). Sperm with average path velocity <5 μm s^{-1} were considered immotile, with average path velocity >20 μm s^{-1} were defined as motile, and 5–20 μm s^{-1} as locally motile. Therefore, in the present study sperm motility includes the percentage of local motile sperm and motile sperm. Motility and velocity of fresh and post-thaw sperm were quantitatively recorded by CASA immediately 10 s after activation, and changes of motility of post-thaw sperm frozen with 15% DMSO were observed every 30 s.

2.1.4 Sperm fertilization and hatching experiments

Fertilization capacity of post-thaw sperm frozen with DMSO (6–24% DMSO) was evaluated. The optimized sperm to egg ratio of 500:1 was selected for the following fertilization trials (Li et al., 2006). The artificial fertilization method was described in detail in Liu et al (2007b). Fertilization rates were evaluated within 6–8 h after insemination by counting the percentage of gastrula-stage embryos in relation to the total number of eggs used. Forty-eight hours after fertilization, the number of hatched larvae was counted in each experiment. The hatching rates were calculated as the percentage of hatched larvae in relation to the total number of eggs used in each experiment.

2.1.5 Ultrastructure

Prior to scanning electron microscopy, sperm were fixed in 2.5% glutaraldehyde diluted in PBS (pH 7.6), dehydrated in a series of increasing concentrations of ethanol, critical-point dried, evaporated with gold, and examined with a scanning electron microscope (KYKY-2800B; KYKY Technology Development Ltd., Beijing, China) For transmission electron microscopy, spermwas prefixed in 2.5% glutaraldehyde, post-fixed in 1% osmium tetroxide, and embedded in Epon 812. Ultrathin sections were prepared, counterstained with 2%uranyl acetate followed by lead citrate, and examined with a transmission electron microscope (HITACHI H- 7000; Hitachi Ltd., Tokyo, Japan), and the number of sperm with various categories (normal, slightly damaged, and seriously damaged sperm) of cryodamage was determined. One-hundred sperm were randomly selected for observation each time; this was repeated three times on different sections (total of 300 frozen-thawed sperm for each male).

2.1.6 Rhodamine 123, propidium iodide and flow cytometry

The staining method used was described in (Liu et al. 2007a). An aliquot of mixed fresh or frozen-thawed sperm with 15%DMSO was incubated for 20 min (in the dark, temperature 4 °C) with 5 mg/mL of Rhodamine 123 (Rh123, Sigma Chemical Co., St. Louis, MO, USA).

Thereafter, sperm were incubated for 45 min in 1.5 mL of Cortland extender. With this staining method, only cells with functional mitochondria were stained, due to the negative potential of the inner membrane of the mitochondria. Samples were diluted and counterstained with 5 mg/mL of propidium iodide (PI, Sigma Chemical Co.). After 10 min, sperm samples were analyzed with flow cytometry (FACSvantage SE flow cytometer; Becton Dickinson, Mountain View, CA, USA) as previously described for trout sperm (Ogier de Baulny, et al, 1997). Sperm populations were identified according to their relative red and green fluorescence (staining with PI and Rh123, respectively). Sperm with red (stained with PI) DNA were interpreted as having a damaged plasma membrane, whereas those that were green (stained with Rh123) were interpreted as having intact mitochondrial function. Sperm that were only red (damaged plasma membrane and lacking mitochondrial function), were localized in Region 1, whereas those that were only green (intact membrane and functional mitochondria), were localized in Region 3. Sperm with both red and green fluorescence (damaged plasma membrane and functional mitochondria) were localized in Region 4, and those with no staining (intact plasma membrane, but no mitochondrial activity) were localized in Region 2.

2.1.7 Statistical analysis

To determine the effects of cryopreservation on sperm motility, fertilization capacity, structure and function, a paired-sample t-test was used to compare fresh versus frozen-thawed sperm. All statistical analyses were performed with SPSS Version 11.0 software (SPSS Inc. Chicago, IL, USA) and $P < 0.05$ was considered significant. All data were expressed as mean±S.D.

2.2 Results

2.2.1 Post-thaw sperm viability

The influence of cryopreservation on sperm motility and velocity was shown in Table 1. Percentages of motile post-thaw sperm frozen with 12–21% DMSO were higher than those with 6% DMSO, 9% DMSO and 24% DMSO. However, the procedure of cryopreservation has no significant ($P>0.05$) influence on the motile sperm velocity 10 s after activation compared with fresh sperm. In addition, the post-thaw sperm frozen with 12-21% DMSO showed similar types of straight trajectories.

Cryoprotectant (%)	Motility parameters		
	Locally motile (%)	Motile (%)	Velocity (μm s^{-1})
Fresh sperm	22.0±7.7	64.7±14.2 c	113.1±10.6 a
6% DMSO	16.6±4.6	26.8±11.4 a	89.1±15.0 a
9% DMSO	21.3±7.2	40.3±9.1 ab	91.9±13.5 a
12% DMSO	17.8±10.6	61.6±8.5 c	95.2±12.3 a
15% DMSO	20.4±6.1	64.8±8.7 c	99.3±11.6 a
18% DMSO	21.8±3.9	62.9±6.2 c	97.7±15.2 a
21% DMSO	16.4±4.3	60.8±5.4 c	90.1±12.3 a
24% DMSO	16.6±7.9	55.7±9.2 bc	95.7±8.9 a

Table 1. The influence of cryopreservation on sperm motility and velocity in *P. major*

The motion characteristics of fresh and post-thaw sperm were evaluated by using computer-assisted sperm analysis 10 s after activation. This table shows the percentages of locally motile (VAP range from 5 to 20μm s-1) and motile sperm (VAP > 20μm s-1) as well as their velocity (VAP) for fresh and post-thaw sperm. Values superscripted by the same letter are not significantly different ($P>0.05$, n=5).

The effect of time after activations on post-thaw sperm motility was shown in Fig. 1. The percentages of total motile sperm of both fresh (87. 2 ± 6.1%) and post-thaw sperm (81.9 ± 6.6%) frozen with 15% DMSO were not ($P>0.05$) different significantly 10 s after activation. However, 30 s after activation the percentage of total motile post-thaw sperm (72.3 ± 6.3%) was ($P<0.05$) lower than that of fresh sperm (82.7 ± 7.2%). Sixty seconds after activation, the percentage of post-thaw sperm motility drastically reduced to 38.7 ± 13.2%.

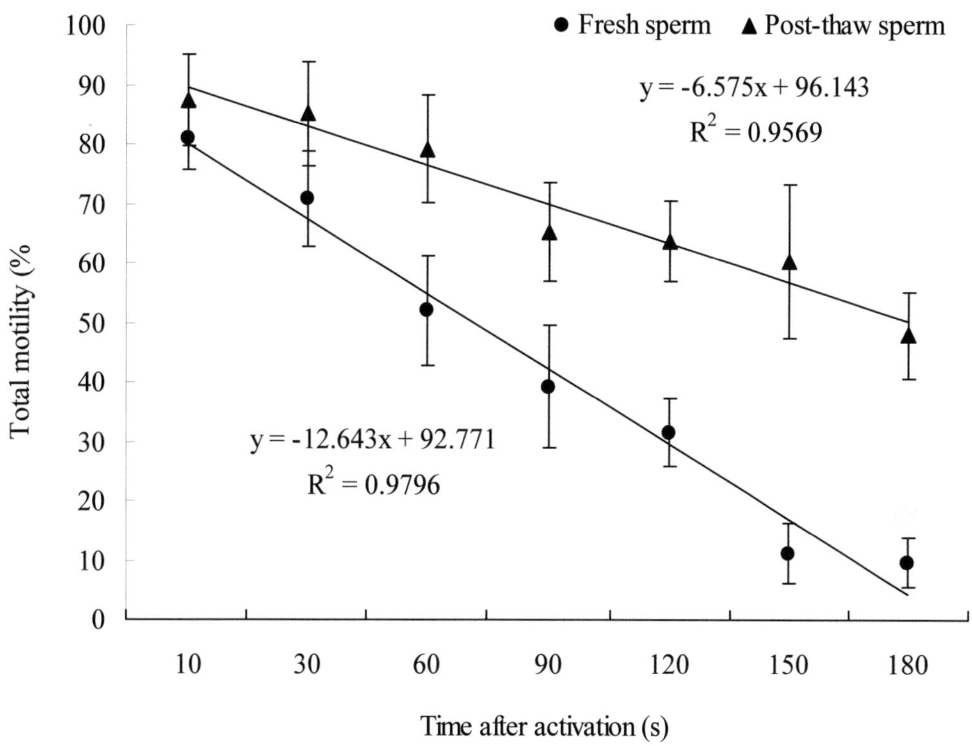

Fig. 1. The influence of time after activation on the motility of fresh and post-thaw sperm in P. major. Ten seconds after activation, the total motilities of fresh and post-thaw sperm frozen with 15% DMSO were observed every 30 s using computer-assisted sperm analysis system. This figure describes the evolution of the total motilities of fresh (▲) and post-thaw sperm (●) after activation respectively (n =5).

2.2.2 Post-thaw sperm fertilizing capacity and hatchability

Fertilization rates and hatching rates of fresh and post-thaw sperm were shown in Fig. 2. The fertilization rates and hatching rates were similar for fresh and post-thaw sperm frozen with 12-21% DMSO. However, lower ($P<0.05$) fertilizing capacity of post-thaw sperm frozen with 6% DMSO, 9% DMSO and 24% DMSO were observed. In addition, the percentages of motile of post-thaw sperm and fertilization rates showed a high positive linear regression (r = 0.876). Similarly, the percentages of motile spermatozoa and hatching rates of post-thaw sperm showed a high positive linear regression (r = 0.878).

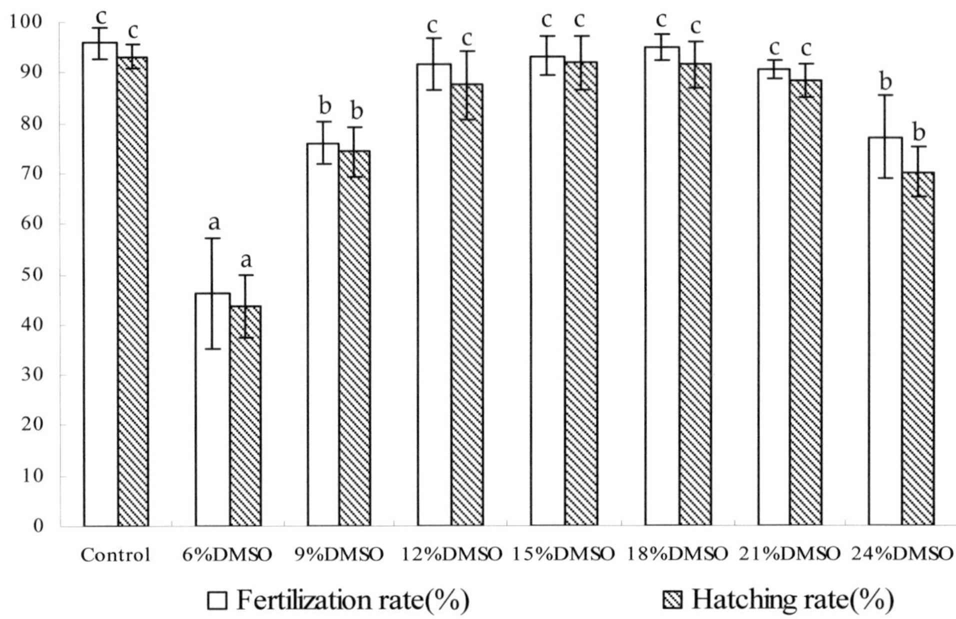

Fig. 2. Fertilization rates and hatching rates of fresh and post-thaw sperm in P. major. Cryopreserved sperm was thawed and activated for the artificial fertilization with sperm to egg ratio 500:1. This figure describes the fertilization rates and hatching rates of post-thaw sperm frozen with 6-24% DMSO. □ For fertilization rates of fresh and post-thaw sperm; ▧ For hatching rates of fresh and post-thaw sperm. Columns marked with the same letter are not significantly different ($P>0.05$, n=5).

2.2.3 Sperm ultrastructure

Ultrastructure of fresh and intact frozen-thawed red seabream sperm are shown in Fig. 1. These sperm had a head, midpiece, and tail. The head was ovoid and contained the nucleus and centriolar complex; the latter consisted of two centrioles. The midpiece was approximately cylindrical and contained mitochondria. The flagellum consisted of nine peripheral doublets and two central microtubules; the axoneme was a typical 9 + 2 structure (Fig. 3 A, B). The proportion of fresh sperm with normal morphology was 77.8 ± 5.6%, whereas after cryopreservation, 63.0 ± 7.2% of the sperm had normal morphology (Fig. 3 C), 20.7 ± 3.1% were partly damaged (e.g. swelling or rupture of head, midpiece and tail region, as shown in Fig. 3 D, as well as damage to mitochondria). Furthermore, 16.4 ± 4.2% were severely damaged; the plasma membranes was completely ruptured and only nuclei, mitochondria, or some fragments of cellular organelles were found (Fig.3 E).

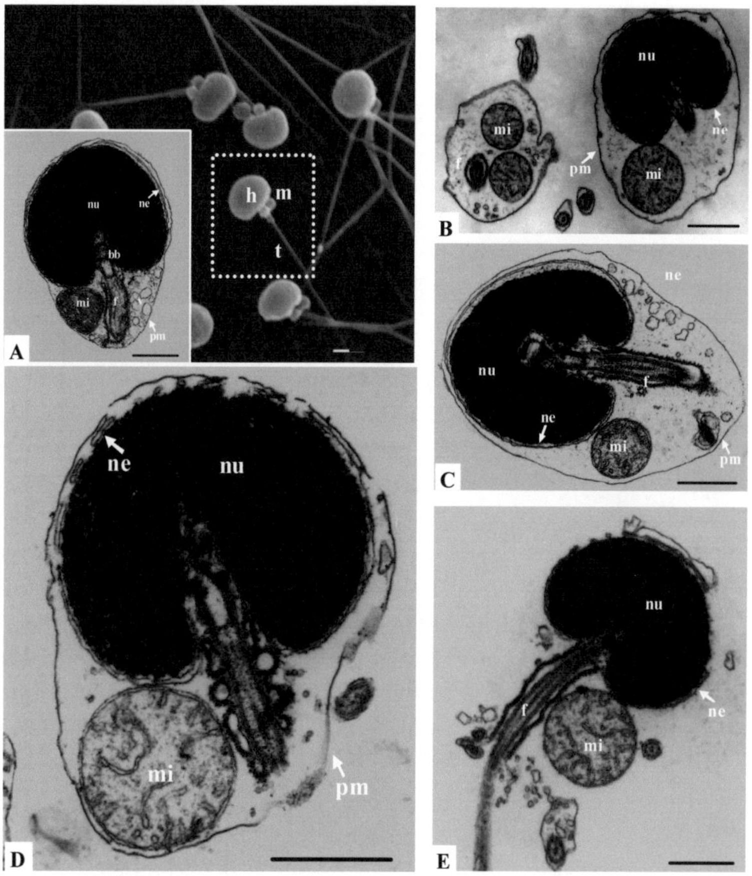

Fig. 3. The morphology and ultrastructure of fresh and normal post-thaw spermatozoa of red seabream. (A) Total view of fresh spermatozoa and the internal structure of head and mid-piece of fresh spermatozoa. (B) Flagellum of fresh spermatozoa. (C) Unchanged spermatozoa cryopreserved with 15% DMSO. (D) Partly damaged spermatozoa. (E) Completely damaged plasmalemma and nuclear envelop. (h, head; m , mid-piece; t, tail region. nu, nucleus; ne, nuclear envelope; bb, basal body; mi, mitochondrion; pm, plasmalemma; f, flagellum; v, vacuole). Scale bar = 0.5 μm.

2.2.4 Fluorescent staining and flow cytometry

Sperm populations were localized into four distinct regions according to their relative green and red fluorescence after staining with PI and Rh123 (Fig. 4). For fresh sperm, 83.9% had an intact membrane and functional mitochondria, 5.1% had nonfunctional mitochondria, 9.8% had nonfunctional mitochondria, and 1.2% had both a damaged membrane and nonfunctional mitochondria; whereas for frozen-thawed sperm, the percentages of sperm localized in four regions were 74.8%(Region 3), 12.7%(Region 4), 9.9% (Region 2), and 2.6%(Region 1), respectively.

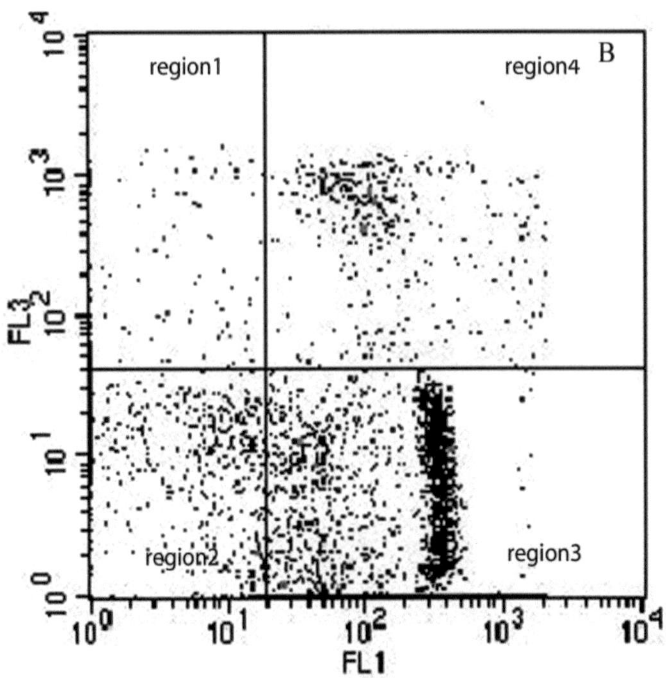

Fig. 4. Flow cytometric dot plots of spermatozoa of red seabream after cryopreservation. Region 1, sperm with a damaged plasma membrane but normal mitochondrial function. Region 2, sperm with an intact plasma membrane but lacking mitochondrial function. Region 3, sperm with an intact plasma membrane and functional mitochondria. Region 4, sperm with a damaged plasma membrane and functional mitochondria.

2.3 Discussion

Motility is an important characteristic for estimating the quality of fresh as well as cryopresrved sperm (Lahnsteiner et al., 1996a). In this study, the freezing-thawing process did not significantly change the main motility pattern and swimming velocity of motile sperm 10 s after activation, and the progressive linear motion was still the dominant pattern. Moreover, for the sperm cryopreserved with 12–21% DMSO, the freezing-thawing process also didn't significantly influence their motility and motility pattern, although it significantly reduced their motility period. However, different results were obtained from the sperm cryopreservation of turbot (Dréanno et al., 1997), which the percentage of motile post-thaw sperm was significantly lower than that obtained from fresh sperm while the velocity and the duration of motion were not significantly modified.

No significant difference in the fertilization rates and hatching rates were observed between sperm cryopreserved with 12–21% DMSO and fresh sperm. However, Lahnsteiner et al. (2003) reported that in cyprinids sperm, the post-thaw fertilization ratios obtained with sperm to egg ratios of $1.3–2.6×10^6$:1 did not reach that of the fresh sperm. Similar results have also been reported in turbot (Chen et al., 2004; Suquet et al., 1998) and flounder (Zhang et al., 2003). These may be due to the species specific or un-ideal protocols used in sperm cryopreservation. In this study, for the post-thaw sperm a high positive correlation was observed between the percentage of motile sperm and fertilizing capacity. This was consistent to the results that obtained from turbot (Dréanno et al., 1999), common carp (Linhart et al., 2000) and African Catfish (Rurangwa et al., 2001).

In the present study, the data from ultrastructural investigation and flow cytometric analysis demonstrates that more that 60% of post- sperm were normal in morphology and mitochondrial function. These results further confirmed the high performance of the protocols established for red seabream. In addition, the high fertilization capacity of post-thaw sperm implies that some of the slightly damaged spermatozoa can still fertilize eggs and develop into larvae. However, it remains to be determined whether the larvae from cryopreserved sperm develop into healthy adults.

During the process of cooling, freezing and thawing, spermatozoa are subjected to a series of damages (Oehninger et al., 2000). In ultrastructural investigation, we found 20.7 ± 3.1% were slightly damaged in some way and 16.4 ± 4.2% were severely damaged. One of the causes may be the ice crystal formation during the freezing process and some researchers agree that intracellular ice formation is the major injury mechanism at rapid cooling rates (Toner et al.,1993; Chao & Liao 2001). Other causes of cryodamages include pH fluctuation, cold shock, osmometric effect, and cryoprotectant toxicity (Chao & Liao 2001). The swelling and rupture of the plasmalemma after thawing may be due to the damage to the unit membrane which is very sensitive to freezing and thawing (Lahnsteiner et al., 1992). Similar morphological changes were reported in post-thaw sperm of ocean pout (Yao et al., 2000), rainbow trout (Lahnsteiner et al., 1996c), and atlantic croaker (Gwo et al., 1991). For example, in grayling sperm (Lahnsteiner et al., 1992), a marked decrease in sperm quality was observed, about 40% to 50% of the spermatozoa were completely damaged, 30% to 40% changed and only 10–20% showed an intact morphology. In this study, flow cytometric analysis, based on membrane integrity and mitochondrial function, was used to assess post-thaw sperm quality. After double staining with Rh123 and PI, we found 74.8% of post-thaw sperm showed membrane integrity and mitochondrial function. In rainbow trout (Ogier de Baulny et al., 1997), the plasma membrane and mitochondrial function were better protected with 10% DMSO.

Plasma membrane integrity and mitochondrial function are the two most important attributes for fertilizing an egg. The damage to membrane integrity and mitochondrial function could destabilize the sperm membrane and affect mitochondrial energy metabolism, thereby affecting spermatozoa viability. However, what interested us most is that although about 30% of spermatozoa were damaged in some way or even totally ruptured, the fertilization capacity of post-thaw sperm were not affected significantly in the standardized artificial fertilization experiment (Oehninger et al., 2000). Three hypotheses can be considered. The first hypotheses is that the sperm that survived freezing-thawing with normal morphology and mitochondrial function as shown in Fig. 3 and Fig. 4 region 3 should be similar to fresh sperm in fertilization capacity. The second is that the sperm cryopreserved with the established method could provide adequate numbers of motile spermatozoa with normal sperm parameters to fertilize the eggs in artificial fertilization experiment. The third is that the process of freezing-thawing may result in a population of partially damaged yet motile spermatozoa, which can fertilize eggs and develop into larva normally. Such a population usually exhibits a certain degree of plasma and mitochondrial membrane leakiness as shown in Fig. 3 and Fig. 4 region 2, 4.

3. Conclusion

In conclusion, the fertilizing capacity and egg hatchability were not significantly reduced by the post-thaw sperm treated with 12-21%DMSO, although the post-thaw sperm quality was influenced during the freezing and thaw process in motility, ultrastructure and mitochondrial function. The cryopreservation protocol used for red seabream sperm should be of great value for the establishment of sperm banks and assessment of ultrastructure and flow cytometry facilitated identification of damaged sperm; However, the exact nature of cryodamage to fish sperm are not yet fully understood. Sperm motility, structure integrity and mitochondrion function were damaged with different extent, although the fertilization capacity of cryopreserved sperm was not changed. There are many questions need to answer, how does the cryodamage reduce the sperm motility duration? If the cryodamages influence the gene expression and the embryo and larvae development? how to improve the post-thaw sperm quality by optimize the cryopreservation method?

4. Acknowledgment

This study was funded by the State 863 High-Technology R&D Project of China (2003AA603510 and 2004AA603310), the Knowledge Innovation Program of the Institute of Oceanology, Chinese Academy of Sciences (Y02507101Q), The National Natural Science Foundation of China (NSFC) (41076100 an31072212) and the earmarked fund for Modern Agro-industry Technology Research System (nycytx-50).

5. References

Babiak, I.; Glogowski, J.; Brzuska, E. & Adamek, J. (1997). Cryopreservation of sperm of common carp, Cyprinus carpio L. Aquaculture Research 28(7): 567–571.

Billard, R.; Cosson, J. & Crim, L. W. (1993). Motility of fresh and aged halibut sperm. Aquatic Living Resources 6(1):67–75.

Blaxter, J. H. S. (1953). Sperm storage and cross-fertilization of spring and autumn spawning herring. Nature 172(4391):1189–1190.

Cabrita, E.; Robles, V.; Alvarez, R. & Herráez, M. P. (2001). Cryopreservation of rainbow trout sperm in large volume straws: application to large scale fertilization. Aquaculture 201(3-4):301–314.

Chao, N. H. & Liao, I,C. (2001). Cryopreservation of finfish and shellfish gametes and embryos. Aquaculture 197(1-4):161–189.

Chen, S. L.; Ji, X. S.; Yu, G. C.; Tian, Y. S. & Sha, Z. X. (2004). Cryopreservation of sperm from turbot (Scophthalmus maximus) and application to large-scale fertilization. Aquaculture 236(1-4):547–556.

Chen, S. L.; Ji, X. S.; Yu, G. C.; Tian, Y. S.& Sha, Z. X. (2004). Cryopreservation of sperm from turbot (Scophthalmus maximus) and application to large-scale fertilization. Aquaculture 236(1-4):547–56.

Christensen, J. M. & Tiersch, T. R. (1997). Cryopreservation of channel catfish spermatozoa: effect of cryoprotectant, straw size, and formulation of extender. Theriogenology 47(3):639–645.

Conget, P.; Ferna´ndez, M.; Herrera, G. & Minguell, J. J. (1996). Cryopreservation of rainbow trout (Oncorhynchus mykiss) spermatozoa using programmable freezing. Aquaculture 143(3-4):319–329.

Donoghue, A. M.; Garner, D. L.; Donoghue, D. J. & Johnson, L. A. (1995). Viability assessment of turkey sperm using fluorescent staining and flow cytometry. Poult Sci 74(7):1191–1200.

Dréanno, C.; Suquet, M.; Quemener, L.; Cosson, J.; Fierville, F.; Normant, Y. & Billard, R. (1997). Cryopreservation of turbot (Scophthalmus Maximus) spermatozoa. Theriogenology 48(4):589–603.

Dzuba, B. B.; Kopeika, E. F. (2002) Relationship between the changes in cellular volume of fish spermatozoa and their cryoresistance. Cryo Letters 23 (6):353-60.

Graham, J. K.; Kunze, E. & Hammerstedt, R. H. (1990). Analysis of sperm cell viability, acrosoma integrity and mitoxhondrial function using flow cytometry. Biol Reprod 43(1):55–64.

Gravance, C. G.; Garner, D. L.; Miller, M. G. & Berger, T. (2001). Fluorescent probes and flow cytometry to assess rat sperm integrity and mitochondrial function. Reprod Toxicol 15(1):5–10.

Gwo, J. C. (2000). Cryopreservation of sperm of some marine fishes. In: T. R. Tiersch & P. M. Mazik , (Ed.), Cryopreservation in aquatic species, World Aquaculture Society, Baton Rouge, Louisiana, USA, pp. 138–60.

Gwo, J. C. 2000. Cryopreservation of sperm of some marine fishes. in: T. R. Tiersch & P. M. Mazik, editors. Cryopreservation in aquatic species. World Aquaculture Society, Baton Rouge, Louisiana, USA, pp. 138–160

Gwo, J. C.; Strawn, K.; Longnecker, M. T. & Arnold, C. R. (1991). Cryopreservation of Atlantic croaker spermatozoa. Aquaculture 94(4):355–375.

He, S. & Woods, III. L. C. (2004). Changes in motility, ultrastructure, and fertilization capacity of striped bassMorone saxatilis spermatozoa following cryopreservation. Aquaculture 236(1-4):677–686.

Henderson-Arzapalo, A.; Colura, R. L. & Maciorowski, A. F. (1994). A comparison of black drum, red drum, and their hybrid in saltwater pond culture. Journal of the World Aquaculture Society 25(2):289–296.

Kime, D. E.; Ebrahimi, M.; Nysten, K.; Roelants, I.; Rurangwa, E.; Moore, H. D. M. & Ollevier, F. (1996). Use of computer assisted sperm analysis (CASA) for monitoring the effects of pollution on sperm quality of fish; application to the effects of heavy metals. Aquatic Toxicology 36(1-4): 223–227.

Kopeika, E.; Kopeika, J.; Zhang, T. (2007) Cryopreservation of fish sperm. Methods Mol Biol.;368:203-17.

Kopeika, J., Kopeika, E.; Zhang, T.; Rawson, D. M. (2003a) Studies on the toxicity of dimethyl sulfoxide, ethylene glycol, methanol and glycerol to loach (Misgurnus fossilis) sperm and the effect on subsequent embryo development. Cryo Letters. 24(6):365-74.

Kopeika, J., Kopeika, E.; Zhang, T.; Rawson, D. M.; Holt, W. V. (2003b) Detrimental effects of cryopreservation of loach (Misgurnus fossilis) sperm on subsequent embryo development are reversed by incubating fertilised eggs in caffeine. Cryobiology46(1):43-52.

Kopeika, J.; Kopeika, E.; Zhang, T.; Rawson, D. M.; Holt, W. V. (2004) Effect of DNA repair inhibitor (3-aminobenzamide) on genetic stability of loach (Misgurnus fossilis) embryos derived from cryopreserved sperm. Theriogenology 61(9):1661-73.

Lahnsteiner, F.; Berger, B. & Weismann, T. (2003). Effects of media, fertilization technique, extender, straw volume, and sperm to egg ratio on hatchability of cyprinid embryos, using cryopreserved semen. Theriogenology 60(5):829–841.

Lahnsteiner, F.; Berger, B.; Horvath, A.; Urbanyi, B. & Weismann, T. (2000). Cryopreservation of spermatozoa in cyprindid fishes. Theriogenology 54(9):1477–1498.

Lahnsteiner, F.; Berger, B.; Weismann, T. & Patzner, R.A. (1996a). Physiological and biochemical determination of rainbow trout, Oncorhynchus mykiss semen quality for cryopreservation. J Appl Aquacult;6:47–73.

Lahnsteiner, F.; Berger, B.; Weismann, T. & Patzner, R.A. (1998). Determination of semen quality of the rainbow trout, Oncorhynchus mykiss, by sperm motility, seminal plasma parameters, and spermatozoal metabolism. Aquaculture 163(1-2):163–181.

Lahnsteiner, F.; Berger, B.; Wiesmann, T. & Patzner, R. A. (1996c). Changes in morphology, physiology, metabolism, and fertilization capacity of rainbow trout semen following cryopreservation. Prog Fish Cult 58(3):149–59.

Lahnsteiner, F.; Weismann, T. & Patzner, R. A. (1992). Fine structural changes in spermatozoa of the grayling, Thymallus thymallus (Pisces:-Teleostein), during routine cryopreservation. Aquaculture 103(1):73–84.

Lahnsteiner, F.; Weismann, T. & Patzner, R.A. (1996b). Cryopreservation of semen of the grayling (Thymallus thymallus) and the Danube salmon (Hucho hucho). Aquaculture 144(1-3):265–274.

Linhart, O.; Rodina, M. & Cosson, J. (2000). Cryopreservation of sperm in common carp Cyprinus carpio: sperm motility and hatching success of embryos. Cryobiology 41(3):241–250

Liu, Q. H.; Chen, Y. K.; Xiao, Z. Z.& Li, J. (2010a). Effect of storage time and cryoprotectant concentration on the quality of cryopreserved sperm in red seabream (Pagrus major). Aquaculture research. 41(9):e89-e95.

Liu, Q. H.; Chen, Y. K.; Xiao, Z. Z.; Li, J. & Xu, S. H. (2010b). Effect of long-term cryopreservation on physiological characteristics, antioxidant activities and lipid peroxidation of red seabream (Pagrus major) sperm. Cryobiology 61(2):189-193.

Liu, Q. H.; Li, J.; Xiao, Z. Z., Ding, F. H.; Yu, D. D. & Xu, X. Z. (2007b). Use of computer-assisted sperm analysis (CASA) to evaluate the quality of cryopreserved sperm in red seabream (Pagrus major). Aquaculture 263(1-4):20–25.

Liu, Q. H.; Li, J.; Zhang S. C.; Ding, F. H.; Xu, X. Z.; Xiao, Z. Z. & Xu, S. H. (2006). An Efficient Methodology for Cryopreservation of spermatozoa of Red Seabream, Pagrus major, with 2-mL Cryovials. Journal of the world aquaculture society 37(3):289-297.

Liu, Q. H.; Li, J.; Zhang, S. C.; Xiao, Z. Z.; Ding, F. H.; Yu, D. D. & Xu, X. Z. (2007a). Flow cytometry and ultrastructure of cryopreserved red seabream (Pagrus major) sperm. Theriogenology 67(6):1168–1174.

Oehninger, S.; Duru, N. K.; Srisombut, C. & Morshedi, M. (2000). Assessment of sperm cryodamage and strategies to improve outcome. Mol Cell Endocrinol 169(1-2):3–10.

Ogier de Baulny, B.; Labbe´, C. & Maisse G. (1999). Membrane integrity, mitochondrial function, ATP content, and motility of the European catfish (Silurus glanis) testicular spermatozoa after freezing with different cryoprotectants. Cryobiology 39(2):177-184.

Ogier de Baulny, B.; Le Vern, Y.; Kerboeuf, D. & Maisse, G. (1997). Flow cytometric evaluation of mitochondrial function and membrane integrity in fresh and cryopreserved rainbow trout (Oncorhynchus mykiss) spermatozoa. Cryobiology 34(2):141-149.

Ohta, H.; Kawamura, K.; Unuma, T. & Takegoshi, Y. (2001). Cryopreservation of the sperm of the Japanese bitterling. Journal of Fish Biology 58(3):670-681.

Richardson, G. F.; Wilson, C. E.; Crim, L.W. & Yao, Z. (1999). Cryopreservation of yellowtail flounder (Pleuronectes ferrugineus) semen in large straws. Aquaculture 174(1-2):89-94.

Rurangwa, E.; Volckaert, F. A. M.; Huyskens, G.; Kime, D. E. & Ollevier, F. (2001). Quality control of refrigerated and cryopreserved semen using computer-assisted sperm analysis (CASA), viable staining and standardized fertilisation in African catfish (Clarias gariepinus). Theriogenology 55(3):751-769.

Segovia, M.; Jenkins, J. A.; Paniagua-Chavez, C. & Tiersch, T. R. (2000). Flow cytometric evaluation of antibiotic effects on viability and mitochondrial function of refrigerated spermatozoa of Nile Tilapia. Theriogenology 53(7):1489-1499.

Suquet, M.; Dréanno, C.; Petton, B.; Norman, Y.; Omnes, M. H. & Billard, R. (1998). Long-term effect of the cryopreservaton of turbot (Psetta maxima) spermatozoa. Aquatic Living Resources 11(1):45-48.

Tiersch, T. R.; Figiel, C. R.; Jr.; Wayman, W. R.; Williamson, J. H.; Carmichael, G. J. & Gorman, O. T. (2000). Cryopreservation of sperm of the endangered razorback sucker. In: T. R. Tiersch & P. M. Mazik, (Ed.), Cryopreservation in aquatic species. World Aquaculture Society, Baton Rouge, Louisiana, USA, Pages 117—122

Tiersch, T. R.; Wayman, W. R.; Skapura, D. P.; Neidig, C. L. & Grier, H. J. (2004). Transport and cryopreservation of sperm of the common snook, Centropomus undecimalis (Bloch). Aquaculture Research 35(3):278-288.

Toner, M.; Cravalho, E. G. & Karel, M. (1993). Cellular response of mouse oocytes to freezing stress: prediction of intracellular ice formation. J Biomech Eng 115(2):169-174.

Van derWalt, L. D.; Van der Bank, F. H. & G. J. Steyn. (1993). The suitability of using cryopreservation of spermatozoa for the conservation of genetic diversity in African catfish (Clarias Gariepinus). Comparative Biochemistry Physiology 106(2):313-318.

Viveiros, A. T. M.; So, N. & Komen, J. (2000). Sperm cryopreservation of African catfish, Clarias garepinus: cryoprotectants, freezing rates and sperm:egg dilution ratio. Theriogenology 54(9):1395-1408.

Warnecke, D. & Pluta, H. J. (2003). Motility and fertility capacity of frozen/thawed common carp (Cyprinus carpio L.) sperm using dimethyl-acetamide as the main cryoprotectant. Aquaculture 215(1-4):167-85.

Yao, Z.; Crim, L. W.; Richardson, G. F. & Emerson, C. J. (2000). Motility, fertility and ultrastructural changes of ocean pout (Macrozoarces americanus L.) sperm after cryopreservation. Aquaculture 181(3-4):361-375.

Zhang, Y. Z.; Zhang, S. C.; Liu, X. Z.; Xu, Y. Y.; Wang, C. L.; Sawant, M. S.; Li, J. & Chen, S. L. (2003). Cryopreservation of flounder (Paralichthys olivaceus) sperm with a practical methodology. Theriogenology 60(5):989-996.

13

Sperm Cryopreservation of Two European Predator Fish Species, the Pikeperch (*Sander lucioperca*) and the Wels Catfish (*Silurus glanis*)

Zoltán Bokor, Béla Urbányi, László Horváth,
Tamás Müller and Ákos Horváth
Department of Aquaculture, Szent István University
Hungary

1. Introduction

Wels catfish (*Silurus glanis*) and pikeperch (*Sander lucioperca*) are two predator fish species cultivated in the traditional Central European pond aquaculture. Their role in the pond ecosystem is the control of the populations of smaller wild fish that enter ponds during their flooding in the Spring and would represent food competition for the cultured cyprinids. In addition, both species are highly priced for their excellent boneless meat, therefore, attempts are made to improve their culture and enhance yields (Horváth et al., 2002).

The process of induced propagation of wels catfish is based on a long-standing technology. However, certain problems can still appear: the method of collecting male gametes is still based on the removal of testes. By the application of this method a particular male can only be used once for propagation. Moreover, differentiation of sexes requires a great deal of experience and the danger of using an immature female with a less developed body structure is still present. As a result of listed problems the success of propagation becomes questionable.

In recent years successful experiments have been carried out in the pikeperch in many farms and research centers for the development of induced propagation in hatcheries. Synchronization of maturation of female individuals is not perfectly developed yet, thus, successful striping requires a constant attention and control. That is why minimizing the presence of males and securing sperm in a most simple way could focus attention on females.

Application of cryopreserved male gametes for fish propagation in hatcheries can serve as a solution for all mentioned difficulties and risks. History of fish sperm cryopreservation dates back to the beginning of 1950s, since then the sperm of more than 200 fish species has been cryopreserved successfully all over the world (Rana, 1995). In spite of this the application of cryopreserved fish sperm is still not very common in aquaculture in contract to e.g. the dairy cattle sector. Most studies done on cryopreservation of fish sperm put the emphasis on optimization of the process, on cryopreservation of a small amounts of male gametes, thus, rendering this technology to the level of laboratory experiments without any basic output for farmers.

By the help of a successful cryopreservation method applicable in fish farms not only the reduction of propagation risks would become possible but also a long-term storage of gametes of substantial breeders as already applied in case of carp. Development of a sperm bank already employed at cattle breeding would also become feasible thus increasing the role and rate of rarely used selection methods of animal breeding in fisheries.

The objective of our work was the development of working protocols for the cryopreservation of wels catfish and pikeperch sperm that can be applied to aquaculture of these species. In both species, this development required a thorough knowledge of culture conditions of the given species, studies of the cryoresistance of sperm and adaptation of cryopreservation methods to hatchery practices.

2. Materials and methods

Experiments on the cryopreservation of sperm of both species were carried out during the years 2005-2008 at various locations in Hungary. Details of experiments including their dates and locations are provided in the descriptions of experiments on each species.

2.1 Experiments on wels catfish sperm

2.1.1 Methods of collection and cryopreservation of male gametes

Gametes applied in the experiments were invariably retrieved directly from the testis removed from the abdomen after the decapitation of male wels catfish (not stripped). After removing it from the abdomen the testis was cut up and squeezed out through a dry gauze into a Petri-dish. After extraction of sperm motility of gametes was examined through a light microscope at 200× magnification.

Ten per cent methanol was used as a cryoprotectant and 6% fructose as a diluent. pH of the cryopreservation medium was adjusted to 7.73 by the help of 1 M $NaHCO_3$ solution. From the diluted sperm treated this way 4 ml was pipetted into a 5-ml straw.

In the process of cryopreservation liquid nitrogen was poured into a polystyrene box on the top of which a polystyrene frame was placed with a height of 3 cm and the straws were laid on it. Samples were stored in a canister storage dewar until use.

Straws were thawed in 40 °C of water for 40 seconds. After thawing the closed ends of straws were cut up and their content was poured into a test-tube or directly onto the eggs used for propagation. Motility of thawed sperm was examined according to the method already described. The method mentioned here was compiled on the basis of former experiments of the cryopreservation group of the Department of Aquaculture on African catfish (Urbányi et al., 1999; Horváth & Urbányi, 2000).

2.1.2 Experiments on cooling time and sperm-egg ratio

Male gametes applied for cryopreservation in 2005 were collected from the farm in Tuka of Szarvas-Fish Kft. and the farm in Szeged of Szegedfish Kft. by joining to their propagation processes. Males were injected with 4 mg/body weight kg carp pituitary in one dose by the assistants of the farm before the extraction of sperm. Length and weight of the body and weight of the testis were measured. Gonadosomatic indices (GSI) were determined from the ratio of testis and body weight. After the squeezing of sperm the motility of gametes was defined as described above. After the collection of gametes the former described cryopreservation method was used with the addition that in case of samples collected in Tuka the effect of cooling time on motility and fertilizing capacity was also tested. Cooling time of samples varied between 3, 5 and 7 minutes. After the cooling period straws were placed into liquid nitrogen.

First propagation tests were completed in the Szajol farm of Fish-Coop Kft. Eggs were gained from fish by a routine propagation process. In the first experiments eggs were divided into 40, 80 and 120 g doses and each dose was propagated with 1 straw of cryopreserved sperm. Fertilized eggs were then incubated in 7 l Zug-jars. Hatching rate was counted after hatching.

The second experimental procedure was performed in the Attala fish farm of Attala Hal Kft., when eggs were divided up to two 150 g doses and one of them was fertilized with one, while the other one with two straws of sperm. In both cases freezing time was 7 minutes. At hatching the ratio of hatching and deformed larvae was determined.

In the experimental procedure the cooling rate was measured, too. A straw was filled with cryopreservation medium. The K type sensor of a Digi-Sense DualLogR digital thermometer (Eutech Instruments, Singapore) was placed into the straw which was then laid onto a 3 cm high polystyrene frame floating on the surface of liquid nitrogen. The thermometer recorded temperature data with 1 second intervals. Temperature data were collected for 6 minutes since storing capacity of the memory of the thermometer allowed the recording of this amount of data.

2.1.3 Cryopreservation and analysis of sperm collected outside of the spawning season in hatchery conditions

In 2006 wels catfish sperm was collected in January and March (aside from the spawning season) in the Köröm farm of Aqua-culture Kft. (Köröm Fish Farm, Local Government of Bőcs at present) from wels catfish kept in a flow-through intensive system. They were kept in tanks in a constant water temperature of 20 °C. The method applied for cryopreservation was the same as already mentioned with the difference that sperm was frozen for 7 minutes on the polystyrene frame before placing it into liquid nitrogen. Motility of sperm was examined both in fresh and cryopreserved samples.

Propagation experiments were performed at the Szajol farm of Fish-Coop Kft. For this, cryopreserved samples originating from Szeged, 2005 and Köröm, 2006 were used. Eggs were collected from female fish by a routine propagation process. In the experimental procedure eggs were distributed into 250 and 350 g dosages and fertilized with one straw of sperm. Fertilized egg doses were then incubated in 7 l Zug-jars and hatching rates were determined after hatching.

2.1.4 Application in hatcheries

The aim of these experiments was to fertilize significant amounts of eggs (150-300 g) with large doses of cryopreserved sperm (5-ml straws) all over the country by joining to propagation work of a given farm. Reliability and repeatability of the method were also examined. In each case one dose of eggs was fertilized with the content of one straw. In the frames of a routine propagation work fertilization with cryopreserved sperm was propagated in five different farms:

- in Attala pond farm of Attala Fish Production and Trading Kft.
- in Köröm fish farm of the Local Government of Bőcs
- in Százhalombatta farm of TEHAG Kft.
- in Ördöngös farm of Aranykárász Co.
- in Szeged farm of Szegedfish Kft.

In the Attala experiment in 15 May 2007 cryopreserved sperm collected ourside of the spawning season was used for the fertilization of 200 g egg doses.

The experiment in Köröm was carried out in 17 May 2007 in which cryopreserved sperm was applied for fertilization also collected off season. Egg doses of 200 g were fertilized both for treated and for control groups.

Following that, egg doses of 200 g were fertilized in the Százhalombatta farm of TEHAG Kft. in 22 May 2007. This time cryopreserved sperm from Szeged was used in the experiments collected in 2006 in the spawning season.

In the fourth experiment in 21 May 2007 egg doses of 200-350 g were fertilized in the Ördöngös farm of Aranykárász Co with cryopreserved sperm collected off season.

The last experiment was performed in the hatchery of Szegedfish Kft. in 23 May 2007. In the research work egg doses of 150 g were fertilized with cryopreserved sperm deriving from Szeged.

In all cases our team personally joined to the propagation work of the farm, thus, we ensured that an adequate amount of cryopreserved sperm was used for the egg doses. Process of propagation was always performed according to the practice of the certain farm. One dose of eggs was fertilized with one straw of thawed sperm. In all experiments hatching percentage of larvae was determined and in the experiment in Attala fertilization rate at 4-8 cell stage was also examined.

2.1.5 Experiments on larval survival

For the analysis of larval survival, growth and survival of feeding larvae in 2007 and non-feeding wels catfish larvae in 2008 were examined.

Analysis of feeding larval stage

When examining feeding larval stage in farm conditions, larvae were produced in Százhalombatta farm of TEHAG Kft. by applying local propagation methods. During fertilization process 1 straw of sperm was added to 200 g of eggs. After fertilization the 200 g doses of eggs were placed into 7 l Zug-jars. For fertilization of individuals in the control group native sperm from males of the farm was used. On the second day after fertilization hatched larvae were counted then after hatching the non-feeding larvae were placed into larva-tanks. On the third day after hatching when larvae started their exogenous feeding the ones devoted for the experiment were counted and placed into troughs. 1000 individuals were placed into a 100 l trough with flow-through water in 3 replcates. The stock was fed with chopped tubifex in every 3 hours. To prevent infections 36% formalin treatment was applied in a concentration of 10 ml/trough in every 4 hours. In each trough velocity of water flow-through was 3 l/minute. Rearing in the trough lasted for 10 days according to the routine practice of the farm.

Laboratory experiments were performed in the Department of Aquaculture. Troughs of the recirculation system applied in the experiment were 40 cm long, 15 cm deep and 10 cm wide (though water depth was 10 cm due to the outlet/stub). Due to photophobia of wels catfish larvae the system was located in a dark room. In the experiment 5×100 individuals (larvae hatched after the application of control and cryopreserved sperm) took part in the treatments. In the first 4 days fish were fed once in every three hours. At the morning and evening feeding they were fed with plankton while at other feeding times experimental fish were fed with Perla Proactiv 4.0 fish diet. Following this 4-day period fish were fed with the above mentioned diet (sometimes also with plankton or Artemia) 4 times a day. Velocity of water flow-through in the troughs was 0.25 l/minute.

In addition to body length and weight condition factor, specific growth rate (S.G.R.) and survival rate (%) was measured.

Examination of non-feeding larval stage

In this developmental stage no farm study was done due to the fact that the trough system was not suitable for the accommodation of such small larvae. Moreover, the hatchery protocol for fry rearing does not use the method of rearing in troughs at this age.

Larvae examined at non-feeding life stage were produced in June 2008 in the hatchery of TEHAG Kft. The method of propagation and utilization of cryopreserved sperm was the same as described at the examination of feeding larval stage. After hatching larvae were transported to the laboratory of the Department of Aquaculture in Gödöllő where the survival of fish was tested using the rearing system built in the previous year. The experiment lasted for 4 days.

2.1.6 Experiments on larval survival

Main effects of cooling time and the quantity of eggs on hatching rate was examined in the Szajol experiment (Part 2.1.2) by the help of two-way analysis of variance ($P<0.05$). In case of testis collected off season at two sampling times as described in Chapter 2.1.3 GSI values were compared by the help of a two-sample t-test. Hatching results gained from fertilization experiments with cryopreserved and control sperm described in Chapter 2.1.4 were analysed with a t-test, too. These statistical tests were performed using the software GraphPad Prism 4.0 for Windows. In the examinations of larval rearing results of survival were compared with Chi^2-test (Kruskal Wallis test), while body length, body weight, condition factors and SGR values with t-tests (Chapter 2.1.5) using the software SPSS 13 for Windows.

2.2 Experiments on pikeperch sperm

2.2.1 Origin of fish

Experiments on pikeperch were conducted two times and in two places: the first place was the Keszthely-Tanyakereszt fish laboratory of Georgikon Faculty of Agriculture at Pannon University and the second one was the Attala hatchery of Attala Hal Kft.

2.2.2 First experiment

The stock of pikeperch taking part in the first experiment originated from Aranyponty Kft. (Sáregres-Rétimajor) (8 females and 10 males: 1424-1870 g). Males were anesthetized by clove oil then sperm was manually stripped and collected with an automatic pipette. Care was taken not to contaminate the gametes with urine or feces.

Motility of fresh sperm was estimated after activating it with water. Motility was examined on slides in a 20× dilution at 200× magnification by the help of a Zeiss Laboval 4 microscope (Carl Zeiss, Jena, GDR). Density of sperm was examined by the Bürker-chamber method in a 1000× dilution.

The following diluents were prepared:

- Glucose diluent (350 mM glucose, 30 mM Tris, pH 8,0)
- KCl diluent (200 mM KCl, 30 mM Tris, pH 8,0)
- Sacharose diluent (300 mM sacharose, 30 mM Tris, pH 8,0)

Methanol and dimethyl-sulfoxide (DMSO) were used as cryoprotectants in a concentration of 10%. All chemicals used in the experiments were purchased from Reanal Zrt. (Budapest, Hungary).

Sperm (200 µl) mixed with a cryopreservation medium (200 µl cryoprotectant, 1600 µl diluent) in a ratio of 1:9 (Horváth et al., 2003; Urbányi et al., 2006) was pipetted into individually marked 0.5-ml straws after 3 minutes of equilibration time. Samples were cryopreserved in a polystyrene box filled with liquid nitrogen. A 3 cm high polystyrene frame was placed onto the top of the nitrogen then straws were laid on this frame where temperature was around -165°C. The time of cryopreservation was 3 minutes. After the freezing process straws were placed into liquid nitrogen and stored there until being used. Thawing was carried out in a 40°C water bath for 13 seconds (Horváth et al., 2003; Horváth et al., 2005). After thawing sperm motility was examined with the same method as described at fresh sperm.

Eggs were distributed to Petri-dishes with a diameter of 5 cm with 200-350 eggs/dose. Fertilization was made with thawed sperm of a half straw (250 µl) right after taking the straw out of the water. Sperm was poured on the egg doses then the gametes were activated with 1 ml of water. Next eggs were allowed to stick to the bottom of the Petri-dish by taking care of the eggs being located in one layer. Fertilization rate was counted at neurula stage.

2.2.3 Second experiment

The second experiment was performed in line with pikeperch propagation in a hatchery. In this research sperm of 4 males and eggs of 1 female were applied. In this case only glucose was used as a diluent with 10% concentration of methanol and DMSO cryoprotectants. Sperm was diluted in a 1:1 and 1:9 ratios. The process of cryopreservation and thawing was the same as the process applied in the first experiment.

Eggs were divided into 10 g doses (about 10 000 eggs according to my counts) and taken into plastic bowls. Each dose was fertilized by a thawed straw of samples (0.5 ml) then these doses were placed into 7 l Zug-jars until hatching. Finally hatched larvae were counted and hatching rate was defined.

2.2.4 Application of cryopreserved sperm to hatchery conditions (preliminary experiment)

This research was done in April 2007 in the hatchery of Attala Hal Kft. Gametes used in the experiment originated from the same farm. Male and female fish were treated by the method applied in the farm in line with local hatchery propagation. In the previous experiment it was difficult to avoid mixing of sperm with urine so in this case stripping was performed by a silicon catheter (inside diameter: 1 mm, outside diameter: 1.5 mm) which was introduced into the sperm duct. Motility estimations were made by the method described earlier. Sperm concentration was examined in a Bürker-chamber in a 1000× dilution.

Sperm originating from 3 males were used in the experiment. It was diluted in a 1:1 ratio with the following composition of cryopreservation medium: 350 mM glucose, 30 mM Tris, pH 8.0 (titrated with ccHCl), methanol with a concentration of 10%. Diluted gametes were pipetted into 0.5-ml straws. Cryopreservation and thawing methods arranged to the method described in the previous chapter. Sperm was stored for 1 week in liquid nitrogen in a canister. After thawing the motility of samples was also examined.

Stripped eggs were divided into 10 and 30 g doses in 3 replicates and into a 50 g dose in one replicate. One dose of eggs was fertilized with one straw. Fresh sperm was used as a control. Each dose was poured into a 7 l Zug-jar for incubation. Hatching rate was counted after hatching.

2.2.5 Applied statistical methods

Results of experiments were evaluated with Graphpad Prism 4.0 for Windows program. Effect of cryoprotectants and diluents on motility and fertilization and the effect of diluent ratio and cryoprotectants on hatching ratio was examined by a two-way analysis of variance (ANOVA) (Chapter 2.2.2).

Results gained from the second experiment (Chapter 2.2.3), namely motility (cryopreserved and fresh sperm) and hatching results (in case of 10 and 30 g egg doses) were analysed by the help of a two-sample t-test ($P \leq 0.05$).

3. Results

3.1 Experiments on wels catfish sperm

The GSI (gonadosomatic index) of male individuals from Tuka was 2 ± 4%. The motility of sperm before cryopreservation was 90%, while after thawing this rate was 0% in case of 3 minutes, 40% in case of 5 minutes and 70% in case of 7 minutes long freezing time. The motility of sperm from Szeged was 80%.

In the experiments carried out in Szajol the highest hatching rate (51 ± 1%) was observed at 7 minutes freezing time and 40 g of dose of eggs, although in the case of 5 and 7 minutes of freezing time a very similar hatching rate (between 40 ± 0% and 51 ± 1%) was observed (Figure 1.). Only the cooling time had a significant main effect ($P < 0.0001$) on the results, considering that 3 minutes long cooling time gave lower hatching rate.

Hatching rate of propagated eggs was 94% in the case of fertilization with a single straw in Attala, while fertilization with two straws resulted 77% hatching rate. The control results were 89% and 81%. It is worth to mention that the ratio of deformed larvae hatched from eggs fertilized with a single straw was only 2.4% (1.8% in control), while in the case of fertilization with two straws it was 11.2% (7.3% in control).

The cooling rate of a straw was approximately -23°C/minute (Figure 2.). It was observed that the temperature of the straw was only -45°C after 3 minutes while after 5 minutes it was -104°C.

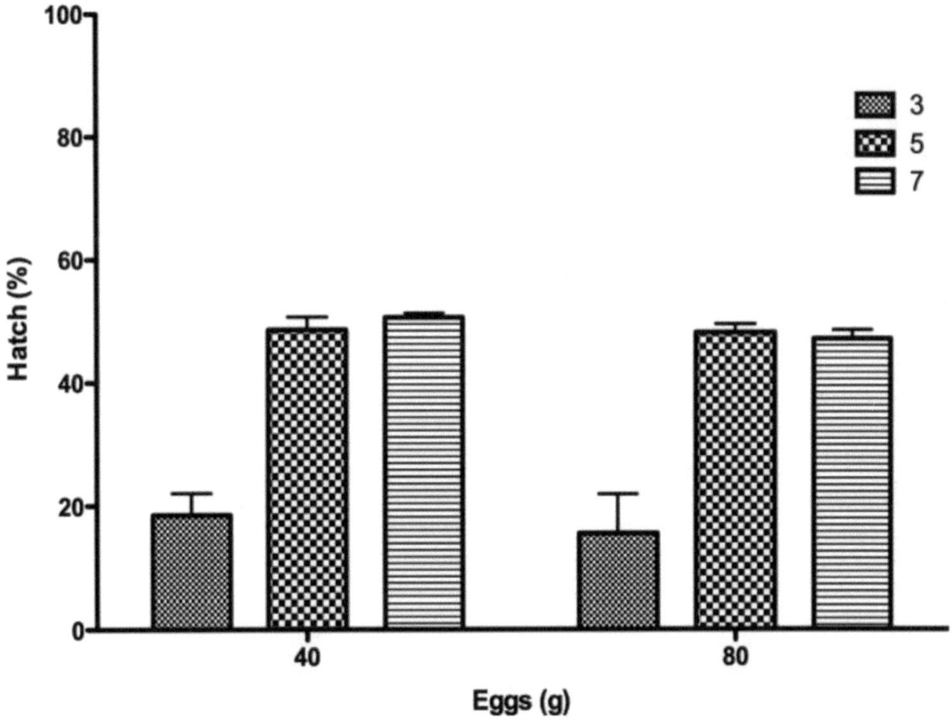

Fig. 1. Hatch rates of wels catfish eggs fertilized with cryopreserved sperm. Each batch of eggs (40 or 80 g) was fertilized with one 5-ml straw of cryopreserved sperm. Columns indicate cooling times employed during cryopreservation (3, 5 or 7 minutes). Data are presented as Mean ± SD (N = 3).

When sperm was collected outside of the spawning season, the average weight of the testes of wels catfish catfish from Köröm was 20.4 g and the average weight of the fish was 2.52 kg, thus GSI rate of them was lower than 1% except one male. This low GSI rate have not had adverse effects on the quality of sperm. No significant difference was observed between GSI rate in January and in March (P = 0.4589). The motility of fresh sperm varied between 50% and 90%. Two of the sperm samples selected for cryopreservation were excluded from further examinations because the motility of these samples was the lowest (50 - 60%). This low rate was caused likely by injuring the cells during squeezing of the testes. Some of the cryopreserved samples were thawed 5 days after freezing and their motility was about 50%.

Sperm frozen in 2005, in Szeged and 2006, in Köröm were used for the propagation experiments. Hatching rate varied between 70 - 80%, except for one sample with 20% of hatching rate. However, according to the head of the farm the fertilization of control group was as bad as the result of the 20% hatching rate. On the basis of the results it was observed that sperm form Köröm (out of spawning season) had similar hatching rate to the sperm from Szeged.

Results of different hatching experiments depended on the propagation method and on the quality of sperm. Fertilization with cryopreserved sperm form Attala resulted 97 ± 1% of fertilization rate while control fertilization rate was 93 ± 1%. There was no significant difference between the two rates ($P = 0.0084$). In the same experiment the hatching rate of the larvae was 95 ± 2% while in the control group this rate was 94 ± 6%. There was no statistically significant difference between the hatching rate of larvae originating from cryopreserved or fresh sperm.

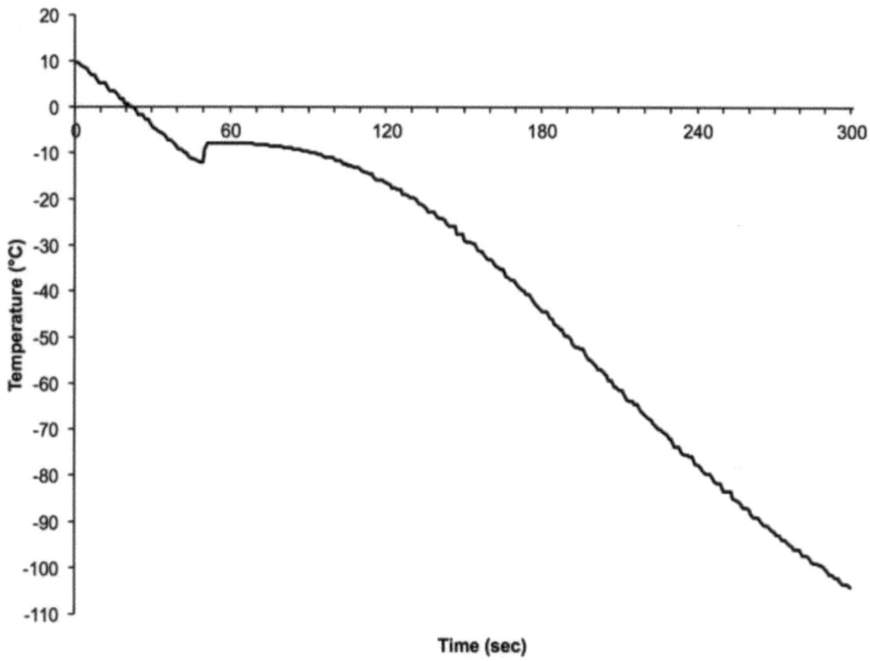

Fig. 2. The cooling profile used with 5-ml straws in the experiments on wels catfish sperm

The hatching rate of larvae originating from cryopreserved sperm in Köröm was 84 ± 5%, while this rate in case of larvae originating from fresh sperm was 69 ± 16%. There was no significant difference between the results. It was observed in the experiments carried out in Százhalombatta that hatching rate of larvae originating from cryopreserved sperm was 50 ± 3%, while the result of the control group was 50 ± 6%. There was also no statistically significant difference between the two groups. Hatching results of the experiments carried out in Ördöngös at the place of Aranykárász Kft. were about 57 ± 22% in case of larvae originating from cryopreserved sperm and 22 ± 18% in case of larvae originating from fresh sperm. In this experiment a significant difference was found ($P = 0.05$) in favor of the cryopreserved group. The hatching rate of larvae originating from cryopreserved sperm in Szeged was 75 ± 3%, while this rate in case of the control group was 83 ± 1%. These results also differ significantly ($P = 0.0249$) but now in favor of the control group.

In the experiments on larval survival, statistically significant difference ($P = 0.034$) was observed on feeding larvae in laboratory conditions regarding 10-day body length. The results showed that larvae originating from cryopreserved sperm had a longer body. During non feeding larval period final body length ($P < 0.001$) and final weight ($P = 0.018$) differed significantly in the two groups in favor of larvae originating from cryopreserved sperm. There was no difference between the larvae originating from cryopreserved or fresh sperm in terms of larval survival.

3.2 Experiments on pikeperch sperm

In the first experiment, in spite of all efforts sperm was mixed with urine during stripping, thus the motility of pikeperch sperm was 50 ± 17%. The motility of the best thawed sample was 28 ± 21 %, which was cryopreserved with glucose diluent and DMSO as cryoprotectant, but statistically significant difference was not be observed among the treatmenrts.

The density (spermatozoa/ml) of pikeperch sperm samples counted in a Burker chamber were the following: 1st male: 0.9375×10^{10}, 2nd male: 1.0100×10^{10}, 3rd male: 0.7037×10^{10}, 4th male: 0.6687×10^{10}.

The highest fertilization rate (43 ± 12%) was observed also in the case of using a combination of glucose diluent and DMSO as cryoprotectant (Figure 3.). During statistical analysis of the data it was found that only the cryoprotectant had a significant effect ($P = 0.0338$) on the ratio of the fertilization.

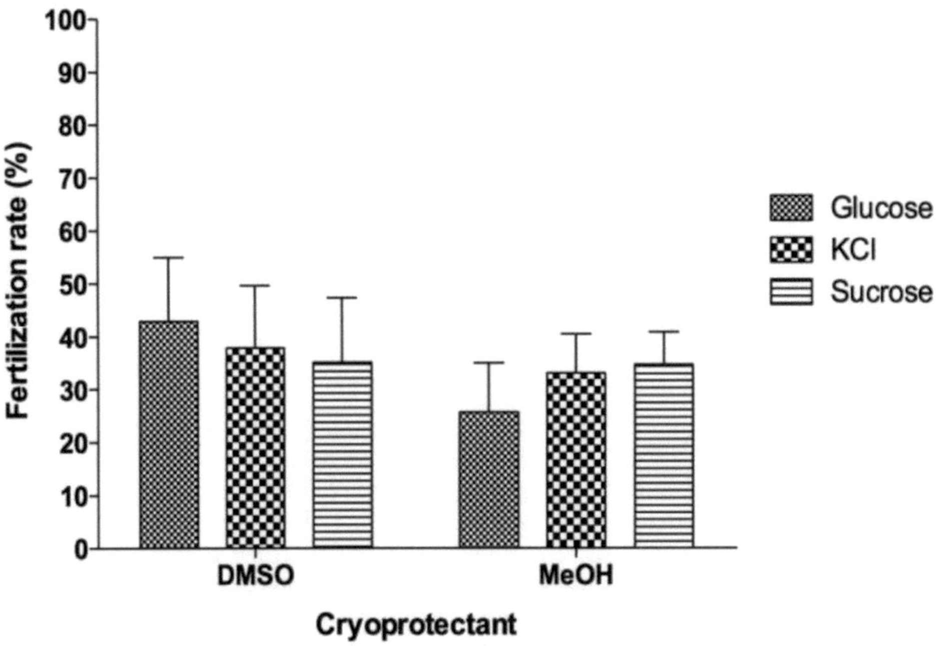

Fig. 3. Fertilization percentages of pikeperch eggs fertilized with cryopreserved sperm in the first experiment. The cryoprotectants dimethyl-sulfoxide (DMSO) and methanol (MeOH) and glucose, KCl or sucrose extenders were compared. Data are expressed as Mean ± SD (N = 3).

The volume of the sperm stripped from pikeperch males in the second experiment was very low (less than 1 ml/individual). The motility of fresh sperm was 45 ± 30%. Similarly to the previous experiment the sperm was mixed with urine again. Motility of thawed pikeperch sperm was very low (0 – 2%) in the samples containin the cryoprotectant DMSO, while motility of sperm frozen in presence of methanol was 40%, independently from rate of dilution. The highest hatching rate (41 ± 22%) was observed with the use of methanol and 1:1 dilution rate, although the statistical analysis has not shown significant differences between hatching rates (Figure 4.).

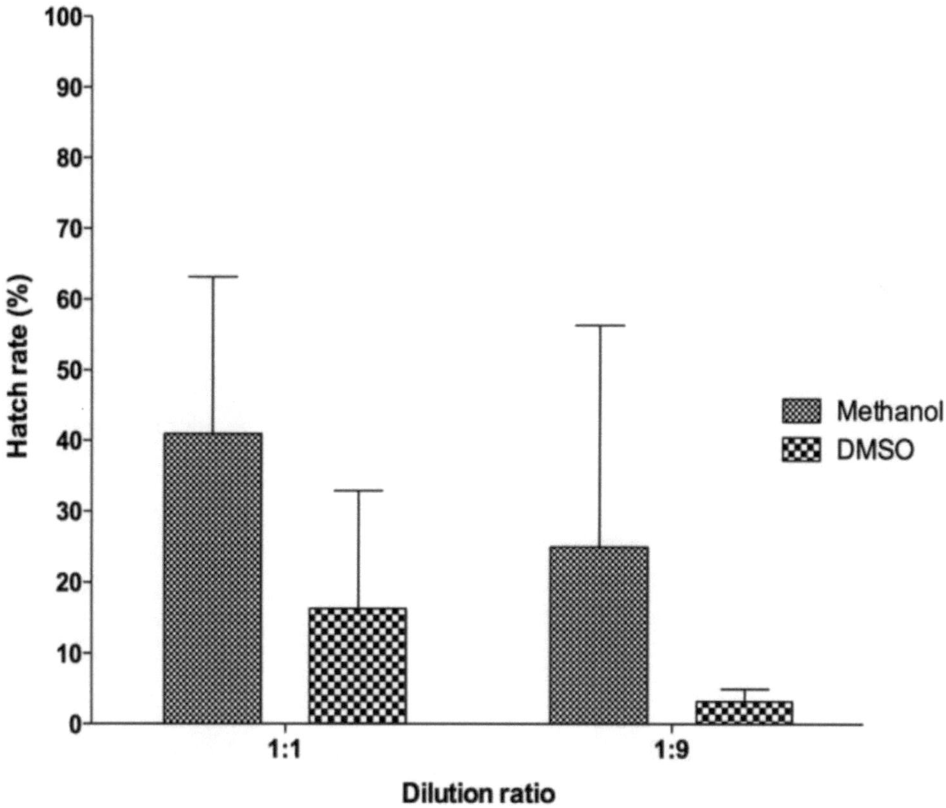

Fig. 4. Fertilization rates of pikeperch eggs fertilized with cryopreserved sperm in the second experiment. The cryoprotectants methanol and dimethyl-sulfoxide (DMSO) and dilution ratios of 1:1 and 1:9 were compared. Data are expressed as Mean ± SD (N = 3).

In the hatchery experiment Stripping sperm with silicon catheter resulted that the motility was 63 ± 10%. Concentration of sperm was $1.8571 \pm 0.1538 \times 10^{10}$, while the number of eggs/g was 1367 ± 54, thus the number of sperm for an egg was 3.396×10^5 in the case of 10-g dose of eggs, 1.132×10^5 in the case of 30-g dose of eggs and 6.792×10^4 in the case of 50-g dose of eggs. Motility of sperm after thawing was 53 ± 5%, thus there was no significant difference ($P = 0.1135$) between the motility of fresh and cryopreserved sperm.

Fertilization of the dose of 10 g of eggs with a single straw resulted 47 ± 4%, while in the case of the dose of 30 g eggs resulted 55 ± 3% hatching rate (Figure 5.). There was no statistically significant difference between the results of the different doses however the result of t-test (P = 0.05701) was very close to the significance level. A hatching rate of 87% was observed in the case of fertilization the dose of 50 g eggs with one thawed straw although in this case there were no replicates in the experiment. It was observed, however, that egg batches of different weight behaved differently in the hatching jars. While egg batches of 10 g stuck together in spite of the attempted elimination of egg stickiness, those in batches of 30 g or 50 g freely rolled on each other, thus improving oxygen supply of fertilized eggs and developing embryos. Thus, it is recommended to use larger batches of eggs for fertilization with cryopreserved sperm, which in turn would facilitate the acceptance of this technology in the aquaculture practice.

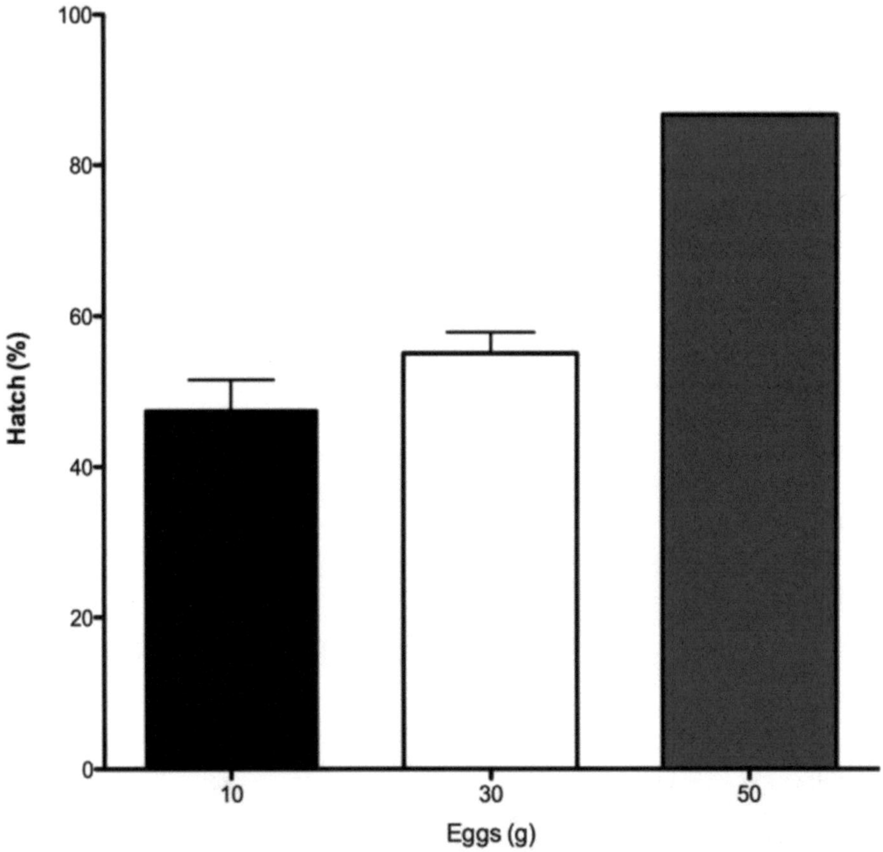

Fig. 5. Hatch rates of pikeperch eggs fertilized with cryopreserved sperm in a commerical hatchery during routine spawning work. Egg batches of 10, 30 or 50 g were used for fertilization (N = 3).

4. Discussion

4.1 Experiments on wels catfish sperm

Hatchery propagation of catfish species including the wels catfish faces several problems. Males of catfish species are typically oligospermic and sperm cannot be stripped but has to be extracted from surgically removed testes (Legendre et al., 1996). As it was mentioned in the Introduction, this can lead to several problems such as shortage of sperm during induced spawning to unnecessary killing of immature females due to their erroneous identification as males.

Cryopreservation of the sperm of catfish species has been studied extensively. Several studies have been published on the cryopreservation of wels catfish sperm (Krasznai & Márián, 1985; Linhart et al., 1993; Ogier de Baulny et al., 1999; Linhart et al., 2005), however, they all reported the use of minute amounts of gametes and did not test practical utilization in the hatcheries.

It can be concluded according to the measured freezing parameters that a longer cooling time is needed for the safe cryopreservation of sperm in 5-ml straws because the temperature is not low enough (-45°C) after 3 minutes. The 7 minutes cooling time, which was used during the experiments, is suitable for these 5-ml straws.

Large amounts of eggs can be fertilised safely with a single 5-ml straw. It was observed in these experiments that the amount of the eggs, fertilised with one straw can be increased because the 2 ml sperm that can be found in a straw contained enough spermatozoa to fertilise 120 g eggs. The use of 5-ml straws has been tested on several fish species including the rainbow trout *Oncorhynchus mykiss* (Wheeler & Thorgaard, 1991; Lahnsteiner et al., 1997; Cabrita et al., 2001) or the paddlefish *Polyodon spathula* (Horváth et al., 2010), however, all previous works report a more or less reduced fertilizing capacity of sperm cryopreserved in these straws as compared to the conventional 0.5-ml French straws. The reaction of sperm to cryopreservation in different straw types seems to be species specific with the sperm of the wels catfish being especially resistant to the incurred cryodamage.

In the experiments no significant decrease was experienced in the quality of sperm after thawing. One of the reasons of this is that proper cooling time was successfully defined, which resulted the best fertilization rate. Thus, it can be said that cryopreserved sperm does not decrease the hatching rate compared to the traditional, routine method.

According to these experiments wels catfish catfish sperm collected outside of spawning season is as suitable for cryopreservation and for fertilization at fish farms similarly to the ones collected in the spawning time.

After the successful cryopreservation and thawing of large amounts of sperm the next step is to carry out safe fertilization with this sperm on large scale. In the experiments doses of eggs between 150 - 350 g were fertilised with a single straw. According to literature data 100 - 200 g eggs can safely be incubated in one 7-l Zug-jar (Szabó, 2000). Cryopreserved sperm showed similar hatching rates to control in every experiment. These results prove that maximum sperm-egg ratio was not reached that might cause a decrease in hatching rate. According to the experiments it can be said that the improved method is suitable for wels catfish fertilization.

After improving the freezing method of sperm the next task was to examine whether the growth and survival of larvae originated from cryopreserved sperm reaches that of larvae originated from fresh sperm. The research was extended to both the feeding and non-feeding larval periods. The results in both cases were that there is no difference in the survival of larvae fertilised with cryopreserved or fresh sperm. In the non-feeding larval period the growth of larvae from cryopreserved sperm exceeded the growth of the control, and in feeding larvae body length was higher compared to the control results.

According to these experiments the survival rate of larvae originating from cryopreserved sperm is as high as in the control and growth level of them in some cases showed better results compared to the control.

4.2 Experiments on pikeperch sperm

During the improvement of the cryopreservation technique of pikeperch sperm in laboratory cryoprotectant DMSO showed better fertilization rates than methanol but fertilization experiments in hatcheries showed opposite results. Literature data can be found on successful usage of both cryoprotectants in several fish species. The objective of this thesis is the usage of cryopreserved pikeperch sperm in hatcheries and according to the results of the experiments in the whole it was concluded that methanol and 1:1 dilution rate is suitable for freezing pikeperch sperm.

A significant variation was observed in motility after thawing and in hatching rate in the first experiments. This variability is caused by mixing of sperm with urine. This problem can successfully be eliminated when the stripping of sperm is conducted with a silicone catheter. According to this method the sperm is stripped with this silicone catheter directly from the testes preventing the mixing of sperm with urine or feces. One year later the use of catheter resulted in substantially better hatching rates.

It was observed that the increasing of the amount of eggs fertilised with a single 0..5-ml straw resulted in improved hatching rates. The reason for this can be that different amounts of eggs behaved differently in Zug-jars. The dose of 10 g of eggs were slightly stuck together, the dose of 30 g of eggs stuck in smaller batches while the dose of 50 g of eggs rolled freely. In spite of the fact that the 50 g of eggs sample had not replicates, these results suggest that fertilisation of larger amounts of eggs result in better hatching rates.

It is supposed that the eggs in the middle of the 10-g batches were more sensitive for oxygen deficiency than the more loose larger egg samples.

Another explanation for these results is that methanol in smaller eggs samples was in higher concentration, thus the toxic effects were more drastic than in larger samples. The lower sperm-egg ratio in larger egg samples had no influence on the results, suggesting that the amount of sperm was in surplus in the case of smaller egg samples.

5. Conclusion

A method has been developed for the cryopreservation of wels catfish sperm that can be used in the practice of fish farms. It is possible to fertilize 150-300 g eggs with sperm cryopreserved in large volumes (5-ml straws). The motility and hatching rate of frozen sperm correspond with the currently used routine method of fertilization with fresh sperm.

The survival and growth parameters of wels catfish larvae originating from cryopreserved sperm was investigated for the first time. According to this study it can be said that survival rate of larvae originating from cryopreserved sperm reaches and in some cases exceeds that of control larvae. This result proves the practical usage of the cryopreservation method.

Pikeperch sperm has successfully been cryopreserved for the first time, and the developed method was tested in hatchery conditions.

6. Acknowledgment

This work was supported by the NKFP 4/006/2004, RET 12/05 and the Bolyai János Fellowship of the Hungarian Academy of Sciences. The authors express their gratitude to the fish farms mentioned in the work for their support.

7. References

Cabrita, E.; Robles, V.; Alvarez, R. & Herráez, M.P., (2001). Cryopreservation of rainbow trout sperm in large volume straws: application to large scale fertilization. *Aquaculture*, Vol.201, pp. 301–314, ISSN: 0044-8486

Horváth, Á. & Urbányi, B. (2000). The effect of cryoprotectants on the motility and fertilizing capacity of cryopreserved African catfish *Clarias gariepinus* (Burchell 1822) sperm. *Aquaculture Research*, Vol.31, pp. 317-324, ISSN: 1355-557X

Horváth, Á.; Miskolczi, E. & Urbányi, B. (2003). Cryopreservation of common carp sperm. *Aquatic Living Resources*, Vol.16, pp. 457-460, ISSN: 0990-7440

Horváth, Á.; Wayman, W.R.; Urbányi, B.; Ware, K.M.; Dean, J.C. & Tiersch, T.R. (2005). The relationship of cryoprotectants methanol and dimethyl sulfoxide and hyperosmotic extenders on sperm Cryopreservation of two North-American sturgeon species. *Aquaculture*, Vol.247, pp. 243-251, ISSN: 0044-8486

Horváth, L.; Tamás, G. & Seagrave, C. (2002). *Carp and pond fish culture, Second edition*, Blackwell, ISBN 0-85238-282-0, Oxford, UK

Krasznai, Z. & Márián, T. (1985). Kísérletek a lesőharcsa (*Silurus glanis* L.) spermájának mélyhűtéses tartósítására és az évszaktól független szaporításra. *Halászat*, Vol.31, pp. 25-28, ISSN: 0133-1922

Lahnsteiner, F.; Weismann, T. & Patzner, R.A., (1997). Methanol as cryoprotectant and the suitability of 1,2 ml and 5 ml straws for cryopreservation of semen from salmonid fishes. *Aquaculture Research*, Vol.28, pp. 471-479, ISSN: 1355-557X

Legendre, M.; Linhart, O. & Billard, R. (1996). Spawning and management of gametes, fertilized eggs and embryos in Siluroidei, *Aquatic Living Resources*, Vol.9, pp. 59-80, ISSN: 0990-7440

Linhart, O.; Billard, R. & Proteau, J.P. (1993). Cryopreservation of European catfish (*Silurus glanis* L.) spermatozoa. *Aquaculture*, Vol.115, pp. 347-359, ISSN: 0044-8486

Linhart, O.; Rodina, M.; Flajshans, M.; Gela, D. & Kocour, M. (2005). Cryopreservation of European catfish *Silurus glanis* sperm: Sperm motility, viability, and hatching succes of embryos. *Cryobiology*, Vol.51, pp. 250-261, ISSN: 0011-2240

Ogier de Baulny, B.; Labbé, C. & Maisse, G. (1999). Membrane integrity, mitochondrial activity, atp content, and motility of the European catfish (*Silurus glanis*) testicular spermatozoa after freezing with different cryoprotectants, Cryobiology, Vol.39, pp. 177-184, ISSN: 0011-2240

Rana, K.L. (1995). Preservation of Gametes. In: *Broodstock Management and Egg and Larval Quality,* N.R. Bromage & R.J. Roberts, (Eds.), 53-75, Blackwell, ISBN 978-0-632-03591-5, Stirling, UK

Szabó, T. (2000). A csuka tógazdasági tenyésztése In: *Halbiológia és haltenyésztés,* L. Horváth, (Ed.), 312 Mezőgazda Kiadó, ISBN 963-9239-45-3, Budapest, Hungary (in Hungarian)

Urbanyi, B.; Horváth, Á. & Dinnyés, A. (1999). Cryopreservation of African catfish (*Clarias gariepinus*) sperm using different cryoprotectants. *Theriogenology,* Vol.51, p. 296, ISSN: 0093-691X

Urbányi, B.; Szabó, T.; Miskolczi, E.; Mihálffy, Sz.; Vranovics, K. & Horváth, Á. (2006). Successful fertilization and hatching of four European cyprinid species using cryopreserved sperm. *Journal of Applied Ichthyology,* Vol.22, pp. 201-204, ISSN: 0175-8659

Wheeler P.A. & Thorgaard G.H., 1991: Cryopreservation of rainbow trout semen in large straws. *Aquaculture,* Vol.93, pp. 95–100, ISSN: 0044-8486

14

Sperm Cryopreservation of Some Freshwater Fish Species in Malaysia

Poh Chiang Chew and Abd. Rashid Zulkafli
Freshwater Fisheries Research Division, FRI Glami Lemi, Jelebu, Negeri Sembilan, Malaysia

1. Introduction

It has been estimated that spermatozoa can last from 200-32,000 years (Stoss & Donaldson, 1983; Suquet et al., 2000). According to Kopeika et al. (2007), several methods of fish sperm storage has been practised including stored in medium with saturated gases, preservation at temperature above zero, in a frozen state as well as by drying. However, the low temperature approaches have been successful in fish sperm cryopreservation. Thus, cryopreservation technology offers the best means for long term storage of fish semen.

To date, successful cryopreservation of fish semen were reported in more than 200 freshwater species and 40 marine species worldwide (Gwo, 2000). Even though in general many successes have been achieved in fish semen cryopreservation, the technique remains as a method that is difficult to be standardized and use in all types of fishes. This is due to the fact that cryopreservation of sperms from different fish species required different conditions, where the protocol needs to be established individually. Even the "general protocol" of cryopreservation of fish sperm summarized by Kopeika et al. (2007) encompassed many variations when different species of fish are involved, particularly in the use of medium ingredients for the cryopreservation.

In view of need to develop individual protocol for successful cryopreservation of fish semen and considering Malaysia has a rich fish fauna with many of them unique to this tropical region, cryopreservation of fish gametes will require detailed study to create new protocol for each fish species intended for semen cryopreservation. To date in Malaysia semen cryopreservation has only been reported for several freshwater fish species, namely *Probarbus jullieni, Tor tambroides, T. deuronensis, Hemibagrus nemurus, Pangasius nasutus, Hypsibarbus wetmorei, Barbonymus gonionotus* and *Clarias gariepinus*. It has been demonstrated that semen cryoperservation plays an important role for the genetic conservation of these fish species.

Cryopreservation technology for fish semen is still not well explored in Malaysia and can be considered as new if compared to the domesticated terrestrial livestocks. Henceforth, this has opened up a new field to be explored with potential applications in aquaculture and in the conservation of the national fisheries genetic resources. Cryopreserved semen could facilitate artificial fertilization especially when mature male fishes are not available or unable to provide viable semen during certain periods of the breeding season. Semen cryopreservation may also be useful for fertilization to produce hybrids of various fish species. It also helps in reducing the cost and labor of maintaining broodstocks under *in situ* condition. In line with the mission of Department of Fisheries (DOF) Malaysia to develop and manage the national fisheries sector in a sustainable manner, the gene bank of freshwater fishes in the form of semen cryobank of Fisheries Research Institute at Glami Lemi was established in 2008. The establishment of the semen cryobank research has achieved the aim of the DOF towards establishing a national semen cryobank (gene bank) in Malaysia for conserving the genetic materials of the threatened or endangered indigenous freshwater fish species and also for those indigenous species which has potential for aquaculture.

The main focus of this chapter will be on the methodology developed for the semen cryopreservation in Malaysia of some fish species mentioned above and the various important steps and several key factors that contributed to the successes in gamete cryopreservation. In addition, the chapter will also present the current status and the challenges of fish semen cryopreservation in Malaysia, especially on the conservation of genetic resources and potential applications of gamete cryopreservation in aquaculture. Challenges in establishment and maintenance of the fish sperm cryobank are also discussed.

2. Semen cryopresevation of freshwater fish species of Malaysia

Malaysia has close to a hundred river systems, two natural lakes (Lake Chini and Lake Bera) and a vast freshwater water bodies and peat swamps (Baluyut, 1983). On fish biodiversity, Malaysia has approximately 616 reported freshwater fish species (Froese & Pauly, 2003). Most of the inland fisheries resources are dominated by the cyprinids and silurids (Khoo et al., 1987). Some endemic species are found in rivers, lakes and peat swamps. In the past decades intensive development activities such as deforestation and land clearing for oil palm plantation or other agricultural uses, indiscriminate logging within and around the catchment areas and construction of dams for hydroelectricity, has led to many riverine fish species suffer high risk of extinction (Khoo et al., 1987; Jackson & Marmulla, 2001). These development activities have resulted in habitat destruction, deterioration of water quality, water pollution and sedimentation, especially during rainy season when runoff is increased. All these certainly have direct great impacts on some fish species and the impacts are irreversible (Ho, 1995). Apart from the environmental degradation resulting from development, other causes of loss in inland fisheries biodiversity are attributed to over-exploitation due to strong market demand, the use of illegal and destructive fishing gears such as poisoning and electro-shocker and the introduction of invasive exotic species (Dudgeon, 2002; Allan et al., 2005). The indigenous species such as Isok barb (*Probarbus jullieni*), Malaysian Mahseer (*Tor* spp.), Mad barb (*Leptobarbus hoeveni*), Hampala barb (*Hampala macrolepidota*), knife fish (*Chitala lopis*), climbing perch (*Anabas testudineus*), snakeheads (*Channa* spp.), Asian arowana (*Scleropages formosus*), the pangasiid catfishes (*Pangasius nasutus, Pangasigodon waandersii*), giant river catfish (*Wallago leerii*), large headed walking catfish (*Clarias macrocephalus*) and giant gouramy (*Osphronemus goramy*) have dwindled in great numbers continuously due to unsustainable fishing activities. At present, most Malaysian fish species could only be conserved probably in the inaccessible or remote areas of the country.

It is no doubt that continuing habitat destruction, overfishing and competition for food from the invading fish species are leading to loss of inland fisheries biodiversity even before much of them could be documented. The lack of data on the indigenous species will subsequently impede efforts to better utilize and manage the nation's inland fisheries resources in a sustainable manner, and can eventually resulting in threatened, endangered or extinction of fish species in Malaysia. With respect to National Biodiversity Policy, it is therefore vital to protect and safeguard the indigenous fisheries resources while the species still exist in the wild. Realizing the risk of depleting fish stocks from natural waters, studies in domestication, management and husbandry of broodstocks, artificial breeding, grow out, nutrition and effort in stock enhancement via restocking of selected indigenous freshwater fish species have been carried out by the Department of Fisheries, Malaysia ever since 1980's. The species which have been studied and documented were the indigenous catfishes (*Clarias macrocephalus* and *C. batrachus*), Tropical bagrid catfish (*Hemibagrus nemurus*), Hampala barb (*Hampala macrolepidota*), Isok barb (*Probarbus jullieni*), Malaysian Mahseer (*Tor tambroides*), local pangasiid catfish (*Pangasius nasutus*) (Pathmasothy & Omar, 1982; Pathmasothy, 1985; Saidin, 1986; Thalathiah et al., 1988; Mohamad-Zaini, 1992; Thalathiah et al., 1992; Ahmad-Ashhar, 1992; Ahmad-Ashhar & Haron, 1994; Ahmad-Ashhar, 1996; Suhairi, 1996; Suhairi et al., 1996; Ahmad-Ashhar, 1998).

In 2007, cryopreservation of semen was implemented as one of the *ex situ* conservation approach, with the exotic species such as the Javanese barb (*Barbonymus gonionotus*) and African catfish (*Clarias gariepinus*) used as the model species to examine the various suitable formulations diluents and skill development in semen cryopreservation. These two species were chosen because they are domesticated species, which are available abundantly in Malaysia and able to breed easily in captivity. In Malaysia, very little work was done on semen cryopreservation in fish species and study on fish semen cryopreservation is still at its infancy. So far only a few indigenous fish species, namely the *P. jullieni, Tor* spp, *H. nemurus, P. nasutus* and *H. wetmorei* were studied. Of which only studies on the *P. jullieni, Tor* spp., and *H. nemurus* were reported (Chew et al., 2010a; Chew et al., 2010b; Muchlisin et al., 2004), while studies on other species remain unpublished. The biology, ecology and reproduction biology of these indigenous fish species are described in the following:

2.1 *Probarbus jullieni*

P. jullieni (English name: Isok barb or Jullien's Golden Carp or seven-striped barb) of family Cyprinidae (Figure 1a) is commonly known as Temoleh or Temelian among Malaysians. This species is listed in the Convention on International Trade in Endangered Species of Wild Fauna and Flora (CITES), Appendix 1 and the IUCN Red List as the endangered species (Hogan et al., 2009; IUCN, 2009). Therefore, conservation of this species is in urgent needed. Isok barb is a migratory species inhabiting river channels with water depth more than 10 m and is endemic to the Pahang River and Perak River in Peninsular Malaysia and the Mekong River basins of Indochina (Roberts, 1992). However, the drastic depletion of the numbers of Isok barb in Perak River was reported ever since the construction of Chenderoh Dam in 1930s. The dam created a physical barrier at the Perak River that permanently blocked the migration path of the Isok barb and increased water level further destroyed the spawning ground of this species (Khoo et al., 1987). Although the Isok barb can grow to reach the weight exceeded 70 kg, like the size of a human being (Baird, 2006), the landing of large size Isok barb (>10kg) has rarely occurred in the past 10 years.

Isok barb is a seasonal-bred species. In Malaysia, induced breeding of this species was carried out successfully in FRI Glami Lemi (formerly known as Freshwater Fisheries Research Centre) in 1990s (Ahmad-Ashhar, 1992; Ahmad-Ashhar & Haron, 1994). In its natural environment, this species spawn in fast flowing deep waters with sandy bottom. Chew et al. (2010a) reported that the spawning behavior of *P. jullieni* in captivity very much associated with the monsoons. Nevertheless, the breeding season of the cultured Isok barb was reported to be 1-2 months earlier (October to January) than the wild populations (December to February).

2.2 *Tor* spp. (*T. tambroides* and *T. deuronensis)*

The Malaysian Mahseer, *Tor* spp. (family: Cyprinidae) or locally known as Kelah, Empurau, Semah or Pelian, is highly sought for its value as a food fish, game fish and ornamental fish (Inger & Chin, 1962; Mohsin & Ambak, 1991; Kottelat et al., 1993). Two valid *Tor* species were identified and described in Malaysia, i.e. *T. tambroides* and *T. douronensis* (Roberts, 1989; Kottelat et al., 1993; Rainboth, 1996; Zhou & Chu, 1996) (Figure 1b and 1c). Of the two *Tor* spp., the *T. tambroides* is more favourable as food fish and can fetch up to about USD 100 - USD 220 per kg and thus is the most expensive freshwater food fish in Malaysia. Malaysian Mahseer occurs in most undisturbed and clear flowing upstream rivers, reservoir systems and lotic habitats throughout the East and Southeast Asia. The major geographical locations of Malaysian Mahseer are Pahang, Perak, Terengganu, Kelantan, Sabah and Sarawak (Ng, 2004; Ambak et al., 2007). Similar to other indigenous fishes in Malaysia, the wild populations of Malaysian Mahseer are declining over the years as the consequences of over-exploitation, natural habitat degradation and water pollution. Therefore, Malaysian Mahseer is now classified nationally as ecologically threatened (Ingram et al., 2005).

The taxonomic status of species within the genus *Tor* has been highly contentious due to plasticity of many external morphological features resulted from considerable geographical and ecological variability (Tsigenopoulos & Berrebi, 2000; Nguyen et al., 2007). Therefore, species identification based on morphological comparison sometimes could be confusing to most people who are not trained in fish taxonomy. Ability to differentiate between the Malaysian Mahseer and the Copper Mahseer (*Neolissochilus* spp.) is another difficult task hampered most of the fish farmers or aqua-culturists. The Copper Mahseer is not a desirable species and thus its value in term of market price is very much lower compared to the Malaysian Mahseer. At present, the seed supply of Malaysian Mahseer is still depending solely on the captive wild stock. As such, those who want to culture this species will have to bear with the risk of getting seed stock that comprises the mixture of these two species. Therefore, a more effective method for species identification by using DNA markers such as the mitochondria DNA (mtDNA) sequences and microsatellte markers are seen as potential in solving this problem (Nguyen et al., 2006a; Nguyen et al., 2006b; Nguyen et al., 2008; Nguyen, 2008).

The first successful hormonal induction spawning of *T. tambroides* under captive pond-reared environment was reported by Ingram et al. (2005). In captive condition, *T. tambroides* seems to spawn all year round. Brood fish of both sexes may strip up to a few times in a year. However, it is reported that geographical reproductive diversity, diet and environment conditions such as changes in temperature, water level, pH, flow velocity, turbidity, rain falls, collectively trigger the Mahseer to spawn (Dobriyal et al., 2000). Malaysian Mahseer is also a large riverine species and that can grow up to 30 – 50kg. In the culture condition, the growth rate of Malaysian Mahseer is relatively slow compared to other aquaculture species. This species usually took up to three years to reach the marketable size of 1.5 - 2.0 kg.

2.3 *Hemibagrus nemurus*

The tropical bagrid catfish, *H. nemurus* (previously known as *Mystus nemurus*) of family Bagridae and with the vernacular name Baung (Figure 1d), is a widely distributed food fish found in most of the inland water bodies in Malaysia (Khan et al., 1995). The occurrence of this species in brackish water was also reported (Inger, 1955). The tropical bagrid catfish is a bottom feeder and fed on a wide range of food from small teleosts, crustaceans, benthic invertebrates to the detritus (Khan, 1987). This species is potential for aquaculture as it receives good acceptance from the local market due to its tender and non-bony meat. This carnivorous species is also a popular species in sport fishing. The artificial spawning of tropical bagrid catfish via hormonal induction using the heteroplastic pituitary extract in combination with human chorionic gonadotropin (HCG) was reported by Thalathiah et al. (1988, 1992). According to Khan et al. (1995), a bimodal male and female gonad somatic index (GSI) pattern throughout a year was shown in this species, suggesting this species could spawn twice in a year.

2.4 *Pangasius nasutus*

The local pangasiid catfish, *P. nasutus* of family Pangasiidae and with local name Patin Buah, (Figure 1e) is a native river catfish endemic to Pahang River in Malaysia. As commented by Hung et al. (2004), catfishes from the family Pangasiidae are of great economical importance in Southeast Asia. *P. nasutus* is one of the favorite food fish due to its white and tender flesh. This omnivorous species commands a high market value (USD 20 per kg) in local market. In the wild, *P. nasutus* fed mainly on the bivalves, mollusks, gastropods and other benthic organisms especialy on sandy bottom rivers (Roberts & Vidthayanon, 1991). Because of its economic value and declining population in the natural waters, this has created popularity and awareness to conserve and culture this species. The first successful induced spawning of *P. nasutus* was reported in 2005. This species is also a batch spawner and exhibits a single-modal GSI pattern throughout the whole year. During the spawning season, multiple releases of eggs from a single female broodfish were observed during the single period (Mohd-Zafri, 2006). The breeding season of this species is associated with rainfalls or monsoons between May - July (Maack & George, 1999).

2.5 *Hypsibarbus wetmorei*

Morphologically *Hypsibarbus wetmorei* is quite similar to the Javanese Barb, *Barbonymus gonionotus* except the appearance of turmeric-like yellow colour on its body (Figure 1f) where the vernacular name Kerai Kunyit is given. This species is of the family Cyprinidae. Kerai Kunyit is highly regarded as a food fish and thus it is a potential aquaculture species. In Malaysia, this species is found to be endemic in Pahang River. Until today, the biology, ecology and repoductive behaviour of this species is not well studied and limited documentation on the species is available. By far, only study on species identification of *Hypsibarbus* spp. (included *H. wetmorei*) using PCR-RFLP method was reported by Jantrarotai et al. (2007).

Fig. 1. Photos of the Malaysian fishes involved in the cryopreservation studies, (a) Isok barb (*P. jullieni*), (b) Malaysian Mahseer (*T. tambroides*), (c) Malaysian Mahseer (*T. deauronensis*), (d) Tropical bagrid catfish (*H. nemurus*), (e) Local pangasiid catfish (*P. nasutus*), and (f) Kerai Kunyit (*H. wetmorei*).

3. Methodology employed for semen cryopreservation of freshwater fishes in Malaysia

As in cryopreservation of other fish species, the cryopreservation of the semen of freshwater fishes from Malaysia to be discussed later falls under the following general procedures (i) collection of semen, (ii) dilution of semen sample, (iii) semen sample packaging, (iv) equilibration, (v) freezing, (vi) cryo-storage, and (vii) thawing. Successful cryopreservation of fish sperm very much depends on a range of factors in each step of the cryopreservation procedures as highlighted by Kopeika et al. (2007).

3.1 Collection of fish semen

Mature and healthy males broodfish were selected and anaesthesized using MS222 or clove oil according to the dosage recommended by the manufacturer. Semen sample was expelled from the male fish by gentle abdominal pressure and collected into a clean and dry tube. Extra precaution should be taken while collecting semen sample. Contamination of sample with blood, water, urine or the feces should be avoided as these contaminants significantly reduced the semen quality and caused poor post-thaw sperm motility. The semen samples were then transferred back to laboratory for quantification of the fresh sperm quality and proceed with further dilution before freezing procedure. The sperm motility rates and sperm concentration of the freshly collected semen sample was evaluated prior to cryopreservation. In order to obtain good post-thaw motility, only semen samples showing sperm motility $\geq$70% were used for cryopreservation.

3.2 Dilution of semen sample

The semen samples with good quality were subsequently diluted at an appropriate sperm to diluent ratio, with suitable extender solution and cryo-protectant. Sperm to diluent ratios ranged between 1:3 to 1:9 were reported to produce best results in fish sperm cryopreservation studies (Scott & Baynes, 1980; Lahnsteiner et al., 1996). Gwo (2000) reported the fish sperm could loss its viability in high dilution ratio especially in marine species. The type of extender solution, cryoprotectant and dilution ratio that were optimum for each Malaysian fish species studied were summarized in Table 1. The type of diluents and dilution ratios workable to preserve sperm motility appear to vary among different fish species. Thus each of these parameters needs to be optimized accordingly via a series of experimentations.

The extender solution helps to maintain sperm viability prior to and during the freezing process. Extender solution is a balanced salt buffer of specific pH and osmotic strength. Apart from salts, sometimes extender solution is prepared with addition of organic compounds such as glucose. The nature of the effect of extenders is based on the control of pH and salt concentration as well as the supply of energy, and can extend the functional life and fertilizing capability of the sperm (Tiersch, 2000). Cryoprotectants function to protect cells from cryo-damage or cryo-injury during freezing and thawing process. The permeating cryoprotectants, namely dimethyl sulfoxide, methanol, ethylene glycol and propylene glycol are among the most frequently used cryoprotective agents among the aquatic organisms (Lahnsteiner et al., 1997; Tiersch, 2000; Tiersch, 2006). However, the permeating cryoprotectants are often toxic to cells, and thus the choice of the types of cryoprotectant and their optimal concentration should be at a balance between protection and toxicity. On the other hand, the non-permeating cryoprotectants such as sucrose, glucose and polymers (e.g. alginate) were often used in combination with the extender solution in the diluents. Sometimes, a combination of different cryoprotectants in certain ratios could help improving the post-thaw motility. The studies of African catfish in our laboratory has shown that a combination of methanol and N, N-dimethylacetamide at ratio 70%:30% produced significant higher post-thaw motility compared to the use of a single cryoprotectant in sample dilution. Tiersch (2000) has also reported that the application of cryoprotectant at concentration between 5% to 20% ususally provides good protection in most fish species. The protective effect and optimal concentration of cryoprotectants could be species specific (Gwo et al., 1991; Suquet et al., 2000; Rideout et al., 2003). Therefore, the optimal concentration needs to be determined individually in each species studied through experimentations.

For Isok barb (*P. jullieni*), Malaysian Mahseer (*Tor* spp.), local pangasiid catfish (*P. nasutus*) and Kerai Kunyit (*H. wetmorei*), a total of 14 extender formulations, five types of cryoprotectants (dimethyl sulfoxide, ethylene glycol, glycerol, methanol and N,N - dimethylacetamide) with concentrations between 5-20% (v/v), semen to diluent ratios (1:1 to 1: 14) were examined as described by Chew et al. (2010a). This may be compared with Muchlisin et al. (2004) who used three extender solutions (the Ringer, physiological saline and saline solution); four types of cryoprotectants (DMSO, ethanol, glycerol and methanol) at three concentrations (5%, 10% and 15%) and three sperm to diluent ratios (1:20, 1:30 and 1:40) in tropical bagrid catfish sperm cryopresevation.

Species	Chemical composition of extender solution	Type of cryoprotectant & concentration (v/v)	Sperm to diluent ratio	Reference
B. gonionotus	Modified from Kurokura et al. 1984 - 62 mM NaCl, 134 mM KCl, 1.5 mM $CaCl_2$, 0.4 mM $MgCl_2$, 2.4 mM $NaHCO_3$ (Horvath et al., 2003)	Methanol 10%	1:4 - 1:6	Unpublished data
C. gariepinus	Ringer solution - 128 mM NaCl, 2.7 mM KCl, 1.4 mM $CaCl_2$, 2.4 mM $NaHCO_3$ (Kurokura et al., 1984)	Methanol + DMA (70:30) 8%	1:6	Unpublished data
P. jullieni	Calcium free Hank's Balance Salt Solution (CF-HBSS) - 152 mM NaCl, 5.9 mM KCl, 0.9 mM $MgSO_4$, 0.36 mM Na_2HPO_4, 0.5 mM KH_2PO_4, 4.6 mM $NaHCO_3$, 6.16 mM Fructose (Tiersch et al., 1994)	Methanol 9 - 10%	1:3 - 1:5	Chew et al., 2010a
Tor spp.	202 mM D(+)-glucose monohydrate, 51.5 mM sodium chloride and 6 mM sodium bicarbonate, with pH 7.1 to 7.8 and osmolality 309 $\pm$ 30 mOsmol/kg	DMSO 10%	1:7	Chew et al., 2010b
P. nasutus	CF-HBSS (Tiersch et al., 1994)	Methanol 9 - 10 %	1:7	Unpublished data
H. wetmorei	Modified Fish Ringer - 111 mM NaCl, 40.2 mM KCl, 2.1 mM $CaCl_2$, 2.4 mM $NaHCO_3$ (Wolf, 1963)	Methanol 9 - 10 %	1:4 - 1:7	Unpublished data
H. nemurus	Modified Fish Ringer - 128 mM NaCl, 2.7 mM KCl, 1.4 mM $CaCl_2$, 2.4 mM $NaHCO_3$, 25.3 mM glucose	Methanol 10%	1:20	Muchlisin et al., 2004

Table 1. A list of extender solution and its composition, type of cryoprotectant and its optimal concentration and sperm to diluent ratio for successful semen cryopreservation of various species of freshwater fishes in Malaysia.

3.3 Packaging and equilibration of diluted semen sample

In this procedure which is performed after the addition of extender solution and cryoprotectant, the diluted semen sample is packed into polyethylene (PE) straws (Chew et al., 2010a; Chew et al., 2010b) or cryo-vials (Muchlisin et al. ,2004). Extended semen sample is subjected to equilibration at temperature <10°C prior to freezing procedure. The duration taken for equilibration is the time required for the cryoprotectant to permeate the cells. Equilibration duration between 15 min to 3 h has been practised for Malaysian fish species sperm cryopreservation and was found to be successful in maintaining a good post-thaw sperm motility (Chew et al., 2010b). In most circumstances, the equilibration duration is set at 15 to 30 min, but it can be varied depending on the type and concentration of cryoprotectant used (Tiersch, 2000).

3.4 Sperm cryopreservation

Cryopreservation involves the removal of excess water from the inside of the cell to the exterior where it can form ice (Tiersch, 2000). A two-step rapid freezing method was used for semen cryopreservation in Isok barb, Malaysian Mahseer, local pangasiid catfish and Kerai Kunyit. After the equilibration procedure, semen samples of these species were subjected to liquid nitrogen vapour exposure for 5-10 min in an insulated styrofoam cooler box filled with liquid nitrogen, with the samples placed between 3 to 4 cm above the liquid nitrogen, after which the samples were directly plunged into liquid nitrogen. For tropical bagrid catfish (Muchlisin et al., 2004), the semen samples were equilibrated on crushed ice (0°C) for 5 min. After the equilibration procedure, the samples were then placed in an ice box containing dry ice (-120°C) for 5 min and subsequently plunged into liquid nitrogen.

Besides the styrofoam cooler and ice box, the vapour shipper is another choice of method could be used for the freezing of fish semen samples. To cryopreserve semen samples by using dry shipper, the extended and packed semen sample was transferred into the fully charged vapour shipper and subsequently transferred and stored in liquid nitogen. The use of the vapour shipper method in freezing as reported in *Tor* spp. (Chew et al., 2010b) was convenient to be used in the field. Such method is simple to use, the cooling rate is more consistent and controllable and it consumes less liquid nitrogen and requires less space compared to the styrofoam cooler box or ice box method.

In our opinion, the use of the sophisticated bench top type of controlled rate programmable freezer is not practical in cryopreservation of fish semen in some laboratories because of difficulty to move this equipment from one location to another. Unfortunately portable type of controlled rate programmable freezer is not user friendly, time consuming and limited number of samples could be processed at a single run. Compared with programmable freezer, the two-step freezing method employing insulated styrofoam cooler box or ice box is simpler, rapid and more cost effective. Nevertheless, the main shortcomings of this simple freezing method is the inconsistency in cooling rates and non-reproducible cryopreservation experiments if performed by different operators.

3.5 Thawing

Thawing is a process to recover the sperm cells from the dormant stage in ultra low temperature. Frozen semen is usually thawed at 40°C, with different optimal durations applied according to the type of packaging and the storage volume as shown in Table 2.

Type of packaging	Volume of semen	Duration of thawing
PE straw	0.25 mL	4s - 6s
	0.5 mL	6s - 8s
	1.2 mL	12s - 15s
Cryo-vial	5 mL	5 min

Table 2. The optimal durations used to thaw the frozen semen samples of freshwater fish species in Malaysia in different types of packaging and storage volume.

3.6 Egg fertilization

A cryopreservation protocol developed for a species is considered success if the semen that cryopreserved according to the certain protocol could successfully fertilize eggs and produced offspring. Fresh semen is usually included in the control treatment. The optimal sperm to egg ratio used should be determined prior to fertilization. The sperm: egg ratio at approximately 250,000:1 is usually sufficient and worked well in most species in our laboratory. However, other sperm to egg ratios from 1000:1 to 500000: 1 were tested by Butts et al. (2009) and 100 000: 1 was found to yield the best fertilization performance in Atlantic cod. The dry fertilization method is favorable in the fertilization test for freshwater fish and thus was applied for all Malaysian fish species studied in our laboratory. In the dry fertilization method, both eggs and sperms were mixed well before hatchery water was later added into the sperm/egg mixture to water-harden the fertilized eggs. After rinsing with hatchery water, fertilized eggs were then incubated in aquarium (with or without using a hatching trough depended on species). Water temperature was kept at room temperature between 23 - 28°C throughout the period of incubation. The duration of embryo development varied between species. Therefore, the duration required for the fertilized egg to hatch is also varied among different species.

4. Discussion on semen cryopreservation of Malaysian freshwater fish species

The successful semen cryopresevation of several Malaysian fish species e.g. *P. jullieni, T. tambroides, T. deuronensis, H. bleekeri, P. nasutus* and *H. wetmorei* using various methods of cryopresevation discussed above may be evaluated via analyses on the sperm motility, fertilization, hatching ability, etc.

4.1 Quality of semen

Sperm motility for the freshly collected semen from the healthy broodfish is usually ranged between 90 - 100% motility provided that the sperm sample was not activated prior to the actual cryopreservation processes or contaminated by water or urine while sampling was carried out. However from our observation, sperm motility of most freshwater fishes dropped to <10% after 24 h if the sample was not extended using suitable extender solution, even though the sample was kept cool in a refrigerator (0- 5°C).

Sperm concentration is one of the important characteristic that determines the sperm quality of an individual male (Billard, 1986; Suquet et al., 1992; Billard et al., 1995). The sperm concentrations of all Malaysian fish species studied were between 2.2×10^8 to 6.2×10^{10} sperm cells per mL and this is in agreement with the studies by Leung & Jamieson (1991). On the other hand, short lifespan after activation is the typical characteristic of spermatozoa of freshwater fish species. The duration of sperm motility of most freshwater fishes is normally <1 min after the sperms are activated (Billard & Cosson, 1992; Lahnsteiner & Patzner, 2008). Duration of sperm motility of *Tor* spp. was about 40-50 s and it slowed down drastically after 20 s of progressive movements (Chew et al., 2010b). For Isok barb, it was about 20 s and slowed down after 10 s of progressive movements. Table 3 shows the range of sperm concentrations and motility duration in each Malaysian species studied at FRI Glami Lemi.

Species	Sperm count (Number of cells per mL)	Duration of motility (*)
Javanese barb (*B. gonionotus*)	$2.52 \times 10^{9} - 1.03 \times 10^{10}$	<15 sec (10 s)
African catfish (*C. gariepinus*)	$5.44 \times 10^{9} - 1.14 \times 10^{10}$	15 - 30 s, (13 s)
Isok barb (*P. jullieni*)	$4.00 \times 10^{8} - 6.24 \times 10^{10}$	20 s, (10 s)
Malaysian Mahseer (*Tor* spp.)	$2.20 \times 10^{8} - 5.98 \times 10^{9}$	40 -50 s, (20 s)
Local pangasiid catfish (*P. nasutus*)	$6.60 \times 10^{8} - 1.36 \times 10^{10}$	25 - 50 s, (15 s)
Kerai Kunyit (*H. wetmorei*)	$7.30 \times 10^{9} - 1.33 \times 10^{10}$	20 -70 s, (20 s)

*Duration of progressive movement before slowed down and finally stopped

Table 3. Sperm concentration and motility duration of each freshwater fish species studied at FRI Glami Lemi.

Osmolality is another critical variable in sperm quality (Honeyfield & Krise, 2000). As seen in many studies, seminal plasma osmolality among males fish is highly variable (Aas et al., 1991). According to Babiak et al. (2001), the extender solution that worked well to cryopreserve spermatozoa of a species should posses the ability to maintain the sperm cell viability by inhibiting sperm motility. The key of success is via the use of an extender solution which is almost isotonic or mimicking the seminal plasma of that particular species. Therefore, it is crucial to know the seminal fluid osmolality of a species before media and diluents for that particular species can be formulated.

In all species of Malaysian fishes studied, sperm motility generally reduced after freezing and thawing process compared to the sperm motility before any freezing procedure. In *P. jullieni*, the motility of cryopreserved semen has reduced by approximately 45% compared to the fresh semen. A reduction of sperm motility by an average of 15% and 30% in *T. tambroides* and *T. deuronensis* respectively was observed. In general, sperm motility reduced between 10 - 70% on average in the species studied (Table 4). These observations are similar to studies in several other species such as *Cyprinus carpio* L. (Wamecke & Pluta, 2003), *Oncorhynchus mykiss* (Lahnsteiner et al., 1996) and *Scophthalmus maximus* (Dreanno et al., 1997).

Species	Motility %	
	Before freezing	Post-thaw
Javanese barb	88 - 97%	15 - 65%
African catfish	60 – 100%	15 – 70%
Isok barb	85 - 100%	30-84% (Mean: 49%)
Malaysian Mahseer	85 -100%	35-89% (Mean: 55%)
Local pangasiid catfish (Patin Buah)	85 -100%	35-70%
Kerai Kunyit	90 - 100%	35-80%
Tropical bagrid catfish (Muchlicin et al. 2004)	80 – 94%	58%

Table 4. Sperm motility (%) of the freshwater fish species studied before freezing and after thawing procedures

For all Malaysian fish species studied, the post-thaw motility rates of the cryopreserved sperm demonstrated similar values ($p>0.05$) even after a year of cryostorage as long as the semen samples are submerged well in the liquid nitrogen and without disturbance during the storage period.

4.2 Fertilization ability and hatching rate

In both Malaysian Mahseer and Isok barb, egg fertilization ability and hatching rates were found significantly lower ($p<0.05$) by using cryopreserved sperm compared with those fertilized using fresh sperm. The speed of embryos development was similar among the fertilized eggs using both cryopreserved and fresh sperm. Besides that, no significant difference ($p>0.05$) was found in the egg fertilization percentages between newly cryo-preserved semen and semen samples after a year of cryo-storage in both species. The performance of egg fertilization and hatching by using cryopreserved semen in four species of freshwater fish species studied in our laboratory is shown in Table 5.

Species	Fertilization (%)	Hatching (%)
Javanese barb	12 – 100	5 – 75
African catfish	21 - 37	19 – 32
Isok barb	1.2 – 10	0.8 – 4.6
Malaysian Mahseer	20 – 55	20 – 53

Table 5. Fertilization and hatching performance using cryopreserved sperm in several freshwater fish species carried out at FRI Glami Lemi

Quantitatively the numbers of hatchings from the cryopreserved semen were low, but qualitatively the hatchlings are visually normal and physically active and healthy as those from the fresh semen. Fingerlings produced from the cryopreserved semen from all four species showed normal development compared with those produced from the fresh semen.

5. Dependence of egg fertilization on semen cryopreservation methods

For the sperm cryopreservation of the fish species such as *P. jullieni* and *Tor* spp., several factors will determine the successful of the egg fertilization. The contributing factors that brought to the success of fish hatching using cryopreserved semen include viability of the post-thaw cryopreserved semen, good quality eggs, handling of the sperm and egg during the fertilization trials and knowledge in culturing the targeted species.

5.1 Viability of the cryopreserved sperm

Semen collection for cryopreservation should be done during the peak of spawning seasons of the particular species because the quality and quantity of spermatozoa are the highest at that particular time. The cryopreserved sperm cells tend to deteriorate very quickly and loss their motilily within 10-30 min after thawing procedure. Therefore, in quantification of the quality of the post-thaw sperm, the post-thaw motility shall be evaluated soon after thawing procedure in order to obtain good results. Similarly in the fertilization trials using cryopreserved semen, the mixture of semen to the eggs should be performed as quickly as possible once the frozen semen is thawed. Semen samples from each male broodfish were cryopreserved separately in order to maintain its inherent variability.

Successful cryopreservation procedures can maintain high quality of cryopreserved semen and this is revealed in high post-thaw sperm motility percentages. Successful cryopreservation procedures balance between the formation of ice crystals within the cells against excessive dehydration, which damages cellular structures (Tiersch, 2000). Each step involved in the cryopreservation procedures that was discussed in section 3 is equally important to produce viable cryopreserved sperm.

5.2 Good quality eggs

Good quality egg is also essential for successful fertilization when cryopreserved semen is used to produce offsprings. Same as sperm quality, egg quality is also the best during the peak of spawning seasons of the particular species. Thus fertilization trial is best to perform during spawning seasons of the fish. This is especially true and applicable for the seasonal bred species. In *P. jullieni*, significant higher fertilization rate (80-95%) was observed during the peak of the spawning season compared to the initial (50-60%) and toward the end of breeding season (30-50%) of the species. Healthy broodfish produced good quality eggs. Good management practice in broodstocks maintenance and good diet (i.e. high protein diet) are the key factors that produced healthy brood stocks.

5.3 Handling of sperm and egg during the fertilization trials

Good control in the timing of egg stripping and thawing of cryopreserved semen is important while using cryopreserved semen to fertilize eggs. Both egg and sperm should be made available at the same time in order to produce high fertilization rates. Spermatozoa of freshwater fishes are usually activated by the hypotonicity of their surrounding media (Morisawa & Suzuki, 1980; Morisawa et al., 1983a; Morisawa et al., 1983b). Most of the time and in majority species, hatchery water is sufficient as the sperm activation media and used in the sperm-egg mixture during fertilization process with cryopreserved semen. However, the use of suitable medium other than hatchery water for sperm activation is sometimes critical in some species (Lahnsteiner et al., 2003). This is observed in *T. tambroides* where specially formulated medium produced significant higher fertilization rate compared to the use of hatchery water as the sperm activation medium (Chew et al., 2010b). The effects of several media on egg fertilization ability had also reported by Billard (1983) on rainbow trout. Fertilization technique used in the fertilization trials is also one of the contributing factors to the success. Our studies showed that fertilization and hatching ability were significantly higher by using dry method compared to wet method.

5.4 Knowledge in culturing of the targeted species

Knowledge on the reproductive biology, broodstock management and husbandry, larva rearing and nursery of the particular fish species to be studied is a prerequisite for successful fertilization using cryopreserved semen. The age of maturity, breeding season (for seasonal bred species), factors that trigger and promote gonad maturation such as the type of nutrition and water quality where the brood fish is maintained are important factors that guaranteed gonad maturation and good health of the broodfish. Besides, well established artificial spawning method of the targeted species is also essential and such knowledge could later help production of fry using cryopreserved semen.

6. Current status of fish sperm cryopreservation in Malaysia –the fish semen cryobank

The current development of fish sperm cryopreservation in Malaysia is mainly focused on the creation of a fish semen cryobank. Such a cryobank for fish semen plays many important roles, especially for conservation of fish stocks and improve aquaculture. The establishment of a fish semen cryobank requires procurement of equipment and facilities for cryopreservation, skill development of the technical operators, choice of species for semen cryobanking, identification of the source for sample collection, developing of suitable cryopreservation procedure for the targeted species, proper record keeping and also proper maintenance of the semen cryobank.

6.1 Fish semen cryopresevation for conservation

Cryopreservation technology provides long term *ex situ* conservation of indigenous and endangered species (Mongkonpunya et al., 2000; Tiersch et al., 2000). Cryopreserved semen could be stored indefinitely without deterioration provided proper maintenance and handling measures are well taken care of all time (Stoss & Donaldson, 1983; Armitage, 1987; Suquet et al., 2000). In Malaysia, fish sperm cryopreservation was employed to save the genetics of the endangered and threatened indigenous species such as *P. jullieni* and *Tor* spp.. Fish species for gene banking and research are chosen on the basis of their threatened status and potential for aquaculture. Species with vulnerable, critically endangered (CR) and endangered (EN) status are prioritized in a conservation programme (Table 6). Species with protandrous behaviour throughout their life cycle such as the *Tenualosa toli* (Terubok) is also protected (MACD, 1996; Blaber et al., 2001; Wong, 2001). Other candidates of indigenous fish species which should be prioritized for semen cryopreservation are the Bala Shark (*Balantiocheilos melanopterus*), the indigenous catfish (*Clarias macrocephalus*) and several *Betta* spp. (*B. chini, B. hipposideros, B. livida, B. tomi* and *B. persephone*).

Semen samples could be stored for years via cryobanking before being used, and thus cryopreserved sperm banks can serve as insurance policies against unforseen catastrophes. However, very strict standardized protocols should be developed so that results are not biased by experimental and treatment variability but only by the inherent variability of the species. Further, these practices should be complemented with habitat conservation procedures (Bart, 2002). As it takes a long time to restore degraded ecosystems, the preserved semen is stored while upgrading of the ecosystems is going on. Both *in situ* and *ex situ* conservation should be applied complimentarily for sustainable management of the indigenous fisheries resources (Harvey, 1998). For an instance in the conservation programme of *P. jullieni,* introduction of closing season (from February to April each year) and conservation zones (12.4 km) at the identified spawning ground of *P. jullieni* at Pahang River have been implemented as the *in situ* conservation approach for the species (Zulkafli et al., 2010).

Conservation of indigenous freshwater fishes is a priority of the Department of Fisheries, Malaysia. Resulting from the success in breeding both *P. jullieni* and *Tor* spp. using cryopreserved semen, cryobanking of the semen of these species was initiated in 2008. However, those samples were collected from the cultured stock. Therefore, effort to further enrich the sperm cryobank through collection from wild stocks for different varieties and populations in the country is the ultimate aim of the Department of Fisheries, Malaysia. At present, semen samples from 88 *Tor* spp., 43 *P. jullieni*, 8 *P. nasutus* and 14 *H. wetmorei* have been collected and cryopreserved in the sperm cryobank of FRI Glami Lemi (Table 7).

The use of cryopreserved semen could support conservation efforts through stock enhancement and repopulation in areas where the species have declined or disappeared. In the breeding and restocking programmes, attempts to save the wild populations have so far largely focused on captive breeding or spawning of wild broodstock and subsequent release of hatchery-reared offsprings into the wild. Hatchery production of fry will support stock enhancement. Consequently, this will hopefully eliminate the need to harvest seed stock from the wild and the translocation of non-indigenous species for such programme.

Species	Status
Balantiocheilos melanopterus (Bala shark)	EN
Betta chini (Chini mouthbrooder)	VU
Betta hipposideros	VU
Betta livida (Emerald-sport fighting fish)	EN
Betta persephone (Black fighting fish)	CR
Betta tomi (Pikehead)	VU
Discherodontus halei (Spot-fin barb)	CR
Eleotris melanosoma	CR
Encheloclarias curtisoma (Soft fin walking catfish)	CR
Encheloclarias kelioides (Soft fin walking catfish)	CR
Encheloclarias prolatus (Catfish)	VU
Parasphromerus harveyi (Harvey's licorice gouramy)	EN
Probarbus jullieni (Isok barb)	EN
Scleropages formosus (Asian Arowana)	EN
Silurus furness	DD
Redigobius bikalanus	CR
Phallostethus dunckeri	VU
Sundoreonectes tiomanensis (Tioman cave loach)	VU
Helicophagus waandersii	EN
Cyclocheilichthys enoplus	EN
Leptobarbus hoeveni (Mad barb)	EN

(Source: DOF, Malaysia; Chong et al., 2010)

Table 6. List of indigenous freshwater fish species in Malaysia with critically endangered (CR), endangered (EN), vulnerable (VU) or data deficient (DD) status.

Species	Number of fish	Volume
Tor spp.	88	150 mL
P. jullieni	43	350 mL
P. nasutus	8	18 mL
H. wetmorei	14	30 mL

Table 7. Current status of cryogenic fish sperm bank of FRI Glami Lemi, Malaysia (since 2008)

6.2 Fish semen cryopreservation for aquaculture

Fish gene banks offer vast potential benefits to hatcheries (Munkittrick & Moccia, 1984; Chao & Liao, 2001). It offers genetic variability to fish hatcheries around the world. The use of frozen semen in breeding programmes offers a means to further broaden the genetic base of the targeted species. Genetic improvement of broodstock or hatchery species for traits such as disease resistant, fast growth rate, salinity tolerance etc. could also make feasible with the establishment of the cryogenic sperm bank. The applications of sperm cryopreservation in aquaculture were also highlighted by Mongkonpunya et al. (2000). In the case of some species, males and females reach maturity over different periods of time, the cryopreserved semen could facilitate artificial fertilization and seed production (Tiersch, 2000). Besides, cryopreserved semen is easier to transport than live fish for culturing. This eliminates the stress to fish. The risk of transmitting diseases is also reduced by using cryopreserved semen.

On the other hand, the use of cryopreserved sperm also provides flexibility in breeding programmes, especially in producing hybrids with favourable characteristics for culturing such as higher viability, intensive growth rate, adaptive flexibility, early sexual maturation etc. (FAO, 1971). Hybridization in fish culture becomes feasible and more manageable with the utilization of cryopreserved semen. For example, in a breeding programme of *H. wetmorei* carried out in our laboratory, *H. wetmorei* was cross-bred with the Javanese barb, *B. gonionotus* successfully via the use of cryopreserved sperm and surrogate egg from *B. gonionotus*. Such breeding procedure was needed because mature female of *H. wetmorei* was not available during the induced breeding programme of the species. Therefore, it is no doubt that the use of cryopreserved sperm provides greater control in breeding programmes.

7. Challenges to the gene banking of fish gamete via cryopreservation

7.1 Procedure optimization

The principle and process behind semen cryopreservation sound rather simple, i.e. the storage of semen samples in ultra-low temperatures and liquid nitrogen (-196°C) is normally used. In avian species (such as chicken, fowl, turkey, goose and duck) and mammal livestock species (such as cattle, horse, boar, sheep and goat), protocols for their semen cryopreservation are well developed and established (Hammerstedt & Graham, 1992; Curry, 2000; Donoghue & Wishart, 2000; Woelders et al., 2003). In fish however, there are no standard protocols that are applicable to all species. Unlike terrestrial animals, it is more difficult to standardize the semen cryopreservation protocols in fish (Tiersch, 2000). This is because different fish species exhibited different responses to the same extenders and cryoprotectants. For an example, the protocol or diluents formulation which served optimal to *P. jullieni* may not be necessarily suitable and served optimum in other species. As such, developing the species-specific and reproducible sperm cryopreservation procedure is thus required for this purpose. These include the choice of the type of extender solution and cryoprotectant, the rates of freezing and thawing etc. In general, the optimization of cryopreservation protocol for a species involved a series of complex interactions among various factors in each step.

7.2 Maintenance and proper recording

It is less costly to maintain a semen cryobank in a long run compared to *in situ* conservation approach such as live genebank. However, some costs need to be allocated for replenishing and topping up of liquid nitrogen in the storage dewars on periodically basis. Obtaining adequate funding and financial support to maintain the semen cryobank is part of the challenges to the institutions owned a semen cryobank. The monitoring of the viability of cryopreserved semen samples is also required from time to time to ensure the viability of stored samples is maintained. Post-thaw sperm motility was assessed for each batch of semen samples a week after cryo-storage and prior to be used for egg fertilization. The viability of the cryopreserved sperm was also evaluated from time to time through egg fertilization tests. A database that is able to provide good records of storage and withdrawals of samples from the semen cryobank is also a long term challenge in genebanking via cryopreservation (Kincaid, 2000). Proper record is essential for ease of samples retrieval in future.

7.3 Technical limitations

There are technical limitations to use cryopreserved semen in fish breeding as it requires involvement of skilled personnels. Therefore, training of operators or technicians in the related discipline is seen required in the technology extension programme. The small volume sample in straws (0.25 mL and 0.5 mL) is sufficient to be used in egg fertilization in laboratory based experiments and in genetic improvement programme of the targeted species. However, it is less practical to use cryopreserved semen for mass production of fry in aquaculture. Adaptation of the current developed protocols for practical application is thus important. The use of bigger straws volume or cryovials (5 mL or 10 mL) should be considered instead.

7.4 Difficulties in obtaining semen samples from wild populations

Difficulties in getting wild stocks are the main constraints in fish semen cryobanking of the indigenous species with threatened or endangered status and those species with high market demand. These limitations have caused difficulties in obtaining the effective population size (N_e), which is very crucial in future restoration efforts for the species. High quality seed is essential to support aquaculture. For many species especially of those riverine species, which their induced breeding method is still not established, their source of seed supply is still depend on the wild caught stocks. In Malaysia, it is also difficult to obtain fry and broodstock from hatchery because the lack of well organised hatchery operation. Each hatchery tends to maintain their own breeders, which is always limited in numbers. As the consequences, this resulted in high inbreeding rates among the hatchery stocks.

In Malaysia, it is now increasingly difficult for fish breeders to locate and collect genetic materials from healthy or relatively undisturbed populations in the wild. The loss of genetic material in fish species can hinder the development of the aquaculture industries, especially fish farms and hatcheries. Many hatcheries often rely on too few breeders to reproduce, resulting in lower production, susceptibility to diseases and poor survival rates in the wild. As wild fish stocks disappear, it becomes even more difficult for hatcheries to find new breeders. At the same time breeding within small populations with limited genetic diversity results in inbreeding depression, i.e. genetic drift, producing small or stunted fish stocks. Therefore, fish genetic resources must be conserved and utilized sustainably because they are the key to maintaining the viability of cultured and natural fish populations. They enable species to adapt to environmental change and also provide the opportunity for genetic improvement programme in aquaculture.

It is observed that large species that breed later in life are more vulnerable to fishing and changes in the environment, particularly in terms of fragmentation of their normal habitats. Indeed, most of the world's largest freshwater fish are at risk according to the IUCN Red List, and over exploitation contributes in a number of these cases. The dragon fish, *Scleropages formosus* is a well known case of over-exploitation. Therefore, giant indigenous species such as freshwater siluroids (*Wallago leeri,*), cyprinids (*Tor* spp., *Probarbus jullieni*), pangasiids (*Helicophagus waandersii*) etc. can be promoted as 'flagship species' or ecosystem ambassadors. At the same time, in terms of preserving biodiversity, by reducing the negative impacts of the continued spread of exotic fish species in the aquatic environment, efforts need to develop indigenous species for use in aquaculture. Those indigenous species that showed good adaptation to pond environment, resistance to handling, possess high growth rate and ability to reach sexually maturity in captivity are worth to be considered and developed as aquaculture species. To achieve the goal, we must safeguard indigenous fish resources both quantitatively and qualitatively from now before it is too late.

8. Conclusion

An overview of the current status of semen cryopreservation of Malaysian freshwater fishes is presented in this chapter. The role of semen cryobank is also discussed. Obviously, semen cryopreservation offers potential applications in *ex situ* conservation and sustainable management of the fisheries genetic resources in Malaysia, especially for those species with rare, vulnerable, threatened or endangered status, those protandry species with sex-changing characteristic over their life time, and also the potential candidates and genetic improved strains for aquaculture development. The successes in semen cryopreservation are very much relied on factors such as having ample knowledge on the biology and reproductive biology of the particular species of interest, trained personnels in various aspects such as gamete cryopreservation, breeding methods, broodstock management and husbandry and larva rearing and nursing of the targeted species.

9. Acknowledgement

We would like to thank the Department of Fisheries Malaysia for supporting this research through a development grant and also thanks to all research staffs of Freshwater Fisheries Research Division, FRI Glami Lemi and the Aquaculture Extension Centre, Perlok who have kindly provided assistance during the project.

10. References

Aas, G.H.; Refstie, T. & Gjerde, B. (1991). Evaluation of milt quality of Atlantic salmon. Aquaculture, 95: 125-132.

Ahmad-Ashhar, O. (1992). Induced spawning of Ikan Temoleh *Probarbus jullieni* Sauvage using human chrionic gonadotropin and heteroplastic crude piscine extract. In: Proceedings of the National IRPA Seminar, Kuala Lumpur, 9-11 January 1992. pp. 243-244.

Ahmad-Ashhar, O. (1996). Breeding and seed production technology of the Jelawat (*Leptobarbus hoevenii*).

Ahmad-Ashhar, O. (1998). Pembiakan Dan Pengeluaran Benih Ikan Jelawat (*Leptobarbus hoevenii*) Bleeker. *Buku Panduan Bil 2/98*. 21pp.

Ahmad-Ashhar, O. & Haron, A. (1994). Pembiakan aruhan Ikan Temoleh *Probarbus jullieni* Sauvage menggunakan ekstrak pituitari dan hormon human chorionic gonadotrophin (H.C.G.). *Proc. Fish. Res. Conf., DOF, Mal. IV:* 253-256.

Allan, J.D., Abell, R., Hogan, Z., Revenga, C., Taylor, B.W., Welcomme, R.L. & Winemiller, K. (2005). Overfishing of Inland Waters. *BioScience* 55(12): 1041-1051.

Ambak, M.A., Ashraf, A.H. & Budin, S. (2007). Conservation of the Malaysian Mahseer in Nenggiri basin through community action.

Babiak, I., Glogowski, J., Goryczko, K., Dobosz, S., Kuzminski, H., Strzezek, J. & Demianowicz, W. (2001). Effect of extender composition and equilibration time on fertilization ability and enzymatic activity of rainbow trout cryopreserved spermatozoa. *Theriogenology* 56: 177–192.

Baird, I. G. (2006). *Probarbus jullieni* and *Probarbus labeamajor*: the management and conservation of two of the largest fish species in the Mekong River in southern Laos. *Aquatic Conservation: Marine and Freshwater Ecosystems* 16(5): 517-532.

Baluyut, E. (1983). Stocking and introduction of fish in lakes and reservoirs in the ASEAN region. *FAO Tech. Pap.* 236. 82 pp.

Bart, A. (2002). Conservation of fish genetic diversity: need for development of a cryogenic genebank in Bangladesh. In: Penman, D.J., Hussain, M.G., McAndrew, B.J. and Mazid, M.A. (eds.) Proceedings of a workshop on Genetic Management and Improvement Strateges for Exotic Carps in Asia, 12-14 February 2002. Dhaka, Bangladesh. Pp 107-110.

Billard, R. (1983). Effects of cleolomic and seminal fluids and various saline diluents on the fertilizing ability of spermatozoa in the Rainbow trout, *Salmo gairdneri*. *J. Reprod. Ferti.*, 68: 77-84.

Billard, R. (1986). Spermatogenesis and spermatology of some teleost fish species. *Reprod. Nutr. Dev.*, 2: 877-920.

Billard, R. & Cosson, M.P. (1992). Some problems related to the assessment of sperm motility in freshwater fish. *J. Exp. Zool.*, 261: 122-131.

Billard, R., Cosson, J., Crim, L.W. & Suquet, M. (1995). Sperm physiology and quality. In: *Brood stock management and egg and larval quality.* (Eds) Bromage, N. R. and Roberts, R.J. Blackwell Science, Oxford. pp. 25-52.

Blaber, S.J.M., Milton, D.A., Brewer, D.T. & Salini, J.P. (2001). The shads (*genus Tenualosa*)of tropical Asia: An overview of their biology, status and fisheries. In: *Proceedings of the International Terubok Conference Sarawak, Malaysia, 11 -12 September 2001, Sarawak.* pp. 9-17.

Butts, I.A.E., Trippel, E.A. & Litvak, M.K. (2009). The effect of sperm to egg ratio and gamete contact time on fertilization success in Atlantic cod *Gadus morhua* L. *Aquaculture* 286 (1-2):89 – 94.

Chao, N.H. & Liao, I.C. (2001). Cryopreservation of finfish and shellfish gametes and embryos. *Aquaculture* 197: 161-189.

Chew, P.C., Hassan, R., Rashid, Z. A. & Chuah, H.P. (2010a). The endangered *Probarbus jullieni* (Sauvage) - Current status on its sperm cryopreservation in Malaysia. *Journal of Applied Ichthyology* 26: 797-805.

Chew, P.C., Rashid, Z. A., Hassan, R., Asmuni, M. & Chuah, H.P. (2010b). Semen cryo-bank of the Malaysian Mahseer (*Tor* spp.) *Journal of Applied Ichthyology* 26: 726-731.

Chong, V.C., Lee, P.K.Y. & Lau, C.M. (2010). Diversity, extinction risk and conservation of Malaysian fishes. *Journal of Fish Biology* 76: 2009-2066.

Curry, M.R. (2000). Cyopreservation of semen from domestic livestock. *Reviews of Reproduction* 5: 46-52.

Dobriyal, A.K., Kumar, N., Bahuguna, A.K. & Singh, H.R. (2000). Breeding ecology of some cold water minor carps from Garhwal Himalayas. Coldwater Fish and Fisheries: 177-186. (Eds) Singh, H.R. and Lakra, W.S. Narenda Publishing House, New Delhi. 337p.

DOF. (n.d.). Endangered freshwater fish species in Malaysia under the IUCN Red List. In: *3rd January 2011. Available from: <http://www.dof.gov.my/c/document_library/get_file?uuid=d8ebcbc5-4d4a-4125-8290-5d5bd78b1a5c&groupId=172176>.*

Donoghue, A.M. & Wishart, G.J. (2000). Storage of poultry semen. *Animal Reproduction Science* 62: 213-232.

Dreanno, C., Suquet, M., Quemener, L., Cosson, J., Fierville, F., Normant, Y. & Billard, R. (1997). Cryopreservation of turbot (*Scophthalmus maximus*) spermatozoa. *Theriogenology* 48: 589-603.

Dudgeon, D. (2002). Fisheries : pollution and habitat degradation in tropical Asian rivers. *Encyclopaedia of Global Environmental Change, Vol. III* (ed. I. Douglas), pp. 316–323. John Wiley & Sons, Chichester, U.K.

FAO. (1971). Seminar/Study Tour in the U.S.S.R. on Genetic Selection and Hybridization of Cultivated Fishes. 19 April - 29 May 1968. Lectures. Rep.FAO/UNDP(TA), (2926): 360 p.

Froese, R., & Pauly, D. (2003). (Eds.). *Fishbase*. In: *World Wide Web electronic publication, 3rd January 2011. Available from: <http://www.fishbase.org>.*

Gwo, J.C. (2000). Cryopreservation of sperm of some marine fishes. In Tiersch, T.R. and Mazik, P.M. (Eds.), *Cryopreservation in Aquatic Species*. World Aquaculture Society, Baton Rouge, L.A. pp. 138-160.

Gwo, J.C., Strawn, K., Longnecker, M.T. & Arnold, C.R. (1991). Cryopreservation of Atlantic croakker spermatozoa. *Aquaculture* 94: 355-375.

Hammerstedt, R.H. & Graham, J.K. (1992). Cryopreservation of poultry sperm: the enigma of glycerol. *Cryobiology* 29: 26-38.

Harvey, B. (1998). An overview of action before extinction. In: Harvey B, Ross C, Greer D, Carolsfeld J (eds) Action before extinction: an international conference on conservation of fish genetic diversity. World Fisheries Trust, Victoria Inger, F.R. and Chin, P.K. 1962. *The Freshwater Fishes of North Borneo*. Chicago Natural History Muzeum, Chicago.

Ho, S.C. (1995). Status of Limnological Research and Training in Malaysia. In: Limnology in Developing Countries (eds B. Gopal and R.G. Wetzel). International Association for Limnology. International Scientific Publications, 50-B., Pocket C., Sidhartha Extension, New Delhi. Pp 163-189.

Hogan, Z., Baird, I.G. & Phanara, T. (2009). Threatened fishes of the world: *Probarbus jullieni* Sauvage, 1880 (Cypriniformes: Cyprinidae). *Environ. Biol. Fish* 84: 291-292.

Honeyfield, D.C. & Krise, W.F. (2000). Measuament of milt quality and factors affecting viability of fish spermatozoa. In: Tiersch, T.R. and Mazik, P.M. (Eds.), *Cryopreservation in Aquatic Species*. World Aquaculture Society, Baton Rouge, L.A. pp. 49-58.

Horvath, A., Miskolczi, E. & Urbányi, B. (2003). Cryopreservation of common carp sperm. *Aquat. Living Resour.* 16: 457–460.

Hung, L. T., Suhenda, N., Slembrouck, J., Lazard, J. & Moreau, Y. (2004). Comparison of dietary protein and energy utilization in three Asian catfishes (*Pangasius bocourti, P. hypophthalmus* and *P. djambal*). *Aquaculture Nutrition* 10: 317-326.

Inger, R.F. (1955). Ecological notes on the fish fauna of a coastal drainage of north Borneo. *Fieldiana Zool.* 37: 47 -90.

Inger, R.F. & Chin, P.K. (1962). The Freshwater Fishes of North Borneo. *Fieldiana Zool.* 45: 1-268.

Inger, R.F. & Chin, P.K. (2002). *The Freshwater Fishes of North Borneo with a supplementary chapter by Chin Phui Kong*. Natural History Publications (Borneo), Kota Kinabalu. 268 pp. + S-78.

Ingram, B., Sungan, S., Gooley, G., Sim, S.Y., Tinggi D. & de Silva S.S. (2005). Induced spawning, larval development and rearing of two indigenous Malaysian Mahseer, *Tor tambroides* and *T. duoronensis*. *Aquacult. Res.* 36(10): 1001-1014.

IUCN (2009). IUCN Red List of Threatened Species. Version 2009.1. <www.iucnredlist.org>. Downloaded on 25 August 2009.

Jackson, D.C. & Marmulla, G. (2001). The influence of dams on river fisheries. In Dams, fish and fisheries: Opportunities, challenges and conflict resolution (ed. Marmulla, G.). *FAO Fisheries Technical Paper* 419: 1-45.

Jantrarotai, P., Sutthiwises, S., Kamonrat, V., Peyachoknagul, S. & Vidthayanon, C. (2007). Species identification of 3 *Hypsibarbus* spp. (Pisces: Cyprinidae) using PCR-RFLP of cytochrome b gene. *Kasetsart J. (nat. Sci.)* 4: 660 - 666.

Khan, M.S. (1987). Some aspects of the biology of ikan baung, *Mystus nemurus* C. & V. with reference to Chenderoh Reservoir. Thesis of MSc. Universiti Pertanian Malaysia, Serdang.

Khan, M.S., Ambak, M.A., Ang, K.J. & Mohsin, A.K.M. (1995). Reproductive biology of a tropical catfish, *Mystus nemurus* Cuvier and Valenciennes, in Chenderoh Reservoir Malaysia. *Aquaculture and Fisheries Management* 21: 173-179.

Khoo, K.H., Leong, T.S., Soon, F.L., Tan, S.P. & Wong, S.Y. (1987). Riverine fisheries in Malaysia. *Archiv. feur Hydrobiologie Beiheft*. 28: 261 - 268.

Kincaid, H.L. (2000). Development of Databases for Germplasm Repositories. In: Tiersch, T.R. and Mazik, P.M. (eds) *Cryopreservation in Aquatic Species.* World Aquaculture Society, Baton Rouge.Pp 323-331.

Kopeika, E., Kopeika, J. & Zhang, T. (2007). Cryopreservation of fish sperm. In: Day, J.G. and Stacey, G.N. (Eds) *Methods Mol. Biol. Vol.* 368: *Cryopreservation and Freeze-Drying Protocols, 2nd edition.* Humana Press Inc., Totowa, NJ. pp. 203 -217.

Kottelat, M., Whitten, A.J., Kartokasari, S.N. & Wirjoratmodjo, S. (1993). *Freshwater Fishes of Western Indonesia and Sulawesi*. Singapore: Berkeley Book Pte Ltd.

Kurokura, H., Hirano, R., Tomita, M. & Iwahashi, M. (1984). Cryopreservation of carp sperm. *Aquaculture* 37: 267-273.

Lahnsteiner, F. & Patzner, R.A. (2008). Sperm morphology and ultrastructure in fish. In: Alavi, S.M.H., Cosson, J.J., Coward, K. and Rafiee, G. (Eds) *Fish Spermatology*. Alpha Science International Ltd., Oxford, UK. pp. 1-61.

Lahnsteiner, F., Patzner, R.A. & Weismann, T. (1996). Semen cryopreservation of salmonid fishes: influence of handling parameters on the post thaw fertilization rate, Aquacult. Res. 27(9): 659–671.

Lahnsteiner, F., Weismann, T. & Patzner, A.R. (1997). Methanol as cryoprotectant and the suitability of 1.2 ml and 5 ml straws for cryopreservation of semen from salmonid fishes. *Aquaculture Research* 28: 471-479.

Lahnsteiner, F., Berger, B. & Weismann, T. (2003). Effects of media, fertilization technique, extender, straw volume, and sperm to egg ratio on hatchability of cyprinid embryos, using cryopreserved semen. *Theriogenology* 60: 829-841.

Leung, L.K.P. & Jamieson, B.G.M. (1991).Live preservation of fish gametes. In: Jamieson, B.G.M. (Ed.), *Fish Evolution and Systematics: Evidence from Spermatozoa.* Cambridge Univ. Press, Cambridge, pp. 245–269.

Maack, G.G. & George, M.R. (1999). Contributions to the reproductive biology of *Encrasicholina punctifer* Fowler, 1938 (engraulidae) from West Sumatra, Indonesia. *Fisheries Research* 44: 113 - 120.

MACD (1996). Life Cyle of Terubok (*Tenalosa toli*) With Recommendations for the Conservation, Management and Culture of The Species. Phase 3 Final Report. A collaborative project of CSIRO Division of Fisheries, Marine Laboratories, Cleveland Queensland, Australia and Inland Fisheries Branch, Ministry of Agriculture & Community Development (MACD), Kuching. Pp.64.

Mohamad-Zaini S. (1992). Pembiakan Aruhan Dan Pengeluaran Benih Ikan Sebarau *Hampala macrolepidota. Buletin Perikanan* 82. 13 pp.

Mohd-Zafri, H. (2006). Morphology and general reproductive stages of *Pangasius nasutus* from Sg. Pahang in Maran District, Pahang, Malaysia. Thesis Msc, UPM.

Mohsin, A.K.M. & Ambak, M.A. (1991). *Freshwater Fishes of Peninsular Malaysia.* Penerbit Universiti Pertanian Malaysia, Serdang, Selangor.

Mongkonpunya, K., Pupipat, T. & Tiersch, T.R. (2000). Cryopreservation of sperm of Asian Catfishes including the endangered Mekong Giant Catfish. In: Tiersch, T.R. and Mazik, P.M. (Eds.), *Cryopreservation in Aquatic Species.* World Aquaculture Society, Baton Rouge, L.A. pp. 108-116.

Morisawa, M. & Suzuki, K. (1980). Osmolality and potassium ion: Their roles in initiation of sperm motility in eleosts. *Science* 210: 1145 - 1147.

Morisawa, M., Suzuki, K. & Morisawa, S. (1983a). Effects of potassium and osmolality on spermatozoan motility of Salmonid fishes. *Journal of Experimental Biology* 107: 105 - 113.

Morisawa, M., Suzuki, K., Shimizu, H., Morisawa, S. & Yasuda, K. (1983b). Effects of potassium and osmolality on motility of spermatozoa from freshwater cyprinid fishes. *Journal of Experimental Biology* 107: 95 - 103.

Muchlisin, Z.A., Hashim, R. & Chong, A.S.C. (2004). Preliminary study on the cryopreservation of tropical bagrid catfish (*Mystus nemurus*) spermatozoa: the effect of extender and cryoprotectant on the motility after short-term storage. *Theriogenology* 62: 25-34.

Munkittrick, K.R. & Moccia, R.D. (1984). Advances in the cryopreservation of salmonid semen and suitability for a production-scale artificial fertilization program. *Theriogenology* 21: 645-59.

Ng, C.K. (2004). *Kings of the Rivers – Mahseer in Malaysia and the region.* Inter Sea Fishery (M) Sdn. Bhd. Selangor.

Nguyen, T.T.T. (2008). Population structure in the highly fragmented range of *Tor douronensis* (Cyprinidae) in Sarawak, Malaysia revealed by microsatellite DNA markers. *Freshwater Biology* 53: 924-934.

Nguyen, T.T.T., Baranski, M., Rourke, M. & McPartlan, H. (2006). Characterisation of microsatellite DNA markers for the mahseer species, *Tor tambroides* and cross-amplification in other four congeners (*T. douronensis, T. khudree, T. putitora,* and *T. tor*). *Molecular Ecology Notes* 7: 109-112.

Nguyen, T.T.T., Ingram, B., Sungan, S., Gooley, G., Sim, S.Y., Tinggi, D. & De Silva, S.S. (2006). Mitochondrial diversity of broodstock of two indigenous mahseer species, *Tor tambroides* and *T. douronensis* (Cyprinidae) cultured in Sarawak, Malaysia. *Aquaculture* 253: 259-269.

Nguyen, T.T.T., Ingram, B., Sungan, S., Tinggi, D., Sim, S.Y., Gooley, G. & De Silva, S. S. (2007). Guidelines for genetic management and conservation of two mahseer species in Sarawak, Malaysia. (Downloaded on 20 October 2009)

Nguyen, T.T.T., Sukmanomon, S., Na-Nakorn, U. & Ziming, C. (2008). Molecular phylogeny and phylogeography of mahseer species (Cyprinidae: Cypriniformes) as inferred from sequences of two mitochondrial gene regions. *Molecular Phylogenetics and Evolution* (in press) (doi:10.1016/j.ympev.2008.01.006).

Pathmasothy, S. (1985). The Effect of Three Diets with Variable Protein Level on Ovary Development and Fecundity in *Leptobarbus hoevenii*. In: Cho, C.Y., D.S. Cowey and W. Watanabe, Finfish Nutrition in Asia; Methodological Approach to Research and Development. Proceedings of the Asian Finfish Nutrition Workshop Held in Singapore 23-2 August, 1983. Ottawa: IDRC., pp.107-112.

Pathmasothy, S., & R. Omar. (1982). The Effect of Four Different Diets on the Growth of *Leptobarbus hoevenii*. *MARDI Research Bulletin* 10(1): 100-113.

Rainboth, W.J. (1996). *Fishes of the Cambodian Mekong. FAO Species Identification Field Guide for Fishery Purposes.* Food and Agriculture Organization (FAO) Publication, Rome.

Rideout, R.M., Litvak, M.K. & Trippel, E.A. (2003). The development of a sperm cryopreservation protocol for winter flounder, *Pseudopie-uronectes americanus* (Walbaum): evaluation of cryoprotectants and diluents. *Aquaculture Research* 34: 653-659.

Roberts, T.R. (1989). *The Freshwater Fishes of Western Borneo (Kalimantan Barat, Indonesia).*California Academy of Sciences, California.

Roberts, T.R. (1992). Revision of the Southeast Asian cyprinid fish genus *Probarbus*, with two new species threatened by the proposed construction of dams on the Mekong River. *Ichthyl. Expl. Freshw.* 3: 37-48.

Roberts, T.R. & Vidthayanon, C. (1991). Systematic revision of the Asian catfish family Pangasiidae, with biological observations and description of three new species. *Proceedings of the Academy of NaturalSciences of Philadelphia* 143: 97-144.

Saidin, T. (1986). Induced spawning of *Clarias macrocephalus* (Gunther). In: Maclean, J.L., Dizon, L.B. & Hosillos, L.V. (eds.) *The First Asian Fisheries Forum.* Asian Fisheries Society, Manila, Philipines. P. 683 - 686.

Scott, A.P. & Baynes, S.M. (1980). A review of the biology, handling and storage of salmonid spermatozoa, J. Fish Biol. 17: 707–739.

Stoss, J & Donaldson, E.M. (1983). Studies on cryopreservation of eggs from rainbow trout (*Salmo gairdneri*) and coho salmon (*Oncorhynchus kisutch*). *Aquaculture* 31: 51 - 65.

Suhairi, A. (1996). Preliminary observations on the culture of the Temoleh (*Probarbus jullieni*) in ponds. Paper presented in *Fisheries Semeniar July 1996*, IPP Batu Maung, Pulau Pinang, Malaysia.

Suhairi, A., Misri, S., Abdul Ghani, H. & Ahmad Ashhar, O. (1996). Kajian awal ternakan ikan temoleh (*Probarbus jullieni*) dalam kolam. *Proc. Fish. Res. Conf., DOF, Mal., 1996*: 278-283.

Suquet, M., Omnes, M.H., Normant, Y. & Fauvel, C. (1992). Assessment of sperm concentration and motility in turbot *Scophthalmus maximus. Aquaculture* 101: 177-185.

Suquet, M., Dreanno, C., Fauvel, C., Cosson, J. & Billard, R. (2000). Cryopreservation of sperm in marine fish. *Aquaculture Research* 31: 231-243.

Thalathiah, S., Abas Fauzi, O. & Ibrahim, T. (1988). First successful attempt to induce breed *Mystus nemurus* (C & V) at Batu Berendam, Melaka. *Proc. 11th Ann. Conf. MSAP:* 53 - 55.

Thalathiah, S., Ibrahim, T. & Mansor, A. (1992). Induced spawning of *Mystus nemurus* (C & V) using heteroplastic pituitary extract, HCG and an analog of LHRH. In: *Proceedings of Fisheries Research Seminar, Malacca, 27 – 29 June 1989. Pp. 185 – 188.*

Tiersch, T.R. (2000). Introduction. In: Tiersch, T.R. and Mazik, P.M. (Eds.), *Cryopreservation in Aquatic Species*. World Aquaculture Society, Baton Rouge, L.A. pp. 19-26.

Tiersch, T.R. (2006). Fish sperm cryopreservation for genetic improvement and conservation in Southeast Asia. *Fish For The People* 4(2): 21-33.

Tiersch, T.R., Goudie, C.A. & Carmichael, G.J. (1994). Cryopreservation of channel catfish sperm: storage in cryoprotectants, fertilization trials, and growth of channel catfish produced with cryopreserved sperm. *Transactions of the American Fisheries Society* 123: 580-586.

Tiersch, T.R., Figiel, C.R., Wayman, Jr., W.R., Williamson, J.H., Carmichael, G.J. & Gorman, O.T. (2000). Cryopreservation of sperm of the endangered razorback sucker. In: Tiersch, T.R. and Mazik, P.M. (Eds.), *Cryopreservation in Aquatic Species.* World Aquaculture Society, Baton Rouge, L.A. pp. 117-122.

Tsigenopoulos, C.S. & Berrebi, P. (2000). Molecular phylogeny of North Mediterranean freshwater barbs (Genus Barbus: Cyprinidae) inferred from cytochrome b sequences: biogeographic and systematic implications. *Molecular Phylogenetics and Evolution* 14: 165-179.

Wamecke, D. & Pluta, H.J. (2003). Motility and fertilizing capacity of frozen/thawed common carp (*Cyprinus carpio* L.) sperm using dimethyl-acetamide as the main cryoprotectant. *Aquaculture* 215: 167-185.

Woelders, H. Zuidberg, C.A. & Hiemstra, S.J. (2003). Applications, limitations, possible improvements and future of cryopreservation for livestock species. In: *Proceedings of the Workshop on Cryopreservation of Animal Genetic Resources in Europe. Paris, 23rd February 2003.* pp 67-76.

Wolf, K. (1963). Physiological salines for freshwater teleosts. *The Progressive. Fish Culturist.* 25: 135-140.

Wong, P. P. H. (2001). The status of the Terubok (*Tenualosa toli*) fishery in Sarawak, Malaysia. In: *Proceedings of the International Terubok Conference Sarawak, Malaysia, 11 -12 September 2001, Sarawak.* pp. 91-99.

Zhang, Y.Z., Zhang, S.C., Liu, X.Z., Xu, Y.Y., Wang, C.L., Sawant, M.S., Li, J. & Chen, S.L. (2003). Cryopreservation of flounder (*Paralichthys olivaceus*) sperm with a practical methodology. *Theriogenology* 60: 989-996.

Zhou, W. & Chu, G.H. (1996). A review of *Tor* species from the Lancangjiang River (Upper Mekong River), China (Teleostei: Cyprinidae). *Ichthyological Exploration of Freshwater* 7: 131-142.

Zulkafli A.R., Chew, P.C. & Johari, I. (2010). Conservation of Freshwater Fishes and Enhancement Programmes in Peninsular Malaysia. 2nd *National Conference on Agrobiodiversity Conservation and Sustainable Utilization (NAC2)-Agrobiodiversity for sustainable economic development, 11-13 May 2010, Tawau, Sabah. Pp 48-50.*

15

Cryopreservation of Brown Trout (*Salmo trutta macrostigma*) and Ornamental Koi Carp (*Cyprinus carpio*) Sperm

Yusuf Bozkurt[1], İlker Yavas[2] and Fikret Karaca[2]
[1]*Mustafa Kemal University, Faculty of Fisheries, Department of Aquaculture, Hatay*
[2]*Mustafa Kemal University, Faculty of Veterinary Medicine, Department of Reproduction and Artificial Insemination, Hatay*
Turkey

1. Introduction

Cryopreservation is considered as one component in an effective strategy to save endangered species by facilitating the storage of their gametes in gene banks (Gausen, 1993; Akcay et al. 2004). Cryopreservation offers several benefits that in this way stocks can be protected from being totally eliminated due to sudden disease outbreak, natural utilization in hatcheries' production and laboratory experiments can be ensured. Stocks can be maintained more economically and experimental materials for advanced studies, such as gene transfer, can be made more accessible (Chao & Liao, 2001; Tekin et al. 2003).

Cryopreservation techniques involve addition of cryoprotectants, freezing and thawing of sperm samples, all of which may result in some damage to the spermatozoa and may decrease egg fertilization rate (Kopeika et al. 2003). Therefore, before cryopreservation of sperm, a through evaluation of different extender solutions, cryoprotectants, straw sizes and thawing rates is essential to develop optimum cryopreservation protocols for various species (Yavas & Bozkurt, 2011).

The species-specific cryopreservation procedure needs a suitable extender, as undiluted semen is not suitable for long-term preservation. Similarly, addition of optimum amount of cryoprotectant reduces cell damages associated with dehydration, cellular injuries and ice crystal formation (Leung, 1991). Although cryoprotectants help to prevent cryoinjuries during freezing and thawing, they can become toxic to cells when the exposure time and concentration are more (Tekin et al. 2007). In addition, type of cryoprotectants is also very specific to many species. There are no universal extenders and cryoprotectants available that can be used across species.

On the other hand, motility is the most commonly used parameter to evaluate sperm quality in fishes (Billard et al. 1995). This parameter is acceptable so that spermatozoa must be motile to achieve fertilization. Furthermore, sperm motility is an important component of a cryopreservation program in order to prevent poor sperm quality semen samples prior to freezing and to estimate the fertility of the stored sperm after thawing (Akçay et al., 2004;

Bozkurt, 2008). Thawing temperature and duration are also critical factors in the survival of cryopreserved sperm cells (Morris, 1981). Optimal freezig/thawing procedures have not been reported for *Salmo trutta macrostigma* sperm. So, in the present study three different thawing temperatures and thawing durations were also tested related to motility.

For this reason, there is a need to improve techniques on gamete storage and evaluation of sperm quality to facilitate optimization of controlled reproduction in fish (Alavi & Cosson, 2005). Important parameters for cryopreservation include type of extenders and cryoprotectants, dilution ratios, freezing/thawing rates and fertilization rates (Bozkurt et al. 2005).

Salmo trutta macrostigma is a salmonid species occurring in inland water habitats of Southern Europe, Western Asia, Northern Africa and Anatolia (Geldiay & Balik, 1988). It is also critically endangered fish species in inland waters because of illegal fishing, overfishing, and other environmental changes, including hydroelectric plants and pollution. For this reason a biological conservation program has been considered for *Salmo trutta macrostigma* in Turkey. On the other hand, ornamental koi carp is evaluated by its colour and have been used in the selective propagation programs. These brightly colored koi carps are the result of selective breeding of wild carp. Over centuries a range of pleasing colors, patterns and shapes have been developed for this valuable species. Therefore, reliable methods for brown trout and koi carp sperm cryopreservation could benefit both aquaculture application and conservation of biodiversity.

Therefore, the present study was conducted in order to examine the effect of ionic extenders combined with different cryoprotectants at different ratios and to test the effect of different thawing temperatures and thawing periods on the post-thaw sperm quality of brown trout (*Salmo trutta macrostigma*) and koi carp (*Cyprinus carpio*) and development of a cryopreservation protocol for sperm of this commercially valuable two species.

2. Materials and methods

2.1 Broodstock management

The experiments were carried out spawning season of the brown trout (*Salmo trutta macrostigma*) and koi carp (*Cyprinus carpio*). In the pre-spawning period the mature bown trouts were kept seperately in small ponds under constant environmental conditions. The water temperature ranged 8-10°C during the spawning period. During the experiment, fish were kept under natural photoperiod. Mean water temperature and dissolved oxygen of the broodstock ponds were 8.7±2.46°C and 9.2±7.2 ppm respectively.

The koi carp broodstock was collected from wintering ponds by seining and transported into the hatchery 48 h prior to gamete collection. In the hatchery, male and female broodfish were held separetely in shadowed tanks (V=1000 L) supplied with continuously (2.5 L min^{-1}) well-aerated water of 24°C. Brown trout and koi carp broodstock were not fed during the experiments.

2.2 Gamete collection

Sperm was collected by gently hand-stripping without anesthesia from mature 10 brown trout males. For koi carp cryopreservation experiments, semen was collected from 5 anesthetized (0.1 g/l MS 222) males by manual abdominal stripping 12 h after a single injection of 2 mg/kg of carp pituitary extract (CPE) at 20-22 °C water temperature. Eggs were collected by hand stripping 10-12 h after a double injection of 3.5 mg/kg of CPE. The first injection, 10% (0.35 mg/kg) CPE was given 10 h before the second (3.15 mg/kg).

For sperm collection, the urogenital papilla's of mature male fishes were carefully dried and sperm was hand-stripped directly into test tubes. Following sperm collection, the tubes containing sperm were placed in a styrofoambox containing crushed ice (4°C). Contamination of sperm with water, urine or faeces was carefully avoided. Sperm was transported to the laboratory within 15 min. For collection of eggs from koi carps, females were wiped dry, stripped by gentle abdominal massage and the eggs from each female were collected in a dry metal bowl. Eggs were checked visually and only those lots of homogenous shape, colour and size were used in the fertilization experiments.

2.3 Determination of fresh sperm quality parameters

Motility was estimated subjectively using light microscope (Olympus, Japan) with a x400 magnification. Samples were activated by mixing 1 µl of sperm with 20 µl activation solution (0.3% NaCl) on a glass slide. The percentage of motility was defined as the percentage of spermatozoa moving in a forward motion every 20% motile increment (i.e., 0, 20%, 40%, 60%, 80%, and 100%) (Vuthiphandchai & Zohar, 1999). Motility measurements were performed within 15 s. after activation. Sperm cells that vibrated in place were not considered to be motile. Sperm motility was estimated with three replicates of samples. For cryopreservation experiments, samples below 80% motile spermatozoa were discarded. Duration of sperm motility was determined using a sensitive chronometer (sensitivity: 1/100 s) by recording the time following addition of the activation solution to the sperm samples.

Spermatozoa density was determined according to the haemacytometric method. Sperm was diluted at ratio of 1:1000 with Hayem solution (5g Na_2SO4, 1g NaCl, 0.5g $HgCl_2$, 200 mL bicine) and density was determined using a 100 µm deep Thoma haemocytometer (TH-100, Hecht-Assistent, Sondheim, Germany) at 400x magnification with Olympus BX50 phase contrast microscope (Olympus, Japan) and expressed as spermatozoa $x10^9$ mL^{-1} (three replicates). Counting chambers were always kept in a moist atmosphere for at least 10 min before cell counting. Sperm pH was measured using indicator papers (Merck, 5.5-9) within 30 min of sampling.

2.4 Experiment 1 - Brown trout (*Salmo trutta macrostigma*)

Collected sperm from 10 males that showing >80 motility was pooled into equal aliquots according to the required semen volume and sperm density to eliminate effects of individual variability of gamete donors. Semen and extenders were kept at 4°C prior to dilution. Pooled semen was diluted at 1:3 ratio (semen/extender) with extender containing 4.68 g l^- NaCl, 2.98 g l^- KCl, 0.11 g l^- $CaCl_2$ and Trizma-HCl 3.15 g l^- in distilled water; pH 9.0 (Billard & Cosson, 1992). The extender contained methanol and egg yolk at ratios of 5%, 10% and 15% separately. Dilution of semen with extender resulted in sperm concentrations of around $2.5x10^9$ cells/ml extender that was enough to avoid damage due to sperm compression during freezing and thawing (Lahnsteiner, 2000). Following sperm suspension was equilibrated for 10 min at 4°C.

Within 1 h after sperm collection, the diluted semen samples were drawn into 0.25mL plastic straws (IMV, France). The open end of straws were sealed with polyvinyl alcohol (PVA). Following, the straws were placed on a styrofoam rack that floating on the surface of liquid nitrogen in a styrofoam box. The straws were frozen in liquid nitrogen vapour 4 cm above of the liquid nitrogen surface (temperature of styroframe surface was about -140°C) for 10 min. Following, the straws were plunged into the liquid nitrogen (-196°C) and stored for several days. For thawing, straws were thawed at 30°C for 10 s by gentle agitation in water bath. Thawed sperm was activated using pond water.

On the other hand, post-thaw sperm quality tests were carred out to evaluate motility rate and duration of motility. For this aim, sperm motility rate and duration of motility values following cryopreservation in the same ionic extender containing 15% egg yolk were determined. Sperm was thawed at 25°C, 35°C or 45°C for 5s, 15s or 25s and activated in 0.3% NaCl and 1% $NaHCO_3$.

2.5 Experiment 2 - Ornamental koi carp (*Cyprinus carpio*)

Collected semen from the 5 males that showing >80 motility was pooled into equal aliquots according to the required semen volume and sperm density needed to eliminate effects of individual variability of the donors. Semen and extenders were kept at 4°C, then diluted at a ratio of 1:3 (semen/extender) with 3 different extenders containing 10% DMSO. Extender 1 contained 5.8 g/L NaCl, 0.2 g/L KCl, 0.22 g/L $CaCl_2$, 0.04 g/L $MgCl_2 6H_2O$, 2.1 g/L $NaHCO_3$, 0.04 g/L $NaH_2PO_4.2H_2O$, 3.75 g/L glycine (Ravinder, et al. 1997). Extender 2 contained 300 mM glucose and 10% egg yolk pH:8 (Tekin et al. 2003) and extender 3 contained 4.68 g/L NaCl, 2.98 g/L KCl, 0.11 g/L $CaCl_2$, 3.15 g/L Tris-HCl, pH:9 (Billard & Cosson, 1992).

The diluted samples were drawn into 0.25 ml plastic straws (IMV, France) and were sealed with polyvinyl alcohol (PVA). Having been diluted, the samples were equilibrated for 10 min at 4°C. After equilibration, the straws were placed on a styrofoam rack that floated on the surface of liquid nitrogen in a styrofoam box. The straws were frozen in liquid nitrogen vapour 3 cm above the surface of liquid nitrogen (-140°C) for 10 min. After 10 min the straws were plunged into the liquid nitrogen (-196°C) and stored for several days. For thawing, the straws were removed from liquid nitrogen and immersed in 30°C water for 10 seconds. Thawed sperm was activated using 0.3% NaCl and observed under microscope for determination of spermatozoa motility and motility durations.

For fertilization experiments, pooled eggs from 3 mature females were used to determine fertilization rates. Egg samples (about 100 eggs) were inseminated in dry Petri dishes with fresh sperm or frozen sperm immediately after thawing at a spermatozoa:egg ratio of 1×10^5: 1. Eggs were inseminated by the dry fertilization technique using a solution of 3 g urea and 4 g NaCl in 1 L distilled water. The sperm and eggs were slightly stirred for 30 min, washed with hatchery water (24°C; 9 mg/l O_2), and gently transferred to labeled Zuger glass incubators with running water (24°C) where they were kept until hatching (3-4 d). Living and dead eggs were counted in each incubator during incubation and dead eggs were removed. When the fertilized eggs developed to embryos at the gastrula stage, the fertilization rate (number of gastrula stage embryos/number of total eggs) was calculated.

2.6 Statistical analysis

Results are presented as means±SE. Differences between parameters were analyzed by repeated analysis of variance (ANOVA). Significant means were subjected to a multiple comparison test (Duncan) for post-hoc comparisons at a level of α=0.05. All analyses were carried out using SPSS 10 for Windows statistical software package.

3. Results

3.1 Fresh sperm quality parameters

In brown trout fresh semen volumes were rather variable and ranged from 9 to 17 ml and mean volume was 12.6±4.28 mL. Motility values were ranged from 75% to 90%. Samples that motility values were below than 80% were not used for the cryopreservation experiments. The mean motility value of fresh sperm samples were 84.5±7.59%. Mean spermatozoa movement duration (s), sperm density x10^9/mL and pH values were achieved as 57.4±3.8 s, 24.8±4.62 x10^9/mL and 7.28±2.46 respectively.

In koi carp mean fresh semen volume, spermatoza motility, motility duration, spermatozoa density and pH values of the collected fresh milt samples were determined as 6.2±4.7 ml, 85.4±2.4%, 125.2±3.5 s, 22.8 x 10^9 mL^{-1} and 7.4±3.7, respectively.

3.2. Experiment 1 - Brown trout (*Salmo trutta macrostigma*)

Post-thaw motility of sperm cryopreserved in ionic extender containing two different cryoprotectants at three different ratios is shown in Table 1. There were significant effect of cryoprotectants on motility rates. Sperm samples cryopreserved in the extenders containing egg yolk yielded greater post-thaw motility rates than methanol containing extenders. Sperm frozen with extender containing 15% egg yolk had the highest post-thaw motility. Differences between the post-thaw motility values were significant (P<0.05).

Methanol (5%)	Methanol (10%)	Methanol (15%)	Egg Yolk (5%)	Egg Yolk (10%)	Egg Yolk (15%)
10.6±4.57Af	15.2±5.80Ae	17.4±4.72Ae	40.5±3.27Aab	42.3±6.1Aa	45.3±4.27Aa
7.5±2.69Ae	9.6±3.37Be	12.3±5.24Ad	30.6±2.86Bb	35.4±4.17Abab	40.2±5.36Aba
5.4±2.73Ae	7.2±3.79Be	10.5±4.27Abd	25.7±4.69BCbc	30.2±5.29Bb	37.8±8.29Ba

Means followed by different superscripts (lowercase for lines and uppercase for columns within the same sperm feature) are different (p<0.05). (mean±SE, n=3).

Table 1. Post-thaw motility (%) of brown trout sperm cryopreserved with different cryoprotectants.

It was observed that a decrease in motility duration occurred following cryopreservation. The longest post-thaw motility longevity was also achieved with extender containing 15%

egg yolk as 54.2±3.46 s. Differences between the means of motility durations were significant ($P<0.05$). (Table 2).

Methanol (5%)	Methanol (10%)	Methanol (15%)	Egg Yolk (5%)	Egg Yolk (10%)	Egg Yolk (15%)
20.3±2.57Ad	24.5±2.47Acd	28.6±3.46Ac	40.2±1.29Aa	42.6±4.57Aa	46.4±2.38Ba
15.23±4.39Abf	19.6±4.39Abef	23.4±3.47Abe	37.2±3.45Ac	45.3±5.39Ab	54.2±3.46Aa
12.3±4.17Bd	15.7±1.28Bd	20.4±8.25Bcd	20.4±4.59Bcd	25.3±2.48Bc	32.5±5.27Cb

Means followed by different superscripts (lowercase for lines and uppercase for columns within the same sperm feature) are different ($p<0.05$). (mean±SE, n=3).

Table 2. Post-thaw longevity (s) of brown trout sperm cryopreserved with different cryoprotectants.

Sperm motility rate (Figure 1) and longevity of motility (Figure 2) values following cryopreservation in the ionic extender containing 15% egg yolk were determined. Sperm was thawed at 25°C, 35°C or 45°C for 5s, 15s or 25s and activated in 0.3% NaCl and 1% $NaHCO_3$.

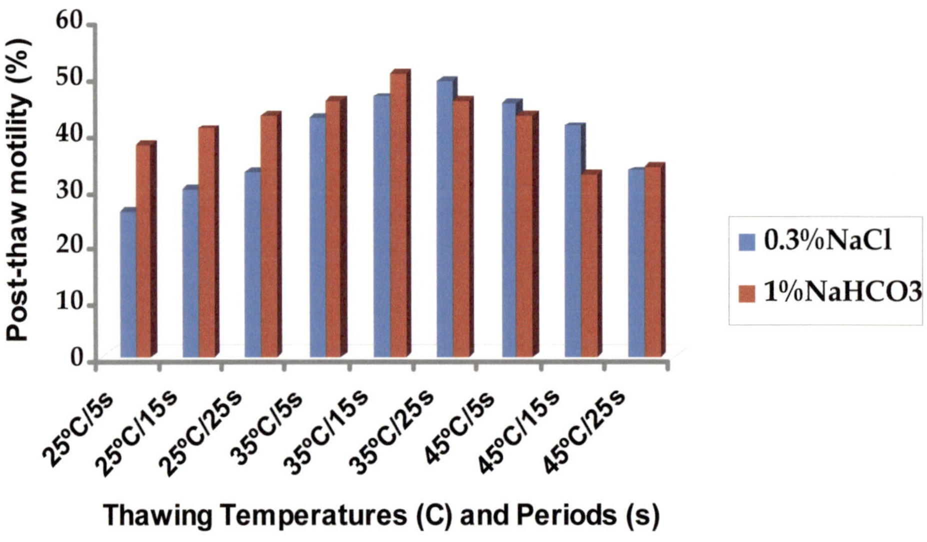

Fig. 1. Post-thaw motility (%) of brown trout sperm thawed at different degrees, periods and activating agents.

Post-thaw sperm motility rates were affected by thawing rates and activation agents and ranged from 25% to 50%. Also, the activating agents affected the duration of motility. All sperm samples triggered in 1% $NaHCO_3$ were motile for a longer period (32-57 s) compared with samples triggered in 0.3% NaCl (24-53 s). Differences between the post-thaw motility and longevity values were significant ($P<0.05$).

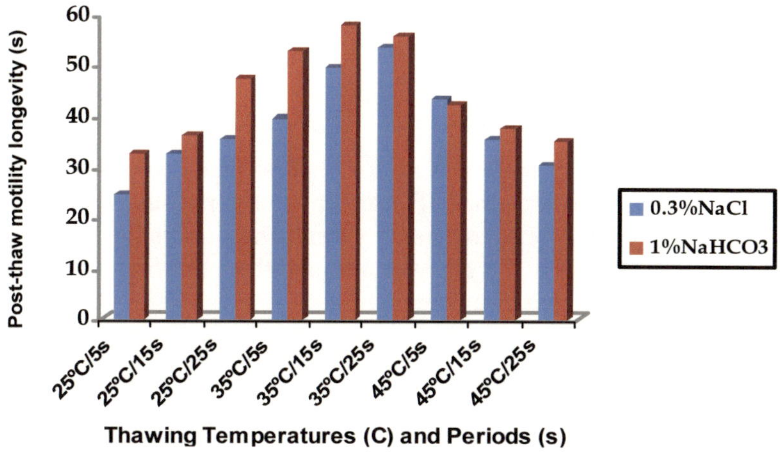

Fig. 2. Post-thaw motility longevity (s) of brown trout sperm thawed at different degrees, periods and activating agents.

3.3 Experiment 2 - Ornamental koi carp (*Cyprinus carpio*)

Effect of three different extenders containing 10% DMSO on the post-thaw motility and movement duration, fertilization and hatching rates of koi carp are shown in Table 3. Mean post-thaw motility of koi carp sperm was 75.3±6.4% while the best motility was determined as 85%. The overall mean fertilization rate was determined as 99.2±0.72 while the best fertilization rate was determined as 100%. The highest hatching rate was determined as 50% in all experimental groups. Motility features and hatching rates of cryopreserved koi carp sperm was statistically different between the experimental groups ($p<0.05$).

Extenders	Post-thaw motility (%)	Post-thaw motlity duration (s)	Fertilization rates (%)	Hatching rates (%)
E1	75.2 ± 0.4^{a}	27.5 ± 1.2^{a}	99.6±0.5	42.5 ± 1.9^{b}
E2	78.6 ± 0.7^{b}	32.9 ± 0.4^{b}	99.7±0.5	46.2 ± 0.7^{c}
E3	72.3 ± 0.2^{a}	25.2 ± 0.6^{a}	98.3±1.2	37.4 ± 0.2^{a}
Control	-	-	99.8±0.2	86.2 ± 0.4^{d}

Means followed by different superscripts are different ($p<0.05$). (mean±SE, n=3).

Table 3. Effect of different extenders on post-thaw motility, fertilization and hatching rates of koi carp sperm.

4. Discussion

Successful cryopreservation of fish spermatozoa depends on a range of factors including the collection of high quality sperm, equilibration conditions, choice of cryoprotectant medium, cooling/thawing regimes, and conditions for fertilization. Even though some general rules can be applied to any fish species, optimization of the protocol is needed for each individual species (Kopeika et al. 2007). Several factors have affected post-thaw quality of cryopreserved sperm from both brown trout (*Salmo trutta macrostigma*) and ornamental koi carp (*Cyprinus carpio*). The results obtained in the present study contribute significantly improve the development protocol of sperm cryopreservation in brown trout and ornamental koi carp at large scale.

4.1 Brown trout (*Salmo trutta macrostigma*)

The results of the present study demonstrate for the first time cryopreservation of brown trout (*Salmo trutta macrostigma*) sperm. In the present study, post-thaw sperm quality was initially evaluated on the basis of sperm motility score and duration of motility for brown trout. For this aim, the effect of two cryoprotectants and three thawing temperatures on the post-thaw sperm quality of brown trout were assessed.

Motility is induced after the spermatozoa released into the aquatic environment during natural reproduction or after transfer to an activation medium during controlled reproduction (Alavi & Cosson, 2006). When salmonid spermatozoa are released into water they have a brief period of sperm activity between 20 and 40s (Morisawa & Morisawa, 1986). A better knowledge of the characteristics of fresh sperm motility is necessary to evaluate sperm quality in commercial hatcheries before artificial reproduction and in laboratories before experiments. Preliminary examination of fresh sperm was carried out in order to determine the relationship between sperm motility and seminal plazma composition of *Salmo trutta macrostigma* sperm (Bozkurt et al. 2011a).

Comparison of different cryoprotectant recipes and freeze-thaw protocols are difficult when each treatment tested for the ability of sperm to fertilise eggs. Cryoprotectants can suppress most cryoinjuries when used higher concentrations but at the same time it can become toxic to the cells. Therefore, a suitable concentration was needed for the development of a cryopreservation protocol. Methanol has been used successfully for sperm cryopreservation in African catfish (*Clarias gariepinus*) (Burchell) (Steyn & Van Vuren, 1987), tilapia (*Sarotherodon mossambicus*) (Peters) (Rana & McAndrew, 1989), bagrid catfish (*Mystus nemurus*) (Muchlisin et al., 2004) and salmonid fish (Lahnsteiner et al., 1996). Mansour et al. (2006) showed that 10% methanol was more effective as a cryoprotectant for Arctic char spermatozoa than 10% DMA or 10% DMSO when used with a glucose diluent. However, the effects of higher levels of methanol cryoprotectant were not investigated. In the present study, brown trout semen in an extender containing 15% egg yolk resulted in the highest overall percentage of sperm motility.

In addition, penetrating cryoprotectants could affect the percentage of motile sperm. In salmonids, some authors reported that higher post-thaw motility from methanol than from DMSO and other cryoprotectants (Mansour et al. 2006). In the present study, methanol and egg yolk have statistically significant effect on the percentage of sperm motility. On the other hand, it should be noted that egg yolk achieved better results than methanol for cryopreservation of brown trout sperm. Extender containing glucose, egg yolk and DMSO described by Alderson and McNeil (1984) gave good results in cryopreservation experiments with large straws. Baynes and Scott (1987) also reported that egg yolk is a valuable component in extenders for salmonid sperm cryopreservation. Furthermore, the addition of egg yolk to the medium interferes with the good visualization of spermatozoa during the motility rate analysis. With this in mind, we have tested several extender/cryoprotectant combinations with the addition of egg yolk that preserve sperm during storage and yet allow good visualization during motility analysis.

On the other hand, thawing temperature also play an important role in the post-freeze semen quality of fish (Wayman et al., 1998). Generally, thawing rates should be high to avoid recrystallization (Lahnsteiner, 2000). Significant post-thaw motility was determined when brown trout sperm was thawed at temperature of 35°C in the present study. According to the results of the present study, it was shown that higher temperatures are necessary to recover membrane stability or metabolism of spermatozoa. Also it appears that either recrystallization and ice crystal formation during thawing were reduced or avoided by this thawing procedure, or enzymatic activities were the best reactivated (Lahnsteiner, 2000). Although thawing from -196°C to 4°C is generally considered as critical phase because of potential recrystallization, the process was similar for all species. Furthermore, the two activating agents (0.3% NaCl and 1% $NaHCO_3$) tested did not affect post-thaw motility rates or quality motility score, although, in general, higher scores were observed when 1% $NaHCO_3$ was used in the present study. Duration of motility was significantly higher when 1% $NaHCO_3$ was used as an activating agent.

4.2 Ornamental koi carp (*Cyprinus carpio*)

The main purpose of the current experiment was to develop an appropriate protocol for ornamental koi carp (*Cyprinus carpio*) sperm cryopreservation to increase sperm availability outside the breeding season. By banking male gametes when they are abundant, most efforts can be devoted to raising healthy female broodstock and obtaining good quality eggs within a short captivity culture period. Through cryopreservation, a sperm repository can also be established for all males in captivity or from the wild. Such repository is important to maintain the genetic diversity to avoid inbreeding or loss of heterozygosity for captive breeding programs as well as possible future stock enhancement in the wild (Cabrita et al., 2009).

In the present experiment, koi carp males gave sperm characterized by good spermatozoa density and percent of motility. Such sperm should be used for cryopreservation experiments when considering minimization of artificial selection and sperm competition during hatchery operations in order to maintain the greatest biodiversity (Campton, 2004).

During the cryopreservation process, one of the important issue is the use of cryoprotectants, which role is to prevent cell damage during the freezing and thawing steps. Several cryoprotectants have been used for fish sperm cryopreservation, including methanol, ethylene glycol and dimethyl sulphoxide (DMSO); however, DMSO is reported to be the most efficient to cryopreserve fish spermatozoa (Anel & Cabrita, 2000) due mainly to its small molecular size, which allows it to enter and exit the spermatic cell easily (Tiersch et al. 1998).

The best fertilization rate obtained with extender II with 99.7% eyeing rate in koi carp. These results can be explained by the presence of 10% DMSO as a cryoprotectant in this extender. It can be concluded that DMSO has higher permeability by permeating into cell, causing reduced ice crystal formation for koi carp sperm. On the other hand, Lahnsteiner et al., (1996) used 10% methanol, 10% DMSO, 10% DMA, 5% glycerol and mixture of 5% DMSO and 1% glycerol for semen cryopreservation of the grayling (*Thymallus thymallus*) and the Danube salmon (*Hucho hucho*), which methanol showed the highest fertilization rates in relation to control 95.3% and 91.1% for grayling and Danube salmon, respectively.

Cryopreservation protocol carried out in the present study with a 1:100,000 egg: spermatozoa ratio, almost the same fertilization efficiency was obtained whether frozen or fresh semen was used. This may be due to differences in extender, cryoprotectant, equilibration, egg quality, or protocol. In the present study, the interaction between the percentage of motile post-thaw sperm and fertilizing capacity was highly positive, similar to results in common carp (Linhart et al., 2000), African catfish (Rurangwa et al., 2001) and grass carp (Bozkurt et al., 2011b).

On the other hand, a wide range of temperatures used to thawed cryopreserved sperm with temperatures from refrigeration (4°C) to 80°C were reported (Lahnsteiner et al., 2000). A fast thawing temperature decreases the recrystallization effect in the spermatic cells and therefore diminishes the membrane damage (Tiersch et al.1998). Higher temperature such as 30°C were also used to thaw cyprinid semen in several studies (Stoss & Hotz, 1983) that similar with the pesent study.

5. Conclusion

It can be concluded that the cryopreservation protocol developed in this study is rather effective and brown trout (*Salmo trutta macrostigma*) and ornamental koi carp (*Cyprinus carpio*) sperm can be successfully cryopreserved. It seems that cryopreservation of brown trout sperm with ionic extenders containing 15% egg yolk is rather effetive on post-thaw sperm quality. In addition, based on the results obtained from this study, it is possible to suggest that sperm cryopreserved with ionic extender containing 10% DMSO packed in 0.25 mL volume straws and thawed at 30°C are the most suitable conditions to retain the sperm quality in koi carp having optimal sperm motility, duration of motility as well as high fertility percentages close to the values obtained with fresh sperm.

This study can help establish a frozen sperm bank for the conservation of genetic material of the brown trout and koi carp. On the other hand, additional research is needed on the effects of cryoprotectants, protective agents and freezing technique in cryopreservation on malformations, survival and condition of progeny produced with cryopreserved spermatozoa of brown trout and koi carp.

6. Acknowledgements

This research was financed by Research Fund of Musafa Kemal University (1005-M-0111). The authors would like to thank the staff of the Fish Production Station of General Directorate of National Parks in Tarsus and State Hydraulic Works (SHW) Fish Production Station in Adana, Turkey.

7. References

Akcay, E.; Bozkurt, Y.; Secer, S. & Tekin, N. (2004). Cryopreservation of mirror carp semen. *Turkish Journal of Veterinary and Animal Sciences*, 28 (5), 837-843.

Alavi, S.M.H. & Cosson, J. (2005): Sperm motility in fishes: (I) Effects of temperature and pH: a review. *Cell Biol. Intern.* 29, 101–110.

Alavi, S.M.H. & Cosson, J. (2006). Sperm motility in fishes: II. Effects of ions and osmolality. *Cell Biol. Int.*, 30, 1-14.

Alderson, R. & Macneil, A.J. (1984). Preliminary investigation of cryopreservation of Atlantic salmon *Salmo salar* and its application to commercial farming. *Aquaculture*, 43: 351-354.

Anel, I. & Cabrita, E. (2000). Effect of external cryoprotectants as membrane stabilizers on cryopreserved rainbow trout sperm. *Theriogenology*, 13: 623-635.

Baynes, S.M. & Scott, A.P. (1987). Cryopreservation of rainbow trout spermatozoa: the influence of sperm quality, egg quality and extender composition on post-thaw fertility. *Aquaculture*, 66:53-67.

Billard, R. & Cosson, M.R. (1992). Some problems related to the assessment of sperm motility in freshwater fish. *Journal of Experimental Zoology*, (261) 122-131.

Billard, R.; Cosson, J.; Crim, L.W. & Suquet, M. (1995). Sperm Physiology and Quality. In: *Broodstock Management and Egg and Larval Quality*. Blackwell Science, Oxford, pp: 25-52.

Bozkurt, Y.; Akçay, E.; Tekin, N. & Secer, S. (2005). Effect of freezing techniques, extenders and cryoprotectants on the fertilization rate of frozen rainbow trout *(Oncorhynchus mykiss)* sperm. *The Israeli Journal of Aquaculture-Bamidgeh*, 57 (2), 125-130.

Bozkurt, Y. (2008). Physical and biochemical properties of *Salmo trutta abanticus* semen. *Indian Veterinary Journal*, 85 (3), 282-284.

Bozkurt, Y; Ogretmen, F.; Kokcu, O. & Ercin, U. (2011a). Relationship between seminal plasma composition and sperm quality parameters of the *Salmo trutta macrostigma* (Dumeril, 1858) semen: with emphasis on sperm motility. *Czech Journal of Animal Science*, 56 (8), 355-364.

Bozkurt, Y.; Yavas, I.; Ogretmen, F.; Sivaslıgil, B.; & Karaca, F. (2011b). Effect of glycerol on fertility of cryopreserved grass carp (*Ctenopharyngodon idella*) sperm. *The Israeli Journal of Aquaculture-Bamidgeh*, IIC:63.2011.635.

Cabrita, E.; Engrola, S.; Conceicao, L.; Pousaoferreira, P. & Dinis, M.T. (2009). Successful cryopreservation of sperm from sex-reversed dusky grouper, Epinephelus marginatus. *Aquaculture*, 287: (1-2), 152-157.

Campton, D. F. (2004). Sperm competition in salmon hatcheries: the need to institutionalize genetically benign spawning protocols. *Transactions of American Fisheries Society*, 133, 1277–1289.

Chao, N.H. & Liao, I.C. (2001). Cryopreservation of finfish and shellfish gametes and embryos. *Aquaculture*, 197:161-189.

Gausen, D. (1993). The Norwegian gene bank programme for Atlantic salmon (*Salmo salar*). *In*: J.C. Cloud and G.H. Thorgaad (eds.), *Genetic conservation of salmonid fishes. Plenum*. New York, USA. pp.181-187.

Geldiay, R. & Balik, S. (1988): *Turkiye tatlısu balıkları* (Freshwater fishes in Turkey). Ege Univ. Fen Fak. Kitaplar Serisi. 97, pp. 519.

Kopeika, J.; Kopeika, E.; Zhang, T. & Rawson, D.M. (2003). Studies on the toxicity of dimethyl sulfoxide, ethylene glycol, methanol and glycerol to loach (Misgurnus fossilis) sperm and the effect on subsequent embryo development Cryo Letters. 24 (6) : 365-374.

Kopeika, E.; Kopeika, J. & Zhang, T. (2007). Cryopreservation of fish sperm . Methods Mol Biol. 368 : 203 - 217.

Lahnsteiner, F.; Weismann, T. & Patzner, R.A. (1996). Cryopreservation of semen of the grayling (*Thymallus thymallus*) and the Danube salmon (*Hucho hucho*). *Aquaculture*, 144:265-274.

Lahnsteiner, F. (2000). Semen cryopreservation in the salmonidae and in the northern pike. *Aquaculture Research*, 31:245-258.

Lahnsteiner, F.; Berger, B.; Horvath, A.; Urbanyi, B. & Weismann T. (2000). Cryopreservation of spermatozoa in cyprinid fishes. *Theriogenology*, 54: 1477–1498.

Leung, L.K.P. (1991). Principles of biological cryopreservation. In. *Fish evolution and systematics: Evidence from spermatozoa*. Jamieson, B.M.G. (Ed). Cambridge University Press, Cambridge. 231-244.

Linhart O.; Rodina M. & Cosson, J. (2000). Cryopreservation of sperm in common carp: sperm motility and hatching success of embryos. *Cryobiology*, 41:241-250.

Mansour, N.; Richardson, G.F. & McNiven, M.A. (2006). Effect of extender composition and freezing rate on post-thaw motility and fertility of Arctic char, *Salvelinus alpinus* (L.), spermatozoa. *Aquaculture Research*, 37: 862- 868.

Morisawa, S. & Morisawa, M. (1986). Acquisition of potential for sperm motility in rainbow trout and chum samon. *The Journal of Experimental Biology*, 126: 89-96.

Morris, G.J. (1981). *Cryopreservation: an introduction to cryopreservation in culture collections.* Cambridge (England): Institute of Terrestrial Ecology.

Muchlisin, Z.A.; Hashim, R. & Chong, A.S.C. (2004). Preliminary study on the cryopreservation of tropical bagrid catfish (*Mystus nemurus*) spermatozoa; the effect of extender and cryoprotectant on the motility after short-term storage. *Theriogenology*, 62, 25–34.

Rana, K.J. & McAndrew, B.J. (1989). The viability of cryopreserved tilapia spermatozoa, *Aquaculture*, 76: 335-345.

Ravinder, K.; Nasaruddin, K.; Majumdar, K. C. & Shivaji, S. (1997). Computerized analysis of motility, motility patterns and motility parameters of carp following short-term storage of semen. *Journal of Fish Biology*, 50: 1309–1328.

Rurangwa, E.; Volckaert, F.A.M.; Huyskens, G.; Kime, D.E. & Ollevier, F. (2001). Quality control of refrigerated and cryopreserved semen using computer-assisted sperm analysis (CASA), viable staining and standardized fertilisation in African catfish (*Clarias gariepinus*). *Theriogenology*, 55:751-769.

Secer, S.; Tekin, N.; Bozkurt, Y.; Bukan, N. & Akçay, E. (2004). Correlation between biochemical and spermatological parameters in rainbow trout (*Oncorhynchus mykiss*) semen. *The Israeli Journal of Aquaculture-Bamidgeh*, 56 (4), 274-280.

Steyn, G.J. & Van Vuren, J.H.J. (1987). The fertilising capacity of cryopreserved sharptooth cattish (*Clarias gariepinus*) sperm. *Aquaculture*, 63: 187-193.

Stoss, J. & Holtz, W. (1983). Successful storage of chilled rainbow trout (*Salmo gairdneri*) spermatozoa for up to 34 days, *Aquaculture*, 31: 269–274.

Tekin, N.; Secer, S.; Akcay, E. & Bozkurt, Y. (2003). Cryopreservation of rainbow trout (*Oncorhynchus mykiss*) semen. *The Israeli Journal of Aquaculture-Bamidgeh,* 55 (3), 208-212.

Tekin, N.; Secer, S; Akcay, E.; Bozkurt, Y. & Kayam, S. (2007). Effects of glycerol additions on post-thaw fertility of frozen rainbow trout sperm, with an emphasis on interaction between extender and cryoprotectant. *Journal of Applied Ichthyology*, 23 (1): 60-63.

Tiersch, T.R.; Williamson, J.H.; Carmichael, G.J. & Gorman, O.T. (1998). Cryopreservation of sperm of the endangered razorback sucker. *Transactions of the American Fisheries Society*, 127: 95-104.

Vuthiphandchai, V. & Zohar, Y. (1999). Age-related sperm quality of captive striped bass, *Morone saxatilis*. *Journal of World Aquaculture Society*, 30: 65–72.

Wayman, W.R.; Tiersch, T.R. & Thomas, W.R. (1998). Refrigerated storage and cryopreservation of sperm of red drum, *Sciaenops ocellatus* L. *Aquaculture Research*, 29: 267-273.

Yavas, I. & Bozkurt, Y. (2011). Effect of different thawing rates on motility and fertilizing capacity of cryopreserved grass carp (*Ctenopharyngodon idella*) sperm. *Biotechnology and Biotechnological Equipment*, 25 (1), 2254-2257.

16

Cryopreservation of the Sperm of the African Catfish for the Thriving Aquaculture Industry in Nigeria

Ofelia Galman Omitogun[1,*], Olanrewaju Ilori[1], Olawale Olaniyan[1], Praise Amupitan[1], Tijesunimi Oresanya[2], Sunday Aladele[2] and Wasiu Odofin[2]

[1]*Biotechnology Laboratory, Department of Animal Sciences, Obafemi Awolowo University (OAU), Ile-Ife,*

[2]*National Centre for Genetic Resources and Biotechnology (NACGRAB), Moor Plantation, Ibadan, Nigeria*

1. Introduction

The production of fish in Nigeria is still very small and cannot sufficiently satisfy the increasing demand of its population of 140 million. To solve the populace's high demand for fish, Nigerians resort to aquaculture which is currently faced with major constraints including lack of fish seed and quality of feed. The scarcity of good broodstock has necessitated the need to conserve the fish genetic resources which are wasted during natural and artificial induced spawning process of fish breeding. One way of expanding aquaculture in Nigeria is by devising a means of preserving genetic resources of our broodstock for all year round supply of fish seed through cryopreservation (Omitogun *et al.*, 2006).

The African catfish *Clarias gariepinus* Burchell, 1822 is one of the most suitable species for aquaculture in Africa. Since the 1970's, it has been considered to hold a great promise for fish farming in Africa. Some other merits of African catfish are: high growth rate reaching market size of 1 kg in 5–6 months under intensive management conditions: highly adaptable and resistant to handling and stress; can be artificially propagated by induced spawning techniques for reliable mass supply of fingerlings; commands a very high commercial value where it is highly cherished as food in Nigerian homes and hotels (Olaleye, 2005.).

The Clariid freshwater fishes belong to the family Clariidae with a wide geographical distribution in Africa consisting of 14 genera (Teugels, 1986a) and 32 species (Teugels, 1986b) in Nigeria. Syndenham (1980) reported that the family consists of 5 subgenera namely: Clarias, Clarioides, Anguilloclarias, Platycephaceloides, and Brevicephaloides. *C.*

* Corresponding Author

gariepinus is the species native to Africa where it is grown although mostly on a subsistence level for food. The fish is hardy and adaptable to diverse environments even with poor water quality with its air breathing ability (Hecht *et al.,* 1996). *C. gariepinus* is a typically non-aggressive stalking predatory omnivore that hunts at night using non-visual primary sense organs especially the senses of touch through the barbels and tactile organs on the mouth and skin (Bruton, 1996).

The availability of gametes throughout the year is important to ensure a constant supply of fish. In captivity (25^{0}C; 12h light per day), *C. gariepinus* gametogenesis is continuous once sexual maturity is reached. However, whereas females can be stripped of eggs after treatment with pituitary extracts, spermatogenesis and male reproductive behavior do not take place spontaneously, even after hormonal therapy. To obtain spermatozoa it is necessary to kill male brood fish or surgically remove the testes. Storing batches of spermatozoa by cryopreservation would significantly improve the reproductive potential of male catfish. The procurement of reliable broodstock (of good genetic quality), fingerlings and as juvenile fish for stocking ponds and fish farms has been a major set back in the development of catfish culture in Nigeria. This is because these cultivable species are not easily obtained from the wild. The development of cryopreservation procedures for sperm of Clarias species will aid in the recovery of threatened and endangered species as well as in the genetic selection and maintenance of lines of selected stocks. Cryopreserved sperm can also benefit commercial aquaculture industry by allowing females to be spawned when males are not available, decreasing the need to hold captive male as broodstock.

1.1 Success of cryopreservation in African catfish in Nigeria

Cryopreservation of fish spermatozoa has been the subject of many investigations. Successful cryopreservation depends not only on the right choice of cryoprotectant and extender, but also on the freezing protocol used. Cryoprotectant and freezing rate together determine the damage to spermatozoa due to intracellular ice crystallization.

The first two years of the NACGRAB- OAU Department of Animal Sciences cryopreservation project (2005-2007) were dedicated to optimization of cryopreservation protocols of the catfish sperm under short-term condition in deep freezer at -10 to -30°C (Oyeleye and Omitogun, 2007) and testing the viability of cryopreserved sperm by studying the ability of these cryopreserved sperm in fertilizing freshly spawned eggs (Omitogun *et al.,* 2006).

The second phase of the project (2008-2010) was dedicated to cryopreservation of the catfish sperm under long term conditions in liquid Nitrogen (-290^{0}C) and testing the motility and ability to fertilize eggs. Evaluation of optimization and economic feasibility of cryopreserved sperm was also carried out (Omitogun *et al.,* 2010). To this end, cryopreserved sperm in liquid nitrogen in Dewar container was further diluted and cryopreserved from 3-8 months, then was taken to an identified and willing commercial catfish farm with the objective of testing the ability of cryopreserved sperm of African giant catfish to fertilize a whole clutch of eggs from a mature female catfish, normally used by commercial farmers and consequently confirm the viability of using cryopreserved sperm in normal commercial hatchery operations. Our hypothesis was that if cryopreseved sperm is practically tested on commercial scale and is proven economically feasible, being a true reflection of what was obtained in the laboratory (Oyeleye and Omitogun, 2007) then this will help to conserve male brood stock (Omitogun *et al.*, 2010) which are normally slaughtered for fry production of catfish, and likewise ensure all-year round artificial propagation, helping the fish farmers in overcoming the problem of scarcity of male catfish breeders which are often encountered in the dry season.

1.2 Background information: Sperm: Egg ratio for optimum fertilization of catfish eggs

Cryopreservation of African catfish semen in liquid Nitrogen (LN_2) will invariably help us to conserve the genetic resources of our desirable male fish breeders for all year round artificial propagation and also help in overcoming the problems of scarcity of desirable male catfish breeders often encountered by the farmers most especially in the dry season and to meet high demand for catfish consumption (Oyeleye and Omitogun, 2007).

Sperm collection in African catfish as mentioned requires killing the male fish in order to excise the testes, it is important to maximize the use of a single male by optimization of sperm: egg insemination ratio. For fresh spermatozoa, the effective insemination ratio was estimated as 245 x 10^3 spermatozoa per egg in *C. gariepinus* (Steyn, 1987) and 50 x 10^3 spermatozoa per egg in *Heterobranchus longifilis* (Otenne *et al.*, 1996.). Because a percentage of spermatozoa die during freezing and thawing processes, the effective insemination ratio for frozen spermatozoa should be higher. In channel catfish, 50 x 10^6 frozen-thawed spermatozoa per 0.5 ml straw enabled fertilization of 250 eggs (200 x 10^3 spermatozoa per egg; Tiersch *et al.*, 1994). In blue catfish, *Ictalurus furcatus*, a minimum of 13,000 x 10^3 frozen-thawed spermatozoa per egg were needed to achieve 54% of control fertilization. In *C. gariepinus*, 49 x 10^3 live frozen-thawed spermatozoa per egg achieved a hatching rate (51.2%) equal to the control (51%).The insemination ratio was within the range 6 to 24 x 10^3 spermatozoa per egg. (Steyn, 1993). During ovulation the belly of the female will swell considerably due to water absorption of the ovary. The speed of the ripening process is dependent upon water temperature and likewise, the development process from fertilized egg to hatching is dependent upon water temperature (Coppens, 2009).

African catfish spermatozoa were first successfully cryopreserved by Steyn *et al.* 1985 who obtained 40% motility 24h after storage in LN_2. Glucose in combination with glycerol has been most widely used cryoprotective solution. Recently, glucose in combination with DMSO was also shown to be effective (Urbanyi *et al.*, 1999). Freezing rates can be rapid (*e.g.*, pellet freezing on dry ice or in LN_2 vapor) or slow (*e.g.*, at fixed rates in programmable freezer (Steyn, 1993). In most cases, sperm quality was only evaluated in terms of motility after thawing. When fertilization was included in the evaluation, sperm: egg ratio was not optimized and was often excessive (Padhi *et al.*,1995). Using excess spermatozoa for fertilization obviously masks the quality of cryopreserved spermatozoa, making comparison of protocols difficult (Viveiros *et al.*, 2000).

Methods for cryopreserving spermatozoa and optimizing sperm: egg dilution ratio in African catfish *Clarias gariepinus* was first developed by Viveiros *et al.*, 2000) where 5 to 25% DMSO and methanol were tested as cryoprotectant, by diluting semen in Ginzburg fish ringer and freezing in 1-milliliter cryovials in a programmable freezer. To avoid an excess of spermatozoa per egg, post-thaw semen was diluted 1:20, 1:200 or 1:2000 before fertilization. Even frozen- thawed spermatozoa with low numbers of live cells yielded adequate hatching rates. They found out that the maximum sperm dilution ratio to achieve hatching rates similar to control was 1:200 without losing fertilization ability. However, at 1:2000 the hatching rates produced with frozen spermatozoa were lower than the control African catfish. Similarly *Heterobranchus longifilis* spermatozoa were diluted 1:3 before freezing and 1:10 after thawing and had the same fertilization ability (78.9%) as the control (81.1%). On the contrary for *Cyprinus carpio,* no spermatozoa survived when diluted higher than 1: 5 before and after freezing (Lubzens *et al.,* 1997).

Cryopreservation of catfish spermatozoa is useful as a routine method of gamete storage and management. However, the economic factor should also be considered. The technology of cryopreservation with the use of liquid nitrogen though desirable but is cost intensive. Therefore there is a need to study how the cryopreserved semen will be maximally utilized with good fertilizing results at the same time cost-effective and affordable for the farmers

To avoid wastage of cryopreserved spermatozoa per egg clutch after dilution with physiological salt solution, fertilization of various measures of egg clutches were tested in the present research with differently cryopreserved spermatozoa for optimization and for cost evaluation. In a second study the concentration of the semen was reduced, *i.e.,* diluted at a dilution ratio of 1:20 and 1:200 to verify the spermatozoa are not in excess and consequently be wasted.

Another study was carried out in order to assist the farmers to determine the approximate amount of egg clutch that will be adequate for a milliliter (ml) of cryopreserved semen without wasting spermatozoa in order to evaluate economic cost and profitability. The aim of this study was to verify the possibility of cryopreserving African catfish under long-term condition in liquid Nitrogen (LN_2) and evaluate the viability and fertility optimization of a specific amount (*i.e.,* 1 ml) of cryopreserved semen of African catfish cryopreserved in LN_2 using various cryoprotective agents with different measures of egg clutches. This paper aimed to establish a standard fertility ratio between a ml of semen and clutch of eggs in order to prevent wastage of semen; be able to maximize the resources and evaluate profitability of cryopreservation in liquid nitrogen by evaluating the effects on the cryopreserved semen as to motility and hatchability, the ability to hatch the eggs from a gravid female and survival of ensuing larvae.

2. Materials and methods

For the benefit of prospective users of cryopreservation of African catfish semen, the whole process is pictorially presented in Figures 1-12.

2.1 Husbandry of the broodstock

The broodstock of the African catfish, *C. gariepinus* were obtained from reliable farms in Ile-Ife and Ibadan, Nigeria and were transported in 25 litre tank opened at the top to the Wet Laboratory of the Department of Animal Sciences, Obafemi Awolowo University, Ile Ife.

The matured male and female broodstock were kept at constant temperature of 27^{0}C in a 1000 litre tank connected to a source of water by a pipe connected from the reservoir plastic tank placed in an elevated stand in the laboratory and its drainage was located at the bottom of the tank for easy flow by gravity.

The broodstock were fed with an imported floating palletized feed *i.e., Coppens^R* feed (42% protein, ISO -170 certified, Netherlands) containing a large percentage of high quality fish meal, which is especially important to facilitate repeated spawning at a maintenance level of 1.5% body weight on a daily basis by gradual hand broadcast. The water quality was regulated through proper monitoring and replaced weekly.

2.2 Selection of broodstock

The sexually matured female was selected according to their swollen, reddish genital papilla and a well distended, swollen soft abdomen. A slight pressure was applied on the abdomen towards the genital papilla after which ripe eggs oozed out which were green-brownish in color and ripe eggs are generally uniform in size. The female broodstock was stocked in the hatchery for about 2 days without feeding so that the alimentary tract was empty at the time of stripping. It is very important that the collected eggs did not get contaminated. Sexually matured male broodstock was selected based on a reddish or pinkish pointed and vascularized genital papilla. The temperature of the broodstock kept in the tank was maintained at 25–27^{0}C.

2.3 Preparation of extender-cryoprotectant

Two extenders were used in this study, phosphate buffered saline (PBS) and Ginzburg Fish Ringer (GFR) with pH of 7.4 and 7.6 respectively. The extenders were prepared as shown in Table 1 with Calcium-free Hanks Balanced Salt Solution (Ca-FHBSS) used in experiment 2. 14; after which they were sterilized for 20 min at 15 lbs/inch^2 using a pressure cooker to avoid contamination and deterioration of the spermatozoa.

2.4 Semen collection

A good quantity of the sperm of African catfish cannot be stripped and sperm can only be obtained after sacrificing it. Sexually matured male weighing 0.8$\pm$ 0.2kg were selected and kept in a different tank of about 50 l capacity for about 18 h prior to the time of sperm collection .The male broodstock was dried with clean towel and then made unconscious by breaking its backbone. The body cavity was carefully opened with a pair of sterilized scissors without damaging the testes after which the two testes were dissected out. It was then removed with a pair of forceps, the blood veins cleared out and rinsed in saline solution. The testes were lacerated with a new and sharp razor blade; the milt was gently squeezed out and collected in a sterilized Petri dish. The whole process was carried out in a disinfected environment to avoid bacterial contamination which can lead to degradation of samples, transfer of pathogens and inaccurate estimation of motility. Sterilized instrument and aseptic techniques for collection of sperm was incorporated to reduce the contamination by bacteria. The volume of the extracted sperm was measured with a 5.0 ml sterilized syringe.

2.5 Cryopreservation of sperm, cryoprotective agents used

Different combination of cryoprotective (CPA) agents are shown in Table 2 as CPA-DP (DMSO+PBS), CPA-DF (DMSO+GFR), CPA-GP (Glycerol+PBS), CPA-GF (Glycerol+GFR), CPA-DGP (DMSO+Glucose+PBS) and CPA-DGF (DMSO+Glucose+GFR). Before cryopreservation of semen, motility of the fresh semen was evaluated in two trials of dilution: 1:1 and 1: 20. In these trials two different extenders were used: Phosphate buffered saline (PBS) and Ginzburg Fish Ringer (GFR). The volume of the extracted semen was measured with a 5.0 ml syringe (DISCARDM(R) NIG) which was diluted with the extender PBS and GFR for first trial on a ratio 1:1 v/v and 1:10 v/v respectively, then mixed evenly with differently prepared cryoprotective agent combinations at a ratio 1:1 (PBS) for fertility and hatchability evaluation.

In the second trial, the two extenders were used but at different sperm dilution ratio for PBS and GFR at a ratio 1:20 for both extenders. The resulting semen-cryoprotective agent solution in each trial after thorough mixing was dispensed into labeled 1ml cryotubes with a 2-step freezing protocol of first freezing on the chilled water blocks at -10^0 C for 10 min before it was finally transferred into the liquid nitrogen for a long-term preservation.

Composition (g/1000 ml)	Phosphate buffer saline (PBS)	Ginzburg Fish Ringer	Calcium-Free HBSS 200mOsmol/kg
NaCl	8.0	7.0	5.26
KCl	0.02	0.28	0.26
$CaCl_2$	-	0.33	-
$NaHCO_3$	0.23	-	0.33
Na_2HPO_4	1.15	-	0.04
KH_2PO_4	0.20	-	0.04
$Mg\ SO_4\ 7H_2O$	-	-	0.13
$C_6H_{12}O_6$	-	-	0.66

Table 1. Composition of the extenders (g/l) tested for cryopreservation of catfish sperm in liquid Nitrogen.

Cryoprotective agent (CPA)	DMSO	Glycerol	PBS	GFR	Glucose/ Sucrose
DP	10	-	90	-	-
DF	10	-	-	90	-
GP	-	10	90	-	-
GF	-	10	-	90	-
DGP	10	-	85	-	5
DGF	10	-	-	85	5

DMSO =Dimethylsulphoxide, PBS =Phosphate buffered saline, GFR=Ginzburg Fish Ringer.: DP = DMSO and PBS; DF = DMSO and GFR; GP = Glycerol + PBS; GF = Glycerol + GFR; DGP = DMSO + PBS + Glucose/Sucrose; DGF = DMSO + GFR + Glucose/Sucrose

Table 2. Composition (%) of the cryoprotective agents used

Fig. 1. The African catfish broodstock.

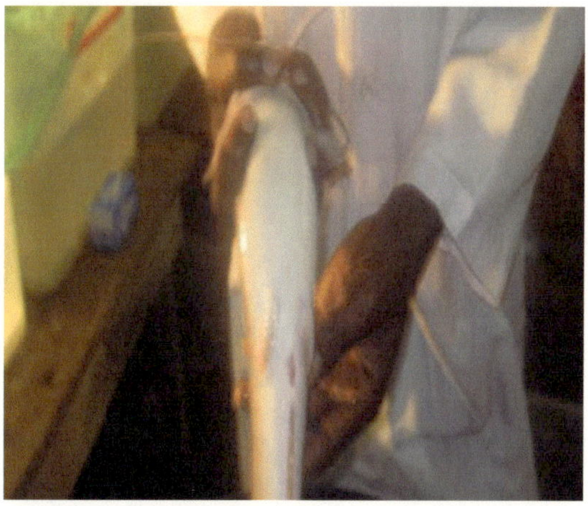

Fig. 2. Male catfish showing the genital openings.

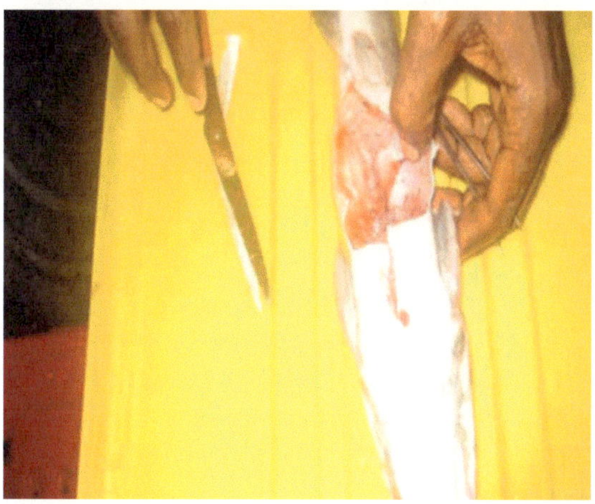

Fig. 3. Removal of testes from male catfish

Fig. 4. Laceration of testes to extract milt

Fig. 5. Motility evaluation using a microscope and haemacytometer.

Fig. 6. Sperm cryopreserved in liquid Nitrogen for 4-8 months

Fig. 7. Removing the cryopreserved sperm and motility evaluated

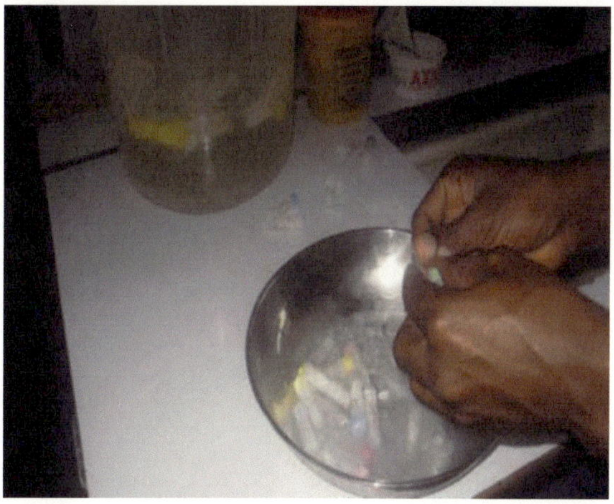

Fig. 8. Rapid thawing process is employed.

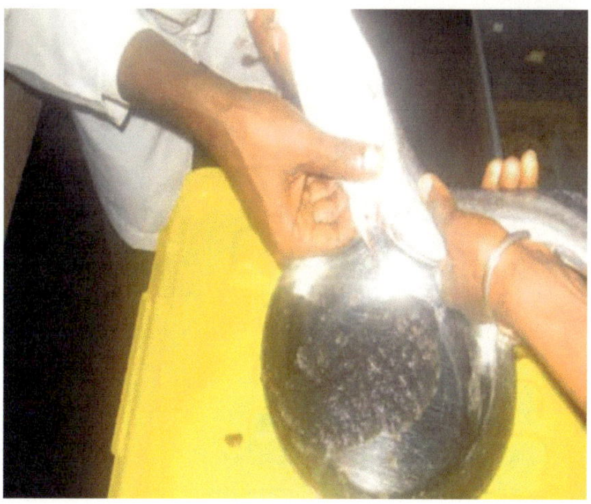

Fig. 9. Stripping the eggs from gravid female catfish after injection with Ovaprim

Fig. 10. Eggs divided into various clutch weights in Petri dishes

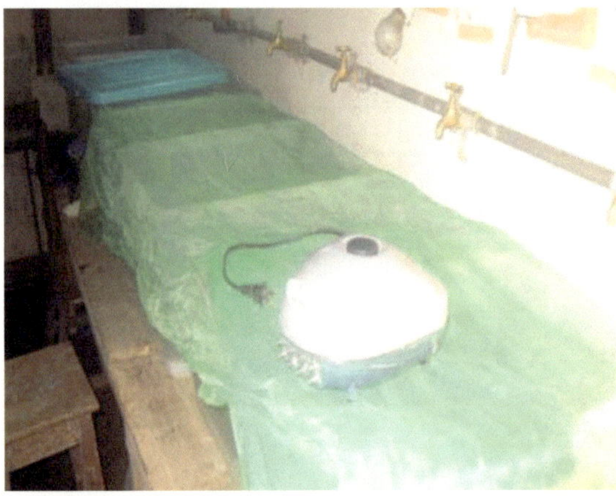

Fig. 11. Fertilized eggs incubated in aerated plastic containers covered with nets

Fig. 12. The temperature was kept at 25-27° C and sometimes covered with black plastic sheets.

2.6 Induced spawning and stripping

The readiness of the female broodstock to be used for breeding was tested by holding it in a head-up vertical position and a slight pressure was applied by pressing its abdomen with a thumb from the pectoral fin towards the genital papilla after which eggs run out freely. The selected broodstock were kept separately in different tanks without feeding them, after they were injected with 0.35 ml Ovaprim ® (Syndel, Canada) per kg live weight (Oyeleye and Omitogun, 2007) and then left for 10-12 hours latency period as a post ovulatory maturation period and to ensure high hatching rates and low proportion of deformed larvae (Hogendoorn, 1979).

2.7 Method of female stripping

The female body surface was gently dried with clean towel. It was tightly held at both ends by two persons with wet towels and stripped by a gentle press on the abdomen with a thumb towards the rear. The first free running eggs obtained at a slight pressing of the induced female broodstock were collected for fertilization (Legendre and Oteme, 1995).

2.8 Egg clutch variation and fertilization

After inducing, the female fish was stripped and the clutch of eggs weighed (about 150–160g/kg of the body weight). The eggs were weighed in various measures of 1.0g, 2.0g, 3.0g, 4.0g and 5.0g based on the level of each experiment. That is for 1.0g, it was weighed seven times with replicates for the different cryoprotective agents and control. After which it was fertilized with the cryopreserved semen in liquid nitrogen after thawing in warm water at 35^0C for 5 min. Fresh semen was used to fertilize same amount of clutches of eggs to serve as the control for both trials.

2.9 Motility evaluation

The motility of the spermatozoa before and after the addition of the cryopreservative agents, CPA and after thawing was evaluated for each trial. The cryopreserved semen was also further evaluated for fertility, hatchability and survivability for each trial. The motility test was done by diluting a drop of post thawed or fresh spermatozoa either with PBS, GFR or 0.9 % saline solution at a ratio 1:100 from which one drop of the solution was put on the hemocytometer and viewed subsequently under the microscope 10X and 40X, low and high power objectives of the microscope. The result arrived at is converted to the total number of spermatozoa per ml by multiplying it by the dilution factor (100) and 10^4 (SIGMA, 1994) as follows:

$$\textit{Total no of spermatozoa per ml} = \textit{Average No. of counted spermatozoa} \times 10^4 \text{ } \textit{of the cryopreserved semen} \quad (1)$$

2.10 Fertility and hatchability evaluation

The development process from fertilized eggs to hatching is dependent upon water temperature while hatching rate is, next to egg quality, dependent on the water quality;

temperature, oxygen level, pH and water hardness. After stripping of the induced female broodstock, the eggs were weighed in grams depending on the on-going experiment *i.e,* 1.0g (600±100 eggs), 2.0g, etc. The various measures of eggs (repeated 12 times together with replicates) were fertilized with cryopreserved semen thawed at 35^{0}C for 5 minutes, and a pair of egg clutches with fresh semen as control experiment. The mixture of eggs and semen was stirred gently for at least 1.0 min to allow contact and adequate fertilization. Within a few minutes after fertilization the eggs absorbed water and could become sticky so the eggs were distributed in a netted basket suspended in the hatching trough (50cm x 35cm x 30cm) containing contaminant-free (passed through a purification system with ultraviolet sterilization at 3000 μW/cm^2) well-aerated water in a single layer so that the eggs get sufficient oxygen during incubation. The hatching troughs were completely covered with mosquito net and black polythene materials placed under 200 Watt bulbs to prevent mosquitoes and other insects laying eggs and to increase level of heat generation. The system was supplied with an electric aerator to increase level of oxygen dissolved in the water.

The incubated eggs were monitored and temperature maintained between 26^{0}C -27^{0}C for incubation between 23- 25 h. Soon after hatching the larvae passed through the net and the dead eggs and shells remain on the net in the basket. The larvae were then simply separated from the unfertilized eggs and eggshells by lifting the basket and the nets out of the hatching trough.

The percentage, % fertility and hatchability were determined subjectively after 12–15 h of fertilization by identifying the healthy developing eggs which were transparent green brownish in colour (Coppens, 2007) while the dead eggs were also estimated:

$$\%\ Fertility = (No.\ of\ fertilized\ eggs\ /\ No.\ of\ inseminated\ egg)\ X\ 100\% \quad (2)$$

$$\%\ Hatchability = \frac{Total\ No.\ of\ fertilized\ eggs\ -\ No.\ of\ unhatcched\ eggs}{Total\ No.\ of\ Fertilized\ eggs} \times 100\% \quad (3)$$

2.11 Post-hatching survivability evaluation

This is done by allowing the newly hatched larvae of all the treatments and that of the control to live on the remains of their yolk sacs for the first 4 days (Heicht *et al.,*1996) after hatching out of the eggs and thereafter carefully removed from the hatching troughs and were fed with Artemia (Inve Aquaculture, USA) on a regular basis (*i.e.,* twice per day). Irregularities in the activities of the fry in terms of feeding, movement in water was observed at the same time taking note of the dead fry which were removed immediately to avoid contamination of water. Survivability evaluation which was observed for a period of about 3 - 4 weeks was done for each stage of the experiment together with fertility and hatchability for fresh (control experiment) and cryopreserved spermatozoa. The post-hatching survivability was evaluated as follows:

$$\%\ Survivability = \frac{Total\ No.\ of\ larvae\ -\ No.\ of\ dead\ larvae}{Total\ No.\ of\ larvae} \times 100\% \quad (4)$$

2.12 The control

The control for both trials was prepared by the use of fresh semen obtained from the lacerated testes from a normal gravid male broodstock with the use of sterilized dissecting knives but activated with saline solution in the ratio of 1:1 v/v and subsequently used to fertilized various measures of egg clutches ranging from 1.0g, 2.0g, 3.0g, 4.0g and 5.0g normal eggs from the same batch of eggs i.e. from the same fish. Control was set up for the evaluation of each parameter for trials *i.e.,* motility, fertility, hatchability and survivability.

The motility evaluation of the post-thawed cryopreserved sperm was evaluated after dilution with the extender on a ratio 1:100 using hemocytometer (SIGMA, 1994).

2.13 Further scaling up for commercial application of cryopreserved semen dilution and egg clutch fertilization

Motility of the fresh semen was evaluated in two different trials of dilution, before any cryopreservation of the semen. Phosphate-buffered-saline (PBS) and ordinary saline water were the two extenders used in the two different trials and were diluted 1:1 and 1:40. The extracted semen volume was measured with a 5.0ml syringe (DISCARDIM(R)NIG) which was diluted with the extender PBS for trial on a ratio 1:20v/v and 1:200v/v respectively and thereafter mixed evenly with the cryoprotective agents at a ratio of 1:1 and cryopreserved for the next seven months in liquid Nitrogen stored in Dewar container

For the second trials, the same cryoprotectant (85% PBS+5% glucose+10% DMSO) was used but at different sperm dilution ratio of 1:40. In both trials, the resulting semen in each experiment after thorough mixing was then dispensed into labeled 1ml cryotubes while a 2-step freezing protocol, e.g. initial freezing onto frozen water (ice) blocks at -10^{0}C for 30min before the final transfer into liquid nitrogen for the next 4 to 7 months.

After induction with Ovaprim, the female broodstock was stripped and the clutch of eggs weighed (which was about 150-200g/kg body weight) and divided into three portions of about 120 g each for the experiment on fertilization with cryopreserved sperm of diluted 1:1 and 1: 40 and for the control.

This procedure was repeated three times in a nearby commercial farm 5kms away from the University to serve as the replicates and to ensure the repeatability of the experiments.

2.14 Experiment on storing sperm in refrigerator

The various extenders used were

- 200mOsmol/kg Ca-F HBSS
- 300 mOsmol/kg Ca-F HBSS
- 400 mOsmol/kg Ca-F HBSS
- RPMI 1640 (SIGMA) Culture Medium in 0.9% NaCl Solution

The three different osmolalities of extender Ca-FHBSS (Calcium-free Hanks Balanced Salt Solution) were prepared according to Riley, 2002. The sperm with the 200, 300 and 400 mOsmol/kg Ca-F HBSS were kept in the refrigerator at 4^{0}C. The semen samples with RPMI and 0.9% NaCl solutions were kept at both room temperature and refrigerator. Two replicates were made for each treatment.

2.15 Statistical analysis

The data collected on the parameters, motility, fertility and hatchability was subjected to standard statistical analysis. The data collected were analyzed using analysis of variance (ANOVA) to find a level of significance at $p < 0.05$. The number of motile sperm counted per square of hemocytometer and percentage of motile sperm obtained after cryopreservation were subjected to Duncan's multiple range test to evaluate effects of types of cryopreservatives as well as egg clutch weights and sperm dilutions and period of cryopreservation on sperm motility. The data collected on motility, fertility and hatchability in the first trial were subjected to 2- way ANOVA at a significance level of $p < 0.05$. The bar charts and line chart showing the relationship between the period of refrigeration and the % motility for the various extenders were also employed for better understanding of the results.

2.16 Cost of production of cryopreserved semen

The costs of chemicals and other consumables used for the study were listed in Table 5. The rates per gram or per ml were calculated to determine the effective cost per ml of the cryopreserved sperm (Table 6).

3. Results and discussion

3.1 Effects of different cryoprotectants on fertility, hatchability, motility and survivability

In the first trial with dilution ratio of 1:1 (sperm: extender) the effect of nature of cryoprotectants on the parameters measured was significant ($p<0.05$). Though dimethylsulphoxide+5% glucose+ PBS (DGP) was higher, DGP and DP gave the best results and was not significantly different ($p>0.05$) from each other but significantly different ($p<0.05$) from other cryoprotectants. It was followed by GP but not significantly different ($p>0.05$) from other cryoprotectants. Fertility also followed the same trend, but DGP was significantly different from DP from their LSD values followed by GP. GF gave the least result but significantly different ($p<0.05$) from other cryoprotectants. Related trend was also observed for other cryoprotectants but that of control was higher ($p<0.05$) than other treatments for each parameter. A similar trend was also observed for motility and survivability.

In Table 3, control (fresh semen), DGP and DP were compared. It is obvious that control has highest mean value (significantly different), this may be expected as the control was not passing through any treatments and processes.

In the second trial when the semen was diluted at ratio 1:20 (sperm: extender) a very close trend was observed which shows that the spermatozoa seems to be too much and probably wasted in the first trial. Besides control which was significantly different ($p<0.05$) for all the parameters, DGP gave the best result before DP, this may be explained by extracellular protection offered by the glucose, but the mean values are not significantly different ($p>0.05$) from each other. Fertility also related to hatchability followed the same trend with hatchability. In fertility, DP, DGP and GP were not significantly different from each other. However, the mean values are significantly different from each other for other parameters in decreasing order of C > DGP > DP > GP > GF. Generally, DGP gave the best followed by DP (without 5% glucose) while Glycerol in combination with Ginsburg fish ringer gave the lowest result.

A close means values were also discovered with trial 1 which indicate the results were better in trial 2, the further diluted semen which could supposedly be explained by addition of extenders.

The differences in fertility and hatchability with control and cryoprotectants tested may be due to the mild damage done to the spermatozoa during the process of lacerating the testes to extract the semen, and also due to the intracellular vitrification (Cryobiosystems, 2009)- a commonly occurring problem in the process of cryopreservation in liquid Nitrogen.

3.2 Effects of egg clutch weight on viability of African catfish gametes

There was a significant effect ($p<0.05$) of egg clutch weight on fertility, hatchability and survivability. In trial 1, although, the hatchability increases with increase in egg weight but the increment at egg weight 4.0g and 5.0g was not significantly different (LSD=88.749, $p>0.05$). There was fertility optimization at 4.0 g of egg clutch weight which though, close to the mean values of 5.0g which is higher but not a uniform increase. The same trend was observed for fertility of eggs but much higher than corresponding hatchability which may be due to loss of eggs to external factors like temperature, contamination and possible error during record taking. There is no significant difference for fertility at egg clutch weight 3.0g and 4.0g but there was significant difference ($p<0.05$) from 5.0g. Survivability was not significantly different from each other except for 5.0g ($p<0.05$).

In the second trial, a similar trend was observed; hatchability was highest ($p<0.05$) at egg clutch weight 4.0g. No significant difference ($p>0.05$) in survivability was observed except for egg weights 3.0g and 4.0g.

Generally, for both trials, egg clutch weight at 4.0 g gave the optimum viability value.

3.3 Effect of type of cryoprotectant and egg clutch weight interaction on hatchability and fertility

From the statistical analysis, it showed that there could also be effect of interaction of both cryoprotective agent (CPA) and egg clutch weight on fertility and hatchability.

The result, as observed shows a significant effect of ($p<0.05$) of interaction of CPA and egg clutch weight on fertility, hatchability for the first trial and only on fertility for the second trial.

Effect of interaction of cryoprotectant on hatchability was not different from the trend of results obtained in previous results. However, the cryoprotective agents were not significantly different ($p>0.05$) from each other but DGP, GP and DP still maintained the higher mean values while control took the highest. DP and DGP were not significantly different ($p>0.05$) from each other for egg clutch weight such as 1.0g to 5.0g. However, GP was significantly different ($P<0.05$) from DGP and DP for egg clutch weight 2.0-5.0g. For GF, DF and DGF, there was also no significant effect ($p>0.05$) with changes in egg weights. The effect of interaction of both CPA and egg weights on fertility was also significant ($p<0.05$).

The results for both trials were also very close, following the same trend, except for control changes in egg clutch weights from 1.0g to 5.0g was generally not significantly different for DP and DGP followed by GP and GF which were not significantly different (p>0.05) from means values of egg clutch weight 4.0g–5.0g for DGP and DP. However, GF gave the lowest value for the two trials.

3.4 Scaling up of the applications of the cryopreserved semen for commercial aquaculture

3.4.1 Effect of dilution ratio on viability of catfish gametes

Table 3 shows the comparison for the parameters measured among the control, dilution ratios 1:1 and 1:40. The fresh semen gave the highest fertility and hatchability rates (P<0.05). It is significantly different from ratio 1:1 and ratio 1:40. Comparing the fertility and hatchability rates of the two different dilution rates, ratio 1:1 gave the highest fertility and hatchability rates which was significant (P<0.05). Survival rate however, followed a different trend in which dilution ratio 1:1 gave the highest survival rate closely followed by ratio 1:40 while the control semen gave the least survival rate (P< 0.05).

Most importantly, the differences in fertility and hatchability may also be attributed to the condition of the farmer's hatchery environment in which many environmental and sanitation conditions were compromised for maximum profit (Amupitan *et al.*, 2010).

DILUTION RATIO	MOTILITY	FERTILITY	HATCHABILITY	SURVIVAL
1:1	55%[b]	30%[b]	35%[b]	15%[a]
1:40	49%[b]	29%[c]	34%[c]	14%[b]
C	72%[a]	54%[a]	62%[a]	13%[c]
LSD	13%	9.2%	25%	13%

Means in the same column with different letter are significantly different at P<0.05, C=Control, LSD = Least Significant Difference

Table 3. Effects of dilution ratio on viability of Catfish semen diluted at ratio 1:1 and ratio 1:40

3.4.2 Effect of dilution ratio on motility of catfish semen and survival of ensuing larvae

The fresh semen gave the highest motility at appreciably high percentage (71.00%) which was significantly different (P<0.05) from cryopreserved semen diluted at ratios 1:1 (50.52%) and 1:40 (49.05%) (Fig. 13). However, there was no significant difference between the two diluted cryopreserved semen (P>0.05). It was evident that the freezing process and cryopreservation decreased sperm motility after cryopreservation. It could be deduced that cryopreserved sperm still needs to be completely activated after thawing in order to fertilize the whole clutch of eggs since there is a direct relationship between motility and fertility.

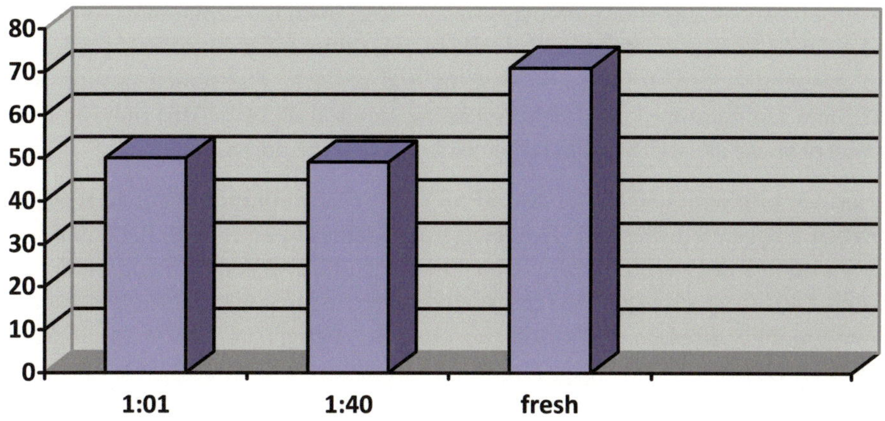

Fig. 13. Motilities of cryopreserved semen at ratios (1:1, and 1:40) compared with the control fresh semen

Cryopreservation in liquid Nitrogen did not have any effect on the survival of *C. gariepinus* larvae produced from cryopreserved semen as shown. However, larvae produced by the cryopreserved semen gave a higher survival, ratio 1:1 (65%), Ratio 1:40 (63%) and fresh semen (50%) ($P>0.05$).

The present study proves that sperm of African catfish cryopreserved aged up to 7 months in liquid Nitrogen and diluted more than 40 times with the extender is viable. The reason for low survivability rate in the control experiment using fresh semen may be attributed to high stocking density (because of the greater number of surviving fry) as practiced by many farmers, *i.e* the quantity of larvae per unit volume of water is less for cryopreserved sperm which in turn was favourable for survival.

3.5 Motility evaluation of refrigerated catfish sperm cells in different extenders

This experiment evaluated the effect of extenders and period of refrigerated storage on the sperm motility of *Clarias gariepinus* sperm cells with the intent to identify a suitable extender for the refrigerated storage of the sperm cells of *Clarias gariepinus.* Semen samples were collected from mature broodstock and were refrigerated with various different extenders at ratio 1:3 namely: Calcium-free Hanks' Balanced Salt Solution (Ca-F HBSS), RPMI 1640 culture medium and 0.9% NaCl. Ca-F HBSS extender was prepared at 3 different osmolalities: 200mOsmol/kg, 300mOsmol/kg and 400mOsmol/kg. Sperm in RPMI 1640 and 0.9% NaCl extenders were also kept at room temperature to assess the effect of refrigeration on motility of catfish sperm cells. Motility was monitored on a 24-hour basis and % motility was evaluated daily. Results showed that sperm cells of *Clarias gariepinus* using 200mOsmol/kg as extender ($p<0.05$) can be stored under refrigeration for 12 days. However, of all the extenders evaluated, RPMI 1640 proved to be the most effective extender ($p<0.05$) retaining higher motility of the refrigerated sperm cells of *Clarias gariepinus.*

3.5.1 Effect of refrigeration on % motility of sperm cells

The semen samples extended with 0.9% NaCl and the RPMI culture solution at room temperature did not have motile sperm cells after 48 hours. The motility of the semen sample with 0.9% NaCl at room temperature dropped from the initial motility of 74.82% to 6.24% (Fig.3.6. 1) after the first 24 h while the semen sample with the RPMI culture solution at room temperature had 0.4% motility at the end of 24 h. The semen sample with 0.9% NaCl at room temperature had a fishy irritating smell after 24 h. This may be due to the production of waste since the sperm cells metabolised at the normal rate.

The semen samples with 0.9% NaCl and RPMI 1640 culture solution retained motility much longer when refrigerated (Fig. 14). The refrigerated semen sample with 0.9% NaCl retained motility for up to 7 days with motility after 24, 48, 72, and 168 hourly being 34.75%, 17.03%, 14.06% and 4.46% respectively. The refrigerated semen sample with RPMI however kept for 9 days with motility after 24, 48, 72, 168 and 216 hours being 53.47%, 37.62%, 25.64%, 8.32% and 5.25% respectively.

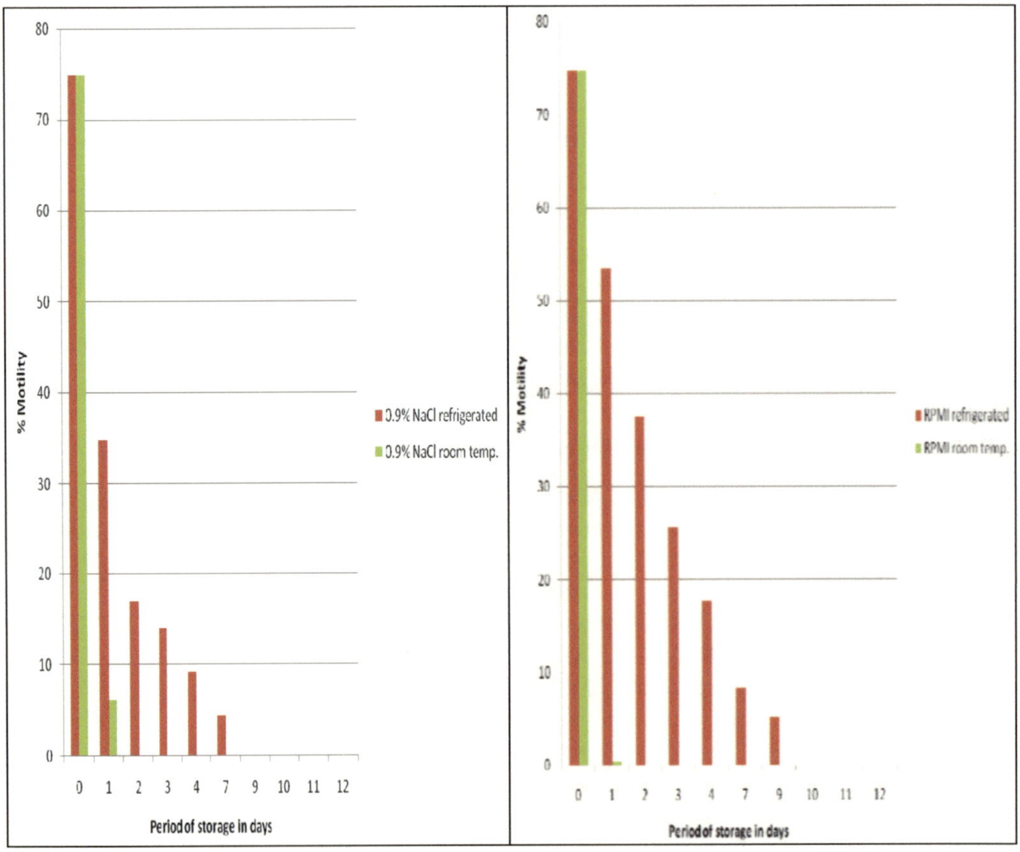

Fig. 14. Effect of refrigeration on motility of semen sample extended with NaCl (left) and RPMI (right). The refrigerated semen sample retained motility till the 7th (in NaCl) and 9th day (in RPMI) whereas semen samples at room temperature only lasted till the first day after storage.

The art of refrigeration provides a low temperature which lowers the metabolic rate of living organisms. The extended semen samples maintained at room temperature proceeded at the normal metabolic rate, hence the sperm cells could not survive up to 48 h. When the sperm cells were still within the fish in the testis, they were supplied with nutrients and the waste they produced are excreted out of the testis, they cannot be supplied with energy or nutrients except provided externally as in tissue culture. Their wastes also accumulate in the solution in which they are suspended in.

High metabolic rate means faster rate of using up available resources by living organism such as nutrients and energy. It also means that waste will be produced at a faster rate thereby causing fast accumulation of waste in the solution in which the sperm cells are suspended. This accumulation will immediately reach a toxic level causing fatality in the sperm cells. Whereas the low temperature provided by the refrigerator to the refrigerated semen sample reduced the metabolic rate of the sperm cells, thereby reducing the rate at which the available nutrients and energy in the semen-extender solution are used up. The nutrient and energy in the semen-extender solution lasted a much longer period when refrigerated, thus keeping the sperm cells alive for a longer period than in the semen samples extended at room temperature.

3.5.2 Effect of osmolality of Ca-F HBSS on % motility

After 24 hours, the motility of the refrigerated semen sample with the 200mOsmol/kg Ca-F HBSS dropped to 34.85% and motility was retained till the 12thday (288 hours) with 0.5% motility. The refrigerated semen sample with 300mOsmol/kg Ca-F HBSS retained motility for 10 days with % motility at 24 h being 31.88 and motility by the tenth day had dropped to 2.28%. The refrigerated semen sample with 400mOsmol/kg Ca-F HBSS retained motility also for 10 days but with a lower % motility at the 10th day (0.4% motility) but with motility at 24 h being 34.46% (Fig. 15).

Based on the length of days for which motility was retained, the 200mOsmol/kg Ca-F HBSS proved to be a good extender since it retained motility for 12 days but with a very low motility (0.5%) However the 300mOsmol/kg, although retained motility for only 10 days, is better since it had the highest motility at the fourth, seventh, ninth and tenth day (i.e. 16.93%, 10.73%, 2.77% and 2.28%).

A good extender should be isotonic to the seminal plasma of the fish. This is to keep the sperm cells immotile until ready for use. Sperm cells are immotile in the seminal plasma and when semen is released in aquatic environment, osmolality goes down and motility is initiated in freshwater species. (Maria *et al.*, 2006). Motility in freshwater species is initiated by exposure of the semen to a hypotonic solution (Morisawa and Suzuki, 1980). Use of extender solutions that are similar in chemical concentration and osmolality are essential to optimizing storage time (Baynes *et al.*, 1981). According to Mansour *et al.*, 2002, motility of *Clarias gariepinus* is completely but irreversibly suppressed in electrolytes and non-electrolytes with an osmolality of 200mOsmol/kg. This statement by Mansour *et al.*, 2002 proves that the osmolality of the seminal plasma of *Clarias gariepinus* is less than or equal to 200mOsmol/kg. This explains why the 200mOsmol/kg of Ca-F HBSS retained motility till the twelfth day as it is closer to being isotonic to the seminal plasma of *Clarias gariepinus*.

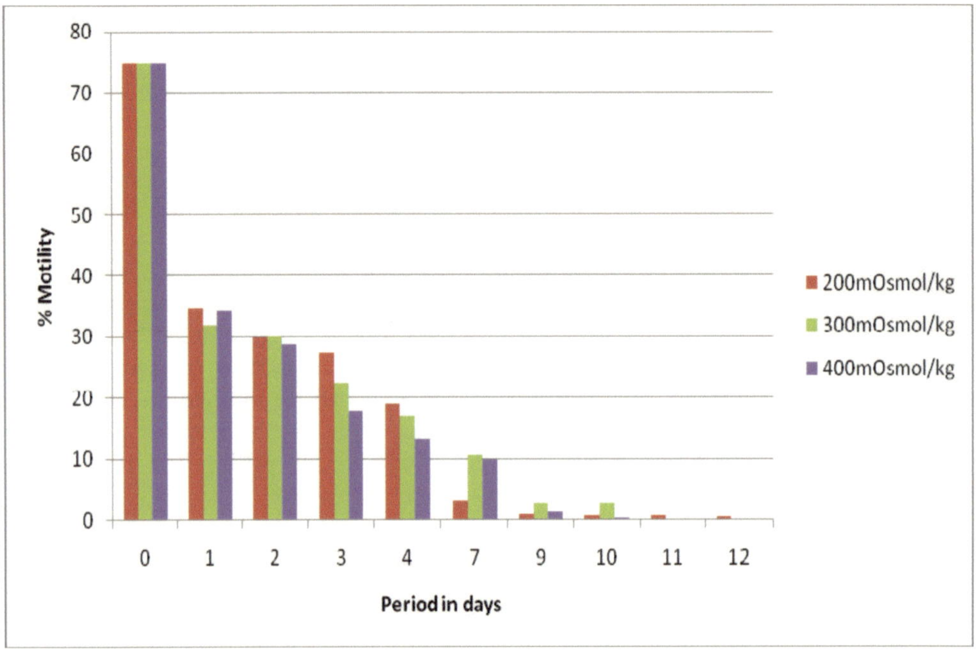

Fig. 15. Effect of osmolality on the sperm motility of refrigerated semen sample in Ca-F HBSS The 200mOsmol/kg Ca-F HBSS extended semen sample retained motility up to day 12 (0.5%)

3.5.3 Effect of extenders on % motility

The refrigerated semen sample with the RPMI culture solution had relatively very high motility at 24 hours after refrigeration; it had a motility of 53.47% as against the 34.85% motility of the semen sample with the 200mOsmol/kg Ca-F HBSS which comes next in rank with it after 24 h. The RPMI extended semen sample also had the highest motility after 48 h of refrigeration (37.62%) with the next in rank being semen sample extended with 300mOsmol/kg Ca-F HBSS with motility of 30.10%. However, the RPMI extended semen sample retained motility only till the fifth day with motility at the fifth day being 5.25% (Fig.16).

The control experiment being semen samples extended with 0.9% NaCl solution retained motility for 7 days with motility at the seventh day being 4.46% and its motility at 24 hours was 34.75% which is exceeded by the semen sample extended with the 200mOsmol/kg Ca-F HBSS which had 34.85% motility at 24 h (Fig. 17).

The relatively high motility retained by the refrigerated semen sample with RPMI culture medium may be due to the additional nutrient supply provided by the RPMI culture medium solution. The RPMI medium culture contains many amino-acids, vitamins and growth factors.

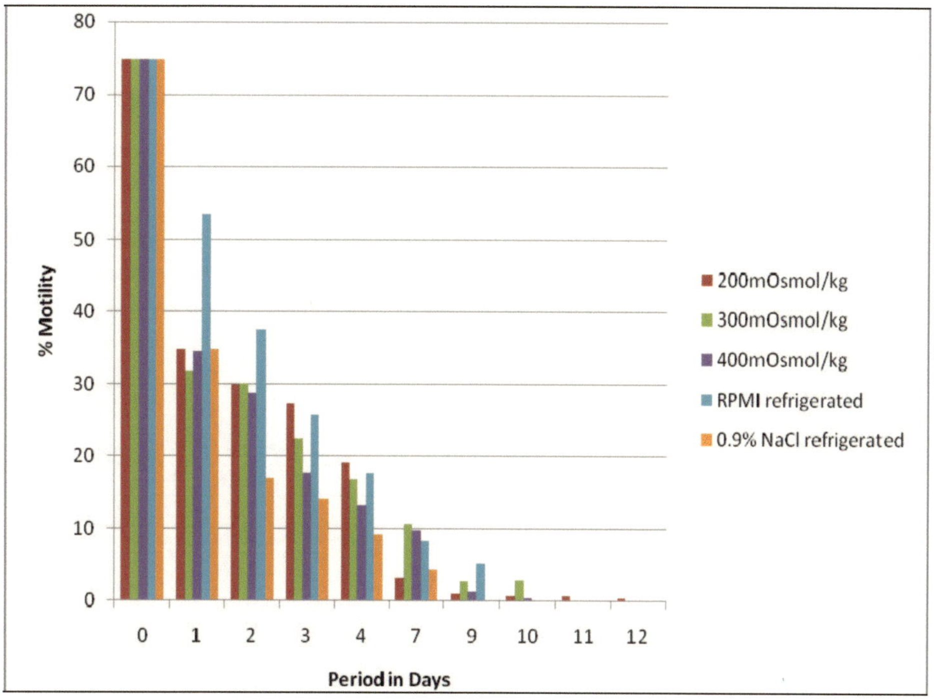

Fig. 16. Effect of the different extenders on the motility of the sperm cells of *Clarias gariepinus*.

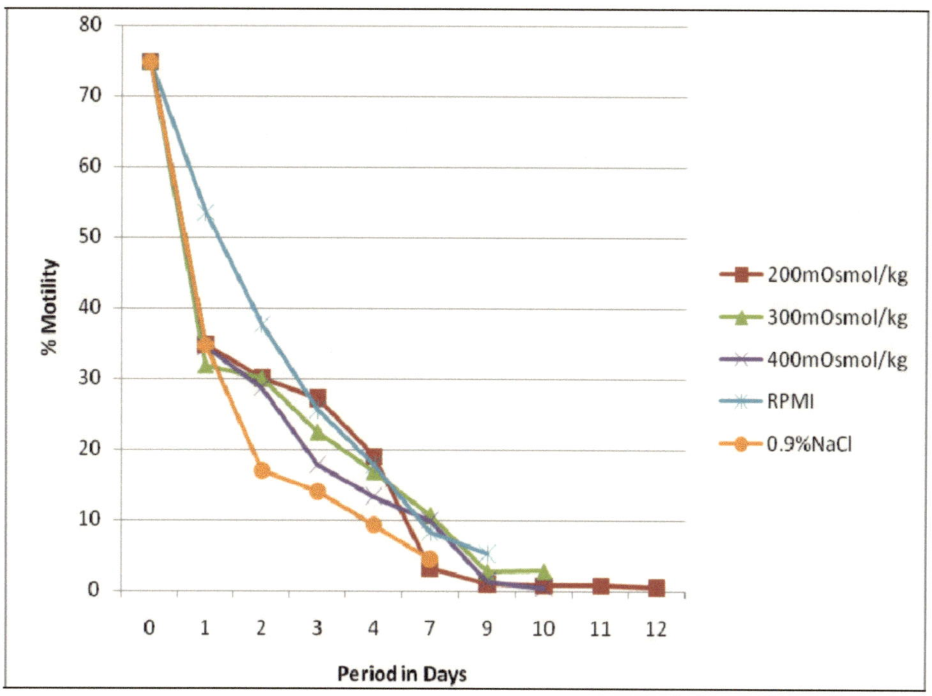

Fig. 17. The decline in motility as the period of storage of refrigerated sperm increased in different extenders.

3.6 Cost of production of cryopreserved catfish sperm for the aquaculture industry

The cost analyses for each reagent and materials used in cryopreservation per ml, g or piece is shown in Table 5. Table 6 shows the cost estimates for the 4-best cryoprotecting agents (DP,

ITEMS	COST (in Naira)	RATE
Liquid Nitrogen LN_2	₦16000/20 l Dewar	₦ 800/l
DMSO	₦ 3700/100/ml	₦ 37/ml
NaCl	₦ 1800/500g	₦ 3.6/g
KCl	₦ 2200/500g	₦ 4.4/g
Na_2HPO_4	₦ 3400/500g	₦ 6.8/g
KH_2PO_4	₦ 2600/500g	₦ 5.2/g
Distilled water	₦100/l	₦ 0.1/ml
$NaHCO_3$	₦ 6028/500g	₦ 12.056/g
$CaCl_2$	₦ 2200/500g	₦ 4.4/g
Glycerol	₦ 2500/2.5l	₦ 1.00/ml
Glucose	₦ 4,100/500g	₦ 8.2/g
Cover Slip	₦175/box (100pcs)	₦ 1.75/slip
Cryovials	₦ 8500/1000	₦8.50/tube
Broodstock (male)	₦ 1200	₦ 1200

Table 5. The cost of reagents and other consumables used to cryopreserve African catfish semen in liquid Nitrogen (LN_2) and listed and the rates per l or ml, g, or per piece are estimated

Ingredients	DP	DGP	GP	GF
Broodstock	1200	1200	1200	1200
Liquid Nitrogen	5000	5000	5000	5000
Glycerol	-	-	1.00	1.00
NaCl	5.76	5.76	5.76	5.76
KCl	0.0176	0.0176	0.0176	0.0176
Na_2HPO_4	1 .564	1 .564	1 .564	-
KH_2PO_4	0.208	0.208	0.208	-
$CaCl_2$	-	-	-	0.2904
$NaHCO_3$	-	-	-	0.5546
Distilled Water	50.00	50.00	50.00	50.00
Glucose	-	4.10	-	-
Cover slip	1.75	1.75	1.75	1.75
Cryovials (80 paces)	680.00	680.00	680.00	680.00
Saline Solution	3.24	3.24	3.24	3.24
Miscellaneous	300.00	300.00	300.00	300.00
TOTAL	7279.5396	7293.6396	7243.5396	7192.1214
Cost/unit of 80 cryovials	7279.5396	7293.6396	7243.5396	7192.1214
Cost per 1ml cryovial	₦ 90.9942	₦ 91, 0455	₦ 90.5442	₦ 89.9015
Proposed Selling price	₦ 100.00	₦ 100.00	₦ 100.00	₦ 100.00
Profit /cryovial	₦ 9.00	₦ 8.95	₦ 9.46	₦ 11.01

Estimation based on 200 ml extender for each cryoprotective agent: DP:DMSO-PBSS, DGP: DMSO-Glucose-PBS; GP: Glycerol-PBS; GF: Glycerol-Fish Ringer .

Table 6. Cost estimates/ml in Naira (₦) for preparation of cryopreserved African catfish semen using different cryoprotectants

DGP, GP and GF. Though, they were not all procured from the same source, all gave almost similar total cost which ranges between ₦7,192. 00 - ₦7,280. 00 and cost/vial ranges between ₦ 89.00 and ₦ 91.00 (1 $ = ₦ 150.00; ₦, naira is the Nigerian currency). Consequently, the proposed economical selling price (in case a cryobank in the University or the collaborating research agency is to be established) considering the present inflation rate is the same for the type of cryoprotectant to be used, but net profit per vial is slightly different.

The selling price for GF is expected to be reduced irrespective of its total cost because of lower viability rate of semen cryopresevеd with it. Profit per vial was also calculated which is highest for GP at ₦100/vial and lowest for GF at ₦95/vial. The selling price ranges from ₦95.00 - ₦100.00 which would encourage the buyers to buy it affordably with reasonable profit for the institute (cryobank).

However, the total cost of GP and GF are lower because of the lower cost price of glycerol, but GF gave the least total cost because of a slight difference in the type and amount of chemical used. Total cost of DGP is higher than that of DP, because of cost of glucose inclusion and higher cost of DMSO.

4. Conclusion

From this study it can be concluded that at further dilution of semen together with different cryoprotecting agents, viability of the semen is still maintained but it varies depending on the type of cryoprotecting agents used. DMSO-dimethylsulphoxide proved to be more efficient than other cryopreservatives in preserving sperm viability. Its potential in cryopreservation can be increased when used in combination with a 5% glucose solution, i.e. DGP and DP proved to be the best even for both trials.

Also, viability of African catfish *C. gariepinus* semen cryopreserved in liquid nitrogen can be maintained for a relatively long time provided ideal protocols are strictly followed. From economic feasibility perspective of cryopreservation of catfish semen, cryopreserved semen is economically feasible and profitable for the cryobank institute or company. The farmers are also assured of the viability of the sperm cells they are buying.

The ability of the African giant catfish (*Clarias gariepinus* Burchell, 1822) semen cryopreserved from 4-8 months with different combinations of extender and cryoprotecting agents, dimethylsulphoxide (DMSO) and glycerol with two extenders: GFR (Ginzburg fish ringer) and PBS (Phosphate buffer saline) to fertilize various egg clutch weights were investigated to evaluate the optimum clutch of egg a milliliter of cryopreserved semen can fertilize. DMSO + glucose with PBS and DMSO+ PBS only proved to be the best cryoprotectant-extender combination in maintaining viability of catfish semen. The optimum viability of the semen was also observed at 4.0-5.0 g of clutch of eggs/ml of semen with little deviation. The first trial was on dilution ratio of 1:1 but in the second trial, the semen was diluted further at a ratio of 1:20 and tested on various egg clutch weights (1, 2, 3, 4, 5 g) to evaluate the viability of cryopreservation even at further dilution. There was a significant effect of different cryoprotecting agents ($p<0.05$) on egg clutch weights. There was a significant effect ($p<0.05$) for hatchability and fertility in the first trial but only fertility in the second trial.

In another experiment we tried higher dilution ratio 1:40 and with bigger clutch of eggs (120g) of a standard female breeder (1.0 kg + 0.2 kg) simulating the practices of many commercial farmers, this time with the cryopreserved sperm. Compared with the control fresh semen that gave the highest motility at appreciably high percentage (71%) which was significantly different ($P<0.05$) from cryopreserved semen diluted at ratios 1:1 (50.52%) and 1:40 (49.05%). However, there was no significant difference between the two diluted cryopreserved semen ($P>0.05$). It was evident that the freezing process and cryopreservation decreased sperm motility after cryopreservation. It could be deduced that cryopreserved sperm still needs to be completely activated after thawing in order to fertilize the whole clutch of eggs since there is a direct relationship between motility and fertility.

In order to assist subsistence fish farmers who may not be able to obtain cryopreserved semen, the motility of sperm cells stored under refrigerated conditions in different extenders was studied. This research evaluated the effect of extenders and period of refrigerated storage on the sperm motility of *Clarias gariepinus* sperm cells with the intent to identify a suitable extender for the refrigerated storage of the sperm cells of *Clarias gariepinus.* Semen samples were collected from mature broodstock and were refrigerated with various different extenders at ratio 1:3 namely: Calcium-free Hanks' Balanced Salt Solution (Ca-F HBSS), RPMI 1640 culture medium and 0.9% NaCl. Ca-F HBSS extender was prepared at 3 different osmolalities: 200mOsmol/kg, 300mOsmol/kg and 400mOsmol/kg. Sperm cells in RPMI 1640 and 0.9% NaCl extenders were also kept at room temperature to assess the effect of refrigeration on motility of catfish sperm cells. Motility was monitored on a 24-hour basis and % motility was evaluated daily. Results showed that sperm cells of *Clarias gariepinus* using 200mOsmol/kg as extender ($p<0.05$) can be stored under refrigeration for 12 days. However, of all the extenders evaluated, RPMI 1640 proved to be the most effective extender ($p<0.05$) retaining higher motility of the refrigerated sperm cells of *Clarias gariepinus.*

The viability of sperm preserved under ordinary refrigerated conditions is possible for a short period of time of 2-7 days depending on the amount of extender used. A culture medium like RPMI 1640 used in this study may give longer life span for sperm cells under refrigerated conditions. Extenders like RPMI 1640 or the cheaper Ca-F HBSS is an alternative that can be recommended for farmers who may have excess of sperm cells from slaughtered male fish for more female gravid eggs that can be sourced within a time period of one week.

The cost of production of a cryotube of sperm (cost of materials, reagents, liquid nitrogen, etc.) was carried out to determine the cost of a milliliter of cryopreserved sperm with a view to selling cryopreserved semen by the research laboratory to farmers who may not be able to afford to buy a male broodstock yielding an affordable cost of ₦100/ml compared to current cost of a male breeder which is ₦ 1000 -1500 each (1 $ = 165 ₦). This also ensures the farmer that the cryopreserved sperm cells they might alternatively buy are viable and will be able to induce the spawning of the female broodstock.

5. Acknowledgements

This research was made possible by research grants from the OAU University Research Committee (11812 AVT), National Centre for Genetic Resources and Biotechnology, Ibadan and the World Bank-Assisted STEP-B (Science and Technology Post-Basic), Abuja. The authors are grateful for the technical assistance of Messrs. Akin Babatunde, Samuel Oladejo, Rotimi Solanke and Wasiu Olaniyi. Our sincere gratitude goes to Prof. B. Solomon of NABDA (National Biotechnology Development Agency) for partially funding the publication charges and Dr. Igor Katkov and Mrs O. Osoniyi for their comments and their expert editing of the manuscript.

6. References

Amupitan, P.S, Ilori, O.I, Aladele, S.E, Odofin, W. T. and Omitogun, O.G. (2010). Scaling up the viability of cryopreserved giant catfish semen in liquid nitrogen for commercial application. In: *34th Annual Conference of the Genetic Society of Nigeria.* NIHORT, Ibadan. Pp 335-343. ISSN 0189-9686

Baynes, S.M., Scott, A. P. and Dawson, A. P. (1981). Rainbow trout spermatozoa: Effect of cations and pH on motility. *Journal of Fish Biology* Vol.19. pp. 259-267.

Bruton, M.N. (1996). Alternative life-history strategies of catfishes. In: Legendre, M. and J.P. Proteau (Eds.).The Biology and culture of Catfishes. *Aquatic Living Resources,* Paris Vol. 9 Hors Series, pp. 35-41.

Coppens International. (2009). www.coppens.eu

Cryobiosystems. (2006). Fundamentals of cryobiology- http://www.Cryobiosystem.imv. com/CBS/Cryobiology/cons.cbs-asp.

Hetch,T., Oellermann, L. and Verheust, L.. (1996). Perspectives and clariid catfish culture in Africa. In: Legendre, M. and Proteau, J-P. (eds.).The Biology and Culture of Catfishes. *Aquatic Living Resources. Paris,* 1996, Vol. 9, Hors Series pp. 197-206.

Hogendoorn H. (1979). Controlled propagation of the African catfish, *Clarias lazera (C & V).*I. Reproductive biology and field experiments. *Aquaculture* Vol. 17 pp. 323-333.

Legendre, M. (1986). Seasonal changes in sexual maturity and fecundity, and HCG-induced breeding of the catfish *Heterobranchus longifilis* Val. (Clariidae) reared in Ebrien lagoon (Ivory Coast). *Aquaculture* Vol. 55 pp. 201-213.

Legendre, M. and Oteme, Z.I. (1995). Effect of varying latency period on the quality and quantity of ova after HCG-induced ovulation in the African catfish, *Heterobranchus longifilis* (Teleostei: Clariidae). *Aquatic Living Resources.* Paris Vol. 8 pp. 309-316.

Lubzens E., Daube, N., Pekarsky, I.., Magnus, Y., Cohen, A., Yusefovich, F. and Feigin, P. (1997). Carp (*Cyprinus carpio* L.). Spermatozoa cryobanks strategies in research and application. *Aquaculture* Vol. 155 pp. 173-178.

Mansour, N, Lahnsteiner, F. and Patzner, R. A. (2002). The spermatozoan of the African Catfish: fine structure, motility, viability and its behaviour in seminal vesicle secretion. *Journal of Fish Biology* Vol. 60 No. 3. pp. 545-560.

Maria A. N., Viveiros, A. T. M., Orfao L. H., Oliveira, A. V. and Morales, G. F. (2006). Effects of cooling and freezing on sperm motility of the endangered fish Piracanjuba *Brycon orbignyanus* (Characiformes, Characidae) *Animal Reproduction* .Vol. 3 No. 1 pp. 55-60.

Morisawa, M. and Suzuki, K. (1980). Osmolality and potassium ions: their roles in initiation of sperm motility in teleosts. *Science* Vol. 210 pp.1145-1146.

Olaleye, V.F. (2005). A review of reproduction and gamete management in the African catfish, *Clarias gariepinus* (Burchell 1822). *Ife Journal of Science* Vol. 7 No.1 pp. 63-70.

Omitogun, O. G., Oyeleye, O.O., Betiku, C.O. Ojiokpota, C., Aladele, S.E. and Sarumi, M.B. (2006). Potentials of short-term cryopreserved sperm of the giant African catfish, *Clarias gariepinus*(Burchell, 1822) for aquaculture in Nigeria. In: Olakojo, S.A., Ogunbodede, B.A. and Akande, S.R (Eds) Proceedings of the 31st*Annual Conference of the Genetic Society of Nigeria* pp. 141-146. NACGRAB, Moor Plantation, Ibadan, Nigeria. Nov. 6-9, 2006. ISSN 0189-9686.

Omitogun, O.G, Olaniyan, O. F., Oyeleye, O.O., Ojiokpota, C. Aladele, S.E. and Odofin, W. T.(2010). Potentials of short term and long term cryopreserved sperm of African giant catfish *(Clarias gariepinus* Burchell 1822) for aquaculture. *African Journal of Biotechnology* Vol. 9 No. 41. pp. 6973-6982, ISSN 1684–5315. Academic Journals.

Oteme, J., Nunez-Rodriguez, J., Kouassi, C.K. Hom, S. and Agnese, J.F. (1996). Testicular structure, spermatogenesis and sperm cryopreservation in the African clariid catfish, *Heterobranchus longifilis* (Valenciennes, 1840). *Aquaculture Research* Vol. 27 pp. 805-813.

Oyeleye, O. O and Omitogun, O. G. (2007). Evaluation of motility of the short-term cryopreserved sperm of African giant catfish (*Clarias gariepinus*). *Ife Journal of Agriculture* Vol. 22 .No.(1) pp.11-16. Ile-Ife, Nigeria. ISSN 0331-6351.

Padhi, B.K. and Mandal, R.K. (1995). Cryopreservation of spermatozoa of two Asian freshwater catfish, *Heteropneutes fossilis* and *Clarias batrachus*. *Journal of Aquaculture in the Tropics* Vol. 10 pp.23-28.

Riley, K. L. P. (2002). Refrigerated storage and cryopreservation of sperm for the production of Red snapper and Snapper hybrids. *M. Sc Thesis*, Agriculture and Mechanical College, Louisiana University. USA.

SIGMA (1994). Cell counting and cell viability. Saint Louis, MO 63178 USA. Pp. 1634-1635.

Steyn, G.J, Van Vuren, J. H., Schoonbee, H.J. and Chao, N. (1985). Preliminary investigations on the cryopreservation of *Clarias gariepinus* (Clariidae,:Pisces) sperm. *Water SA*. Vol. 11 No.1. pp.15-18.

Steyn, G.J. (1987). The fertilizing capacity of cyopreserved sharptooth catfish, *Clarias gariepinus* sperm. *Aquaculture* Vol. 6 pp. 187-193.

Suquet, M., Dreanno C., Fauvel, C., Cosson, J. and Billard, R. (2000) Cryopreservation of sperm in marine fish. *Aquaculture Research* Vol. 31 pp. 231-243.

Sydenham, D.H.J. (1980). New species of *Clarias* from West Africa. Rev. Zool. *Afr., Vol .XCIV, Fasc.,* Vol. 31 pp. 659-677.

Teugels, G.G. (1986a). Clariidae. In: Daget, J., Gosse, J-P., Thys van den Audenaerde,D.F.E. (editors). Check-list of the Freshwater fishes of Africa, Brussels, MRAC, Tervuren, *ORSTOM,* Paris, pp.66-101.

Teugels, G.G. (1986b). A systematic revision of the African species of the genus Clarias (Pisces:Clariidae). *Annals Musee. Revue. Africain. Contributions.* Vol. 247 pp. 1-199.

Tiersch, T. R., Goudie, C.A. and Carmichael, G.J. (1994). Cryopreservation of channel catfish sperm. Storage in cryoprotectants, fertilization trial and growth of channel catfish produced with cryopreserved sperm. *Transactions of the American Fisheries Society* Vol. 123 pp. 580-586.

Urbanyi B., Horvath, A., Varga, Z. and Horvath, L. (1999). Effect of extenders on sperm cryopreservation of African catfish, *Clarias gariepinus* (Burchell). Aquaculture Research. Vol.30 pp.145-151.

Van der Walt, L.D., Van der Bank, F.H. and Steyn, G.J. (1993). The suitability of using cryopreservation of spermatozoa for the conservation of genetic diversity in African catfish (*Clarias gariepinus*). *Comparative Biochemistry and Physiology.* Vol. 106A pp.313-318.

Viveiros, A.T., So, N. and Komen, J. (2000). Sperm cryopreservation of African catfish, *Clarias gariepinus;* Cryoprotectants, freezing rates and sperm: egg dilution ratio. *Theriogenology* Vol. 54 pp. 1395-1408.

Part 5

Cryopreservation of Plants

17

Comparison of Cryopreservation Methods of Vegetatively Propagated Crops Based on Thermal Analysis

Jiří Zámečník, Miloš Faltus, Alois Bilavčík and Renata Kotková
Crop Research Institute
The Czech Republic

1. Introduction

There is a trend to preserve the plant germplasm by not only conventional *ex situ* methods or *in vitro* techniques, but also, more recently, by cryopreservation. Cryopreservation techniques are based on the storage of plant samples at very low temperature at which practically no chemical reactions occur and consequently, neither aging nor genetic changes of plant material. There has been a great development progress of cryopreservation methods during last years. Cryopreservation becomes a highly utilized technique for germplasm conservation. Generally the cryopreservation is storage of the samples. The samples can be e.g. organs and shoots tips from *in vitro* culture, or from the field, such as mature, immature bulbils, cloves of garlic or dormant buds of fruit trees, at the ultra-low temperature (mainly -196 °C, the temperature of liquid nitrogen).

Although the technique was introduced for plants in the '70s, it has never been applied on a wide scale due to the high cost of cryo-freezers; indeed, it was used in order to escape the formation of lethal intracellular ice crystals, time-consuming and laborious slow-cooling procedures. A new cryogenic - vitrification technique is now available, aiming at the direct immersion of plant specimens from tissue cultures in liquid nitrogen, without resorting to an expensive apparatus for slow cooling and with a considerable simplification of the procedures (Benson, 2008). The vitrification method simplifies cryogenic process and makes possible an increased application of cryopreservation on wide-range plant genetic resources. The glassy state is the objective status of cryopreservation methods named vitrification.

The aim of this study is a comparison of different cryopreservation methods based on the vitrification achieved by dehydration and glass transition temperature (T_g), and their efficiency towards optimal regeneration of vegetatively propagated plants. The thermal characteristics, evaluation of frozen water content, and the glass transition temperature were measured by a differential scanning calorimeter.

2. Importance of cryopreservation of vegetatively propagated plants

Some of vegetatively propagated plants are not able to reproduce by seeds e.g. garlic plant (*Allium sativum* L.). The only way how to propagate it is to use its cloves or bulbils for seeding plants for further growing. The vegetatively propagated plant germplasm is endangered by abiotic and biotic factors in the field conditions. Although the production area of many vegetatively propagated plants has been decreasing, many local cultivars and varieties remain. In the presence of decreasing cultivar variability in production areas, diminishing of old orchards, as well as appearance of diseases close to field collection areas, the question of safely maintaining the broad genetic potential of fruit trees is arising.

Two safe methods ensure vegetatively propagated plant germplasm maintenance with a low risk of loss: slow-growth *in vitro* culture and the cryopreservation methods. Advantages of *in vitro* collection are aseptic and stable conditions of the cultivation and availability of the material during the year. A disadvantage is the necessity of sequential plant multiplication. Advantages of cryo-collection are low costs for its long-term maintenance and material stability. Disadvantages are a longer time for the plant to recover from stored material and a rather high input costs of the cryopreservation procedure. The best way how to maintain germplasm is the combination of both methods. The base collection should be maintained by *in vitro* collection that provides the material in case of requirements. Core collection of the most valuable material, should be backed-up by cryo-collection for long-term storage, and plants are recovered just in case the genotype is lost from the base collection. For that reason, important vegetatively propagated plant collections have started to introduce accessions to slow-growth *in vitro* cultures and simultaneously in cryo-collection in liquid nitrogen (Gonzalez-Arnao *et al.*, 2008; Keller *et al.*, 2008; Kim *et al.*, 2006).

3. Cryopreservation methods

The latest results from the field of low temperature biology suggest that the main factor influencing the success of the cryopreservation method is the maintenance of a glassy state in plant samples and the avoidance of ice nucleation. The danger of ice nucleation and subsequent ice crystallization leading to frost damage during cooling and rewarming of samples is considered as a critical point of cryopreservation. That is the reason why many of the new progressive methods use and involve a glassy state in plant material intended for cryopreservation. Knowledge of the glass transition temperature is useful not only for improving methods involving glassy state in plant shoots tips. It also provides information essential for the long-term storage of shoot tips.

This biotechnology is based on the induction of the vitrification status – glass induction by dehydration, addition of cryoprotectants and a very fast decrease in temperature. Vitrification can be achieved in a number of ways (Sakai & Engelmann, 2007) but they usually all have the results of increased solute concentration to a critical viscosity. Low water content minimizes the ice crystallization that is potentially dangerous for plant cells and increases the temperature of glass transition. Supposing that the change of water status in the certain range is not limiting for plant regeneration. Plant Vitrification Solutions (PVS) marked with numbers according to the specific mixture of basic cryoprotectants and their concentrations are usually used for osmotic dehydration. Another cryopreservation method used, is based on desiccation in the air-flow cabinet. It is defined with the flow rate, temperature and humidity or on desiccation over various saturated salt solutions with steady-state activity of water.

3.1 Cryoprotectants involved in vitrification method

The cryopreservation method using a vitrification solution was first described by (Luyet, 1937). The vitrification solutions were firstly named, according to the first author of the publication and later the vitrification solutions have abbreviated names from Plant Vitrification Solution (PVS) with a number according to the time of their first appearance in the literature. The main ones are Luyet (1937), Fahy (1985), Steponkus (Langis & Steponkus, 1990), PVS1 (Uragami *et al.*, 1989; Towill, 1990), PVS2 (Suzuki *et al.*, 2008), PVS3, PVS4, PVS5 (Nishizawa *et al.*, 1993), VS6 (Liu *et al.*, 2004a), PVSL (Liu *et al.*, 2004b) VSL (Suzuki *et al.*, 2008), with different concentration and combination of the main four components: dimethylsulfoxide, sucrose, glycerol and ethylene glycol. The increased efficiency of vitrification methods was achieved by treating plants in the pre-cultivation step before cryopreservation of plant shoot tips in so called Loading Solution (LS) (Dumet *et al.*, 2002; Matsumoto & Sakai, 1995; Sakai *et al.*, 1991; Sakai & Engelmann, 2007). The cryoprotective substances should fulfil several basic parameters, such as cell permeability, viscosity, toxicity and the minimum concentration necessary for the vitrification, which eliminates the formation of ice crystals.

Cryoprotective substances help to ensure the stability of membranes and enzymes in the subsequent dehydration by vitrification solutions and to avoid the formation of ice crystals (Kartha & Leung, 1979; Kim *et al.*, 2006). The samples are exposed to a several hour-long treatment by some cryoprotective substances, and then they are plunge-frozen in liquid nitrogen. The effect of cryoprotective solution composition for plant regeneration was studied in different plant species (Ellis *et al.*, 2006; Kim *et al.*, 2004; Kim *et al.*, 2009; Tanaka *et al.*, 2004).

In the most recent approaches to the garlic cryopreservation, vitrification method can be induced by treating the shoot tips of plantlets with a highly concentrated a mixture of glycerol and sucrose. (Nishizawa *et al.*, 1993) developed Plant Vitrification Solution 3 (PVS3) with 50% glycerol (w/v) and 50% sucrose (w/v) in water. It is noteworthy that, following these procedures, the plant specimens can be directly plunged into liquid nitrogen, where they can be stored for an indefinite period of time without undergoing the risks of contamination or genetic alterations.

3.2 Methods based on dehydration

Potato (*Solanum tuberosum* L.) is a plant species sensitive to frost temperatures. Cryoprotocol for potato has to solve the problem of how to overcome temperature between 0 °C and -130 °C during cooling and warming without ice crystal growth and cell damage. Cold acclimation is not appropriate as pre-cultivation for potato plant (Hirai & Sakai, 1999; Schafer-Menuhr *et al.*, 1996; Kaczmarczyk, 2008). The only method for potato vitrification is a water content decrease in samples, and than the rapid cooling and warming rate. Water content decrease is achieved by preculturing explants with osmotic compounds, air desiccation or vitrification. On bases, vitrification (Sarkar & Naik, 1998), droplet (Schafer-Menuhr *et al.*, 1996) and recently vitrification-droplet (Halmagyi *et al.*, 2004; Schafer-Menuhr *et al.*, 1996) methods were developed or adapted for potato.

3.3 Encapsulation-dehydration

One of the other cryopreservation methods is encapsulation-dehydration. The shoot tips were encapsulated in an alginate gel. Experiments with dynamic dehydration studies demonstrated the necessity of meristems encapsulation (Benson *et al.*, 1996; Grospietsch *et al.*, 1999; Hirai & Sakai, 1999). The encapsulation of shoot tips prolongs the dehydration period up to seven hours at a low relative humidity. The alginate beads without shoot tips had approximately the same dehydration-time curve. On the contrary no encapsulated shoot tips were completely dehydrated up to 1hour. The static dehydration of shoot tips was done over the various saturated salt solutions.

4. Cryoprotocols

4.1 Garlic

The unripe topsets were surface sterilized by chloramines and 75% ethanol and from this point, all preparations were performed under sterile conditions using sterile instruments and culture media in a laminar flow box. Opening the surface sterilized topset in sterile condition, all the inside structures were sterile (Fig. 1a). The sterile unripe bulblets were removed and cuts were made to the clusters or clumps of 3-8 bulbils. The bulbils varied in the thickness (approximately 2 mm) and in the length (3-5 mm) depending on the genotype and stage of ripening. Inside the unripe bulbils shoot tips were (Fig. 2a) with meristematic tissue.

Pre-culture of unripe clusters of bulbils was done on the MS culture medium (Murashige & Skoog, 1962) with 0,2 mg L^{-1} BAP and 0,02 NAA mg L^{-1} with 10 % sucrose for 20-24 h at 22 °C and 16 h light in Petri dishes sealed with Parafilm.

Fig. 1. Plants grown in *in vitro* conditions ready for dissection of shoot tips: (a) Garlic. Scale bar, 5 mm; (b) Potato. Scale bar, 1 mm; (c) Hop. Scale bar, 5 mm and (d) Apple tree. Scale bar, 10 mm.

Fig. 2. The size and shape of shoot tips used for cryopreservation. (a) Garlic. Scale bar, 1 mm; (b) Potato. Scale bar, 0,25mm; (c) Hop. Scale bar, 0,25 mm and (d) Apple tree. Scale bar, 1 mm.

Cryoprotocol Steps	The Procedure
Unripe bulbils dehydration	
	Immersion in the loading solution (13,7 % (w/v) sucrose + 18,4 % (w/v) glycerol in the liquid medium) (Sakai *et al.*, 1991) for 20 minutes
Cryopreservation	
	Dehydration by PVS3 (Nishizawa *et al.*, 1993) at a laboratory temperature for 2 hours
	Removing and adding fresh PVS3 before freezing
	Aluminum foil stripes with 5-10 clusters of unripe bulbils plunged directly into liquid nitrogen at least for one hour in liquid nitrogen (Sarkar & Naik, 1998)
Thawing	
	Rapid warming immersion into a 40 °C water bath for 30-120 seconds for thawing
Survival and regeneration evaluation	
	Sub-culture on MS medium supplemented with 0,2 mg L^{-1}BAP and 0,02 NAA mg L^{-1} with 3 % sucrose for seven days in the dark
	Evaluation of survival after two weeks (Fig. 12a.)
	Evaluation of regeneration after 8-10 weeks

Table 1. Cryopreservation steps of garlic

4.2 Potato

Potato explants (Fig. 1b.) were multiplied by nodal cuttings in plastic boxes (Vitro Vent container, Duchefa) on 100 ml modified MS medium (Grospietsch *et al.*, 1999) with 7 g L^{-1} agar and 30 g L^{-1} sucrose, without myo-inositol and phytohormones, with a decreased amount of nitrogen at pH 5,5. Nodal cuttings were cultivated at 22 ± 1 °C, 80 µmol m^{-2} s^{-1} and photoperiod 16/8 h light/dark (L/D) (Fig. 1b). Subculture interval was 3-4 weeks.

Nodal cuttings were planted in the same conditions as the pre-cultured plants but only 50 ml medium was used per one box. After 4 day pre-culture, lateral buds elongated to at least 1 mm (Fig. 2b). Subsequently, 25 ml 2 M sucrose was added into each container and explants were cultivated at the same conditions for the next 5-6 days.

Cryoprotocol Steps	The Procedure
Shoot tips dehydration	
	Shoot tips (Fig. 2b.) (1-2 mm) were transferred onto a filter paper moistened with 14 ml 0,7 M sucrose and phytohormones of the same composition as in a recovery medium (0,5 mg L^{-1} IAA + 0,5 mg L^{-1} Kinetin + 0.2 mg L^{-1} GA3) at 22 ± 1 °C and photoperiod 16/8 h (L/D) and shielded with a leaf of paper for overnight. The second day the shoot tips were transferred (20 tips per foil) onto aluminum foils (20 x 6 x 0.05 mm) Dehydration above silicagel for 1,75 – 2 h.
Cryopreservation	
	Plunging aluminum foils into liquid nitrogen Stored in cryovials (two foils with 20 shoot tips per vial).
Thawing	
	Alluminum foils plunged rapidly into the water bath at a laboratory temperature. Transfer immediately onto the recovery medium (Grospietsch *et al.*, 1999) with the same composition as the pre-culture medium but with phytohormones (0,5 mg L^{-1} IAA + 0,.5 mg L^{-1} Kinetin + 0,2 mg L^{-1} GA3)
Survival and regeneration evaluation	
	Survival was defined by shoot tip growth and by green color and recovery as a new explant development (Fig. 12b.) Plant regeneration was evaluated 2 and 8 weeks after cryopreservation.

Table 2. Cryopreservation steps of potato

4.3 Hop

Maternal plants were cultivated on a multiplication medium without phytohormones at 22 ± 1 °C, 80 µmol m^{-2} s^{-1}, photoperiod 16/8 h (L/D); subculture interval was 8 weeks. Modified solid medium (Murashige & Skoog, 1962) without casein and myoinositol, with decreased amount of nitrogen (25 % (w/v) of NH_4NO_3 and 50 % (w/v) of KNO_3 of the original Murashige and Skoog medium), with 40 g L^{-1} glucose, pH 5,5 without phytohormones was used as the multiplication medium (MSH).

Nodal cuttings were planted in the same conditions as during explant multiplication but only 50 ml medium was used per one box (Fig 1c). After 7-10 d pre-culture, lateral buds elongated to 1-2 mm. Then the explants were transferred into cold acclimation conditions at 4 °C for 7-10 days. Subsequently 25 ml 0,7 M sucrose was added into each container and explants (Fig. 1c.) were cultivated at the same conditions for the next 7-10 days.

Cryopreservation steps for hop shoot tips were the same as for potato (see Tab. 2).

4.4 Apple tree

In vitro plants (Fig. 1d) were cultivated in 100 ml Ehrlenmeyer flasks with 20 ml of MS medium, 3% (w/v) sucrose, 6 g L^{-1} agar, supplemented with GA3 1 mg L^{-1}, BAP 1 mg L^{-1}, IBA 1 mg L^{-1}, at 20 ± 1 °C, 8/16 (L/D) photoperiod of light intensity 100 μmol m^{-2} s^{-1}.

Cryoprotocol Steps	The Procedure
Shoot tips dehydration	
	One to two week subcultivation on fresh MS medium,
	Cold hardening for 4-6 weeks at +4± 1 °C, short photoperiod (8/16, L/D) of light intensity 25 μmol m^{-2} s^{-1}
	Petri dish with 16 ml of MS medium and poured with 16 ml of 2M sucrose solution for 48 hours
	Encapsulation of shoot tips (Fig. 2d.) with meristematic cells in 3% w/v dipped into alginate in 0,75M sucrose for 10 min and then drop into 0,1 M CaCl2 in 0,6 M sucrose for 10 min to net the alginate and form a bead
Cryopreservation	
	Beads are gently dried on a sterile filter paper
	Additional dehydration in the laminar flow box at laboratory temperature and dehydrated for different time up to 4 -5 hours, Placed in 2 ml cryovials and plunged in liquid nitrogen.
Thawing	
	Shoot tips with beads were warmed up by plunging cryovials in 40 °C water
Survival and regeneration evaluation	
	Survival was defined by shoot tip growth and by green color and regenerated as a new explant development (Fig. 12d.)

Table 3. Cryopreservation steps of apple tree

5. Thermal analysis

Differential scanning calorimetry (DSC) belongs to thermal methods that can be used for measurement and determination of phase and glass transitions for cryopreservation. In principle, the DSC measures the temperatures and heat flows associated with transitions in plant material as a function of time and temperature. It gives information about endothermic or exothermic changes or changes in heat capacity. The obtained data can be used for determination of glass transition, temperature of ice nucleation, melting, boiling, crystallization time and kinetic reaction - the most important characteristics useful for cryopreservation (Zámečník & Faltus, 2009). The danger of nucleation and subsequent intracellular ice crystallization leading to frost damage during cooling and rewarming of the samples is considered a critical point of plant survival at ultra-low temperatures.

The differential scanning calorimetry method is based on the regulated decrease/increase temperature of the sample and reference and the measurement of temperature and heat flow corresponding to the sample. There are two different types of the differential scanning calorimeters. The power compensation DSC type directly measures heat release/uptake from the sample and the heat flow type measures differences of temperature between reference and sample and recalculates the differential heat flux. The most common cooling/heating rate of the sample is 10 °C min^{-1}.

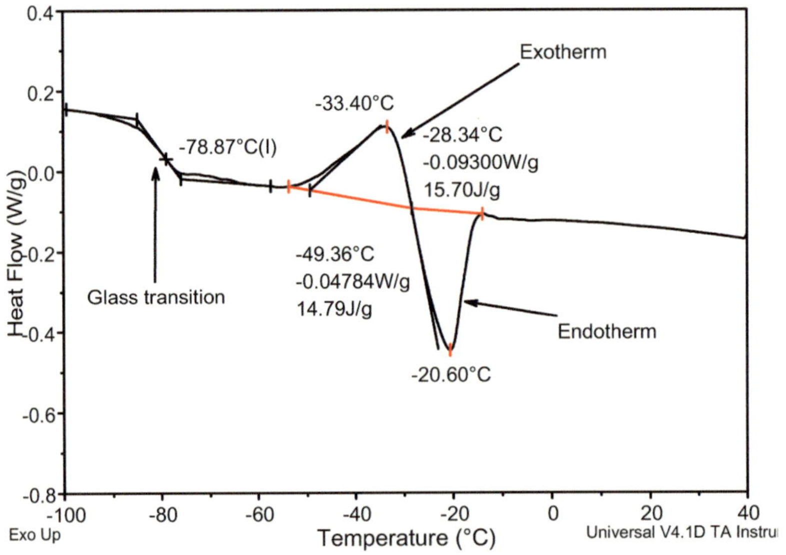

Fig. 3. Example of apple tree shoot tips heat flow response to the temperature. Cooling and warming rate was 10 °C min^{-1}. Exo up.

In our experiments we used shoot tips of *in vitro* cultures of apple tree. Samples with different water content were obtained by air dehydration of alginate encapsulated shoot tips in the flow box or by dehydration at 4 °C of *in vitro* cultures. For DSC measurement dissected shoot tips were placed in aluminum sample pans and measured by Differential Scanning Calorimeter TA2920. Samples were cooled down to -120 °C (rate of 10 °C min^{-1}). The data were collected during heating to 20 °C (rate of 10 °C min^{-1}). The purge gas was either nitrogen or helium.

A Differential Scanning Calorimeter is used as a main tool in cryobiology to assist cryopreservation protocol development, to store thermograms as a documentation of cryo-protocols in the use and to keep information about stored samples and their thermal properties before, during and after cryopreservation (Benson *et al.*, 1996; Faltus & Zámečník, 2009; Šesták & Zámečník, 2007; Zámečník *et al.*, 2007)

There is an example (Fig. 3) of measured thermal characteristics of shoot tips of apple tree *in vitro* culture cv. Greensleeves by the DSC. Samples of an approximate weight of 10 mg were crimped in an aluminium sample pan and cooled from room temperature to -120 °C. Cooling and heating rate was 10 °C min^{-1}. The glass transition, exothermic and endothermic characteristics were analyzed in detail during heating. Thermal characteristics were measured by DSC TA 2920 (TA Instruments) and evaluated by Universal Analysis 2000 for Windows (TA Instruments).

6. Water content and glass transition

6.1 Garlic

Water content during dehydration of garlic cv. Djambul 2 clusters of shoot tips by PVS3 (Fig. 4). Total amount of water (solid line) and amount of crystallized water (dash line) in shoot tips treated with PVS3 rapidly descend during the first 1,5 hours and further is constant. Crystallized water reaches minimum after 1,75 hours of PVS3 treatment. In this case the decrease of water content in the unripe bulbils is probably so low that it can have no further influence on the glass transition change. In comparison with the measurements in this study on the apple tree shoot tips (see below), the glass transition temperature increases with decreases of water after dehydration.

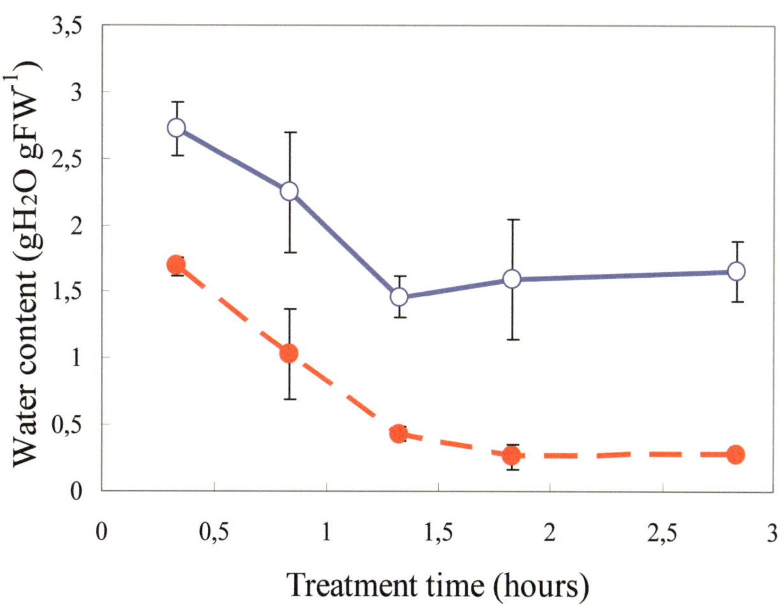

Fig. 4. Unripe garlic bulbils water content (empty circle) and the part of frozen water (full circles) during PVS3 treatment. Note: The unripe bulbils were in the loading solution first 20 minutes than they were immersed in to the PVS3. The bars are standard deviation of mean.

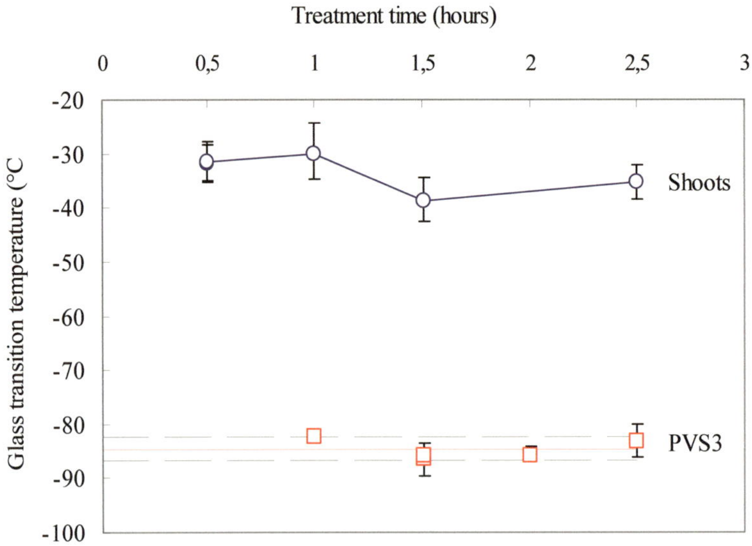

Fig. 5. Glass transition temperature of *Allium* shoot tips after moisture loss by dehydration in the Plant Vitrification Solution 3 (PVS3). Circles show the glass transition midpoint in shoot tips and bars show the onset and endset of glass transition. Squares show the glass transition of PVS3 in shoots. The full line is for the glass transition of PVS3 without shoots. The dashed line above is the endset of glass transition and below the onset of glass transition for PVS3.

Glass transition of garlic shoot tips was measured after different times of treatment - unripe bulbils in PVS3 at 23 °C (Fig. 5). At each curve, there were two S-shape heat flow changes during warming of the samples, typical for glass transition. The lower glass transition temperature on unripe bulbils heat flow curves coincides within the range of onset and endset of the glass transition temperature of PVS3 measured after unloading unripe bulbils. This glass transition temperature can be of PVS3 coating on the surface of the shoot tips immersed in PVS3.

The high glass transition temperature corresponds to glass transition of the shoot tips because at this range of temperature there were no thermal events on the PVS3 temperature dependent curve. There is no significant difference in the change of shoot tip glass transition changes from 0,5 to 2,5 hours of PVS3 treatment. The detectable glass transition was found between -30 °C and -39 °C. The average glass transition temperature is -33,5 °C after 0,5 hour. From these results it is obvious that the glass transition at higher temperature is for shoot tips saturated with PVS3. So, for the survival of shoot tips after thawing from liquid nitrogen, the second glass transition which occurs at higher temperatures is important (Zamecnik *et al.*, 2011).

6.2 Potato

Nodal cuttings were pre-cultured on medium with added sucrose solution. The final sucrose concentration in medium was 0,7 M. The importance of sucrose pre-treatment before potato cryopreservation proved by Grospietsch *et al.*, (1999) and Halmagyi *et al.*,

(2004). Halmagyi *et al.*, (2004) showed the highest plant regeneration after cryopreservation following a pre-treatment with 0,5 M sucrose. Similarly, Sarkar and Naik (1998) found a slightly negative effect of 0,7 M sucrose pre-treatment compared in comparison with 0,5 M or 0,3 M sucrose pre-treatment. In the present study the injury of potato explants was not observed after 0,75 M sucrose treatment.

Total water content in the shoot tips after nodal cutting pre-culture was approximately 5 g of H_2O per 1 g of dry mass (gH_2O g DW^{-1}) (Fig. 6.). Frozen water content in shoot tips was 4,3 gH_2O g DW^{-1} and the unfrozen 0,7 gH_2O g DW^{-1}. Subsequently shoot tips were isolated and loaded with 0,7 M sucrose in a Petri dish on filter paper for overnight. Total water content of shoot tips decreased to 2,1 gH_2O g DW^{-1}, from which 1,4 gH_2O g DW^{-1} represents the frozen water fraction and 0,7 gH_2O g DW^{-1} the unfrozen water fraction. Because the total water content and frozen water fraction decreased but the unfrozen fraction did not change, the ratio of frozen/unfrozen water content (WC_f/WC_u) decreased from 6 to 1,9. The following air dehydration resulted in a decrease of total water content due to both water fractions decrease. After 1,5h air dehydration above silicagel the total water content in shoot tips was 0,49 gH_2O g DW^{-1} from which the frozen water content was 0,09 gH_2O g DW^{-1}, and the unfrozen 0,4 gH_2O g DW^{-1}. Resulting WCf/WCu ratio decreased to 0,22. The prolonged dehydration decreased both water fractions. After 2h air dehydration above silicagel the total water content in shoot tips was 0,28 gH_2O g DW^{-1} from which 0,006 gH_2O g DW^{-1} belonged to the frozen fraction and 0,276 gH_2O g DW^{-1} to the unfrozen fraction. The WC_f/WC_u ratio decreased to 0,12 after 2h air dehydration of shoot tips above silicagel, which represents 2 % crystallized water of the total water content.

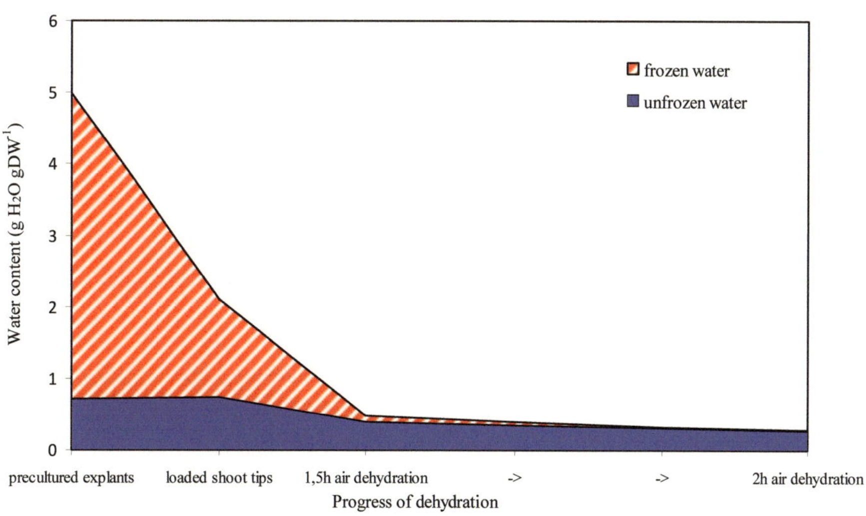

Fig. 6. The progress of dehydration of potato explants (cv. Désirée) after specific steps of cryoprotocol. Explants were pre-cultured on medium with 0,7 M sucrose. The isolated shoot tips were loaded with 0,7 M sucrose solution for overnight. The loaded shoot tips were dehydrated by dry air above silicagel for 2 hours. The amount of frozen and unfrozen water was determined by the DSC analysis.

The decrease in percentage of crystallized water in shoot tips during 1,75 to 2h air dehydration is illustrated in Fig. 6. The crystallized water content decreased from approximately 9% to 2%. Dehydration of shoot tips was connected to the glass transition temperature increase from -38 to -32 °C. The optimal water content of potato shoot tips was approximately 0,4 gH_2O g DW^{-1} that was obtained between 1,5h and 2h air dehydration above silicagel according to the size of particular genotype shoot tips. The temperature of glass transition was approximately -35 °C and the amount of frozen water was very small but still detectable (Fig. 7). Decrease in water content and onset of melting temperature was also found after dehydration by PVS2 solution or 10 % DMSO (Kaczmarczyk, 2008, Kaczmarczyk *et al.* 2011). However the T_g found by these cryoprotectants was lower than -100 °C. The higher temperature of glass transition found in this study indicated a higher stability of material stored at ultra-low temperatures.

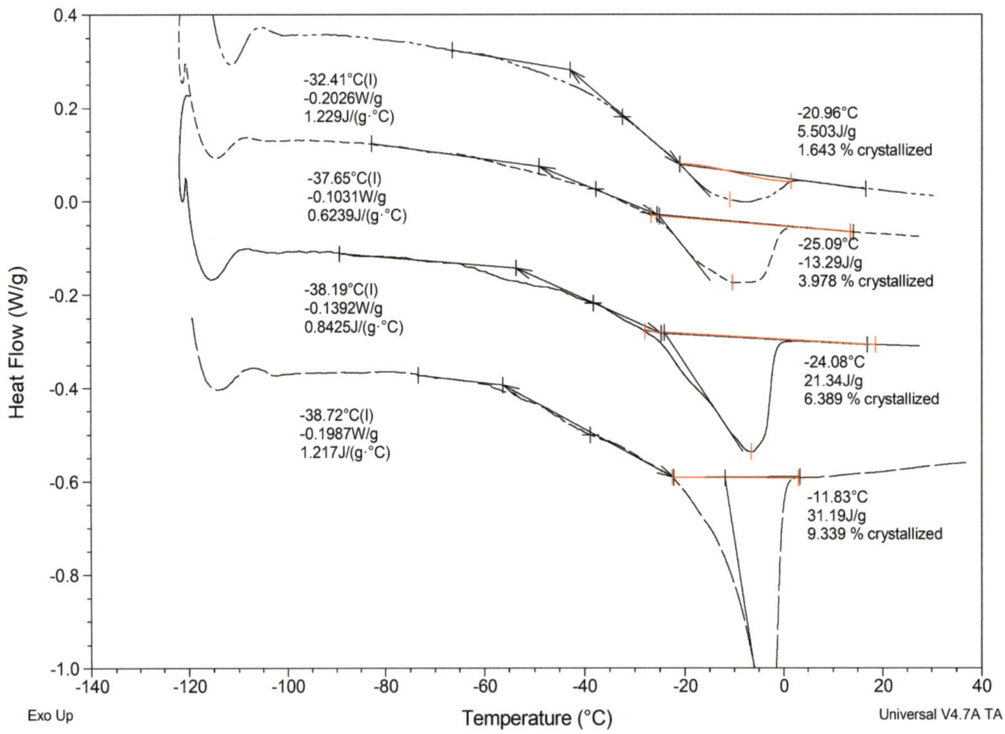

Fig. 7. DSC curves of air dehydrated potato shoot tips (cv. Désirée) air-dehydrated above silicagel for 1,75 to 2 hours. Heat flow was evaluated during warming the samples from -130 to 30 °C by ramp temperature 10 °C min^{-1}. Glass transitions were defined by the temperature of glass transition, change of heat flow per g of sample and change in specific heat capacity (C_p). Melting exotherms are defined by the onset temperature of melting, enthalpy change of thermal event, and crystallinity of water. Curves are shifted along y-axis for clarity according to crystallized water.

The most valuable accessions from sub-collection of old potato cultivars of the Czech origin were selected from the potato in vitro-bank at the Crop Research Institute (CRI) to store them by cryopreservation method. A new cryopreservation method based on nodal cutting osmotic pre-treatment, shoot tips sucrose loading and their air dehydration on aluminum foils was used for storage of 58 selected potato. All plant accessions prepared for storage in cryo-bank were virus-free. Average post-thaw recovery of hop and potato was 36 % and 25 %, respectively. Recovery of new plants was successful in all tested genotypes.

6.3 Hop

Isolated hop shoot tips (cv. Saazer) were dehydrated by air above silicagel (Fig. 8). Water content was 2,4 g water per 1g dry mass before air dehydration. The highest water decrease was measured during the first 30 minutes of dehydration. Water content of hop shoot tips was 0,68 $gH_2O\ gDW^{-1}$ after 32 minutes of dehydration. Shoot tips water content decreased below 0,5 $gH_2O\ gDW^{-1}$ after 70 minutes of dehydration and reached 0,4 $gH_2O\ gDW^{-1}$ after 100 minutes of dehydration. After 120 minutes the shoot tips water content was 0,37 $gH_2O\ g\ DW^{-1}$. The plant regeneration depended on the time of dehydration, which was influenced by the shoot tips water content. The highest explant regeneration was achieved after 90 minutes of dehydration at a water content close to 0,4 $gH_2O\ g\ DW^{-1}$.

In a former study, a decrease in the endothermic peak was found during air dehydration by encapsulation-dehydration method used for hop cryopreservation (Martinez *et al.*,1998; Martinez *et al.*, 1999; Martinez & Revilla, 1998). A negligible amount of freezable water was detected in shoot tips after the water content decreased to 18 % and no freezable water was found at a water content of 14 %. The glass transition temperature was found at a water content of 18 % and lower. The temperature of glass transition increased with a decrease of water content (Fig. 9).

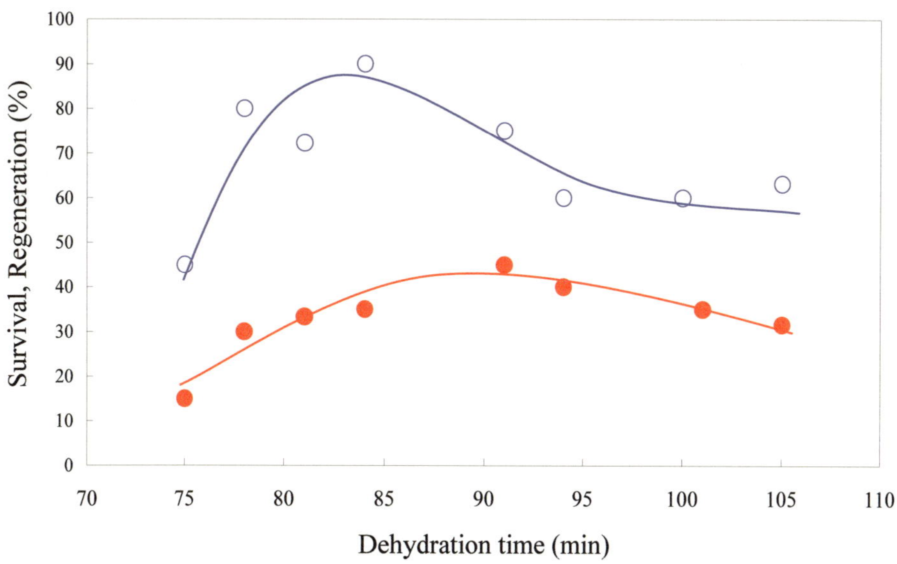

Fig. 8. Survival (empty circles) and regeneration (full circles) of hop explants during dehydration (cv. Saazer).

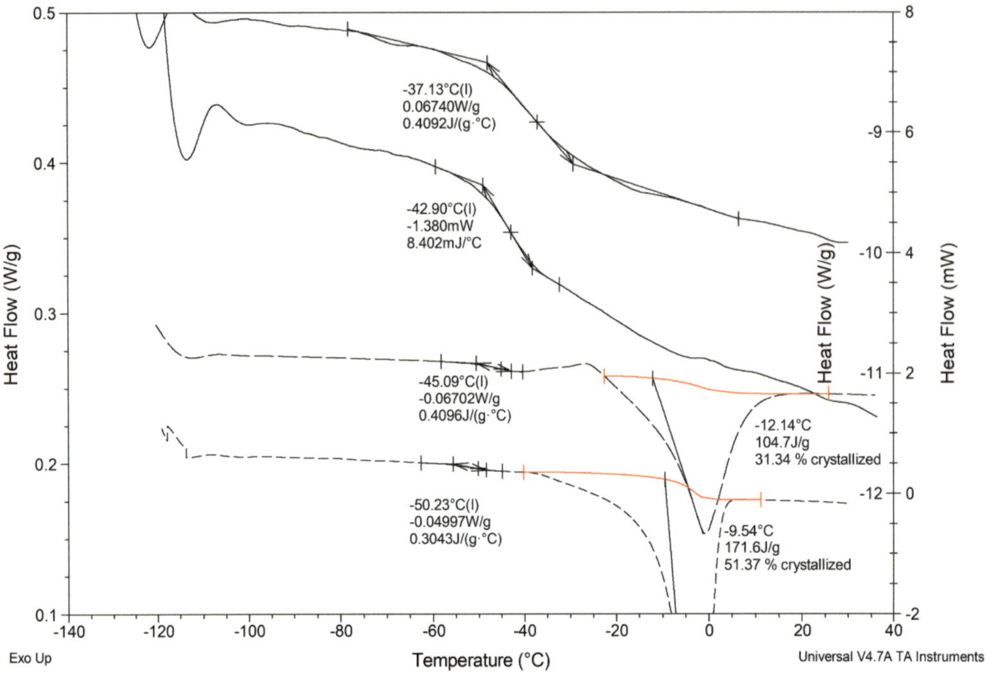

Fig. 9. Heat flow response to the temperature of hop shoot tips. Glass transition temperature of hop shoot tips after moisture loss by dehydration above silicagel. Cooling and warming rate was 10 °C min^{-1}. Curves are shifted along y-axis for clarity.

6.4 Apple tree

In Fig. 10, there is an example of measured thermal characteristics of encapsulated shoot tips of apple tree *in vitro* culture cv. Greensleeves by the DSC. Samples of approximate weight of 10 mg were crimped in an aluminium sample pan and cooled from room temperature to -120 °C. The cooling and heating rate was 10 °C min^{-1}. The glass transition, exotherm and endotherm characteristics were analysed in detail during heating. Thermal characteristics were measured by DSC TA 2920 (TA Instruments) and evaluated by Universal Analysis 2000 for Windows (TA Instruments).

The course of dehydration of encapsulated shoot tips of *in vitro* cultures of apple tree cv. Greensleeves in an open Petri dish exposed to air flow in laminar flow hood at laboratory temperature is demonstrated in Fig. 10. The determination of water content of 20 encapsulated shoot tips placed in the Petri dish was done by weighing the shoot tips during dehydration after approximately 4 hours of drying to the constant weight in an oven (105 °C). The water content was calculated as a proportion of g of water to g of dry matter. From measurement it was evident, that the dehydration consists of two parts; a faster one at the beginning and a slower one after approximately 2 hours of dehydration. From 2 h of dehydration, the level of water content in encapsulated apple tree shoot tips is almost constant. In Fig. 10, there are thermal characteristics (during heating) of encapsulated apple tree shoot tips during their dehydration in the air flow in flow hood. The more dehydrated the samples, the higher the glass transition temperature (characterised by the inflex point I) was measured and no endothermal events representing water in ice crystal form were detected below 0,4 gH_2O g DW^{-1}. The value of 0,4 gH_2O g DW^{-1} dehydration level corresponds to the levels recommended also by other authors (Gupta & Reed, 2006; Martinez *et al.*, 1999; Wu *et al.*, 1999). The integration of endotherm areas of shoot tip and alginate confirms the importance of dehydration to the levels when ice crystals are not present in shoot tip tissues (Figs. 10,11). The energy counted as integration of the endothermic peak corresponded to the amount of frozen water; the less energy, the smaller amount of ice crystals in the sample. These thermal results led us to dehydrating encapsulated shoot tips below 0,4 gH_2O gDW^{-1}.

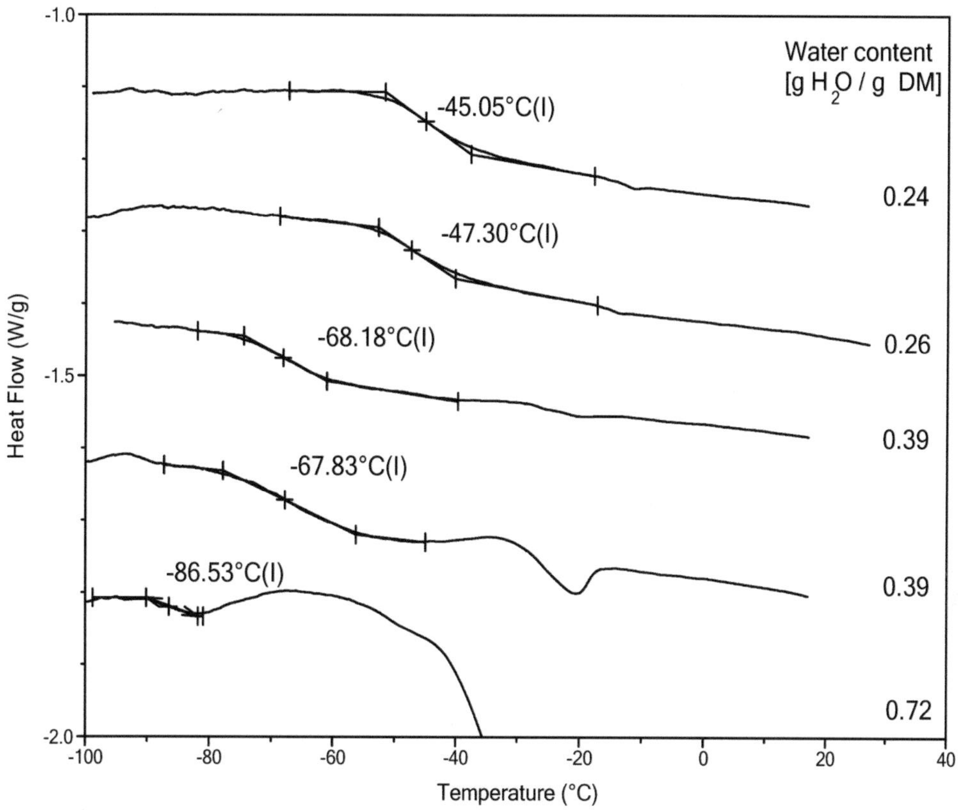

Fig. 10. Glass transition temperature as an inflection point of heat flow change of encapsulated apple tree shoot tips after water loss expressed as final water content (figures behind the end of the separate curves) by dehydration in the air flow. Curves are shifted along y-axis for clarity.

Dehydration curves corresponding to the loss of water from encapsulated *in vitro* shoot tips were measured (Figs. 10,11). During the dehydration procedure of cryopreservation The proper time/level of dehydration must be taken into consideration for successful cryopreservation.

The less water in plant tissues, the less probable damage from ice crystal formation and growth. On the other side plant tissues withstand only certain dehydration. The most appropriate level of dehydration is determined by DSC by measurement of frozen and unfrozen water (generally it is possible to say water in glassy state) (Fig. 11). After 4h dehydration of encapsulated *in vitro* shoot tips the water content decreased below 0,3 gH_2O gDW^{-1}. Water content decrease slows down markedly at the level of 0,6 gH_2O gDW^{-1}. From this level of dehydration, both exotherms and endotherms start to disappear which corresponded to the end of ice crystal formation and the start of glass transitions with high change of heat capacity (Fig. 11). The survival and regeneration of cryopreserved apple tree shoot tips, cultivar Greensleeves, were 75 % and 53 % respectively after 4h dehydration. Non-dehydrated shoot tips neither survived nor regenerated. Dehydration of shoot tips to the level of glass formation is a crucial factor for their survival at ultralow temperatures.

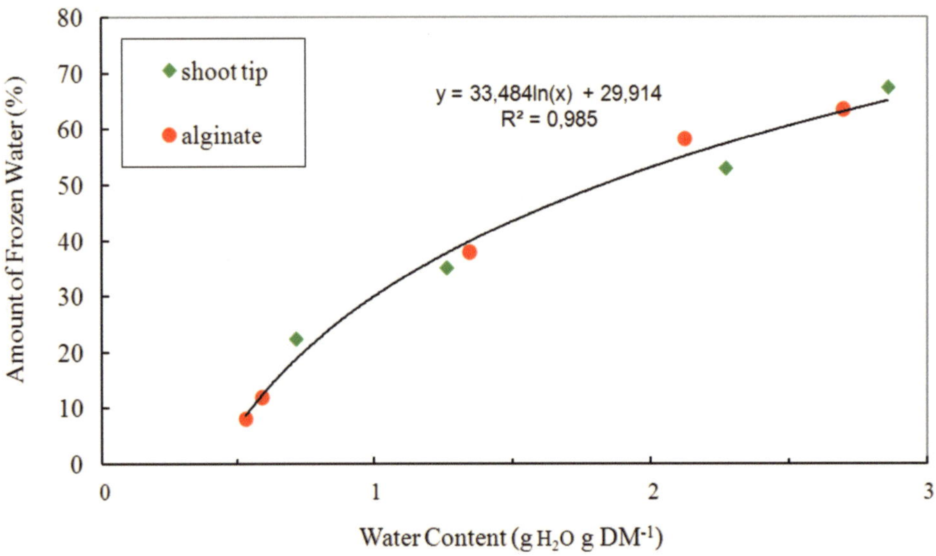

Fig. 11. The amount of frozen water of apple tree shoot tips and alginate beads during dehydration was as the integration of endotherm areas of shoot tip and alginate. Samples of an approximate weight of 3-10 mg were crimped in an aluminum sample pan and cooled from room temperature to -120 °C. The cooling and heating rate was 10 °C min^{-1}. The enthalpy counted as an integration of the endothermic peak corresponded to the amount of frozen water; the less enthalpy the smaller amount of ice crystals in the sample.

7. Regeneration of plants after cryopreservation

The regeneration rate of unripe garlic bulbils was close to 100 % in comparison with the lower regeneration rate of ripe bulbils. The results for ripe bulbils were done on 173 accessions (the measurements on ripe bulbils were not presented, Grospietsch unpublished). The average regeneration rate of ripe bulbils was 40 % and unripe bulbils near 100%. The optimized droplet-vitrification protocol was successfully applied to bulbil primordia of garlic varieties also with high regeneration percentages ranging between 77,4-95% (Engelmann, 2011; Kim *et al.*, 2006)

Fig. 12. Plants regenerated into new plants after immersion in and thawing from the liquid nitrogen: (a) Garlic. Scale bar, 5 mm; (b) Potato. Scale bar, 2 mm; (c) Hop. Scale bar, 1 mm and (d) Apple tree. Scale bar, 5 mm.

The average recovery after cryopreservation of fifty potato cultivars was 24,8% and average hop recovery was 30,5%. Plant recovery eas improved due to the cryoprotocol and media modifications and the average recovery of potato and hop in the year 2007 was 29,1% and 35,5%, respectively (Fig. 13). The highest frequency of plant recovery was near to the average recovery in both crops.

To improve the stability and safety of potato collection, the cryo-collection of the Czech potato germplasm was established. The sub-collection of old potato varieties of the Czech origin was selected as the most important part of potato germplasm kept in the Czech In Vitro Bank of Potato. Fifty eight selected genotypes were cryopreserved by a new method based on osmotic adjustment of explants with sucrose and following air-dehydration. Currently these 58 old potato cultivars of the Czech origin were backed up in cryo-collection at the CRI in Prague.

The differences in plant survival and regeneration exist either among species or cultivars. Example of survival and regeneration of apple tree *in vitro* cultures cryopreserved by the encapsulation-dehydration method are shown in (Tab. 4). The differences can be caused either by different reaction of cultivars in cryopreservation protocol or cold hardening conditions or *in vitro* cultivation. For example, the cultivar McIntosh belongs to very cold resistant cultivars and also has very high regeneration after cryopreservation. On the contrary, the very cold tender cultivar Zvonkové had no survival in laboratory frost tolerance test on dormant buds (data not shown) but *in vitro* cultures were able to survive the cryopreservation procedure, although with very weak regeneration.

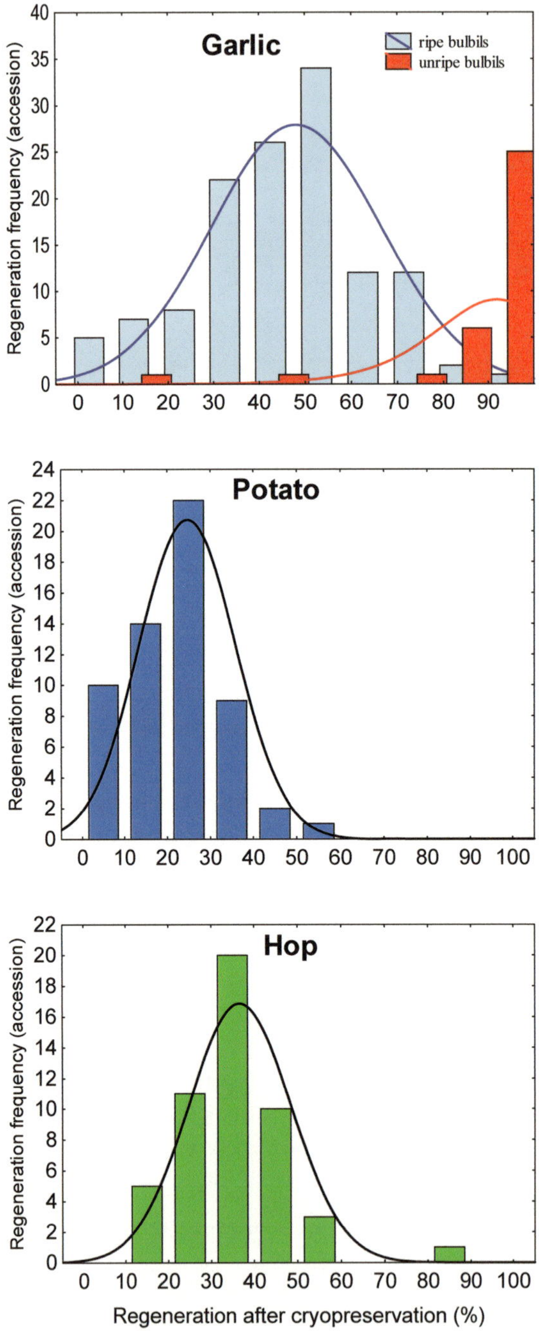

Fig. 13. Regeneration frequency of genotypes (accessions) of garlic, potato and hop stored in the Cryobank of vegetatively propagated crops. Altogether, 129 accessions were evaluated in ripe garlic bulbils, 34 accessions in unripe garlic bulbils, 58 accessions in potato and 50 in hop.

There is clear evidence for the necessity of physiological and biochemical adaptations of cryopreservation procedures according to the different demands of used cultivars to fulfill the needs for successful cryopreservation of apple tree *in vitro* germplasm.

Apple tree cultivar	Survival [%]	SD	Regrowth [%]	SD	n	Number of freezing
Alkmene	6 [a]	0,4	6 [a]	0,4	32	2
Golden Delicious	75 [bc]	27,9	55 [bcd]	24,7	39	3
Greensleeves	75 [bc]	15,0	53 [bcd]	7,5	30	2
Chodské	63 [bc]	17,6	46 [abcd]	13,8	84	5
Idared	34 [abc]	13,5	34 [abc]	13,5	45	2
Jonagold	63 [bc]	13,4	44 [abcd]	26,2	50	3
McIntosh	85 [c]	15,0	85 [d]	15,0	20	2
Prima	28 [ab]	15,0	21 [ab]	8,0	20	2
Rubín	78 [bc]	11,0	78 [cd]	11,0	32	2
Zvonkové	44 [abc]	18,7	4 [a]	5,4	56	4
Average	75	21,7	43	21,2	41	3

[a-d] average followed by the same index did not significantly differ at P<0,05 (analysis of variance - Duncan's test) SD - standard deviation (P < 0,05).

Table 4. Survival and regrowth of apple tree cultivars after encapsulation–dehydration cryopreservation protocol. (n = total number of shoot tips used for evaluations of regeneration).

In vitro shoot tips of apple tree were cryopreserved also by vitrification (PVS2) and pre-culture dehydration methods, but the results were not adequate (Tab. 5). Thus the basic approach of cryopreservation of fruit trees in our laboratory was the encapsulation-dehydration method. The average survival and regeneration of evaluated apple tree cultivars were 75 % and 53 %, respectively (Tab. 4). Similar values of regeneration were obtained by (Condello *et al.*, 2011) with the droplet-vitrification method in two apple tree cultivars. On the other hand, (Wu *et al.*, 2001) reached regeneration of up to 86 %. They recommended a prolonged subcultivation of mother plants and their cold acclimation, which decreased water content in shoot tips and subsequently increased the regeneration in all cryopreservation procedures they evaluated. According to our unpublished data and in concordance with other authors, (Wu *et al.*, 2001; Condello *et al.*, 2011), the adaptation of the *in vitro* mother plants appears to be one of the important steps for improving regeneration of cryopreserved cultivars with lower survival. The encapsulation-dehydration cryopreservation method is a suitable tool to conserve a broader spectrum of apple tree germplasm.

8. Comparison of different cryopreservation methods

The bases for achieving high regeneration after plant cryopreservation using a vitrification method needs to observe the time fulfilment in dehydrating procedures and prevention of tissue damage by chemical toxicity of the cryoprotectants. Otherwise, plant parts in solutes can be injured by the cryoprotectant, by strong osmotic stress during the cryoprotective solution treatment. The time for a high regeneration rate of plants after cryopreservation must be optimized. During this time the explants are dipped in the vitrification solutions, at the optimal temperature, which is involved in the procedure (Condello *et al.*, 2011; Faltus & Zamecnik, 2009; Sakai & Engelmann, 2007; Zámečník *et al.*, 2007).

The temperature-induced glasses-the point at which this occurs, is called the glass transition temperature (T_g) - molecular motion nearly ceases and the liquid becomes a glassy solid. Vitrification of cells and tissues is a physical process which avoids intracellular ice crystallization during ultra-rapid freezing by the transition of the aqueous solution of the symplast into an amorphous glassy state of the cells. As a consequence of the vitrification process, plant tissues are protected from the damage and remain viable during their long-term storage at –196 °C.

Experimental determination of glassy state in plants is complicated by the endothermic reaction overlapping with the glass transition. Ice crystallization as the first-order reaction has discontinuous change in heat capacity contrary to glass transition, which is characterized by heat capacity change. The main problem is to distinguish the endothermic reaction and the glass transition temperature during the measurement of the thermal events. Thermal analysis methods of glass transition temperature and temperature of ice crystals melting in plant tissues were determined. Standard Differential Scanning Calorimetry (DSC) method and Temperature Modulated Differential Scanning Calorimetry (TMDSC) were usually used (Condello *et al.*, 2011; Zámečník & Faltus, 2009).

	Cryopreservation Method		
Plant	Droplet -vitrification (PVS3)	Ultra-rapid freezing	Encapsulation -dehydration
Garlic	**	NT	*
Potato	*	**	*
Hop	NT	**	NT
Apple tree	*(§)	*	**

Table 5. Cryopreservation methods used in this study. Results obtained by cryopreservation methods with a high regeneration rate were presented only: ** - high regeneration rate; *- tested; NT- not tested; (§)-PVS2 was used.

The vitrification method, based on involvement of biological glass, requires a highly concentrated solution of cryoprotectant (from 5 to 8 M), at which the cells are osmotically dehydrated to a certain level. This level of dehydration is characterized by no frost-heaving of water and with a minimum, or no production of ice crystals. The cells treated by cryoprotectant are then vitrified before or during immersion into liquid nitrogen (Sakai *et al.*, 1991). Vitrification belongs to new well-developed procedures supplying frost dehydration of cells pursued at a low temperature. Namely, it removes most of the freezable water through the exposure of the plant shoot tips to a highly concentrated vitrification solution at temperatures above the freezing point.

Several procedures of cryopreservation of fruit trees germplasm, especially apple tree were evaluated, including two-step cryopreservation with controlled ice nucleation (Niino *et al.*, 1992; Niino & Sakai, 1992; Seufferheld *et al.*, 1999; Zhao *et al.*, 1999), and the type of vitrification method of extirpated *in vitro* cultures. The procedures were adjusted and corrected according to the thermal methods for determination of ice nucleation (Tyler *et al.*, 1988), glass transitions in plant material, and exothermic and endothermic characteristics of plant buds. A combination of encapsulation-dehydration cryoprotocol was chosen as the most appropriate system of *in vitro* cultures (Chang & Reed, 2001). The reaction of selected cultivars was different on *in vitro* sub-cultivation and subsequent cryopreservation protocol.

Many plants from the temperate and tropical region were successfully cryopreserved by the encapsulation-dehydration method, which belongs to cryopreservation methods. The ecapsulation-dehydration cryopreservation procedure is based on encapsulating shoot tips of pretreated *in vitro* plants with subsequent dehydration either in sterile air flow (Benson *et al.*, 1996; Dereuddre *et al.*, 1991) or above silicagel (Grospietsch *et al.*, 1999). Dehydrated beads with encapsulated shoot tips are in most cases plunged directly into liquid nitrogen or slowly frozen in programmable freezers (Fabre & Dereuddre, 1990); Zhao *et al.*, 2001). Rewarming proceeds either slowly by placing beads on Petri dishes or the cryotubes are placed in a water bath of temperature ranging from 25 to 45 °C for several minutes (Gupta & Reed, 2006; Matsumoto & Sakai, 1995). Survival and regeneration of shoot tips are evaluated after placing the re-warmed beads on cultivation medium which can be compound modified by phytohormones (their combination and concentration), to stimulate proliferation or eliminate the phenolic compounds (Paulet *et al.*, 1993). The medium can be softer in some cases to allow an easier regeneration from beads (Reed *et al.*, 2006) or even the explants can be extracted from the beads (Niino & Sakai, 1992). The importance of direct explant regeneration to the new plants without callus inter phase is important for the genetic stability after usage of this method.

9. Conclusion

The four main crops (garlic, potato, hop and apple tree) have been cryopreserved in the Czech Plant Cryobank. The methods of cryopreservation are based on cryoprotocol of the vitrification procedures (encapsulation-dehydration, dehydration by vitrification solution and a modified ultra-rapid freezing method based on preconditioning of the plant shoot tips on an osmotic solution). Cryopreservation is well advanced for vegetatively propagated species, and techniques are ready for large-scale experimentation in an increasing number of cases (Engelmann, 2011). We have started the routine cryopreservation with Czech potato genotypes in collaboration with Potato Research Institute Ltd., Havlíčkův Brod. The other three crops are supported by specialized companies (*Allium* genera by the Gene Bank Olomouc branch, the Research and Breeding Institute of Pomology at Holovousy and the Hop Research Institute). The Czech Plant Cryobank operates as a safe duplicate to repositories of germplasm kept in field or *in vitro* conditions. In this way the Czech Plant Cryobank in Prague joined the effort of the potato cryobank in Germany (Keller *et al.*, 2008), and Korea (Kim *et al.*, 2006). Currently three of the EU countries (The Czech Republic, Germany and Poland) involved in the maintenance of national vegetatively propagated *Allium* collections are developing the methodology for cryopreservation (project

EURALLIVEG, EU) and established the Tripartite *Allium* Cryobank to store meristematic explants in liquid nitrogen.

10. Acknowledgment

This work was partially supported by project 0002700604 MSMT and QH71228 MZe. The authors would like to thank M. Grospietsch, Jana Chittendenová, Aneta Babická and Adéla Hreňuková for their help and Dr. V. Skládal for critical review of this manuscript.

11. References

Benson, E.B.; Reed, B.M.; Brennan, R.M.; Clacher, K.A. & Ross, D.A. (1996). Use of thermal analysis in the evaluation of cryopreservation protocols for Ribes nigrum L germplasm. *Cryo-Letters*, Vol.17, No.6, (November 1996), pp. 347-362, ISSN0143-2044

Benson, E.E. (2008). Cryopreservation of phytodiversity: A critical appraisal of theory practice. *Critical Reviews in Plant Sciences*, Vol.27, No.3, (2008), pp. 141-219, ISSN0735-2689

Chang, Y.J. & Reed, B.M. (2001). Preculture conditions influence cold hardiness and regrowth of Pyrus cordata shoot tips after Cryopreservation. *Hortscience*, Vol.36, No.7, (December 2001), pp. 1329-1333, ISSN0018-5345

Condello, E.; Caboni, E.; Andre, E.; Piette, B.; Druart, P.; Swennen, R. & Panis, B. (2011). Cryopreservation of Apple in Vitro Axillary Buds Using Droplet-Vitrification. *Cryoletters*, Vol.32, No.2, (March 2011), pp. 175-185, ISSN0143-2044

Dereuddre, J.; Hassen, N.; Blandin, S. & Kaminski, M. (1991). Resistance of Alginate-Coated Somatic Embryos of Carrot (Daucus-Carota L) to Desiccation and Freezing in Liquid-Nitrogen .2. Thermal-Analysis. *Cryo-Letters*, Vol.12, No.3, (May 1991), pp. 135-148, ISSN0143-2044

Dumet, D.; Grapin, A.; Bailly, C. & Dorion, N. (2002). Revisiting crucial steps of an encapsulation/desiccation based cryopreservation process: importance of thawing method in the case of Pelargonium meristems. *Plant Science*, Vol.163, No.6, (December 2002), pp. 1121-1127, ISSN0168-9452

Ellis, D.; Skogerboe, D.; Andre, C.; Hellier, B. & Volk, G. (2006). Implementation of garlic cryopreservation techniques in the national plant germplasm system. *Cryoletters*, Vol.27, No.2, (March 2006), pp. 99-106, ISSN0143-2044

Engelmann, F. (2011). Use of biotechnologies for the conservation of plant biodiversity. *In Vitro Cellular & Developmental Biology-Plant*, Vol.47, No.1, (February 2011), pp. 5-16, ISSN1054-5476

Fabre, J. & Dereuddre, J. (1990). Encapsulation Dehydration - A New Approach to Cryopreservation of Solanum Shoot-Tips. *Cryo-Letters*, Vol.11, No.6, (November 1990), pp. 413-426, ISSN0143-2044

Fahy, G.M. (1985). Cold Shock Injury Is A Significant Factor in Freezing-Injury - A Position Against. *Cryobiology*, Vol.22, No.6, (1985), pp. 628-628, ISSN0011-2240

Faltus, M. & Zámečník, J. (2009). Thermal Characteristics of Some Vitrification Solutions. *Cryoletters*, Vol.30, No.5, (September 2009), pp. 389-389, ISSN0143-2044

Gonzalez-Arnao, M.T.; Panta, A.; Roca, W.M.; Escobar, R.H. & Engelmann, F. (2008). Development and large scale application of cryopreservation techniques for shoot and somatic embryo cultures of tropical crops. *Plant Cell Tissue and Organ Culture*, Vol.92, No.1, (January 2008), pp. 1-13, ISSN0167-6857

Grospietsch, M.; Stodůlková, E. & Zámečník, J. (1999). Effect of osmotic stress on the dehydration tolerance and cryopreservation of Solanum tuberosum shoot tips. *Cryo-Letters*, Vol.20, No.6, (November 1999), pp. 339-346, ISSN0143-2044

Gupta, S. & Reed, B.M. (2006). Cryopreservation of shoot tips of blackberry and raspberry by encapsulation-dehydration and vitrification. *Cryoletters*, Vol.27, No.1, (January 2006), pp. 29-42, ISSN0143-2044

Halmagyi, A.; Fischer-Kluver, G.; Mix-Wagner, G. & Schumacher, H.M. (2004). Cryopreservation of Chrysanthemum morifolium (Dendranthema grandiflora Ramat.) using different approaches. *Plant Cell Reports*, Vol.22, No.6, (January 2004), pp. 371-375, ISSN0721-7714

Hirai, D. & Sakai, A. (1999). Cryopreservation of in vitro-grown meristems of potato (Solanum tuberosum L.) by encapsulation-vitrification. *Potato Research*, Vol.42, No.2, (1999), pp. 153-160, ISSN0014-3065

Kaczmarczyk, A. (2008). Physiological, biochemical ,histological and ultrastructural aspects of cryopreservation in meristematic tissue of potato shoot tips (2008), ISSN 978-3-89574-694-9

Kaczmarczyk, A., Rokka, V-M & Keller E. R. J. (2011). Potato Shoot Tip Cryopreservation. A Review Source: *Potato research*, Vol. 54, No. 1, (March 2011),pp. 45-79, ISSN0014-3065

Kartha, K.K. & Leung, N.L. (1979). Cryopreservation of Plant Meristems As A Means of Germ Plasm Storage. *Cryobiology*, Vol.16, No.6, (1979), pp. 582-583, ISSN0011-2240

Keller, E.R.J.; Kaczmarczyk, A. & Senula, A. (2008). Cryopreservation for plant genebanks - A matter between high expectations and cautious reservation. *Cryoletters*, Vol.29, No.1, (January 2008), pp. 53-62, ISSN0143-2044

Kim, H.H.; Cho, E.G.; Baek, H.J.; Kim, C.Y.; Keller, E.R.J. & Engelmann, F. (January 2004). Cryopreservation of garlic shoot tips by vitrification: Effects of dehydration, rewarming, unloading and regrowth conditions. *Cryoletters*, Vol.25, No.1, (January 2004), pp. 59-70, ISSN0143-2044

Kim, H.H.; Lee, J.K.; Yoon, J.W.; Ji, J.J.; Nam, S.S.; Hwang, H.S.; Cho, E.G. & Engelmann, F. (2006). Cryopreservation of garlic bulbil primordia by the droplet-vitrification procedure. *Cryoletters*, Vol.27, No.3, (May 2006), pp. 143-153, ISSN0143-2044

Kim, H.H.; Lee, Y.G.; Shin, D.J.; Ko, H.C.; Gwag, J.G.; Cho, E.G. & Engelmann, F. (2009). Development of Alternative Plant Vitrification Solutions in Droplet-Vitrification Procedures. *Cryoletters*, Vol.30, No.5, (September 2009), pp. 320-334, ISSN0143-2044

Langis, R. & Steponkus, P.L. (March 1990). Cryopreservation of Rye Protoplasts by Vitrification. *Plant Physiology*, Vol.92, No.3, (March 1990), pp. 666-671, ISSN0032-0889

Liu, H.Q.; Yu, W.G.; Dai, J.X.; Gong, Q.H.; Yang, K.F. & Lu, X.Z. (2004a). Cryopreservation of protoplasts of the alga *Porphyra yezoensis* by vitrification. *Plant Science*, Vol.166, No.1, (January 2004a), pp. 97-102, ISSN0168-9452

Liu, Y.G.; Wang, X.Y. & Liu, L.X. (2004b). Analysis of genetic variation in surviving apple shoots following cryopreservation by vitrification. *Plant Science*, Vol.166, No.3, (2004b), pp. 677-685, ISSN0168-9452

Luyet, B. (1937). On the mechanism of cellular death by high pressure cytological modifications accompanying death in yeast. *Comptes Rendus Hebdomadaires Des Seances De l Academie Des Sciences*, Vol.204, (1937), pp. 1506-1508, ISSN0001-4036

Martinez, D. & Revilla, M.A. (1998). Cold acclimation and thermal transitions in the cryopreservation of hop shoot tips. *Cryo-Letters*, Vol.19, No.6, (November 1998), pp. 333-342, ISSN0143-2044

Martinez, D.; Revilla, M.A.; Espina, A.; Jaimez, E. & Garcia, J.R. (1998). Survival cryopreservation of hop shoot tips monitored by differential scanning calorimetry. *Thermochimica Acta*, Vol.317, No.1, (July 1998), pp. 91-94, ISSN0040-6031

Martinez, D.; Tames, R.S. & Revilla, M.A. (1999). Cryopreservation of in vitro-grown shoot-tips of hop (Humulus lupulus L.) using encapsulation/dehydration. *Plant Cell Reports*, Vol.19, No.1, (November 1999), pp. 59-63, ISSN0721-7714

Matsumoto, T. & Sakai, A. (1995). An Approach to Enhance Dehydration Tolerance of Alginate-Coated Dried Meristems Cooled to -196-Degrees-C. *Cryo-Letters*, Vol.16, No.5, (September 1995), pp. 299-306, ISSN0143-2044

Murashige, T. & Skoog, F. (1962). A Revised Medium for Rapid Growth and Bio Assays with Tobacco Tissue Cultures. *Physiologia Plantarum*, Vol.15, No.3, (1962), pp. 473-&, ISSN0031-9317

Niino, T. & Sakai, A. (1992). Cryopreservation of Alginate-Coated Invitro-Grown Shoot Tips of Apple, Pear and Mulberry. *Plant Science*, Vol.87, No.2, (1992), pp. 199-206, ISSN0168-9452

Niino, T.; Sakai, A.; Yakuwa, H. & Nojiri, K. (1992). Cryopreservation of In vitro-Grown Shoot Tips of Apple and Pear by Vitrification. *Plant Cell Tissue and Organ Culture*, Vol.28, No.3, (March 1992), pp. 261-266, ISSN0167-6857

Nishizawa, S.; Sakai, A.; Amano, Y. & Matsuzawa, T. (1993). Cryopreservation of Asparagus (*Asparagus-Officinalis* L) Embryogenic Suspension Cells and Subsequent Plant-Regeneration by Vitrification. *Plant Science*, Vol.91, No.1, (1993), pp. 67-73, ISSN0168-9452

Paulet, F.; Engelmann, F. & Glaszmann, J.C. (1993). Cryopreservation of Apices of In-Vitro Plantlets of Sugarcane (*Saccharum* Sp Hybrids) Using Encapsulation Dehydration. *Plant Cell Reports*, Vol.12, No.9, (July 1993), pp. 525-529, ISSN0721-7714

Reed, B.M.; Schumacher, L.; Wang, N.; D'Achino, J. & Barker, R.E. (2006). Cryopreservation of bermudagrass germplasm by encapsulation dehydration. *Crop Science*, Vol.46, No.1, (January 2006), pp. 6-11, ISSN0011-183X

Sakai, A. & Engelmann, F. (2007). Vitrification, encapsulation-vitrification and droplet-vitrification: A review. *Cryoletters*, Vol.28, No.3, (May 2007), pp. 151-172, ISSN0143-2044

Sakai, A.; Kobayashi, S. & Oiyama, I. (1991). Survival by Vitrification of Nucellar Cells of Navel Orange (*Citrus-Sinensis* Var Brasiliensis Tanaka) Cooled to 196-Degrees-C. *Journal of Plant Physiology*, Vol.137, No.4, (February 1991a), pp. 465-470, ISSN0176-1617

Sarkar, D. & Naik, P.S. (1998). Cryopreservation of shoot tips of tetraploid potato (*Solanum tuberosum* L.) clones by vitrification. *Annals of Botany*, Vol.82, No.4, (October 1998), pp. 455-461, ISSN0305-7364

Schafer-Menuhr, A.; Muller, E. & Mix-Wagner, G. (1996). Cryopreservation: an alternative for the long-term storage of old potato varieties. *Potato Research*, Vol.39, No.4, (1996), pp. 507-513, ISSN0014-3065

Sestak, J. & Zamecnik, J. (2007). Can clustering of liquid water and thermal analysis be of assistance for better understanding of biological germplasm exposed to ultra-low temperatures. *Journal of Thermal Analysis and Calorimetry*, Vol.88, No.2, (May 2007), pp. 411-416, ISSN1388-6150

Seufferheld, M.J.; Stushnoff, C.; Forsline, P.L. & Gonzalez, G.H.T. (1999). Cryopreservation of cold-tender apple germplasm. *Journal of the American Society for Horticultural Science*, Vol.124, No.6, (November 1999), pp. 612-618, ISSN0003-1062

Suzuki, M.; Tandon, P.; Ishikawa, M. & Toyomasu, T. (2008). Development of a new vitrification solution, VSL, and its application to the cryopreservation of gentian axillary buds. *Plant Biotechnology Reports*, Vol.2, No.2, (June 2008), pp. 123-131, ISSN1863-5466

Tanaka, D.; Niino, T.; Isuzugawa, K.; Hikage, T. & Uemura, M. (2004). Cryopreservation of shoot apices of in-vitro grown gentian plants: Comparison of vitrification and encapsulation-vitrification protocols. *Cryoletters*, Vol.25, No.3, (May 2004), pp. 167-176, ISSN0143-2044

Towill, L.E. (1990). Cryopreservation of Isolated Mint Shoot Tips by Vitrification. *Plant Cell Reports*, Vol.9, No.4, (August 1990), pp. 178-180, ISSN0721-7714

Tyler, N.; Stushnoff, C. & Gusta, L.V. (1988). Freezing of Water in Dormant Vegetative Apple Buds in Relation to Cryopreservation. *Plant Physiology*, Vol.87, No.1, (May 1988), pp. 201-205, ISSN0032-0889

Uragami, A.; Sakai, A.; Nagai, M. & Takahashi, T. (1989). Survival of Cultured-Cells and Somatic Embryos of Asparagus-Officinalis Cryopreserved by Vitrification. *Plant Cell Reports*, Vol.8, No.7, (October 1989), pp. 418-421, ISSN0721-7714

Wu, Y.J.; Engelmann, F.; Zhao, Y.H.; Zhou, M.D. & Chen, S.Y. (1999). Cryopreservation of apple shoot tips: Importance of cryopreservation technique and of conditioning of donor plants. *Cryo-Letters*, Vol.20, No.2, (March 1999), pp. 121-130, ISSN0143-2044

Wu, Y.J.; Zhao, Y.H.; Engelmann, F.; Zhou, M.D.; Zhang, D.M. & Chen, S.Y. (2001). Cryopreservation of apple dormant buds and shoot tips. *Cryoletters*, Vol.22, No.6, (November 2001), pp. 375-380, ISSN0143-2044

Zámečník, J. & Faltus, M. (2009). Evaluation of Thermograms from Differential Scanning Calorimeter. *Cryoletters*, Vol.30, No.5, (September 2009), pp. 388-388, ISSN0143-2044

Zámečník, J, Faltus, M. & Bilavčík, A. (2007). Cryoprotocols used for cryopreservation of vegetatively propagated plants in the Czech cryobank. *Advances in Horticultural Science*, Vol.21, (2007), pp. 247-250,

Zamecnik, J.; Faltus, M.; Kotkova, R. & Hejnak, V. (2011). Glass transition determination in Allium shoot tips after dehydration. Acta Hort. (ISHS) 908 (April 2009), pp. 33-38, ISSN0567-7572

Zhao, Y.H.; Wu, Y.J.; Engelmann, F. & Zhou, M.D. (2001). Cryopreservation of axillary buds of grape (*Vitis vinifera*) in vitro plantlets. *Cryoletters*, Vol.22, No.5, (September 2001), pp. 321-328, ISSN0143-2044

Zhao, Y.H.; Wu, Y.J.; Engelmann, F.; Zhou, M.D.; Zhang, D.M. & Chen, S.Y. (1999). Cryopreservation of apple shoot tips by encapsulation-dehydration: Effect of preculture, dehydration and freezing procedure on shoot regeneration. *Cryo-Letters*, Vol.20, No.2, (March 1999), pp. 103-108, ISSN0143-2044

18

Cryopreservation of Tropical Plant Germplasm with Vegetative Propagation – Review of Sugarcane (*Saccharum* spp.) and Pineapple (*Ananas comusus* (L.) Merrill) Cases

Marcos Edel Martinez-Montero[1], Maria Teresa Gonzalez Arnao[2] and Florent Engelmann[3,4]
[1]*University of Ciego de Avila/Bioplantas Centre*
[2]*Universidad Veracruzana,*
[3]*IRD, UMR DIAPC*
[4]*Bioversity International*
[1]*Cuba*
[2]*Mexico*
[3]*France*
[4]*Italy*

1. Introduction

Sugarcane (*Saccharum* sp. hybrids) is a crop of major importance, which is cultivated on a large scale in tropical and subtropical regions primarily for its high sucrose content. Cultivated pineapple (*Ananas comosus* (L.) Merrill, which is now called Ananas comosus var comosus) belongs to the family Bromeliaceae. It is economically the fourth most important crop worldwide in terms of tropical fruit production and follows banana, mangoes and citrus. One of the main drawbacks faced by sugarcane and pineapple agriculture worldwide is the vegetative (i.e. asexual) nature of its conventional propagation. The consequence is that plants in the field must be replaced at intervals ranging from 1 to 5 years, a process that is costly, tedious and time-consuming. Furthermore, if the planting material is of low quality, yields decrease and more tillage is needed. The crops are exposed to natural disasters, while the propagation system leads to systemic disease transmission, and natural selection and plagues also take their toll. Moreover, the industry is in dramatic need of planting material, which cannot be produced in sufficient quantities to meet the demand using classical macropropagation techniques.

In vitro culture techniques have been extensively developed and applied for several thousand plant species including sugarcane and pineapple. Their uses are of high interest for multiplication, conservation and transformation of plant germplasm. Indeed, they allow the multiplication of plant material with high multiplication rates in an aseptic environment, reduction of space requirements, genetic erosion is reduced under optimal storage conditions, and minimized of the expenses in labour costs. Moreover, tissue culture systems greatly facilitate the international exchange of germplasm as the size of the samples is drastically diminished and they can be shipped in sterile conditions. Different *in vitro* conservation methods are employed, depending on the storage duration requested. For short- and medium-term storage, the aim is to reduce growth and to increase the intervals between subcultures. For long-term storage, cryopreservation, i.e. storage at ultra-low temperature, usually that of liquid nitrogen (−196°C), is the only current method ensuring long-term storage of germplasm from vegetatively propagated species. At this temperature, all cellular divisions and metabolic processes are stopped; therefore, plant material can thus be maintained without alteration or modification. Moreover, cultures are stored in a small volume, protected from contamination, requiring very limited maintenance.

This Chapter comprises two main sections focusing on the establishment, optimization and application of cryopreservation techniques to different tissues of *in vitro* sugarcane and pineapple cultures. The first part presents the cryopreservation protocols developed for sugarcane apices isolated from *in vitro* grown plants, embryogenic calluses and somatic embryos, as well as some analytical techniques (electrolyte leakage, protein content and lipid peroxidation products), used to describe the impact of the successive steps of the protocol on the physiological state of the cultures, which are also useful to refine the cryopreservation protocol. The effect of cryopreservation on the phenotypical development, both *in vitro* and in the field, of sugarcane plants regenerated material will be also presented. The second section presents the studies performed to set up and refine a cryopreservation protocol for apices of pineapple *in vitro* plantlets. The protocol established following the vitrification approach was successfully applied for the first time to shoot tips of three pineapple varieties, and then extended to nine pineapple accessions belonging to the *in vitro* collection of Bioplantas Centre in Cuba. In addition, we present the preliminary assays developed using callus of two pineapple cultivars. In the conclusion, we discuss the possibilities and prospects of utilisation of cryopreservation techniques for the long-term storage of other vegetatively propagated tropical plant species.

2. Cryopreservation protocols for sugarcane

Several review papers have been published, which provide lists of species which have been successfully cryopreserved (Cyr, 2000; Engelmann, 1997; Engelmann & Takagi, 2000; Sakai et al., 2002). For vegetatively propagated species, cryopreservation has a wide applicability in terms of species coverage, since protocols have been successfully established for roots and tubers, fruit trees, ornamentals, forestry species and plantation crops from both temperate and tropical origin (Engelmann, 2004; Kaczmarczyk et al., 2008; Engelmann, 2010; Engelmann, 2011).

In the case of sugarcane, cryopreservation protocols have been developed for various materials: apices of *in vitro* plantlets using the encapsulation-dehydration technique (Gonzalez-Arnao et al., 1993; Paulet et al., 1993); cell suspensions (Finkle & Ulrich, 1979; Gnanapragasam & Vasil, 1990) and embryogenic callus using classical freezing protocols (Eksomtramage et al., 1992; Gnanapragasam & Vasil, 1992; Jian et al., 1987) and simplified cryopreservation protocols (Martinez-Montero et al., 1998). Recently, it was published the cryopreservation procedure based on vitrification techniques for somatic embryos (Martinez-Montero et al., 2008).

2.1 Cell suspensions

The first experimental research on sugarcane cell suspension was accomplished at the end of 1970 and at the beginning of 1980 decades (Finkle & Ulrich, 1979; Finkle & Ulrich, 1982; Ulrich et al., 1979; Ulrich et al., 1984). These authors demonstrated that resistance of cells to freezing to -23°C and -40°C was possible with little decrease in survival by using mixtures of glucose, dimethylsulfoxide and polyethylene glycol as cryoprotectants. However, no plants were recovered from the cryopreserved cell suspension.

Later on, Gnanapragasam & Vasil (1990) reported that efficient plant regeneration was obtained from a cryopreserved embryogenic cell suspension of one commercial sugarcane hybrid established from leaf derived callus. They observed pregrowing the cells for three days in Murashige & Skoog (1962) basal medium supplemented with 0.33 M sorbitol was essential to the process. A regeneration efficiency of 92% was obtained and plants regenerated from cryopreserved cells, and grown to maturity in the greenhouse, were morphologically identical to regenerated control plants. Later, there were not detected differences at molecular level using RFLP technique comparing plants regenerated from cryopreserved and control cells for three sugarcane hybrids (Chowdhury & Vasil, 1993).

2.2 Embryogenic callus

The first success for cryopreservation of sugarcane embryogenic callus was obtained by Ulrich et al., (1979) for the hybrid H50-7209. It was a pretreatment using a combination of 10% polyethylene glycol, 8% glucose and 10% DMSO, freezing rate of 2°C.min^{-1} until a first transfer temperature of -40°C and freezing rate of 5°C.min^{-1} until second transfer temperature of -80°C. However, the recovery of cryopreserved callus was achieved only with root regeneration. Ulrich et al., (1984) obtained after modifications of the same protocol a limited number of albino plantlets from cryopreserved calluses.

Later on, high survival rates (ca. 90%) and recovery of whole plants were obtained by Jian et al., (1987), Eksomtramage et al., (1992) and Gnanapragasam & Vasil (1992). The conditions defined were different from that used by Ulrich et al., (1979, 1984). For cryoprotective treatment, a mixture of sorbitol and DMSO was used by Jian et al., (1987) and Gnanapragasam & Vasil (1992); Eksomtramage et al., (1992) employed a mixture of sucrose and DMSO. Freezing conditions were also different: 1°C.min^{-1} from 0°C to -10°C, and kept for 15min at the same freezing rate from -10°C down to -40°C and kept for 1-5 h, and finally immersed into liquid nitrogen (Jian et al., 1987); or 0.5°C.min^{-1} down to -40°C or -45°C with no plateau at the end of the controlled freezing sequence (Eksomtramage et al., 1992 and Gnanapragasam & Vasil, 1992). Moreover, the technique developed by Eksomtramage et al., (1992) was successfully applied to calluses of 10 varieties.

These authors have followed the strategy known as dehydration by extracellular freezing, which uses a controlled freezing regime (Withers & King, 1980). However, this procedure requires expensive and sometimes complex programmable freezing devices, limiting its use to laboratories specializing in cryopreservation (Ashmore, 1997; Reed, 2001). Furthermore, their research has been focused on the cryopreservation of sugarcane calli obtained from segments of immature leaves belonging to *in vitro* cultured plants; however, such explants are known to have a limited morphogenetic capacity (Krishnaraj & Vasil, 1995) and it is widely acknowledged that immature embryos, as well as young inflorescences, are physiologically better explants for calli production because they retain their embryogenic capacities (Merkle et al., 1995).

2.2.1 Optimization of methodology for sugarcane callus

Our research team (Martinez-Montero et al., 1998, 2006), using the cryo-research for sugarcane callus described above as starting point, published the results for establishing step by step a methodology for the cryopreservation of sugarcane calli with embryogenic structures obtained from immature inflorescence (Figure 1). We optimized the following aspects according to the *in vitro* survival and regeneration (plants per 500 mg of calli) percentages for: Selection of the cooling procedure, the effect of the cooling procedure and of the type of alcohol, the effect of the induction time of extracellular ice crystals, the effect of post-subculture time, the effect of sucrose and dimethylsulfoxide concentration in the cryoprotective medium, and the effect of the pre-freezing time.

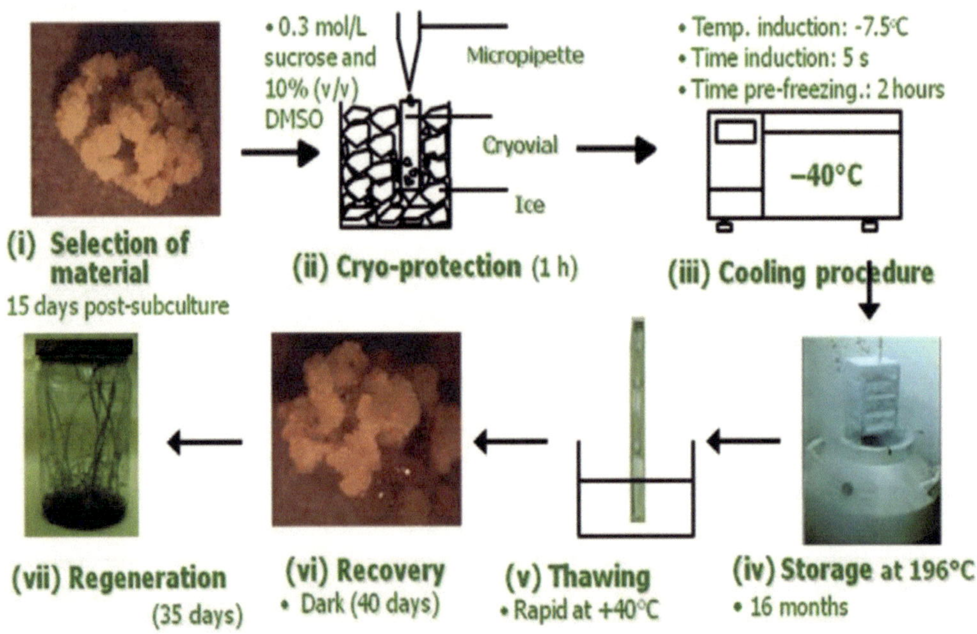

Fig. 1. Optimized methodology for the cryopreservation of sugarcane calli with embryogenic structures.

Firstly, we based on the results carried out by Maddox et al. (1983) and Withers (1985) who successfully used uncomplicated freezing procedures for cellular suspensions of *Nicotiana* and *Musa*, respectively (Figure 2). We evaluated the application of one of these devices as an alternative to establish techniques for the cryopreservation of sugarcane calli with embryogenic structures (i.e., cooling rate controlled by a computer-coupled programmable freezer).

As results was detected a survival after storage in liquid nitrogen for both cooling procedures implying the existence of a protective dehydration process that allows the vitrification of some cells without the formation of intracellular ice crystals. However, from a practical point of view, seeding ice crystals is much more difficult when using the procedure proposed by Maddox et al., (1983).

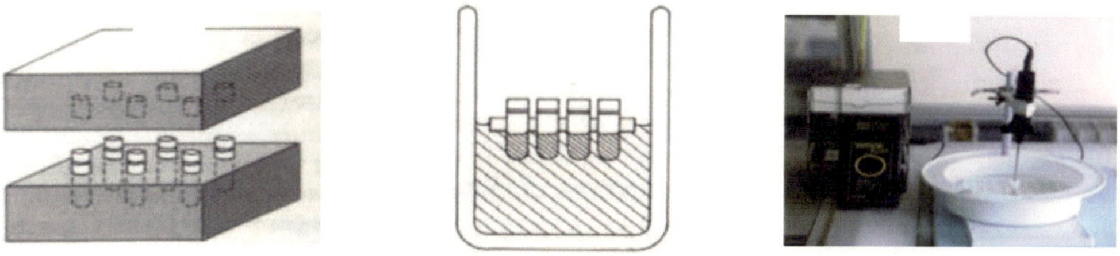

a b c

Fig. 2. Devices used by Maddox et al. (1983) (A), Withers (1985) (B) and our research group (C) for the establishment of cryopreservation procedures.

In general, the "classical" cryopreservation protocols provide insufficient detail on seeding ice crystals step (Martinez-Montero et al., 2006). For instance, although the analysis of this parameter must have been unavoidable for the development of the protocols of Jian et al., (1987) and Gnanapragasam & Vasil (1992), these analyses are not described in their articles; and even Eksomtramage et al., (1992), who first mentioned the need for this step when freeze sugarcane calli, provide little detail on its implementation.

Optimizing the effect of post-subculture time on the survival and regeneration of plants from cryopreserved calli the results reported by Jian et al., (1987) and Martinez-Montero et al., (2006) coincided. We founded that survival after cryopreservation is associated with the selection step during the post-subculture period, reaching a maximum at 15 days post-subculture. Moreover, we correlated this finding to the physiological state of the calli before cryopreservation and measured the growth of the calli. These results are the basis for a rational selection of the material to be cryopreserved, since several authors have shown that there is a correlation for different species between the phase of active growth of the calli and its performance upon cryopreservation (Reinhoud et al., 2000; Withers, 1985; Yoshida et al., 1993). It has been proven that the morphology of the cultured cells has a marked influence on cryotolerance. In most species, only small cells with a highly dense cytoplasm, usually found in small cellular aggregates in the periphery of the callus, survive after cryopreservation; whereas large, vacuolated cells are damaged during freezing (Kristensen et al., 1994; Withers, 1985).

We also founded the decrease in survival and regeneration when using sucrose concentrations higher than 0.3 M. The importance of sucrose tolerance within this setting is determined by the role of this disaccharide in the regulation of the hydric potential of the cells (Tetteroo, 1996); sucrose has also occasionally been considered an inducer of cellular division and differentiation (Feher, 2003). Furthermore, there is evidence suggesting that sucrose functions as a genetic regulatory signal for genes coding for enzymes and proteins involved in transport and storage (Lunn & MacRae, 2003). Additionally, Ausborn et al., (1994) and Turner et al., (2001) detected that sucrose stabilizes the lipid bilayers on the membranes by forming disaccharide-lipid hydrogen bonds, whereas Niu et al., (1997) founded that the right amount of intracellular sucrose can protect a number of enzymes from ion-mediated toxicity.

According to our results, the sugarcane calli did not survive the cryopreservation procedure when dimethylsulfoxide was omitted from the cryoprotective mixture; and both the survival and plant regeneration percentages rose steadily with increasing dimethylsulfoxide concentrations, up to 10%. However, in clear contrast with the results obtained when testing different amounts of sucrose, there are no differences in survival between the cryopreserved and non-cryopreserved samples at concentrations higher than the optimum (10% v/v in this case), and the contrast is even starker when comparing plant regeneration rates, where the cryopreserved material performs even better than the non-cryopreserved calli. These results agree with those of Finkle et al., (1985) for rice cells, who concluded that the effects achieved by using dimethylsulfoxide are paradoxical, since although this substance is clearly toxic, but inhibits the growth of ice crystals during cryopreservation.

Our data obtained during the experiment for dimethylsulfoxide tolerance are coherent with the reports for other biological systems, associated to the high degree of toxicity of dimethylsulfoxide (Arakawa et al., 1990; Fahy et al., 1990). Kartha et al., (1988) detected that dimethylsulfoxide produces an inhibition of 35 to 42% on the growth of embryogenic cultures of white spruce when used at a concentration higher than 5% (v/v), and Klimaszewska et al., (1992) reported a 28% reduction in the growth of embryogenic tissues from black spruce when treated with 15% (v/v) dimethylsulfoxide; this effect, according to the microscopic observations of these authors, is due to the induction by this substance of a strong plasmolytic effect at the cellular level.

However, and in spite of these findings, dimethylsulfoxide has been, and still is used as a cryoprotectant during storage at ultra-low temperatures. According to Engelmann, (2000) this apparent paradox is due to the fact that dimethylsulfoxide is always used as part of a cryoprotective mixture, rather than individually. Arakawa et al., (19990) have provided evidence that the toxicity of dimethylsulfoxide in isolated proteins is mediated by hydrophobic interactions, which are favored at increasing temperatures; in this context, this effect is minimized by the use, during preculture, of sucrose at 0°C, which induces the biosynthesis of proteins that neutralize the toxic effects of this agent (caused by its interaction mainly with lysine residues) (Anchordoguy et al., 1991; Klimazewska et al., 1990; Swan et al., 1999).

Although the exact cryoprotective mechanism of dimethylsulfoxide at ultra-low temperatures remains unknown, it is widely acknowledged that it depends on the colligative properties of this penetrating compound; that is, dimethylsulfoxide affects the formation of ice crystals by decreasing the equilibrium freezing point of the solution, in direct dependence on its molar concentration (Kinoshita et al., 2001; McGann, L.E. & Walterson, 1987). Dimethylsulfoxide, as a cell-penetrating agent, also decreases the intracellular concentration of toxic electrolytes on unfrozen cells (Finkle et al., 1985).

Anchordoguy et al., (1991) suggested, furthermore, that there is another, not colligative mechanism for dimethylsulfoxide-mediated cryoprotection, which involves ionic interactions between the oxygen atom from this molecule and phospholipid bilayers. Such a mechanism would stabilize the cell membranes during the freeze-thaw cycle. The findings by us related wit better plant regeneration percentages from cryopreserved calli as compared to calli which had not been cryopreserved could be similar to those of Aronen et al., (1999) in embryogenic cultures of *Abies cephalonica*. We later founded that the storage in liquid nitrogen eliminates a high proportion of cells which had been previously damaged by dimethylsulfoxide, since only small, meristematic cells survive this treatment. Furthermore, the use of cryoprotective mixtures containing other agents greatly minimizes the inherent toxicity of dimethylsulfoxide.

On the other hand, it is recognized that the process of apoptosis (programmed cell death) is not circumscribed to animals, but also occurs in plants, where it is used for the selective elimination and suicide of unwanted cells (Krishnamurthy et al., 2000). According to Joyce et al., (2003), among the cells undergoing this process are those which have sustained high levels of *in vitro* stress, which can compromise their physiology. Such a mechanism might, therefore, be involved in dimethylsulfoxide-mediated toxicity.

After using the simple freezing procedure proposed by Martinez-Montero et al. (1998), it was determined that, apparently, the best dehydration levels are reached by the sugarcane calli when kept for 2 or 3 hours at -40 °C. It should be noted that the survival percentages achieved in this study were comparable to the best values obtained by Jian et al., (1987) and Eksomtramage et al., (1992). Survival rates did not increase with longer pre-freezing times, probably due to excessive dehydration of the material. Studies based on the use of nuclear magnetic resonance spectroscopy in *Catharanthus roseus* cells (Chen et al., 1984), vegetative apple buds (Tyler et al., 1988) and different tissues from *Rhododendron japonicum* (Ishikawa et al., 2000) have determined that the optimum pre-freezing time for a specified pre-freezing temperature depends on the amount of water still remaining inside the cells. Tyler et al., (1988) proved the need for the pre-freezing step when they showed that the incubation of samples at an intermediate negative temperature before immersion in liquid nitrogen would result in a better performance of the cryopreserved material after thawing.

Finally, the optimized protocol carried out by our team took into account the *in vitro* survival and regeneration (plants per 500 mg of sugarcane calli) percentages and was validated for: a) three varieties (CP52-43, C1051-73, C91-301) (Table 1) ; b) explants obtained either from immature inflorescences or immature leaves from *in vitro* plants; c) calli stored for up to 16 months under liquid nitrogen, belonging to the CP52-43 variety (Table 2).

Variety		Survival (%)	Regeneration (plants per 500mg of calluses)
-LN	CP52-43	98,8 a	230 a
	C91-301	69,5 c	72 c
	C1051-73	44,1 d	55 d
+LN	CP52-43	89,0 b	150 b
	C91-301	38,8 d	42 e
	C1051-73	22,2 e	25 f
Typical Error		0,190	1,421

Table 1. Effects of optimized cryopreservation protocol on survival and plant regeneration produced from control (-LN) and cryopreserved (+LN) sugarcane embryogenic calluses (varieties CP52-43, C91-301 and C1051-73). *Means within columns followed by the same letter are not significantly different (ANOVA p < 0,05 Tukey,). Data were transformed for statistical analysis in accordance with* $x' = 2 \arcsin((x/100)^{0,5})$ *and with* $x' = (0,5 + x)^{0,5}$ *for percentage of survival and plant regeneration, respectively.*

	Time (months)	Survival (%)	Calluses that regenerated plants (%)	Regeneration (plants per 500mg of calluses)
-NL	1	100 a	100,0 a	225,2 a
	4	98,6 a	100,0 a	224,3 a
	8	96,7 ab	66,7 b	77,0 c
	12	97,5 ab	7,1 c	19,1 d
	16	11,5 c	0,0 d	0,0 e
+NL	1	90,6 b	96,7 a	149,3 b
	4	87,6 b	97,8 a	142,9 b
	8	88,0 b	93,3 a	135,1 b
	12	86,4 b	97,0 a	141,7 b
	16	90,0 b	97,5 a	140,3 b
Typical Error		0,189	0,253	1,312

Table 2. Effect of extended storage duration on the survival and plantlet produced from control (-LN) and cryopreserved (+LN) sugarcane embryogenic calluses (variety CP52-43). *Means within columns followed by the same letter are not significantly different (ANOVA $p < 0,05$ Tukey,). Data were transformed for statistical analysis in accordance with $x' = 2$ arcsine $((x/100)^{0,5})$ and with $x' = (0,5 + x)^{0,5}$ for percentage of survival and plant regeneration, respectively.*

2.3 Apices

The first attempt to freeze sugarcane apices were carried out by the group of Bajaj et al., (1987) using apices from *in vivo* plants. This material was pretreated with a mixture of 5% (v/v) of each DMSO, sucrose and glycerol during 45min. Then the freezing was accomplished by rapid immersion in liquid nitrogen of samples. However, the apices recovery was very scarce and with only small callus formation without plant regeneration.

Later on, research for the development of a cryopreservation protocol for sugarcane apices was carried out in the framework of collaborative program involving IRD (Institut de recherche pour le développement, Montpellier, France), CIRAD (Centre de coopération internationale en recherche agronomique pour le développement, Montpellier, France), CNIC (Centro Nacional de Investigaciones Cientificas, Havana, Cuba), IPGRI (International Plant Genetic Resources Institute, Rome, Italy) and FAO (Food and Agriculture Organisation of the United Nations, Rome, Italy).

A cryopreservation process using encapsulation/dehydration was set up for apices sampled on *in vitro* plantlets of sugarcane by Paulet et al., (1993) in CIRAD, Montpellier, France. After dissection, apices were cultured for one day on standard medium and then encapsulated in medium with 3% alginate. Optimal conditions comprised preculture for 2d in liquid medium with 0.75M sucrose, desiccation for 6h under the laminar flow or for 10–11 hours with silicagel followed by rapid freezing and slow thawing. Survival after freezing in liquid nitrogen ranged between 38 and 91% for the 5 varieties experimented.

Later on, Gonzalez-Arnao et al. (1993) in CNIC, Havana Cuba investigated the effect of sucrose concentration during the pregrowth treatment and of freezing procedure on the survival of encapsulated apices of six sugarcane varieties. The optimal sucrose concentration was 0.75 M during 24h. We showed that encapsulated apices of sugarcane could withstand freezing in liquid nitrogen using various freezing procedures. Growth recovery of apices after thawing was very rapid and direct, due to the fact that most cells of the apical region had been only slightly harmed, as revealed by histological examination.

Moreover, our group studied apices sampled on *in vitro* plantlets of different varieties and could be cryopreserved using the encapsulation-dehydration technique and stored for one year at the temperature of liquid nitrogen without modification in their recovery percentage (Gonzalez-Arnao et al., 1999). By contrast, apices placed at -70°C or -25°C lost viability very rapidly. There are several explanations for this result: even though vitrification of internal solutes has been observed during freezing of these materials, including sugarcane devitrification and recrystallization processes, which are detrimental to 'cellular integrity, take place at these temperatures (Gonzalez-Arnao et al., 1996). These contrasting results might be linked to the presence of higher levels of residual free water in the latter systems, which would recrystallize rapidly at these temperatures and result in the death of the explants. At lower temperatures comprised between -135 and -196"C, no differences were noted in the regrowth capacity of all materials mentioned above whatever the storage duration tested.

It is interesting to note that, even though the two protocols set up were slightly different, the average results obtained on a total of 15 sugarcane varieties (8 frozen with the CNIC protocol, 7 with the CIRAD protocol) were similar (Table 3). It should also be noted that different varieties showed different sensitivities to preculture and desiccation, and to preculture, desiccation and freezing. However, there was only one case (Ja 60-5) where the difference between control and cryopreserved samples was very high, 70% and 24% survival, respectively. Both protocols are thus potentially applicable to a large range of varieties without any need for further adaptation.

Recently, *in vitro* shoot tips of two clones were successfully cryopreserved using encapsulation-dehydration according to Gonzalez-Arnao et al., (1993) and droplet-vitrification with two vitrification solutions, PVS2 and PVS3 (Barraco et al., 2011). For both clones, encapsulation-dehydration induced significantly higher recovery, reaching 60% for clone H70-144 and 53% for clone CP68-1026, compared with droplet-vitrification in which recovery was 33–37% for clone H70-144 and 20–27% for clone CP68-1026. Optimal conditions included preculture of encapsulated shoot apices for 24 h in liquid medium with 0.75 M sucrose and dehydration with silica gel to 20% moisture content (fresh weight basis) before direct immersion in liquid nitrogen. With both protocols employed, regrowth of cryopreserved samples, as followed by visual observation, was always rapid and direct.

Variety	Survival	
	-LN	+LN
C 87-51 (*)	90	70
C 266-70 (*)	90	86
B 34104 (*)	60	67
B 4362 (*)	88	74
Ja 60-5 (*)	70	24
IAC 5448 (*)	50	38
POJ 2878 (*)	56	60
CP 70-1133 (*)	80	60
CP 68-1026 (**)	100	64
B 69566 (**)	100	91
Co 6415 (**)	80	64
Co 740 (**)	80	38
IAC 5118 (**)	50	38
My 5514 (**)	83	75
Q 90 (**)	100	82

Table 3. Survival of control (-LN) and cryopreserved (+LN) apices using the encapsulationdehydration technique according to the protocol described by Gonzalez-Arnao et al. (1993, *) and Paulet et al. (1993, **).

2.4 Somatic embryos

In Cuba, a micropropagation protocol based on the artificial seed technology has been established for sugarcane, which uses somatic embryos produced on semi-solid medium, an alternative which allows mass multiplication of plants from elite varieties (Nieves et al., 2001; Tapia et al., 1999). However, this protocol has an important limiting factor, which lies with the necessity of safely storing somatic embryos for the long-term (Benson, 2008). Establishing a cryopreservation protocol for somatic embryos would allow solving this problem.

In this sense our group (Martinez-Montero et al., 2008) compared three vitrification-based cryopreservation techniques, viz. vitrification, encapsulation-vitrification and droplet-vitrification for cryopreserving sugarcane somatic embryos. No viability was achieved using the vitrification procedure. The comparison of the recovery pattern of cryopreserved somatic embryos showed that droplet-vitrification procedure was more efficient than encapsulation-vitrification (Table 1) based on the presence of green colour in somatic embryos and on the percentage of clumps with embryos converted into plants. Moreover, the presence of callus together with converted plants was observed with the encapsulation-vitrification procedure.

Protocol	Presence of green colour in SE (%)	Clumps with SE converted into plants (%)	Presence of callus together with converted plants (%)
Encapsulation-vitrification *	8 b	5 b	67 a
Droplet-vitrification**	25 a	18 a	0 b
SEM	0.22	0.21	0.31

Means with different letters are statistically different (*t*-test, $P < 0.05$). The data were transformed before the analysis using $x' = 2 \arcsin((x/100)^{0.5})$. (Number of samples = 144).

*. The selected samples were encapsulated with 3 % (w/v) Na-alginate and loading solution (2 M glycerol + 0.4 M sucrose). The beads formed were dehydrated at 0°C for 10 min in PVS2 before direct immersion in liquid nitrogen.

** The selected samples were loaded in 2 M glycerol + 0.4 M sucrose for 20 min at 25°C. Then, they were placed on the filter paper, transferred in cryotubes and dehydrated at 0°C for 80 min with PVS2. The samples in cryotubes were transferred in droplets of PVS2 solution (10 µl) placed on aluminium foil strips and the strips were immersed in liquid nitrogen.

Table 4. Effect of the cryopreservation protocol on the recovery pattern of sugarcane somatic embryos (SE).

Untreated embryos were white (Fig. 3A & B). Cryopreserved embryos were white to yellow when they were placed on recovery medium; viable embryos turned yellow to green after about 2 weeks; they converted to plants within an additional 2 week period and produced green shoots and roots (Fig. 3 C & D). Callus formation was not observed in germinated embryos and no secondary embryos were produced after the droplet-vitrification procedure (Fig. 3 C). However, callus appeared together with germinated embryos after encapsulation-vitrification (Fig. 3D).

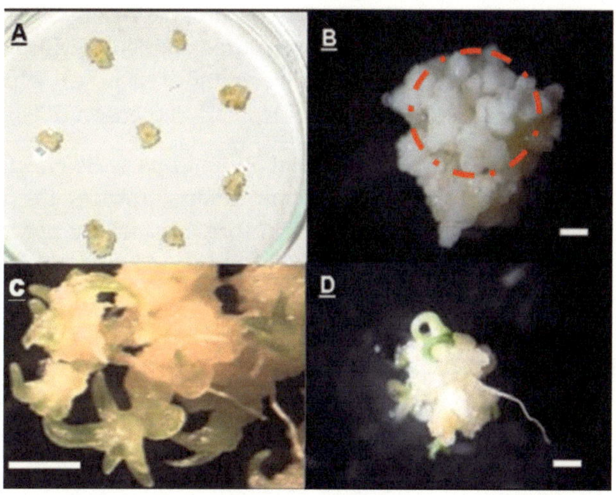

Fig. 3. Initial embryogenic sugarcane callus (A); clumps of somatic embryos selected for cryopreservation experiments (dashed line) (B); recovered clumps of somatic embryos after cryopreservation and 4 weeks after transfer to MS medium under light conditions (C, using droplet-vitrification procedure; D, using encapsulation-vitrification procedure) (bar = 1mm).

The obtained results by Martinez-Montero et al., (2008) contrasted with what is generally observed in the literature, as vitrification is the most frequently employed vitrification-based procedure and it has been applied to a large number of species (Panis & Lambardi, 2006; Sakai & Engelmann, 2007). However, the number of successful reports of application of the droplet-freezing and encapsulation-vitrification techniques is increasing steadily (Engelmann, 2011).

Sugarcane somatic embryos proved very sensitive to PVS2, even though the PVS2 treatment was performed at 0°C, which usually reduces the toxicity of the cryoprotectant solution (Benson, 2008). This high sensitivity rendered the utilization of the vitrification procedure impossible and alternative options had to be sought.

One of the options tested for cryopreservation of sugarcane somatic embryos was the encapsulation-vitrification technique, as developed by Matsumoto et al., (1995). These authors suggested that the toxicity of the PVS2 solution could be reduced by encapsulating the explants in alginate beads. Encapsulation also made the manipulation of the material easier. The positive effect of employing this technique was confirmed by the results, as some viability was achieved after cryopreserving sugarcane embryos using encapsulation-vitrification.

We also tested the droplet-vitrification technique with sugarcane embryos (Martinez-Montero et al., 2008). Droplet vitrification combines the procedure called droplet-freezing, which has been established with cassava (Kartha & Engelmann, 1994) and applied notably to potato (Schäfer-Menuhr et al., 1997) and asparagus shoot tips (Mix-Wagner et al., 2000), in which explants are cooled in a droplet of cryoprotectant solution with the vitrification procedure (Sakai et al., 1990), since explants are cooled in a droplet of PVS2 solution. Droplet-vitrification is relatively easy to implement and generally ensures high recovery after cooling (Sakai & Engelmann, 2007). One of the advantages of this technique is the high cooling and warming rates achieved, compared with others procedures (Benson, 2007; Panis et al., 2005). These high cooling/warming rates ensure complete vitrification during cooling and reduce the risks of devitrification during warming of samples, which is important to avoid the lethal effects of intracellular ice crystal formation (Benson, 2008).

Moreover, Volk & Walters (2006) concluded that PVS2 imparts its effect in the previtrified solution, and at lower temperature the cryoprotectant restricts the mobility of water molecules, so that they are unable to nucleate and ice crystals are not allowed to growth. Benson (2008) empathized that cryoprotection using droplet-vitrification involves a somewhat different principle, due to the behavior of water molecules contained in micro-droplets of vitrification solution. If the biophysical conditions are optimal the droplets can become vitrified on direct exposure to liquid nitrogen.

3. Use of analytical techniques for sugarcane cryopreservation protocols

In the past, cryopreservation protocols have generally been developed using an empirical approach. However, considerable advances have been made in recent years in the use of analytical tools to enhance our current knowledge of the damages induced in biological tissues by cryopreservation (Engelmann, 2011). Various biophysical, biochemical and histo-cytological techniques are available for this purpose (Harding, 1999). Such analytical tools allow the detection of those components of a cryopreservation method which cause the most damage. Usually, these studies are correlated with survival responses and viability testing. However, the application of analytical tools for plant cryopreservation studies is still very scarce and in some cases they are costly to implement and complex to evaluate (Verleysen et al., 2004). Apart mention need the excellent review by Benson (2008) in which it is exposed that contemporary cryopreservation research is now supported by advanced biomolecular or 'omics' technologies, creating a new knowledge base which will hopefully help to solve some of the more difficult cryobiological challenges. However, it will become increasingly so as stakeholders invest in areas commonly interested in low temperature research. Therefore, our research experience is only limited to use non costly and complex analytical techniques yet.

3.1 Effect of cryopreservation on the structural and functional integrity of cell membranes of sugarcane embryogenic callus

Cell membranes are one of the main targets of numerous stressing events, including cryopreservation (Benson, 2007; Fahy et al., 1984; Engelmann, 2011). Various markers, including electrolyte efflux, lipid peroxidation products and cell membrane protein content, reflect the structural and functional integrity status of cell membranes after exposure to such stressing events (Harding, 1999; Verleysen et al., 2004).

Measurement of electrolyte leakage has been used notably for studying the desiccation and cryopreservation sensitivity of various recalcitrant seed species (Sun, 1999). Lipid peroxidation profiles have been used as markers of cell membrane damage during freezing of rice cell suspensions and of the coenocytic alga *Vaucheria sessilis* (Benson et al., 1992; Fleck et al., 1999). Watanabe et al., (1999) have shown that the acquisition of tolerance to cryopreservation of rice cells was related to changes in protein metabolism. An increasing number of proteins and peptides that might contribute to freezing tolerance by reducing the effects of dehydration associated with freezing have been identified (Thomashow, 1999). In the same way, Thierry et al., (1999) have observed in carrot somatic embryos the over-accumulation of boiling-stable proteins, which seems to be related to an increase in tolerance to cryopreservation. Besides, some enzymes, which are induced by low temperature, such as fatty acid desaturase and sucrose phosphate synthase, also contribute to freezing tolerance (Guy, 1999).

Our research group studied the effect of cryopreservation on the structural and functional integrity of cell membranes of sugarcane embryogenic calluses by measuring electrolyte leakage, lipid peroxidation products and membrane proteins (Fig. 4). Firstly, we showed (Martinez-Montero et al., 2002a) that survival and plantlet production were lower with cryopreserved sugarcane embryogenic calluses in comparison with unfrozen control calluses. However, the differences observed between control and cryopreserved calluses in the parameters studied to evaluate membrane structural and functional integrity, including electrolyte leakage, total cell membrane protein content, malondialdehyde and other aldehyde content were only transitory. Indeed, they had all disappeared within 3-4 days after freezing.

Electrolyte leakage, measured to evaluate the overall effect of cryopreservation on the semi-permeability of plasma membranes, revealed a partial loss of membrane semipermeability in callus cells. The transitory character of the electrolyte efflux observed indicates that no dramatic mechanical cell membrane injuries were caused by cryopreservation; rather only reversible lesions were induced by this treatment. As part of this dynamic process, the electrolytes released by damaged cells may have been taken up by living cells.

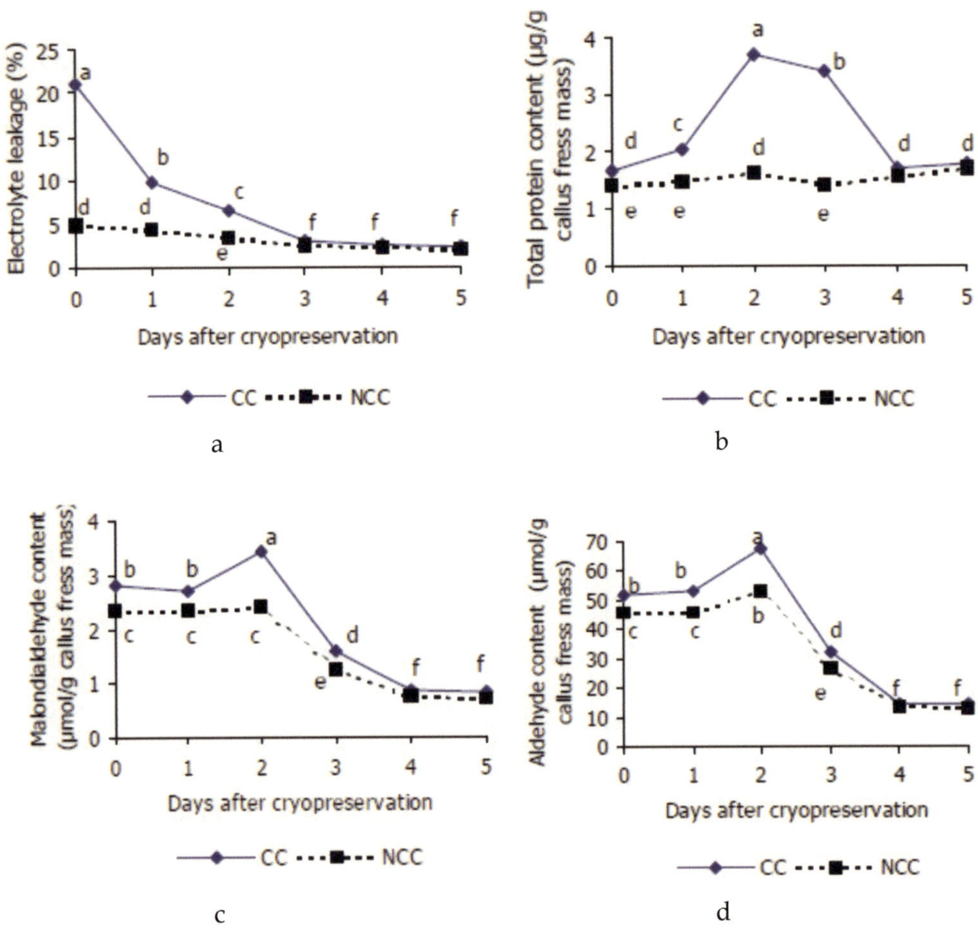

Fig. 4. Effect of cryopreservation on the structural and functional integrity of cell membranes of sugarcane embryogenic calluses by measuring electrolyte leakage (A), total proteins (B), malondialdehyde content (C) and other aldehyde content (D). *Data followed by the same letter are not statistically different (ANOVA, Duncan, $p<0.05$). CC: cryopreserved callus. NCC: non-cryopreserved callus.*

Freezing injury induces the production of free radicals, mainly reactive oxygen species (ROS) (Benson et al., 1992). The ROS signaling pathway is mainly controlled by the production of, and balance between, pro- and antioxidants and the perturbation of ROS homeostasis (Mittler et al., 2004). These changes are perceived by various proteins, enzymes and receptors which influence different developmental, metabolic and defense pathways. Free radicals then attack the lipid fraction of membranes, resulting in the formation of unstable lipid peroxides. These compounds breakdown to form toxic secondary oxidation products (Esterbauer et al., 1988) such as aldehydes, including malondialdehyde and other aldehydic products.

According to our results the main factors affecting sugarcane callus cell membrane damages and electrolyte efflux might thus be oxygen reactive species instead of malondialdehyde and aldehydes themselves, since the highest concentration of these compounds was reached later than the highest level of electrolyte leakage. However, it is also possible that the damages noted after cryopreservation could have been caused by the loss of cellular integrity due to the formation of ice crystals and to the cryoprotectants employed, which could damage the membranes.

We also showed that the content in malondialdehyde and other aldehydes in the microsomal fraction were higher for cryopreserved calluses than unfrozen controls, but only during the first three days after cryopreservation. Benson et al., (1992) have obtained similar results for malondialdehyde with cryopreserved rice cell suspensions. Therefore, they suggested that freezing stress could have caused disruption and uncoupling in some metabolic pathways as reported by Fleck *et al*. (1999) and Dumet et al., (2000) with other biological systems. This could have led to the production of free radicals, thus promoting lipid peroxidation in the cellular membranes of calluses at a very early post-thaw recovery stage.

Variations were also observed in control calluses, concerning mainly electrolyte leakage and lipid peroxidation. The significantly increased levels of malondialdehyde and aldehydes measured during the first 3 days in control calluses might be a result of mechanical membrane damage caused by cutting when preparing the starting material. Fleck et al., (1999) described an increase in lipid peroxidation products after cutting algae filaments into sections. In addition, transfer of material to fresh medium itself is another stress source that may cause an increase in malondialdehyde and aldehydes (Benson, 2007).

The concentration of lipid peroxidation products decreased from the second day onwards and reached a constant value on the fourth day in both frozen and control calluses (Martinez-Montero et al., 2002a). This decrease must have been caused by the activation of antioxidant defense mechanisms. Plants produce antioxidant molecules and have scavenging systems (ß-carotenes, tocopherol isomers, ascorbic acid, glutathione) and enzymatic free radical processing systems (superoxide dismutase, catalase, glutathione reductase, ascorbate peroxidase and various other enzymes) as a protective response to stresses (Leprince et al., 1993). Those antioxidant systems are directly activated by oxidative stress and, consequently, diminish the levels of ROS and thiobarbituric acid reactive substances in cells. Martinez-Montero et al., (2002a) suggested additional experiments to be performed to measure the concentration of such antioxidant molecules and the activity level of the above-cited enzymes in sugarcane embryogenic calluses in relation to cryopreservation.

An increase in cell membrane-related proteins has been described as a response to dehydration and freezing stress (Ausborn et al., 1994). Such proteins are produced as a protective mechanism to preserve membrane structure, ion sequestration and chaperon-like functions (Thierry et al., 1999). According to Martinez-Montero et al., (2002a) the total microsomal fraction protein content was higher in cryopreserved sugarcane calluses during the first 3 d after thawing. This increase was concomitant with an increase in malondialdehyde and aldehyde concentration. They hypothesized that some of the proteins induced by the freeze-thaw cycle may play a role in decreasing the malondialdehyde and aldehyde levels, in addition to the other functions mentioned above.

Even though electrolyte leakage, malondialdehyde, aldehyde and cell membrane protein contents became similar in control and cryopreserved samples 4 d after cryopreservation, cryopreservation consistently affected callus survival and plantlet regeneration (Martinez-Montero et al., 2002a). However, lipid peroxidation products (such as malondialdehyde and aldehydes) might have impaired various cell functions in the sugarcane embryogenic calluses by cross-linking to macromolecules such as DNA and proteins to form mutagenic compounds as reported by Yang & Scaich, (1996) for animal cells. Moreover, the free radicals induced by freezing stress are considered both cytotoxic and genotoxic because they are capable to modify protein structure, to form complexes with DNA and enzymes and to inhibit nucleic acid synthesis (Esterbauer et al., 1988; Grune et al., 1997, Spiteller, 1996). Such impairments might have affected the totipotency of these callus cells.

3.2 Use of electrolyte leakage technique for sugarcane somatic embryos

To allow a quick, reliable prognosis of the experiments performed and to refine the optimal conditions for cryopreservation of somatic embryos, viability was estimated using an electrolyte efflux test by our research group (Martinez-Montero et al., 2008). Firstly, four dissected clumps were either immediately incubated in 20 ml double-distilled water or immersed directly in liquid nitrogen before incubation. The percentage of living cells (cell viability) was calculated by making mixtures of cooled and non cooled clumps (ca. 20 mg total fresh weight). Conductivity of the water was measured before (C0) and after 5, 15 or 25 h of imbibition (Cx). Samples were then autoclaved (30 min at 112°C, 107 kPa) and cooled down to room temperature for 4 h to determine the total conductivity (Ctotal). The percentage of electrolyte leakage was calculated from the ratio: (Cx – C0)*100/(Ctotal – C0). Lastly, a regression analysis was made between the results of the electrolyte leakage test and the cell viability (both were expressed in percentages). The best model that was suitable to represent the experimental data constitutes a standardized curve for analysis of cell viability during all experimentations.

The effect of the imbibitions duration in double-distilled water on the leakage of electrolytes in mixtures of cooled and non-cooled clumps by regression analysis demonstrated that there was a significant linear relationship ($\alpha = 0.05$) between the electrolyte efflux and the viability of somatic embryos and high coefficients of determination ($R^2 > 0.88$) were obtained (Fig. 5). It was clearly observed that a greater functional relationship existed in the lineal equation for imbibitions periods of clumps between 15 and 25 h. Our results confirmed that an equilibration period was necessary for accurate measurement of cell leakage and the electrolyte leakage was practically complete (85% of the electrolytes leaked in a sample with 100% of cooled clumps) after 15 h only. Moreover, the values of electrolyte leakage indicated that close to 10% of the somatic embryos were damaged after dissection for the cryopreservation experiments. Therefore, our test proved useful and precise as it was not only good for distinguishing between living and dead tissues but also for quantifying small differences in the amount of viable tissues.

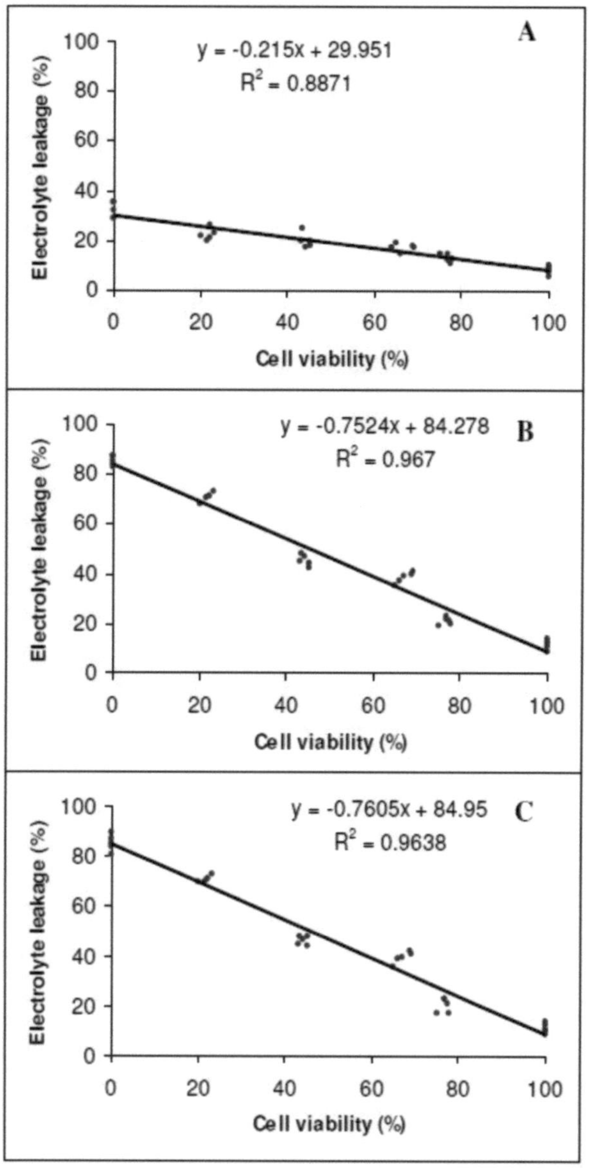

Fig. 5. Effect of the imbibition duration in double-distilled water on the leakage of electrolytes in mixtures of cooled and non-cooled clumps by regression analysis to determine the cell viability of sugarcane somatic embryos. A) after 5 h; B) after 15 h; C) after 25 h.

3.2.1 Optimization of methodology for somatic embryos

The loading treatment is an essential step to achieve high post-rewarming survival of cryopreserved sugarcane somatic embryos because it induces or enhances tolerance of samples to PVS2 treatment (Panis & Thinh, 2001). A loading solution including 2 M glycerol and 0.4 M sucrose is the most commonly employed in cryopreservation protocols (Sakai & Engelmann, 2007). The results obtained by Martinez-Montero et al., (2008) showed that modifying the composition of the loading solution improved viability according to electrolyte efflux test for cryopreserved sugarcane embryos.

This indicates the importance of carefully studying each step of the cryopreservation protocol to optimize survival. We hypothesized that increasing the number of OH groups present in the loading medium progressively decreased viability of control sugarcane somatic embryos, whereas there was an optimum in their number to achieve highest viability after cryopreservation (Fig. 6). It has been suggested that OH groups of sugars/polyalcohols replace water and interact with phospholipids forming hydrogen bonding with membrane phospholipids (Turner et al., 2001). This helps stabilizing cellular membranes during dehydration and cooling and helps maintaining membrane integrity and function through minimizing bilayer disruption and damages (Benson, 2007).

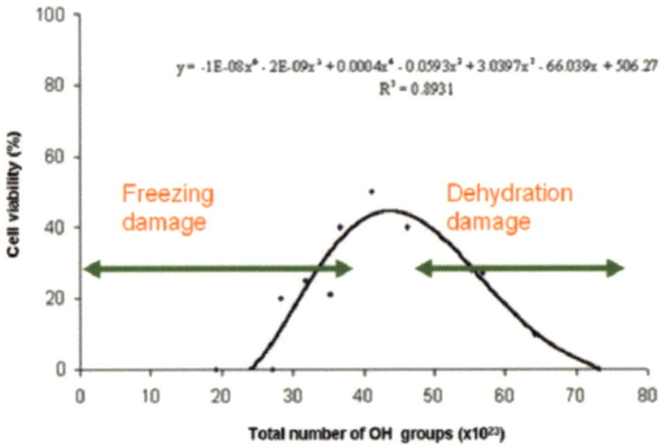

Fig. 6. Effect of the total number of OH groups of glycerol and sucrose on the cell viability of the cryopreserved sugarcane somatic embryos by droplet-vitrification procedure.

From a thermodynamic, kinetic, and structural point of view, the physico-chemical mechanism by which glycerol plus sucrose as co-solvent system can modulate the functionality of a given protein is very important (Baier & McClements, 2005; Ruan et al., 2003). However, the stabilization mechanism of these agents has been attributed to a protein preferential hydration mechanism, as proposed by Timasheff (1993) or to an osmotic stress (Parsegian et al., 1995) where, mathematically, the two mechanisms cannot be distinguished (Parsegian et al., 2000).

In a very well documented paper, Parsegian et al., (2000) indicated that there has been much confusion about the relative merits of different approaches, osmotic stress, preferential interaction (i.e. preferential hydration), and crowding, to describe the indirect effect of solutes on macromolecular conformations and reactions. The two first mechanisms (and crowding) cannot be distinguished as they are derived from the same solution theory. In the preferential hydration model proposed by Timasheff (1993), both the chemical nature and the size of the solute determine water exclusion from the protein surfaces. The osmotic stress emphasizes the role of the water that is necessarily included if solutes are excluded

(Parsegian et al., 2000), dealing also with the movement of water molecules (Parsegian et al., 1995; Stanley et al., 2008).

Duan et al., (2001) observed that the hydroxyl groups present in the glucose units could contribute to protein-sugar interactions in aqueous solutions. These newly formed dipole-dipole interactions could form a hydrophilic layer around the protein units and therefore increase the dispersability of the protein through protein hydration and/or alter the intramolecular interactions in such a way that folding and even dissociation may be favoured (Semenova et al., 2002).

4. Stability assessment

Before using cryopreservation as an additional tool in the overall conservation strategy for any plant material, it is essential to verify that the cryopreservation protocol developed does not have any destabilizing effect and that the plants produced from cryopreserved explants are true to type (Harding, 2004). This author firstly provided a definition for "Cryobionomics" - a novel term describing the remodeled concept of genetic stability and the re-introduction of cryopreserved plants into the environment. Later, Cryobionomics is proposed as an approach to explore links between cryoinjury and genetic instability during and after cryopreservation (Harding et al., 2005, 2008b). There are an increasing number of reports indicating that no changes are observed in the material regenerated from cryopreservation (Engelmann, 1997). However, most of these experiments have been performed very soon after cryopreservation on a small number of individuals, often using material still cultured *in vitro* or after a very short period of growth in vivo and they concern mainly in vitro growth characteristics, or a limited number of biochemical or molecular markers. Only in a limited number of cases (e.g. Côte et al., 2000; Engelmann, 1997; Schäfer-Menuhr et al., 1997) have plants been grown in the field for a long period allowing the assessment of agronomic characteristics.

In the case of sugarcane, numerous experiments have been conducted to study the field behaviour of micropropagated plants (Feldmann et al., 1994; Flynn & Anderlini, (1990); Jackson et al., 1990; Lorenzo et al., 2001; Pena & Stay, 1997), uncovering the occurrence of rejuvenation phenomenons and of epigenetic changes. By contrast, only limited information is available concerning the stability of plants regenerated from cryopreserved material. RFLP analysis did not reveal any difference that could be attributed to cryopreservation between plants of one sugarcane variety produced from control and cryopreserved calluses (Eksomtramage et al., 1992) or cell suspensions (Chowdhury & Vasil 1993). Plants produced from control and cryopreserved shoot tips of one variety were similar as regards pattern of two isoenzymatic systems (Paulet et al., 1993) and six agronomic traits observed during their early growth in vivo (Gonzalez-Arnao et al, 1999).

Moreover, we published data on the field performance of sugarcane plants originating from cryopreserved material (Martinez-Montero et al., 2002b). The field performance of plants produced from embryogenic calluses of one sugarcane commercial hybrid cv. CP52-43 (CP43-64 x CP38-34, Canal Point, USA) cryopreserved using the protocol developed by Martinez-Montero et al. (1998) was evaluated over a period of 27 months by observing several agronomic parameters (Fig. 7). Similar observations were carried out simultaneously for comparison on plants produced from the same callus cultures, but which were not cryopreserved and on plants of the same variety originating from classical macropropagation.

Fig. 7. Regenerated plants from cryopreserved embryogenic sugarcane callus a) during acclimatization (42 days); b, c) after 12 months in the field of the stool; d) after 15 months in the field of the first ratoon.

Treatments were distributed following a randomised block experimental design including four repetitions per treatment. Experimental plots were 7.5 m long with 5 rows each. Intra-row spacing was 0.5 m. The commercial hybrid C266-70 (Co281 x POJ2878, INICA, Cuba) was planted as experiment border. Fertilization at the time of planting included 75 kg/ha urea; 50 kg/ha P_2O_5 and 50 kg/ha K_2O. Additionally, 75 kg/ha urea were supplied 3 months after planting. The following measurements were performed on 100 plants originating from cryopreserved calluses, control calluses and macropropagated plants:

- After 6 months of stool field growth: number of stems per clump (i.e. number of suckers produced from the original plant); stem diameter (cm); stem length (m).
- After 12 months of stool field growth: number of stems per clump; stem diameter (cm); stem length (m); fibre percentage (w/w); juice brix (i.e. mass of sugar (g dry matter) per 100 g of juice, expressed in Brix degree) ; pol (total sugar content) percentage in juice; pol percentage in cane; juice apparent purity tons of pol/ha.
- After cutting of stools and 15 months of field growth of the first ratoon: number of stems per clump; stem diameter (cm); stem length (m); single stem fresh mass (kg); fibre percentage (w/w); juice brix; pol percentage in juice; pol percentage in cane; juice apparent purity; tons of pol/ha.

The results of the evaluation of field grown sugarcane plants after different periods showed significant differences between treatments only during the first six months of field growth of sugarcane stools (Table 4). Stems produced from *in vitro* cultured materials, irrespective of their cryopreservation status, had a smaller diameter and a shorter height than those produced from macropropagated buds. These differences disappeared during the course of the experiment as they were not observed anymore after 12 months of stool field growth.

Parameter measured	Origin of stools			Typical Error
	Cryopreserved calluses	Control calluses	Buds isolated	
Stem diameter (cm)	1,51 b	1,45 b	1,82 a	0,12
Stem length (m)	0,57 b	0,49 b	0,85 a	0,10
Number of stems per meter	5,20 a	5,45 a	4,22 b	0,35

Table 4. Evaluation of several agronomic parameters after six months of field growth of sugarcane stools originating from cryopreserved calluses, control (non-cryopreserved) calluses and buds isolated from macropropagated plants. *Data in rows followed by the same letter are not statistically different (ANOVA, Tukey test, $p < 0,05$).*

No significant difference between treatments was observed for any of the parameters studied after 12 months of stool field growth and 15 months of field growth of the first ratoon (Tables 5 and 6). This study has demonstrated that the differences observed for several agronomic characters between stools originating from cryopreserved and control calluses, and macropropagated material after 6 months of field growth disappeared progressively with time, as no differences could be uncovered in stools after 12 months nor after 15 months of field growth of the first ratoon.

Only very few published reports on this topic deal with comparable number and duration components. No differences have been noted in the vegetative and floral development of several hundreds of palms regenerated from control and cryopreserved oil palm embryogenic cultures, but no detailed account of the observations made has been published (Engelmann, 1997). The most comprehensive study is the comparison of the field behaviour of banana plants regenerated from control and cryopreserved cell suspensions (Côte et al., 2000), which showed that two out of the eleven descriptors analyzed differed between control and cryopreserved material during the first culture cycle but that, similarly to our observations, these differences disappeared during the second culture cycle.

Parameter measured	Origin of stools			Typical Error
	Cryopreserved calluses	Control calluses	Buds isolated	
Stem diameter (cm)	2,58	2,51	2,66	0,09
Stem length (m)	1,85	1,90	1,70	0,11
Stem fresh weight (kg)	1,37	1,33	1,49	0,09
Stem number per meter	9,80	10,55	9,32	1,50
Agriculture recovery (t/ha)	111,88	116,93	115,72	2,62
Juice brix (°Brix)	22,07	21,22	24,69	1,87
Pol percentage in juice (%, w/w)	19,17	18,37	19,35	1,93
Fibre percentage (%, w/w)	11,55	11,23	11,52	0,43
Pol percentage in cane (%, w/w)	16,95	16,30	17,12	1,90
Agro-industrial recovery (t/ha)	18,96	19,06	19,81	1,51

Table 5. Evaluation of several agronomic parameters after 12 months of field growth of sugarcane stools originating from cryopreserved calluses, control (non-cryopreserved) calluses and buds isolated from macropropagated plants. *No statistical differences were found (ANOVA).*

An interesting result in our study is that plants regenerated from control and cryopreserved calluses displayed the same differences in comparison with those originating from macropropagated material. These differences are therefore not induced by cryopreservation but are due to the fact that both groups of plants originate from in vitro cultured material. It is indeed a well known phenomenon that tissue culture induces temporary changes in the behaviour of *in vitro* cultured plants during their early in vivo growth phase (Swartz 1991). Such changes can be induced by *in vitro* culture conditions including notably low light intensity, high humidity, limited gas exchanges, presence of high sucrose concentrations and growth regulators in the medium.

In the case of sugarcane, changes in the field behaviour of plants have been frequently observed after *in vitro* culture. Several authors have reported an increase in the number of new stems per clump (Flynn & Anderlini, 1990; Jackson, 1990; Perez et al., 1999), which generally induces a reduction in the stem diameter and mass. Peña & Stay (1997) stated that, with sugarcane *in vitro* culture stimulated growth and vigour, induced rejuvenation and generally improved agricultural performance. Many authors (Jimenez et al., 1991; Lorenzo et al., 2001; Lourens. & Martin 1987; Taylor et al., 1992) indicate that such differences disappear during the course of field growth and the first clonal multiplication, as observed in our experiments.

Parameter measured	Origin of stools			Typical Error
	Cryopreserved calluses	Control calluses	Buds isolated	
Stem diameter (cm)	2,62	2,59	2,65	0,10
Stem length (m)	1,93	1,98	2,04	0,16
Stem fresh weight (kg)	1,50	1,51	1,54	0,21
Stem number per meter	9,77	9,65	10,09	2,20
Agriculture recovery (t/ha)	122,10	121,42	129,48	5,12
Juice brix (°Brix)	23,22	23,02	22,98	1,79
Pol percentage in juice (%, w/w)	20,45	20,07	19,95	1,85
Fibre percentage (%, w/w)	12,9	12,17	12,5	0,35
Pol percentage in cane (%, w/w)	17,81	17,62	17,45	1,80
Agro-industrial recovery (t/ha)	21,75	21,35	22,59	1,51

Table 6. Evaluation of several agronomic parameters after 15 months of field growth of the first sugarcane ratoon originating from cryopreserved calluses, control (non-cryopreserved) calluses and buds isolated from macropropagated plants. *No statistical differences were found (ANOVA).*

In conclusion, the results obtained in our study validate the cryopreservation protocol developed by Martinez-Montero et al. (1998) for embryogenic calluses. Cryopreservation of embryogenic calluses will thus be incorporated in the scheme established by the Centro de

Bioplantas for mass production of *in vitro* sugarcane plants by means of somatic embryogenesis.

5. Cryopreservation protocols for pineapple

Pineapple is vegetatively propagated and crosses between varieties produce botanical seeds. However, these seeds are highly heterozygous and therefore of limited interest for the conservation of specific gene combinations. Cryopreservation of apices is the most relevant strategy for long-term conservation of vegetatively propagated crops, since true to type; virus free plants can be regenerated directly from cryopreserved apices (Lynch et al., 2007).

Vitrification and encapsulation-dehydration techniques have been widely applied for successfully cryopreserve apices of a large number of different crops which do not require sophisticated equipment for freezing and produce high recovery rates with a wide range of materials (Engelmann, 2010, 2011). Most vitrification protocols use a loading treatment and a stringently timed dehydration with PVS (Benson, 2008b). This two-step procedure has allowed tissues to be more tolerant to osmotic stress and to resist the chemical toxicity induced by the highly concentrated cryoprotective solutions. Exposure duration to PVS is usually not longer than 2 h (Thinh et al., 1999).

The encapsulation-dehydration method for cryopreservation is based on the fact that encapsulation protects the explants and preculture in medium enriched with osmoticum makes them tolerant to air drying (Fabre & Dereuddre, 1990). Preculture in 0.75 M sucrose and desiccation to about 25% water content in beads (fresh weight basis) are the most common conditions used (Gonzalez-Arnao et al., 1996). The application of vitrification solutions to dehydrate encapsulated cells or shoot tips has also been successfully applied to several species (Engelmann, 2004).

5.1 Apices

The first successful result related with the cryopreservation of pineapple (*Ananas comosus*) apices was reported by our group in Cuba (Gonzalez-Arnao et al., 1998b). The encapsulation and vitrification techniques were experimented for freezing apices of pineapple *in vitro* plantlets. Positive results were achieved using vitrification only (Fig. 8). Optimal conditions included a 2 day preculture of apices on medium supplemented with 0.3M sucrose, loading treatment for 25 min in medium with 0.75M sucrose + 1M glycerol and dehydration with PVS2 vitrification solution at 0°C for 7 h before rapid immersion in liquid nitrogen. This method resulted in ranged survival (25-65%) depending of the genotypes. Recovery of cryopreserved apices took place directly, without transitory callus formation.

The negative results after cryopreservation of pineapple apices by encapsulation-dehydration technique can be related to the high sensitivity of pineapple apices to sucrose and dehydration. Pregrowth in media with sucrose concentrations higher than 0.5M was detrimental to survival and a prolonged treatment in 0.5M sucrose was required to improve survival after desiccation, but it was not sufficiently to obtain survival of apices after freezing. The viability loss observed after freezing may be due to the crystallization of remaining freezable water upon freezing. This detrimental crystallization might be avoided by slowly freezing of the encapsulated apices, which would result in removing the remaining freezable water by means of freeze-induced dehydration. Several plant materials cryopreserved by encapsulation-dehydration technique have required slow freezing regime to achieved optimal survival (Engelmann, 2010).

In another set of experiments, our group obtained positive results again only with vitrification (Martinez-Montero et al., 2005). The best protocol comprised a 2-d preculture on semi-solid MS medium supplemented with 0.3 M sucrose, a loading treatment in liquid medium containing 0.4 M sucrose + 2 M glycerol for 25 min, and dehydration for 7h at 0°C with PVS3 before immersion into liquid nitrogen. The highest survivals of apices were: Smooth Cayenne (45%), Cabezona (33%) and Red Spanish (25%).

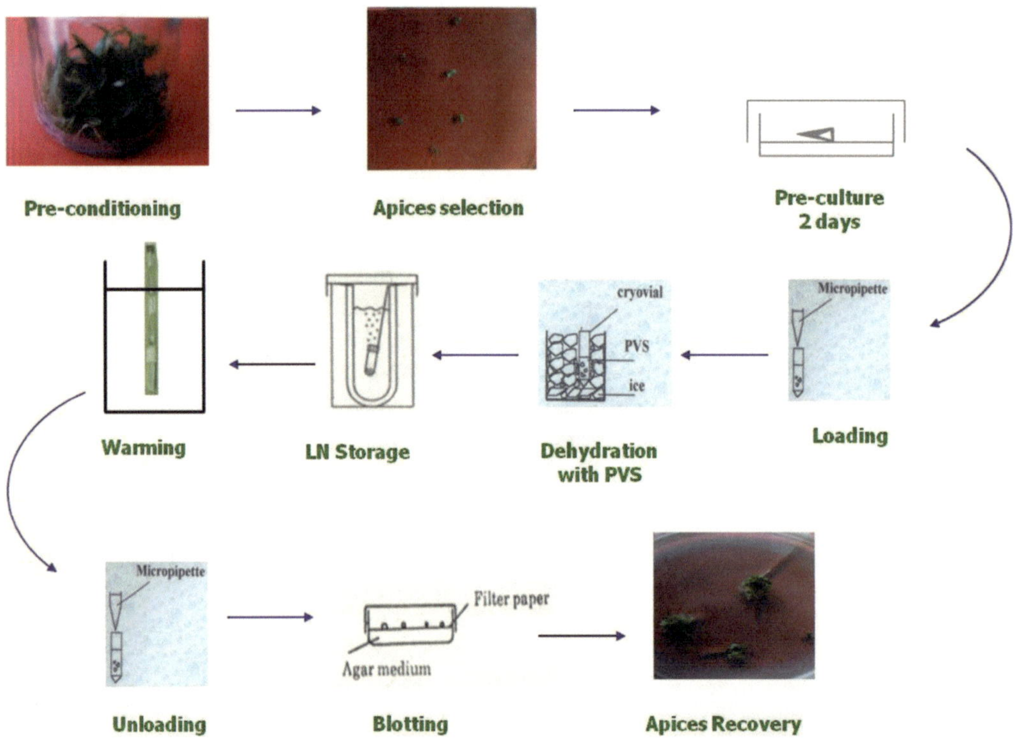

Fig. 8. Cryopreservation protocol established for pineapple apices using vitrification technique.

However, contrary to most vitrification reports, pineapple apices required a prolonged exposure (7 h) to the vitrification solutions (Engelmann, 2010). This result is probably due to the large size, and compact structure of the pineapple apices employed in our experiments with success (Fig. 9): the apices were around 3mm long, and comprised the apical dome tightly covered by 2-3 leaf primordial with a very thick cuticle. Long treatment durations were needed for the vitrification solution to sufficiently dehydrate these very compact structures.

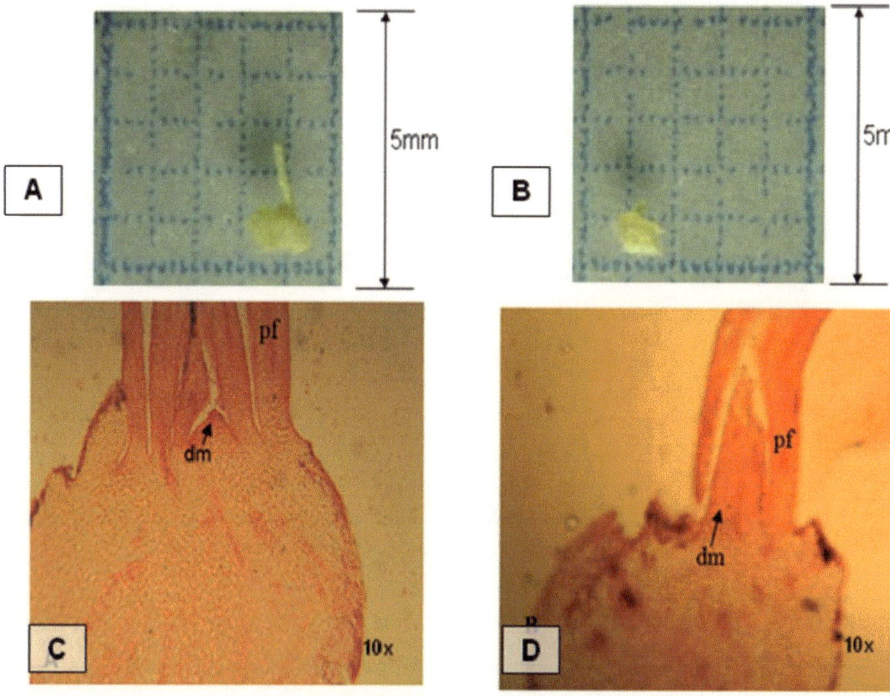

Fig. 9. Dissected pineapple (*Ananas comosus* L. Merrill cv. MD2) shoot tips as viewed by stereo-microscope (A, B); and by light microscopy 10x (C, D). Dissected apices type I (A, C) with apical dome (dm) and 3-4 primordial leaves (pf) and mechanical damaged apices type II (B, D) with one primordial leaf used as controls.

5.1.1 Extension of vitrification protocol

In the case of pineapple, it is believe as symbol in the province of Ciego de Avila (Cuba) due to great cultivated areas dedicated to this crop. In our Institution (Bioplantas Centre) is located the unique field collection of pineapple germplasm in the country. However, this field genebank is prone to disease, or damage through natural disaster and need very high maintenance. For this reason, the cryopreservation of apices obtained from vitroplants could constitutes the most relevant strategy for long-term conservation of pineapple germplasm, since true to type and virus-free plants can be regenerated directly from cryopreserved apices (Martinez-Montero et al., 2005).

The successful application of the vitrification protocol for nine accessions of the *in vitro* collection at Bioplantas Centre was accomplished with the following conditions: type of shoot tip (consisted in meristematic dome area and 3-4 primordial leaves with 2,5 - 3 mm in size); 0,3 mol.L^{-1} sucrose preculture during 2 days; application of the loading solution (0,4 mol.L^{-1} sucrose + 2 mol.L^{-1} glycerol) during 25 min at 25°C; dehydration with plant vitrification solution number three (PVS3: 50% w/v glycerol + 50% w/v sucrose) during 7 hours at 0°C. The results per accessions expressed as percentage of recovery before (-LN) and after (+LN) cryopreservation for six apices per replicate, four replicates per treatment and each experiment was repeated three times are showed in Table 7.

	Survival (%)	
	-LN	+LN
Cayenne of Puerto Rico	80.2	65.5
Perolera	49.9	33.8
Smooth Cayenne of Serrana	50.3	25.3
Cabezona	61.5	27.9
Piña Blanca of Caney	57.9	24.7
P3R5	53.1	20.0
Red Spanish of Caney	45.5	12.1
MD2	80.1	60.2
Bromelia sp.	33.1	6.3

Table 7. Effect of vitrification protocol on survival of apices from eight pineapple accessions and one related specie (*Bromelia* sp.) before (-LN) and after cryopreservation (+LN).

5.1.2 Optimization of methodology for pineapple apices

Further modifications to the procedures might be useful in order to reduce the exposure duration to PVS and achieve higher survival after cooling. Therefore, the objective was to develop a more effective cryopreservation protocol using both vitrification and encapsulation/vitrification. As previously we reported (Gonzalez-Arnao et al., 1998b; Martinez-Montero et al., 2005), pineapple apices are sensitive to sucrose and dehydration exposures. A progressive treatment increasing sucrose concentrations was effective at enhancing their tolerance to dehydration and cooling.

Proline has been shown to have a beneficial effect in several cryopreservation protocols (Luo & Reed, 1997; Rasmunsen et al., 1997; Rudolf & Crowe, 1985; Thierry et al., 1999). In our experiments we confirmed that a 2-d progressive preculture in a mixture of sucrose and proline improved the results obtained after cooling in comparison with sucrose alone. This modification in pregrowth also considerably reduced the required dehydration time in PVS. As previously reported, apices of Puerto Rico variety treated for 2 d in 0.3 M sucrose needed an extended exposure (7 h) to PVS2 at 0°C to achieve high levels of survival (65 %) after cryopreservation. However, following the same vitrification approach, higher survival (72%) was obtained using the best pretreatment in sucrose-proline and only 30 min of exposure to PVS2. As regards PVS3, we also demonstrated that 30 min or 1 h of dehydration were also enough to obtain higher survival (76 %) after cryopreservation.

The role of proline has been associated with its ability to act as source of nitrogen and carbon for reparative post-stress processes (Rabbe & Lova, 1984), to increase the non-freezable fraction of water (Rasmussen et al, 1997), to inhibit membrane mixing and to stabilize proteins during dehydration and freezing (Rudolf & Crowe, 1985). On the other hand, the combination of chemical cryoprotectants may also improve the response of tissues to cryopreservation in comparison with the application of one chemical alone. As reported for wasabi apices, a mixture of glycerol with sucrose was more effective at enhancing their tolerance to dehydration and deep cooling than a preculture in sucrose alone (Sakai et al, 2000).

Dehydration at 0°C instead of 25°C for both vitrification solutions gave better results, as previously reported for pineapple apices by our group. This low temperature reduces the toxicity of the vitrification solutions and increases the potential period of exposure (Withers & Engelmann, 1997). In all our cryopreservation experiments, dehydration with PVS3 at 0°C gave higher recovery rates compared with PVS2, even for diverse genotypes. The encapsulation-vitrification method using PVS3 gave greater survival than the vitrification procedure. Additionally, the manipulation of encapsulated apices permits handling large quantities of material that from the practical point of view is more convenient. These results corroborated that encapsulation-vitrification may also be very useful for cryopreserving desiccation-sensitive germplasm such as pineapple, that could not be successfully cryopreserved using an encapsulation-dehydration approach (Gonzalez-Arnao et al., 2003).

The cryopreservation protocol presented here improved survival and shortened the process compared with previous protocols from our group (table 8). The optimal conditions involved the encapsulation of pineapple apices in calcium alginate (3%) beads, followed by a 2-d progressive preculture in liquid medium with 0.16 M sucrose + 0.3 M proline for 24 h, then 0.3 M sucrose + 0.3 M proline for 24h, a loading treatment for 25 min in 0.75 M sucrose + 1M glycerol solution at room temperature and dehydration for 60 min with PVS3 at 0°C before rapid immersion into liquid nitrogen.

		Regrowth (%)			
Variety	Vitrification Solution	Encapsulation-Vitrification		Vitrification	
		-LN	+LN	-LN	+LN
MD-2					
	PVS2	75±2.7 [a]	39±1.9 [d]	68±1.0 [b]	46±2.0 [c]
	PVS3	76±1.9 [a]	54±2.8 [c]	72±2.0 [b]	48±1.1 [d]
Puerto Rico	PVS2	91±1.3 [a]	83±1.8 [c]	88±1.5 [b]	72±2.0 [d]
	PVS3	93±2.2 [a]	83±1.1 [b]	88±1.4 [a]	76±0.7 [c]

Different letters within rows imply significant differences according to Tukey test ($P<0.05$) and each PVS.

Table 8. Comparison of the encapsulation-vitrification and vitrification procedures on regrowth (%) of pineapple apices after dehydration with PVS2 or PVS3 solutions at 0°C.

5.2 Embryogenic callus

The results of some studies indicated *Fusarium subglutinans* isolates cause fusariose which constitutes the most serious pineapple disease and causes losses as high as 80% of marketable pineapple fruit. It produces phytotoxins in culture that were phytotoxic on calluses (Jin et al., 1996; Kaur et al., 1987) and the correlation between pineapple variety and the toxicity of culture filtrates suggests that filtrates could be used to screen in vitro for disease resistance (Borras et al., 2001). Therefore, the cryopreservation of pineapple calluses can provide a means of effective source when in vitro screening of germplasm for fusariose disease would be attempted. Storing calluses in liquid nitrogen could preserve their regeneration capacity and limits the risk of somaclonal variation, which increases with culture duration.

A simplified freezing protocol mentioned before for sugarcane embryogenic calluses (Martinez-Montero et al., 1998) was used for pineapple calluses of the genotypes "Smooth Cayenne" and "Perolera" (Martinez-Montero et al., 2005). For cryopreservation experiments, 15 to 20 day-old calluses, about 3 to 6 mm in diameter, were employed. Calluses were pretreated with a cryoprotective solution containing 5 to 20 % (v/v) dimethylsulfoxide (DMSO) and 0.5M sucrose for 1h at 0°C. After freezing calluses were transferred directly to recovery medium (MS medium supplemented with dicamba:BAP (2.5:0.5 $mg.L^{-1}$) and citric acid (0.1 $mg.L^{-1}$)). The survival, evaluated 45 days after thawing, corresponded to the percentage of calluses which had increased in size during the recovery period.

As results, the survival of calluses after pretreatment was high and similar for the two genotypes studied (Table 8). It decreased only with 20% of DMSO. After freezing in liquid nitrogen, survival was achieved with 10 and 15% DMSO only and was highest with 15% DMSO (57-67%). Re-growth of successfully cryopreserved calluses was very rapid and they increased in size during the recovery period. With this work, the application of a simplified freezing protocol achieved survival from used pineapple genotypes and this confirm that our previously simplified freezing protocol for sugarcane (Martinez-Montero et al., 1998) can be used wider to others species.

		Survival (%)			
		Smooth Cayenne		Red Spanish	
DMSO (%)	Sucrose (M)	-LN	+LN	-LN	+LN
5	0.5	85±11 a	0 c	87±6 a	0 c
10		81±8 a	25±3 b	85±6 a	30±4 b
15		75±12 a	57±7 a	80±6 a	65±6 a
20		55±6 b	0 c	60±4 b	0 c

Table 9. Effect of cryoprotective solution (DMSO + sucrose) on the survival rate (%) before (-LN) and after (+LN) application of cryopreservation protocol for calluses of two pineapple genotypes. *Values represent means of 50 samples from three replicate experiments, ± SE. Means within columns followed by the same letter are not significantly different (ANOVA p<0,05 Tukey,). Data were transformed for statistical analysis in accordance with* $x' = 2 \arcsin((x/100)^{0,5})$ *for percentage of survival.*

6. Conclusions

For vegetatively propagated species, cryopreservation has a wide applicability both in terms of species coverage, since protocols have been successfully established for root and tubers, fruit trees, ornamentals and plantation crops, both from temperate and tropical origin and in terms of numbers of genotypes/varieties within a given species. With a few exceptions, vitrification-based protocols have been employed. It is also interesting to note that in many cases, different protocols can be employed for a given species and produce comparable results. Survival is generally high to very high. Regeneration is rapid and direct, and callusing is observed only in cases where the technique is not optimized. Different reasons can be mentioned to explain these positive results. The meristematic zone of apices, from which organised growth originates, is composed of a relatively homogenous population of small, actively dividing cells, with little vacuoles and a high nucleocytoplasmic ratio. These characteristics make them more susceptible to withstand desiccation than highly vacuolated and differentiated cells. As mentioned earlier, no ice formation takes place in vitrification-based procedures, thus avoiding the extensive damage caused by ice crystals which are formed during classical procedures. The whole meristem is generally preserved when vitrification-based techniques are employed, thus allowing direct, organised regrowth. By contrast, classical procedures often lead to destruction of large zones of the meristems, and callusing only or transitory callusing is often observed before organised regrowth starts. Other reasons for the good results obtained are linked with tissue culture protocols. Many vegetatively propagated species successfully cryopreserved until now are cultivated crops, often of great commercial importance, for which cultural practices, including *in vitro* micropropagation, are well established. In addition, *in vitro* material is "synchronized" by the tissue culture, and pregrowth procedures and relatively homogenous samples in terms of size, cellular composition, physiological state and growth response are employed for freezing, thus increasing the chances of positive and uniform response to treatments. Finally, vitrification-based procedures allow using samples of relatively large size (shoot tips of 0.5 to 2–3 mm), which can regrow directly without any difficulty. Cryopreservation techniques are now operational for large-scale experimentation in an increasing number of cases. In view of the wide range of efficient and operationally simple techniques available, any vegetatively propagated species should be amenable to cryopreservation, provided that the tissue culture protocol is sufficiently operational for this species.

7. Acknowledgment

The authors thank the Food and Agriculture Organization of the United Nations (FAO), the International Plant Genetic Resources Institute (IPGRI, actually Bioversity International) and the International Foundation for Science (IFS) for partly funding the research programme.

8. References

Anchordoguy, T.J.; Cecchini, C.A.; Crowe, J.H.; Crowe, L.M. (1991) Insights into the cryoprotective mechanism of dimethyl sulphoxide for phospholipid bilayers. *Cryobiology* 28(5), (October 1991), 467-473. ISSN: 0011-2240

Arakawa, T.; Carpenter, J.F.; Kita, Y.A.; Crowe, J.H. (1990) The basis for toxicity of certain cryoprotectants: An hypothesis. *Cryobiology* 27(4), (August 1990), 401-15. ISSN: 0011-2240

Aronen, T.S.; Krajnakova, J.; Haggman, H.M., Ryynannen, L.A. (1999) Genetic stability of cryopreserved embryogenic clones of white spruce (*Picea glauca*). *Plant Cell Rep.* 18, 948-953. ISSN: 0721-7714

Ashmore, S.E. (1997) *Status reports on the development and application of in vitro techniques for the conservation and use of plant genetic resources.* International Plant Genetic Resources Institute, ISBN 92-9043-339-6, Rome, Italy.

Ausborn, M.; Schreier, H.; Brezesinski, G.; Fabian, H.; Meyer, H.W.; Nuhn, P. (1994) The protective effect of free and membrane bound cryoprotectants during freezing and freeze-drying of liposomes. *J Control Release* 30, 105-116. ISSN: 0168-3659

Baier, S.K. & McClements, D.J. (2005) Influence of cosolvent systems on the gelation mechanism of globular protein: Thermodynamic, kinetic, and structural aspects of globular protein gelation. *Comprehensive Reviews in Food Science and Food Safety* 4, 43-54. ISSN: 1541-4337

Barraco, G.; Sylvestre, I.; Engelmann, F. (2011) Comparing encapsulation-dehydration and droplet-vitrification for cryopreservation of sugarcane (*Saccharum* spp.) shoot tips. *Scientia Horticulturae*, 130(1), (August 2011), 320-324. ISSN: 0304-4238

Benson, E.E. (2008a) Cryopreservation of phytodiversity: A critical appraisal of theory and practice. *Critical Reviews in Plant Sciences* 27(3), 141-219. ISSN: 0735-2689

Benson, E.E. (2008b) Cryopreservation theory. In: *Plant Cryopreservation: A Practical Guide.* Reed, B.M. Ed., pp. 15–32. Chapt. 2. Springer, ISBN 978-0-387-72275-7, New York, USA.

Benson, E.E., Lynch, P.T., Jones, J. (1992) Variation in free radical damage in rice cell suspensions with different embryogenic potentials. *Planta* 188: 296–305. ISSN: 0032-0935

Borras, O.; Santos, R.; Matos, A.; Cabral, R.; Arzola, M. (2001) A first attempt to use a *Fusarium subglutinans* culture filtrate for the selection of pineapple cultivars resistant to fusariose disease. *Plant Breed.* 120: 435-438. ISSN: 0179-9541

Chen, T.H.H.; Kartha, K.K.; Constable, F.; Gusta, L.V. (1984) Freezing characteristics of cultured *Catharanthus roseus* (L.) G. Don cells trated with dimethylsulfoxide and sorbitol in relation to cryopreservation. *Plant Physiol.* 75, 720-725. ISSN: 0032-0889

Chowdhury, M.K.U. & Vasil, I.K. (1993) Molecular analysis of plant regenerated from embryogenic cultures of hybrid sugarcane cultivars (*Saccharum* spp.). *Theor. Appl. Genet* 86, 181-188. ISSN: 0040-5752

Côte, F.X.; Goué, O.; Domergue, R.; Panis, B.; Jenny, C. (2000) In-field behavior of banana plants (*Musa* spp.) obtained after regeneration of cryopreserved embryogenic cell suspensions. *Cryo-Letters* 21, 19-24. ISSN: 0143-2044

Cyr, D.R. (2000) Cryopreservation: roles in clonal propagation and germplasm conservation of conifers. In: *Cryopreservation of tropical plant germplasm - Current research progress and applications.* Engelmann, F.; Takagi, H., Eds. (261-268) Tsukuba: JIRCAS; Rome: IPGRI. ISBN 92-9043-428-7, Rome, Italy.

Duan, X.Q.; Hall, J.A.; Nikaido, H.; Quicho, F.A. (2001) Crystal structures of the maltodextrin/maltose-binding protein complexed with reduced oligosaccharides: flexibility of tertiary structure and ligand binding. *Journal of Molecular Biology* 306:5, (9 March 2001), 1115-1126. ISSN: 0022-2836

Dumet, D.; Block,W.; Worland, R.; Reed, B.M.; Benson, E.E. 2000. Profiling cryopreservation protocols for *Ribes ciliatum* using differential scanning calorimetry. *Cryo-Letters* 21: 367-378. ISSN: 0143-2044

Eksomtramage T.; Paulet F.; Guiderdoni E.; Glaszmann J.C.; Engelmann F. (1992) Development of a cryopreservation process for embryogenic calluses of a commercial hybrid of sugarcane *(Saccharum* sp.) and application to different varieties. *Cryo-Letters* 13, 239-52. ISSN: 0143-2044.

Engelmann, F. & Takagi, H. Eds. (2000) *Cryopreservation of tropical plant germplasm - Current research progress and applications*. Tsukuba: JIRCAS; Rome: IPGRI; ISBN 92-9043-428-7, Rome, Italy.

Engelmann, F. (1997) Importance of desiccation for the cryopreservation of recalcitrant seed and vegetatively propagated species. *Plant Genet. Res. Newsletter* 112, 9-18. ISSN: 1020-3362

Engelmann, F. (1997) *In vitro* conservation methods. In: *Biotechnology and Plant Genetic Resources Conservation and Use.* BV Ford-Lloyd, JH Newburry & JA Callow (Eds). pp. 119-161. ISBN: 0851991424, CABI, Wallingford.

Engelmann, F. (2000) Importance of cryopreservation for the conservation of plant genetic resources. In: *Cryopreservation of tropical plant germplasm - Current research progress and applications*. Engelmann, F.; Takagi, H., Eds. (8-20) Tsukuba: JIRCAS; Rome: IPGRI; ISBN 92-9043-428-7, Rome, Italy.

Engelmann, F. (2004) Plant cryopreservation: progress and prospects. *In Vitro Cell Dev Biol Plant* 40, 427–433. ISSN 1054-5476

Engelmann, F. (2010) Use of biotechnologies for the conservation of plant biodiversity. *In Vitro Cell Dev Biol Plant* DOI 10.1007/s11627-010-9327-2, ISSN 1054-5476

Engelmann, F. (2011) Cryopreservation of Embryos: An Overview. In: *Plant Embryo Culture: Methods and Protocols, Methods in Molecular Biology*. Trevor, A.T. & Yeung, E.C. Eds. Vol. 710, DOI 10.1007/978-1-61737-988-8_13, ISBN 978-1-61737-987-1, Springer Science+Business Media, LLC 2011.

Esterbauer, H.; Zollner, H.; Schauer, R.J. (1988) Hydroxyalkenals: Citotoxics products of lipid peroxidtion. In: *ISI Atlas of Sciences: Biochemistry* 311-317. ISBN. 0-941708-00-4

Fabre, J. & Dereuddre, J. (1990) Encapsulation-dehydration: A new approach to cryopreservation of *Solanum* shoot tips. *Cryo-Letters* 11, 413 – 426. ISSN: 0143-2044

Fahy, G.M.; Lilley, T.H.; Linsdell, H.; Douglas, M.S.J.; Meryman, H.T. (1990) Cryoprotectant toxicity and cryoprotectant toxicity reduction: in search of molecular mechanisms. *Cryobiology* 27(3), (June 1990), 247-68. ISSN: 0011-2240

Fahy, G.M.; McFarlane, D.R.; Angell, C.A.; Meryman, H.T. (1984) Vitrification as an approach to cryopreservation. *Cryobiology*, 21(4), (August 1984), 407-426. ISSN: 0011-2240

Feher, A.; Pasternak, T.P.; Dudits, D. (2003) Transition of somatic cells to an embryogenic state. *Plant Cell, Tissue and Organ Culture* 74, 201-228. ISSN: 0167-6857

Feldmann, P.; Sapotille, J.; Gredoire, P.; Rott, P. (1994) Micropropagation of sugarcane. In: *In vitro culture of tropical plants*. C Teisson (ed). CIRAD, ISBN 2-87614-162-0, Montpellier, pp. 15-17.

Finkle, B.J. & Ulrich, J.M. (1979) Effect of cryoprotectants in combination on the survival of frozen sugarcane cells. *Plant Physiology* 63, 598-604. ISSN: 0032-0889

Finkle, B.J. & Ulrich, J.M. (1982) Cryoprotectant removal as a factor in the survival of frozen rice and sugarcane cells. *Cryobiology* 19(3), (June 1982) 329-335. ISSN: 0011-2240

Finkle, B.J. ; Zavala, M.E.; Ulrich, J.M. (1985) Cryoprotective compounds in the viable freezing of plant tissues. In: *Cryopreservation of plant cells and organs.* Kartha, K.K. Ed., 243-267, ISBN 0-8493-6102-8, CRC Press, Boca Raton. USA.

Fleck, R.A.; Day, J.G.; Clarke, K.J.; Benson, E.E. (1999) Elucidation of the metabolic and structural basis for the cryopreservation recalcitrance of *Vaucheria sessilis, Xanthophyceae. Cryo-Letters* 20, 271–282. ISSN: 0143-2044

Flynn, L. & Anderlini, T. (1990) Disease incedence and yield performance of tissue culture-generated seed cane over the crop cycle in Lousiana. *Journal of American Society of Sugar Cane Technologists* 10, 113. ISSN 1991-8178

Gnanapragasam, S. & Vasil, I.K. (1990) Plant regeneration from a cryopreserved embryogenic cell suspension of a commercial sugarcane hybrid (*Saccharum* sp.) *Plant Cell Rep.* 9, 419-423. ISSN: 0721-7714

Gnanapragasam, S. & Vasil, I.K. (1992) Cryopreservation of immature embryos, embryogenic callus and cell suspension cultures of gramineous species. *Plant Science* 83, 205-215. ISSN: 0168-9452

Gonzalez-Arnao, M.T.; Engelmann F.; Huet C.; Urra C (1993) Cryopreservation of encapsulated apices of sugarcane: Effect of freezing procedure and histology. *Cryo-Letters* 14, 303-308. ISSN: 0143-2044

Gonzalez-Arnao, M.T.; Engelmann. F.; Urra-Villavicencio. C.; Morenza, M.; Rios, A. (1998a) Cryopreservation of citrus apices using the encapsulation-dehydration technique. *Cryo-Letters* 19,177-182. ISSN: 0143-2044

Gonzalez-Arnao, M.T.; Juarez, J.; Ortega, C.; Navarro, L.; Duran-Vila, N. (2003) Cryopreservation of ovules and somatic embryos of citrus using the encapsulation-dehydration technique. *Cryo-Letters* 24 (2), 85-94. ISSN: 0143-2044

Gonzalez-Arnao, M.T.; Moreira, T.; Urra, C. (1996) Importance of pregrowth with sucrose and vitrification for the cryopreservation of sugarcane apices using encapsulation-dehydration. *Cryo-Letters* 27, 141-148. ISSN: 0143-2044

Gonzalez-Arnao, M.T.; Ravelo, M.M.; Urra-Villavicencio, C.; Martinez-Montero, M.M.; Engelmann, F. (1998b) Cryopreservation of pineapple (*Ananas comosus*) apices. *Cryo-Letters* 19, 375-382. ISSN: 0143-2044

Gonzalez-Arnao, M.T.; Urra, C.; Engelmann, F.; Ortiz, R.; de la Fe, C. (1999) Cryopreservation of encapsulated sugarcane apices: Effect of storage temperature and storage duration. *Cryo-Letters* 20, 347-352. ISSN: 0143-2044

Grune, T.; Michel, P.; Sitte, N.; Eggert, W.; Albrecht-Nebe, H.; Esterbauer, H.; Siems, W.G. (1997) Increased levels of 4-hydroxynonenal modified proteins in plasma of children with autoimmune diseases. *Free Radical Biology and Medicine* 23 (3), (November 1996), 357-360. ISSN: 0891-5849

Guy, C. (1999) Molecular responses of plants to cold shock and cold acclimation. *J Mol Microbiol Biotechnol.* 1(2), 231-242. ISSN: 1464-1801

Harding, K. (1999) Stability assessments of conserved plant germplasm. In: *Plant Conservation & Biotechnology*. Chapt. 7. Benson, E.E. Ed., pp. 97–107, Taylor and Francis Ltd., ISBN 0-7484-0746-4, London, UK.

Harding, K. (2004) Genetic integrity of cryopreserved plant cells: A review. *Cryo-Letters 25, 3-22.* ISSN: 0143-2044

Harding, K.; Johnston, J.; Benson, E. E. (2005) Plant and algal cell cryopreservation: issues in genetic integrity, concepts in 'Cryobionomics' and current European applications. In: *Contributing to a Sustainable Future*. Benett, I.J.; Bunn, E.; Clarke, H.; McComb, J. A., Eds., (112–119) Proc. Australian Branch IAPTC & B, Perth, Western Australia.

Ishikawa, M.; Ide H.; Price, W.S.; Arata, Y.; Kitashima, T. (2000) Freezing behaviors in plant tissues as visualized by NMR microscopy and their regulatory mechanisms. In: *Cryopreservation of tropical plant germplasm - Current research progress and applications.* Engelmann, F.; Takagi, H., Eds. (22-35) Tsukuba: JIRCAS; Rome: IPGRI. ISBN 92-9043-428-7, Rome, Italy.

Jackson, W.; Wagnespack, H.; Richard, C.; Garrison, D.; Lester, W. (1990) CP65-357 Kleentek test in Lousiana. *Journal of the American Society of Sugar Cane Technologist* 21, 10-13. ISSN 1991-8178

Jian, L.C.; Sun, D.L.; Sun, L.H. (1987) Sugarcane callus cryopreservation. In: *Plant cold hardiness.* Li, P.H. Ed., 323-337, Alan R. Liss, Inc., ISBN 0-8451-1084-8, New York.

Jiménez, E.; Pérez, J.; Gil, V.; Herrera, J.; García, Y.; Alfonso, E. (1995) Sistemas para la propagación de caña de azúcar. In: *Avances de la Biotecnología Moderna* Estrada M, Riego E, Limonta E, Tellez P, Fuentes J (Eds).. Elfos Scientiae. Cuba 3:11.2

Jin, H.; Hartman, G.L.; Nickell, C.D.; Widholm, J.M. (1996) Phytotoxicity of culture filtrate from *Fusarium solani*, the causal agent of Soybean Sudden Death Syndrome. *Plant Dis.* 80: 922-927. ISSN: 0191-2917

Joyce, S.M.; Cassells A.C.; Jain, M.S. (2003) Stressand aberrant phenotypes *in vitro* culture. *Plant Cell Tissue and Organ Culture* 74, 103-21. ISSN:0167-6857

Kaczmarczyk, A.; Shvachko, N.; Lupysheva, Y.; Hajirezaei, M.R.; Keller, E.R.J. (2008) Influence of altering temperature preculture on cryopreservation results for potato shoot tips. *Plant Cell Rep.* 27: 1551-1558. ISSN: 0721-7714

Kartha, K.K. & Engelmann F. (1994) Cryopreservation and germplasm storage. In: *Plant cell and tissue culture.* Vasil, I.K. & Thorpe, T.A. Eds. (195–230) Kluwer, ISBN 978-0-7923-2493-5, Dordrecht.

Kartha, K.K.; Fowke, L.C.; Leung, N.L.; Caswell, K.L.; Hakman I. (1988) Induction of somatic embryos and plantlets from cryopreserved cell cultures of white spruce (*Picea glauca*). *J. Plant Physiol* 132, 529-39. ISSN: 0176-1617

Kaur, G.; Singh, U.S.; Garg, G.K. (1987) Mode of action of toxin isolated from *Fusarium oxysporum* f. sp. ciceri. *Indian Phytopathol* 40: 76-84. ISSN 0367-973X

Kinoshita, K.; Li, S.J.; Yamazaki, M. (2001) The mechanism of the stabilization of the hexagonal II (HII) phase in PE membranes in the presence of low concentrations of dimethyl sulfoxide. *Eur Biophysics J* 30, 207-220. ISSN 0175-7571

Klimazewska, K.; Ward, C.; Cheliak, W.M. (1992) Cryopreservation and plant regeneration from embryogenic cultures of larch (*Larix* x *eurolepis*) and black spruce (*Picea mariana*). *J Exp Bot* 43, 73-79. ISSN: 0022-0957

Krishnamurthy, K.V.; Kishnaraj, R.; Chozavendam, R.; Samuel, C.F. (2000) The programme of cell death in plants and animals. A comparison. *Curr Sci* 79, 1169-1181. ISSN: 0011-3891

Krishnaraj, S. & Vasil, I.K. (1995) Somatic embryogenesis in herbaceous monocots. In: *In vitro embryogenesis in plants. Current Plant Science and Biotechnology in Agriculture,*

Thorpe, T.A. Ed. Kluwer Academic Publishers; 20, 417-469. ISBN: 0-7923-3149-4, Dordrecht , Netherlands.

Kristensen, M.M.H.; Find, J.I.; Floto, F.; Moller, J.D.; Norgaard, J.V.N.; Krogstrup, P. (1994) The origin and development of somatic embryos following cryopreservation of an embryogenic suspension culture of *Picea sitchensis*. *Protoplasma* 182, 65-70. ISSN: 0033-183X

Leprince, O.; Hendry, G.A.F.; McKersie, B.D. (1993) The mechanisms of desiccation tolerance in developing seeds. *Seed Science Research*, 3: 231-246. ISSN: 0960-2585

Lorenzo, J.C.; Ojeda, E.; Espinosa, A.; Borroto, C. (2001) Field performance of temporary bioreactor-derived sugarcane plants. *In Vitro Cell Dev. Biol.-Plant*. 37, 803-806. ISSN 1054-5476

Lourens, A.G. & Martin, F.A. (1987) Evaluation of *in vitro* propagated sugarcane hybrids for somaclonal variation. *Crops Science* 27, 793-796. ISSN: 0011-183X

Lunn, J.E. & MacRae, E. (2003) New complexities in the synthesis of sucrose. *Current Opinion in Plant Biology* 6, 208-224. ISSN: 1369-5266

Luo, J. & Reed, B.M. (1997) Abscisic acid-responsive protein bovine serum albumin and proline pretreatments improve recovery of *in vitro* currant shoot tips and callus cryopreserved by vitrification. *Cryobiology* 34:240-250. ISSN: 0011-2240

Lynch, P.T.; Benson, E.E.; Harding, K. (2007) Climate change: the role of ex situ and cryo-conservation in the future security of economically important, vegetatively propagated plants. *Journal of Horticultural Science & Biotechnology* 82 (2), 157–160. ISSN: 1462-0316

Maddox, A.D.; Gonsalves, F.; Shields, R. (1983) Successful preservation of suspension cultures of three *Nicotiana* species at the temperature of liquid nitrogen. *Plant Science Letters* 28, 157-162. ISSN: 0304-4211

Martinez-Montero, M.E.; Gonzalez-Arnao, M.T.; Borroto-Nordelo, C.; Puentes-Diaz, C.; Engelmann, F. (1998) Cryopreservation of sugarcane embryogenic callus using a simplified freezing process. *Cryo-Letters*, 19, 171-176. ISSN: 0143-2044

Martinez-Montero, M.E.; Lorenzo, J.C.; Ojeda, E.; Quiñones, J.; Mora, N.; Sánchez, M.; Iglesias, A.; Martínez, J.; Castillo, R. (2006) Methodology for the cryopreservation of calli with embryogenic structures for the culture of sugarcane. *Biotecnología Aplicada* 23(4), 360-375. ISSN: 1027-2852

Martínez-Montero, M.E.; Martínez, J.; Castillo, R.; Engelmann, F.; González-Arnao, M.T.; (2005) Cryopreservation of pineapple (*Ananas comosus* (L.) Merr) apices and calluses. *Acta Hort. (ISHS)* 666, 127-131. (ISSN 0567-7572

Martinez-Montero, M.E.; Martinez, J.; Engelmann, F. (2008) Cryopreservation of sugarcane somatic embryos. *Cryo-Letters*, 29(3), 229-242. ISSN: 0143-2044

Martinez-Montero, M.E.; Mora, N.; Quiñones, J.; Gonzalez-Arnao, M.T.; Engelmann, F.; Lorenzo, J.C. (2002a) Effect of cryopreservation on the structural and functional integrity of cell membranes of sugarcane embryogenic calluses. *Cryo-Letters* 23, 237-244. ISSN: 0143-2044

Martinez-Montero, M.E.; Ojeda, E.; Espinosa, A.; Sánchez, M.; Castillo, R.; Gonzalez-Arnao, M.T.; Engelmann, F.; Lorenzo, J.C. (2002b) Field performance of cryopreserved callus-derived sugarcane plants. *Cryo-Letters* 23(1), 21-26. ISSN: 0143-2044

Matsumoto T, Sakai A, Takahashi C y Yamada K (1995) Cryopreservation of *in vitro*-grown apical meristems of wasabi (Wasabia japonica) by encapsulation-vitrification method. CryoLetters. 16: 189-196. ISSN: 0143-2044

Matsumoto, T.; Sakai, A.; Takahashi, C.; Yamada, K. (1995) Cryopreservation of *in vitro*-grown apical meristems of wasabi (*Wasabia japonica*) by encapsulation-vitrification method. *Cryo-Letters* 16, 189-196. ISSN: 0143-2044

McGann, L.E. & Walterson, M.L. (1987) Cryoprotection by dimehyl sulfoxide and dimethyl sulfone. *Cryobiology* 24(1), (February 1987), 11-16. ISSN: 0011-2240

Merkle, S.A.; Parrot, W.A.; Flinn, B.S. (1995) Morphogenicaspects of somatic embryogenesis. In: *In vitro embryogenesis in plants. Current Plant Science and Biotechnology in Agriculture,* Thorpe, T.A. Ed. Kluwer Academic Publishers; 20, 155-204. ISBN: 0-7923-3149-4, Dordrecht, Netherlands.

Mittler, R.; Vanderauwera, S.; Gollery, M.; Van Breusegem, F. (2004) Reactive oxygen gene network of plants. *Trends Plant Sci.* 9, 490–498. ISSN: 1360-1385

Mix-Wagner, G.; Conner, A.J.; Cross, R.J. (2000) Survival and recovery of asparagus shoot tips after cryopreservation using the "droplet" method. *New Zeal. J. Crop & Hort Sci.,* 28, 283-287. ISSN: 0114-0671

Murashige, T.; Skoog, F. (1962) A revised medium for rapid growth and bioassays with tobacco tissue cultures. *Physiol. Plant.* 15: 473-497. ISSN: 0031-9317

Nieves, N.; Martinez-Montero, M.E.; Castillo, R.; Blanco, M.A.; González-Olmedo, J.L. (2001) Effect of abscisic acid and jasmonic acid on partial desiccation of encapsulated somatic embryos of sugarcane. *Plant Cell Tissue and Organ Culture* 65, 15-21. ISSN: 0167-6857

Niu, D.K.; Wang, M.G.; Wang, Y.F. (1997) Plant cellular osmotica. *Acta Biotheoretica* 45(2), 161-169. ISSN: 0001-5342

Panis, B. & Lambardi, M. (2006) Status of cryopreservation technologies in plants (crops and forest trees). In: *The role of Biotechnology in Exploring and Protecting Agricultural Genetic Resources,* Ruane, J. & Sonnino, A. Eds. (61-78) FAO, ISBN 92-5-105480-0, Rome, Italy.

Panis, B. & Thinh, N.T. (2001) *Cryopreservation of Musa germplasm,* INIBAP Technical Guideline 5. Escalant, J.V. & Sharrock, S. Eds., International Network for the Improvement of Banana and Plantain, ISBN: 2-910810-45-3, Montpellier, France. IPGRI, Rome, Italy.

Panis, B.; Piette, B.; Swennen, R. (2005) Droplet vitrification of apical meristems: a cryopreservation protocol applicable to all *Musaceae. Plant Sci.* 168, 45–55. ISSN: 0168-9452

Parsegian, V.A.; Rand, R.P.; Rau, D.C. (1995) Macromolecules and water: Probing with osmotic stress. In: *Methods in Enzymology: Energetics of Biological Macromolecules.* Johnson, M.L. & Ackers, G.K. Eds. 259, (43-94), ISBN: 978-0-12-182160-9, Academic Press Inc., New York.

Parsegian, V.A.; Rand, R.P.; Rau, D.C. (2000) Osmotic stress, crowding, preferential hydration, and binding: A comparison of perspectives. *Proceedings of the National Academy of Sciences USA,* 3987-3992. ISSN-0027-8424

Paulet, F.; Engelmann, F.; Glaszmann, J.C. (1993) Cryopreservation of apices of *in vitro* plantlets of sugarcane (*Saccharum* sp. hybrids) using encapsulation-dehydration. *Plant Cell Rep.* 12, 525-529. ISSN: 0721-7714

Pena, M. & Stay, M. (1997) Ventajas de la propagacion in vitro. *Canaveral* 3(24), 473–497. ISSN: 1026-0781

Pérez, J.N.; Jiménez, E.; Gómez, R. (1999) Mejora genética de la caña de azúcar mediante la inducción de mutaciones y la selección in vitro. In: *Biodiversidad y Biotecnología de la Caña de Azúcar.* A Arencibia & MT Cornide (eds.). pp. 79-92, ISBN: 959-235-015-9, Elfos Scientiae, Cuba.

Rabe, E. and C. J. Lovatt (1984). De novo arginine biosynthesis in leaves of phosphorus-deficient Citrus and Poncirus species. *Plant Physiology* 76(3): 747-752. ISSN: 0032-0889

Rasmussen, P.H.; Jorgensen, B. & Nielsen, J. (1997) Cryoprotective properties of proline in cod muscle studied by differential scanning calorimetry. *Cryo-Letters* 18, 293-300. ISSN: 0143-2044

Reed, B.M. (2001) Implementing cryogenic storage of clonally propagated plants. *Cryo-Letters* 22, 97-104. ISSN: 0143-2044

Reinhoud, P.J.; Van Iren, F.; Kijne, J.W. (2000) Cryopreservation of undifferentiated plant cells. In: *Cryopreservation of tropical plant germplasm - Current research progress and applications.* Engelmann, F.; Takagi, H., Eds. (91-102) Tsukuba: JIRCAS; Rome: IPGRI. ISBN 92-9043-428-7, Rome, Italy.

Ruan, K.; Xu, Ch.; Li, T.; Li, J.; Lange, R.; Balny, C. (2003) The thermodynamic analysis of protein stabilization by sucrose and glycerol against pressure-induced unfolding: The typical example of the 33-kDa protein from spinach photosystem II. *European Journal of Biochemistry* 270, 1654–1661. ISSN: 0014-2956

Rudolph, A.S. & Crowe, J.H. (1985) Membrane stabilization during freezing: the role of two natural cryoprotectants, trehalose and proline. *Cryobiology* 22, 367-377. ISSN: 0011-2240

Sakai, A. & Engelmann, F. (2007) Vitrification, encapsulation-vitrification and droplet-vitrification: a review. *Cryo-Letters* 28, 151–172. ISSN: 0143-2044

Sakai, A.; Kobayashi, S.; Oiyama, I. (1990) Cryopreservation of nucellar cells of navel orange (*Citrus sinensis* Osb. var brasiliensis Tanaka) by vitrification. *Plant Cell Rep.* 9, 30-33. ISSN: 0721-7714

Sakai, A.; Matsumoto, T.; Hirai, D.; Charoensub, R. (2002) Survival of tropical apices cooled to -196°C by vitrification. In: *Plant cold hardiness, gene regulation and genetic engineering.* Li, P.H.; Palva, E.T. Eds., 109-119, Kluwer Academic/Plenum Publishers, ISBN: 0306472864, New York.

Schäfer-Menuhr, A.; Schumacher, H.M.; Mix-Wagner, G. (1997) Long-term storage of old potato varieties by cryopreservation of shoot-tips in liquid nitrogen. *Plant Genet. Res. Newsletter* 111, 19-24. ISSN: 1020-3362

Semenova, M.G.; Antipova, A.S.; Belyakova, L.E. (2002) Food protein interactions in sugar solutions. *Current Opinions in Colloid Interface Science* 7, 438–444. ISSN: 1359-0294

Spiteller, G. (1996) Enzymic lipid peroxidation-α consequence of cell injury? *Free Radical Biology and Medicine* 21(7), 1003-1009. ISSN: 0891-5849

Stanley, C.; Krueger, S.; Parsegian, V.A.; Rau, D.C. (2008) Protein structure and hydration probed by SANS and osmotic stress. *Biophysical Journal* 94, 2777–2789, (April 2008), ISSN: 0006-3495

Sun, W.Q. (1999) State and phase transition behaviors of *Quercus rubra* seed axes and cotyledonary tissues: relevance to the desiccation sensitivity and cryopreservation of recalcitrant seeds. *Cryobiology*, 38(4), (June 1999), 372-385. ISSN: 0011-2240

Swan, T.W.; O´Hare, D.; Gill, R.A.; Lynch, P.T. (1999) Influence of preculture conditions on the post-thaw recovery of suspension cultures of Jerusalem artichoke (*Helianthus tuberosus* L.). *Cryo-Letters* 20, 325-336. ISSN: 0143-2044

Swartz, H.J. (1991) Post culture behaviour: genetic and epigenetic effects and related problems. In: *Micropropagation. Technology and Application.* PC Debergh & RH Zimmermann (eds.). pp. 95-121, Kluwer Academic Publishers, ISBN: 9780792308195, Dordrecht, Netherlands.

Tapia, R.; Castillo, R.; Blanco, M.A.; González, J.L.; Sánchez, M.; Rodríguez, Y. (1999) Induction, maduration and encapsulation of sugarcane somatic (*Saccharum* spp) cv CP5243. *Biotecnología Aplicada* 16, 20-23. ISSN: 1027-2852

Taylor, P.; Ko, H.; Adkins, S.; Rathus, C.; Birch, R. (1992) Establishment of embriogenic callus and high protoplast yielding suspension cultures of sugarcane (*Saccharum* spp. hybrids) *Plant Cell Tissue and Organ Culture* 28: 69-78. ISSN: 0167-6857

Tetteroo, F.A.A. (1996) *Desiccation tolerance of somatic embryoids.* PhD Thesis, Wageningen Agricultural University, The Nederlands. ISBN 90-5485-517-7

Thierry, C.; Florin, B.; Petiard, V. (1999) Changes in protein metabolism during the acquisition of tolerance to cryopreservation of carrot somatic embryos. *Plant Physiol. Biochem.* 37, 145-154. ISSN: 0981-9428

Thinh, N.T.; Takagi, H.; Yashima, S. (1999) Cryopreservation of in vitro-grown shoot tips of banana (*Musa* spp.) by vitrification method. *Cryo-Letters* 20(3), 163-174. ISSN: 0143-2044

Thomashow, M.F. (1999) Plant cold acclimation: Freezing tolerance genes and regulatory mechanisms. *Annu. Rev. Plant Physiol. Plant Mol. Biol* 50, 571-599. ISSN:1040-2519

Timasheff, S.N. (1993) The Control of Protein Stability and Association by Weak Interactions with Water: How Do Solvents Affect These Processes? *Annual Review of Biophysical and Biomolecular Structure* 22, 67-97, (June 1993), ISSN: 1056-8700

Turner, S.; Senaratna, T.; Touchell, D.; Bunn, E.; Dixon, K.; Tan, B. (2001) Stereochemical arrangement of hydroxyl groups in sugar and polyalcohol molecules as an important factor in effective cryopreservation. *Plant Science* 160, 489-497. ISSN: 0168-9452

Turner, S.R.; Krauss, S.L.; Bunn, E. Senaratna, T.; Dixon, K.W.; Tan, B.; Touchell, D.H. (2001) Genetic fidelity and viability of *Anigozanthos viridis* following tissue culture, cold storage and cryopreservation. *Plant Science* 161, 1099-1106. ISSN: 0168-9452

Tyler, N.; Stushnoff, C.; Gusta, L.V. (1988) Freezing of water in dormant vegetative apple buds in relation to cryopreservation. *Plant Physiol* 87, 201-205. ISSN: 0032-0889

Ulrich, J.M.; Finkle, B.J.; Moore, P.H. (1984) Frozen preservation of cultured sugarcane cells. *SugarCane 3*, 11-14.

Ulrich, J.M.; Finkle, B.J.; Moore, P.H.; Ginoza H. (1979) Effect of a mixture of cryoprotectants in attaining liquid nitrogen survival of callus culturesof a tropical plant. *Cryobiology* 16, 550-556. ISSN: 0011-2240

Verleysen, H.; Samyn, G.; Van Bockstaele, E.; Debergh, P. (2004) Evaluation of analytical techniques to predict viability fter cryopreservation *Plant Cell Tissue and Organ Culture* 77, 11-21. ISSN: 0167-6857

Volk, G.M. & Walters, C. (2006) Plant vitrification solution 2 lowers water content and alters freezing behaviour in shoot tips during cryoprotection. *Cryobiology* 52, 48–61. ISSN: 0011-2240

Watanabe, K.; Kuriyama, A.; Kawai, F.; Kanamori, M. (1999) Effect of cryoprotectant treatment and post-thaw washing on the survival of cultured rice (*Oryza sativa* L.) cells after cryopreservation. *Cryo-Letters* 20, 377-382. ISSN: 0143-2044

Withers, L.A. & King, P.J. (1980) A simple freezing unit and routine cryopreservation method for plant cell cultures. *Cryo-Letters* 1, 213-220. ISSN: 0143-2044

Withers, L.A. (1985) Cryopreservation of cultures plant cells and protoplasts. In: *Cryopreservation of plant cells and organs.* Kartha, K.K. Ed., 243-267, CRC Press, ISBN 0-8493-6102-8, Boca Raton. USA.

Yang, M.H. & Schaich, K.M. (1996) Factors affecting DNA damage caused by lipid hydroperoxides and aldehydes. *Free Radical Biology and Medicine* 20(2), (July 1995), 225-236. ISSN: 0891-5849

Yoshida, S.; Hattanda, Y.; Suyama, T. (1993) Variationsin chilling sensitivity of suspension cultured cells of mung bean (*Vigna radiata* (L.) Wilczek) during the growth cycle. *Plant Cell Physiology* 34, 673-679. ISSN 0032-0781

19

Somatic Embryogenesis and Cryopreservation in Forest Species: The Cork Oak Case Study

Conceição Santos
CESAM & Department of Biology, University of Aveiro, Aveiro
Portugal

1. Introduction

1.1 General concepts

It is widely accepted that propagation and conservation strategies are crucial in any forest breeding program and that plant biotechnology has powerful tools to propagate and preserve selected genotypes (Jain, 1999; Park et al., 1988; Park, 2002). In particular, in vitro techniques have demonstrated to be essential in a large number of agricultural and forest breeding programs, allowing large scale clonal propagation of elite genotypes (Pinto et al., 2008) or of endangered forest species (Brito et al., 2009). In vitro techniques are also essential in breeding programs involving genetic improvement (e.g. using techniques such as somatic hybridization or genetic transformation) or involving long term germplasm conservation by cryopreservation (Benson, 2008; Fernandes et al., 2008).

Somatic embryogenesis offers advantages in improving forest species over other in vitro propagation methods (Park, 2002), namely: a) somatic embryos simultaneously possess both the shoot and root meristems; so a distinct rooting stage, which usually involves stressing procedures, is not necessary; b) during somatic embryogenesis, embryos/clusters are often formed faster and potentially at extremely high numbers per explant; c) somatic embryogenesis is amenable to automation, which means it may become cheaper than other clonal propagation techniques (Pinto et al., 2002; 2008); d) finally, a robust and efficient protocol of somatic embryogenesis will allow that embryogenic clonal lines can be preserved for long periods (e.g., in liquid N_2) while corresponding plants are transferred to field conditions and are monitored for their characteristics.

The combination of somatic embryogenesis and conservation strategies in breeding programs allows that one may develop high-value forest clonal varieties merely by recovering from N_2 those genotypes/clones that showed best characteristics in the field. These selected genotypes can later be used for advanced breeding programs and commercial forestry (Park, 2002).

From the exposed advantages, it is generally accepted among breeders that breeding programs using in vitro clonal strategy require as a pre-requisite that a cloning technique (preferably somatic embryogenesis) is well established. Then it also requires the optimization of long-term genetic testing methodologies of clonal lines and long term conservation. The combination of these strategies will then enable large-scale production and disposition of tested clonal lines in industrial forest management. For example, micropropagation processes are well refined for spruce and larch species to support their commercial application (http://cfs.nrcan.gc.ca/ factsheets/conifersomatic). However, for most pine species, it is much more difficult to obtain somatic embryogenesis, though interesting advances are in course (e.g. Park et al., 2010). Similarly, the application of this technology to most forest dicotyledonous species, as is the case of *Quercus* genus, has demonstrated to be difficult due to species general recalcitrance to in vitro culture (Santos, 2008).

1.2 *Quercus suber* in vitro cloning: A reliable protocol?

Cork oak (*Quercus suber* L.) belongs to Fagaceae, an important family of forest trees in the Northern hemisphere dominating temperate forests and Mediterranean ecosystems. In particular, cork oak is an abundant species in the Atlantic and West Mediterranean countries where it is an important component of Mediterranean ecosystems (Pinto et al., 2002). Cork is the bark of the oak, which is a natural, renewable and sustainable raw material product of economic interest for a range of applications. Due to its enormous economical importance, intense research has been focused on cork valuable material and, more recently, on cork oak germplasm-conservation programs (Fernandes et al., 2008).

In the last decades, studies were done to improve protocols of cork oak in vitro micropropagation and conservation (namely cryopreservation). In particular, the currently available rates of success in cork oak plant regeneration by somatic embryogenesis (Fernandes et al., 2011; Lopes et al., 2006; Loureiro et al., 2005; Pinto et al., 2001; 2002) and in cork oak material cryopreservation (Fernandes et al., 2008; Fernandes, 2011) are highly encouraging for researchers and breeders to consider the integration of these strategies in breeding and conservation programs of this species (Figure 1).

Recently, it was also established a Portuguese consortium to identify and characterize cork oak ESTs Gene responses to several biotic and abiotic stresses as well as to developmental conditioning are currently being screened and data will be of upmost importance to cork oak researchers and breeders (http://www.fct.pt/apoios/projectos/consulta/projectos.phtml.en).

For long it has been assumed that *Quercus* species have, at some extent, recalcitrant responses to micropropagation in general, and to somatic embryogenesis in particular. Most common strategies for cork oak micropropagation use stem cuttings or use juvenile material or leaves for somatic embryogenesis (Santos, 2008; Fernandes et al., 2008). When utilizing material selected from adult field trees as explant sources, the use of greenhouse forced sprouts instead of directly collected field material is strongly advised.

The developed micropropagation by stem cutting is efficient with juvenile and, less, mature genotypes. Briefly, after disinfection with sodium hypochloride, explants (with 1-2 apical and/or lateral buds) are inoculated on WPM ("Woody Plant Medium", Lloyd and McCown, 1980) medium containing benzylaminopurine (BAP 0.5 mg/L) and naphthalene acetic acid (NAA 0.1 mg/L). After multiplication and elongation, shoots are exposed to an indol-butiric acid shock for rooting. Plants in this stage are then ready for acclimatization (Pires et al., 2003; Figure 2a).

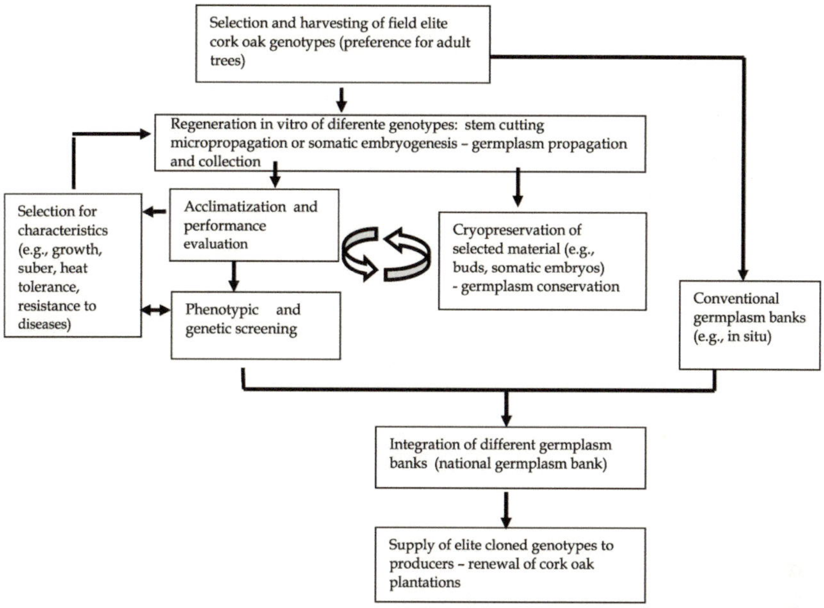

Fig. 1. Schematic representation of a proposed strategy of integrating cork oak micropropagation and cryopreservation technologies in Portuguese breeding programs of this species (adapted from Santos, 2008).

As reported above, somatic embryogenesis is a regeneration strategy with enormous potential for breeding programs. Somatic embryos were developed in *Q. canariensis* (Bueno et al., 1996; 2000), *Q. rubra* (Vengadesan & Pijut, 2009), *Q. serrata* (Ishii et al., 1999; Takur et al., 1999), *Q. robur* (Cuenca et al., 1998; Endemann et al., 2002; Wilhelm et al., 1999), *Q. acutissima* (e.g., Kim, 2000) and *Q. petrea* (Chalupa, 2005). However, not only most studies use juvenile sources of explants (e.g., zygotic embryos and seedlings), but also plant conversion frequencies are still low, supporting the recalcitrance of these species.

Q. suber somatic embryogenesis was obtained first from juvenile plants (e.g., Bueno et al., 1996; 2000; Pinto et al., 2001) and later from leaf explants of mature plants (e.g., Hernandez et al., 2003; Lopes et al., 2006; Pinto et al., 2002; Santos et al., 2007) (Figure 2b,c).

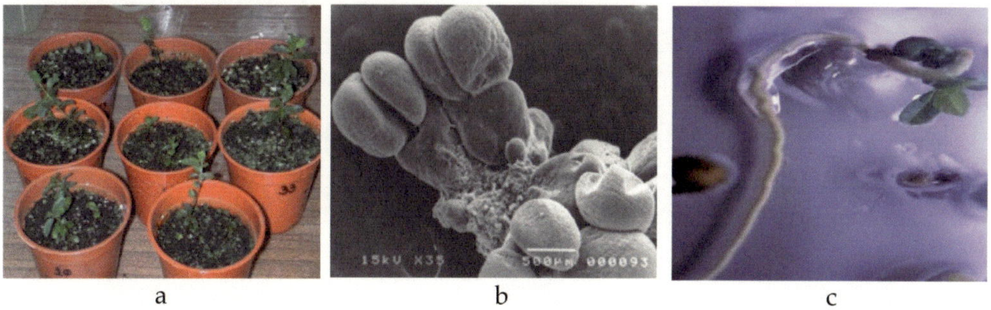

Fig. 2. Micropropagation from field mature cork oak trees: a) Acclimatized plants micropropagated by stem cuttings; b) Scanning electon microscopy of two cotyledonary somatic embryos; c) converted embling (Adapted from Pires et al., 2003; Pinto et al., 2002; Santos, 2008).

Fig. 3. Enhanced protocol of our group for cork oak somatic embryogenesis. MS medium - Murashige and Skoog, 1962; MSWH - MS medium with no growth regulators. (Adapted from Fernandes, 2011).

The initially protocol developed by Santos and collaborators (Pinto et al., 2002; Lopes et al., 2006) was, however, not sufficiently efficient for large scale propagation and for immediate transfer to industrial breeding programmes of cork oak. Meanwhile, those and other authors reported the deficient maturation of somatic embryos during somatic embryogenesis, as the main cause for low conversion rates in this species (Chalupa, 2005; Fernandes et al 2011; Hernández et al., 1999).

Hernández et al. (2003) highlighted that an adequate reserve deposition in the embryo tissues seems to be necessary for their adequate maturation. Efforts were made since then to manipulate physical conditions in order to promote the adequate accumulation of reserves (Fernandes, 2011; Santos, 2008). Fernández-Guijarro et al. (1995) reported that somatic embryos from cork oak young seedlings increased maturation under light followed by storage at 4 °C, and that controlled starvation could benefit synchronization.

Santos (2008) and Fernandes (2011) compared the accumulation profiles of carbohydrate, lipid and protein reserves during the maturation of cork oak somatic embryos and the zygotic counterparts. Assuming that the accumulation of reserves that occurs in zygotic embryos may be ideal for embryos maturation, Fernandes (2011) also compared the accumulation profiles of somatic embryos exposed to different conditions such as polyethylene glycol (PEG), abscisic acid (ABA) and cold, and defined the condition that led to an accumulation profile closer to the one of the zygotic embryos.

From those analyses, the authors proposed an improvement to the initial somatic embryogenesis protocol developed by Pinto et al. (2002) (see Figure 3). In the improved protocol clusters of somatic globular embryos are isolated and transferred to MS medium (Murashige & Skoog, 1962) with PEG. After maturation, cotyledonar embryos are transferred to MS medium and submitted to chilling (4° C). Conversion is then achieved on woody plant medium (WPM) medium supplemented with BAP 0.5 mg/L and NAA 0.1 mg/L. After some weeks plants are acclimatized with success (Fernandes, 2011).

In conclusion, the inclusion of cold and osmotic stress in the protocol improved somatic embryos maturation and consequent conversion in approximately 70% of the genotypes. However, it was evident a genotype dependence in this process, with responsiveness ranging from very-good/in most genotypes to null, in few genotypes (Fernandes, 2011) (Figure 3).

1.3 Current challenges for the SE process

The loss of embryogenic competence is one of the major drawbacks of long term micropropagation potocols (e.g., Brito et al., 2009). In particular, embryogenic masses were maintained for long periods may dedifferentiate and lose their embryogenic potential. In cork oak this phenomena has originated two types of calluses under the same conditions: embryogenic (EC) and non-embryogenic (NEC) and these last calluses rarely regain embryogenic ability (Santos, 2008).

In vitro functional changes during embryogenesis imply changes in explant cells from differentiated and quiescent (G_0) stage to dedifferentiated dividing (G_1-S-G_2/M) stages, and later an evolution to embryogenic states. All these transitions imply changes in gene expression, and in cell cycle dynamics, where growth regulators, namely auxins and cytokinins, are crucial players (Gahan, 2007).

Using embryogenic and non embryogenic calluses of adult cork oak genotypes, our group reported differential distribution of cells staged in G_1, S and G_2 phases according to callus and growth regulators type (Fernandes, 2011) confirming that cell cycle dynamics during somatic embryogenesis suffers exogenously-induced alterations, and in particular, it is conditioned by growth regulators (Gahan, 2007). We also found that using the two different somatic embryogenesis protocols available (Fernandes, 2011; Pinto et al., 2002), not only cell cycle dynamics changed with time during the process, but also genotypes with different somatic embryogenic competences had different cell cycle dynamics (Fernandes, 2011). Curiously, responsive genotypes showed cell cycles with similar progression profiles (Fernandes, 2011).

Considering the key players regulating cell cycle dynamics, cyclins are among the most important. In a broad sense, D-type cyclins are thought to regulate the G_1-to-S transition, A-type cyclins, the S-to-M phase control, and B-type cyclins regulate both the G_2-to-M transition and intra-M-phase control (Gahan, 2007). The cyclin D (CYCD)/retinoblastoma pathway is believed to be involved in controlling both the commitment of cells to the mitotic cell cycle and decisions involving cell growth, differentiation, and cell cycle exit (Dissmeyer et al., 2009; Cools & Veylder, 2009). Key genes for growth and cell division are regulated by E2F transcription factors, which are inactive when bound by retinoblastoma. The phosphorylation of retinoblastoma is initiated by CYCD-containing cyclin-dependent kinases (CDKs) and is completed by cyclin E–CDK2, resulting in the dissociation of retinoblastoma from E2F factors, triggering the passage of cells from G_1- to S-phase (Gahan, 2007). This key role for the G_1 exit pathway results in it being the primary and predominant cell cycle control point. However, cyclin E–CDK2 is rate-limiting for entry into S-phase and can trigger S-phase in the absence of RB phosphorylation. CYCD3;1 are the best studied examples and expression of their genes is regulated by extrinsic signals, such as sucrose availability. CYCD3;1 expression is also regulated by plant hormones (Dewitte et al., 2003; for review see Dewitte & Murray, 2003).

The E2F family plays a critical role in organizing cell cycle progression by coordinating early cell cycle events with the transcription of genes required for entry into S-phase (e.g., Inzé, 2000). Two major classes of genes possess characteristic E2F binding sites, the first class encodes essential enzymes in the pathways for nucleotide and DNA synthesis that are co-ordinately up-regulated in late G_1. The second class corresponds to genes for regulators of cell cycle progression. Genes from both classes respond to ectopic expression of E2Fs from the first sub-group, namely those that can induce entry into S-phase (Dewitte & Murray, 2003; Gahan et al., 2007).

The cell cycle involves a complex network of regulating molecules. So the control of all these classes of checkpoints regulators is under study in cork oak embryogenic (EC) and non-embryogenic (NEC) tissues (Santos, 2011, unpublished data). Understanding and controlling these checkpoints will become a powerful tool to both better understand the embryogenic per se and to manipulate the developmental stages of embryogenic process.

2. Cryopreservation

2.1 General principles

Public and private efforts have been made to protect and conserve germplasm by preserving the genetic material of selected genotypes (e.g., Engelmann, 2000). As in other species, cork oak germplasm preservation can be done in situ (in the field and in natural environment), which demands large areas and is susceptible to environmental hazards. Alternatively preservation may be done ex situ (for general review see Li & Pritchard et al., 2009). In particular, in vitro preservation allows that in a small area, large amounts of genotypes are multiplied and maintained under controlled conditions where environmental influences are minimal. However, precocious ageing as well as somaclonal variation and genetic instability may arise after long term culture (Brito et al., 2009).

Alternatively, cryopreservation is the storage of living materials at extremely low temperatures using usually liquid nitrogen (−196 °C), and is an ideal strategy for plant germplasm preservation (Benson, 2008; Feng et al., 2011; Wang & Perl, 2006). This preservation strategy allows not only the preservation of material in small volumes (involving low maintenance requirements) but also, by reducing to residual values the cell metabolism, it allows that cells are stored for long periods, with low probability of genetic instability occurrence (Feng et al., 2011). Cryopreservation therefore allows: the conservation of plant material minimizing occurrences of genetic instability, contaminations and diseases; the preservation of endangered, rare or selected genotypes. Cryopreservation is already being applied to several plant species including forest woody species (e.g., Sakai et al., 2008). Also different plant materials have been used in this preservation strategy: shoot tips, cell cultures, embryos and seeds (Feng et al., 2011).

For cryopreservation to be useful in breeding programs, it is necessary to develop the cryogenic technique *per se*, and to ensure that robust and efficient regeneration protocols are available. Freezing and thawing stages require that cells are structurally and functionally cryoprotected. This may happen naturally (e.g. some naturally dehydrated material) but usually it is induced artificially with treatment with cryoprotectants that influence ice formation and activity of electrolytes present in the solution. Ideally, cryoprotectants should have low or no cytotoxicity. Cryoprotectants may be: a) permeating compounds, such as dimethylsulphoxide (DMSO, used usually in the range of 5-10%) that has a rapid entrance rate and so requires short incubation periods; another permeating compound is glycerol (used often in the range of 10-20%); b) non-permeating compounds such as sugars, sugar alcohols, polyethylene glycol (PEG). Often mixtures of cryoprotectant compounds are preferred to improve their efficacy (e.g., combinations of PEG : glucose : DMSO). Finally other strategies as cold hardening or ABA treatment may increase the freezing resistance and survival rates of cells.

Plant cryopreservation strategies may include slow or rapid freezing approaches. The first is based on physico-chemical changes during the process, namely associated with apoplastic ice crystal formation while cytoplasm may remain free from intra-cellular ice formation. Slow freezing decreases therefore the osmotic potential of the cytoplasm contributing to the cell desiccation. Rapid freezing is achieved by immersion of the cryoprotectant-treated samples in N_2, leading to an ultra-fast cooling that prevents the formation of ice crystals inside the cell (e.g., Sakai et al., 2008).

Some variants involve vitrification that includes a cell dehydration step prior to storage in N_2 (Sakai et al., 2008). This may rely on the ability of concentrated solutions of cryoprotectants to become viscous to very low temperatures, without ice formation. Consequently, during the vitrification process plant cells are dehydrated and the cytoplasm vitrified during freezing, which allows that ice crystals are rarely formed. Vitrification has already been applied to a large number of species (Panis et al., 2005; Sakai et al., 2008).

Nonetheless, during this process, usually complex and toxic solutions with high osmotic potential, such as the PVS2, are used for cryoprotection (Fernandes et al., 2008). Also the duration of the successive steps of a vitrification protocol is in general very short, hampering the simultaneous treatment of a large number of samples.

Some alternatives to vitrification-based techniques were developed, namely encapsulation-dehydration strategies (e.g., Engelmann et al., 2008). Encapsulation-dehydration is based on concepts related with artificial seeds. Concisely, plant tissues, such as shoot tips or somatic embryos, are covered by for example alginate. Then, they are dehydrated (using exposures to highly concentrated solutions, and/or to air in a flow chamber), before being transferred to N_2. This strategy has the advantage of using less toxic compounds such as glycerol than in other vitrification methods, thus minimizing stress conditions (e.g. Volk et al., 2006). It also has the advantage of easy and inexpensive manipulation, not requiring expensive instruments, as occurs in controlled freezing (Fernandes et al., 2008).

This method has been applied to many species, such as, for example, mulberry (Niino & Sakai, 1992), *Prunus* sp. (Brison et al., 1992), sweet potato (Feng et al., 2011; Hirai & Sakai, 2003), persimmon (Matsumoto et al., 2001), apple (Niino & Sakai, 1992; Paul et al., 2000), lily (Bouman & Klerk, 1990) and even grapevine embryogenic cell suspensions (Wang & Perl, 2006) or pear (Niino & Sakai, 1992; Scottez et al., 1992). In several assays as in those with *Robinia pseudoacacia*, it was demonstrated that encapsulation-dehydration originated better results than vitrification (Verleysen et al., 2005). Recently we have also demonstrated that encapsulation-dehydration was the most efficient method of cryopreservation of *Quercus suber* somatic embryos (Fernandes et al., 2008).

2.2 *Quercus suber* cryopreservation

Propagation of cork oak presents several drawbacks as it has a high heterozigocity, often leading to individuals with high probability of instability and genetically distinct from parents, therefore leading to high numbers of undesired genotypes (Lopes et al., 2006). Moreover, seeds are only stored for short periods as they rapidly loose viability. This recalcitrance jeopardises the development of conservation and improvement programs in this species. As in most forest species, *Quercus suber* conservation approaches consist mainly in agro-forest sustainable systems, and scarce strategies using biotechnological approaches have already succeeded. Valladares et al. (2004) highlighted that highly interesting individuals may be maintained with vegetative propagation.

The cryopreservation of seeds or embryos seems therefore to have huge potential as an innovative preservation strategy, in particular in species with recalcitrance. González-Benito et al. (2002) examined different factors included in the cryopreservation protocols for *Quercus ilex* and *Q. suber* embryonic axes. The authors demonstrated that temperature of in vitro incubation played an important role, mostly for *Q. ilex* axes. *Q. suber* axes were sensitive to desiccation and cooling.

With respect to *Quercus* sp. somatic embryos, Martinez et al. (2003) and Valladares et al. (2004) successfully cryopreserved embryogenic cultures of *Q. robur* and *Q. suber*, using the vitrification method. As reported above, highly toxic cryoprotectants are used in most classical vitrification processes. To overcome these negative effects, recently our group (Fernandes et al., 2008) used a less toxic variation to the classical vitrification technique, called the encapsulation-dehydration method, to cryopreserve *Q. suber* material.

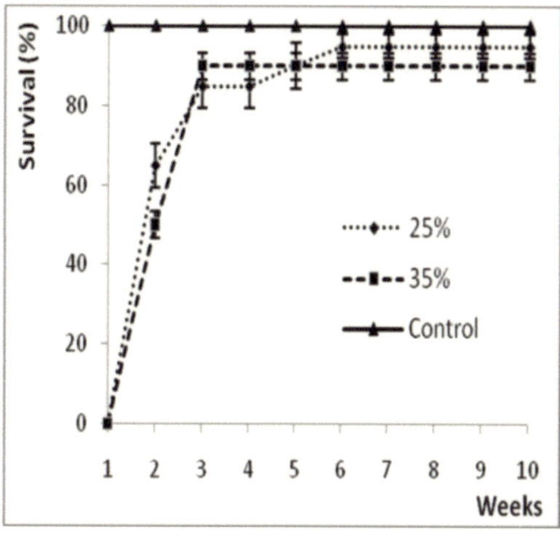

a

b

Fig. 4. a) Percentage of survival of cork oak cells after 10 weeks (for two protocols of cryopreservation used CRY25 and CRY35, corresponding to the final % content of water). b) Acclimatized plant obtained from cryopreserved somatic embryos that were recovered and cultivated on MSWH for maturation and conversion (adapted from Fernandes et al., 2008).

In this standard protocol for cork oak, somatic embryos or embryogenic clusters derived from mature trees and previously maintained on MS medium without growth regulators (MSWH) are used as samples. For encapsulation, embryos/clusters were separated and loaded in the alginate plus $CaCl_2$ solutions, forming the beads (3–4 mm), each one contained one embryo/cluster. They were then pre-cultivated on sucrose-enriched standard liquid medium (MS_{WH} with 0.7M sucrose) for 3 days. Beads were then desiccated by drying in the airflow of a laminar flow cabinet, and carefully weight loss was monitored for water content calculation. Two final water content (WC) values were assessed: 25% (CRY25) and 35% (CRY35). Afterwards, beads were placed in cryotubes (10 per vial), and immersed in N_2 for 24 h. Samples' thawing was done by incubating the cvryotubes at 38 °C (2 min) and incubating in detoxification solution (1 h). Beads were transferred to solid standard medium (MS_{WH}) for regeneration (Figure 4a, b; Fernandes et al., 2008).

This cryopreservation technique developed by Fernandes et al. (2008) for cork oak somatic embryos is simple, effective and non-toxic for the species. Also, survival rates in encapsulated-dehydrated (but non-frozen) cork oak samples achieved 90%. We also demonstrated that cryopreserved somatic embryo derived clones were able to recovery, leading to plants morphologically normal and that had genetic stability.

3. Screening of genetic variation in cloned and cryopreserved material

In vitro regenerated plants may exhibit somaclonal variation as a result from genetic or epigenetic modifications (Fourré et al., 1997; Isabel et al., 1996). It is generally accepted that morphological, cytological and molecular variations may be generated by the imposed stress during in vitro or cryopreservation processes. These induced variations are conditioned by the genotypes used and/or by the techniques/protocols used. Theoretically, any protocol of plant cloning or germplasm conservation should lead to no somaclonal variation.

Genetic variability of in vitro material or of cryopreserved samples may be assessed by several techniques. Molecular and genetic techniques when used individually give a limited perspective of the occurrence of somaclonal variation, but if combined may provide an interesting and complementary toolbox of markers for "true-to-typeness" evaluation (e.g., Santos et al., 2007).

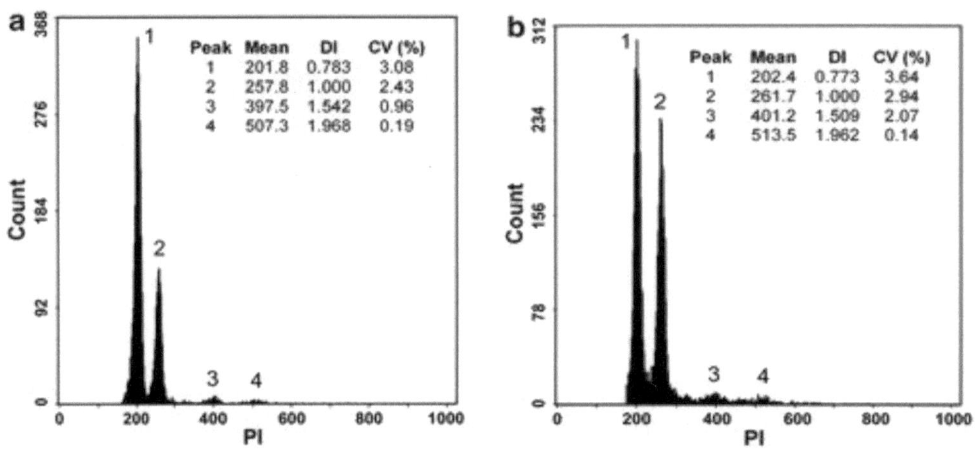

a b

Fig. 5. Flow cytometry histograms of relative propidium iodide (PI) fluorescence intensity obtained after simultaneous analysis of nuclei isolated from *Quercus suber* and from the reference standard *Glycine max* (2C=2.50 pg DNA): **a)** leaves of the mother plant; **b)** somatic embryo. Histogram *peak 1:* nuclei at G_0/G_1 phase of *Q. suber*; *peak 2:* nuclei at G_0/G_1 phase of *G. max*, *peak 3* nuclei at G_2 phase of *Q. suber*, *peak 4* nuclei at G_2 phase of *G. max* (adapted from Loureiro et al., 2005).

Gross genetic variations of genomic origin affect mostly the number of chromosomes and ploidy level and can be detected by chromosome counting or flow cytometry (Loureiro et al., 2005). Chromosome mutations (e.g., inversion, translocation) and genic mutations are screened by molecular markers that detect DNA sequence modifications. Among these, the most popular are RFLPs (restriction fragment length polymorphisms), RAPDs (randomly amplified polymorphic DNAs), AFLPs (amplified fragment length polymorphisms) or microsatellites/SSRs (simple sequence repeats). RFLPs and AFLPs are highly reproducible techniques but are costly and may be time-consuming. For easy application of SSRs, it is required that the microsatellite loci and its flanking primers are readily available for a given species, while RAPDs is simple but may have a lack of reproducibility.

Several genetic and molecular markers have been used to assess somaclonal variation in micropropagated oak plants. Loureiro et al. (2005), using flow cytometry, found no ploidy or DNA content variations in cork oak embryogenic tissues or among somatic embryos (Figure 5).

Also, Fernandes et al. (2008) using cryopreserved material, confirmed by flow cytometric analyses that the two cryopreservation procedures (CRY25 and CRY35) provided genetic fidelity, for the parameters used: in all samples, both ploidy and DNA content were in concordance with literature data: 2C= 1.90 pg DNA (Loureiro et al., 2005; Santos et al., 2007), and changes in DNA content of non-cryopreserved and cryopreserved samples were minimal ($\leq$ 0.01 pg/2C) (Figure 6; Table 1).

Flow cytometric techniques, despite highly rapid and robust, may however not detect minor changes in DNA content, so it must be emphasised that the putative occurrence of small changes in DNA content in these kinds of assays should not be discarded.

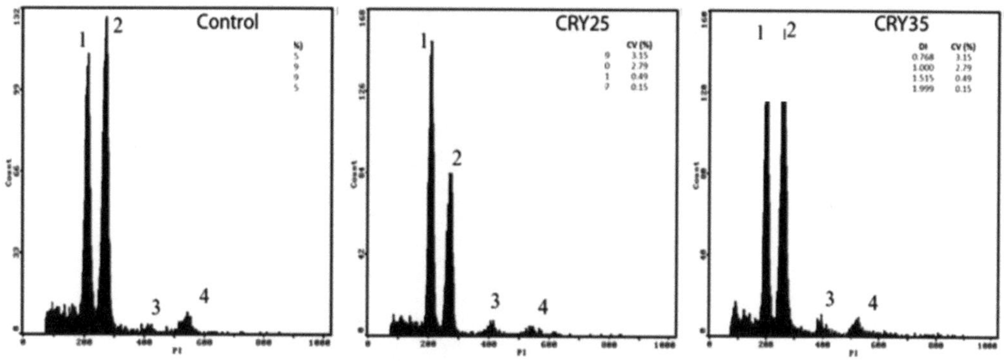

a b c

Fig. 5. Flow cytometric estimation of genome size of *Q. suber* samples of control (a) and after cryopreservation: CRY25 (b) and CRY35 (c). In all graphics peaks 1 and 3 correspond to 2C and 4C nuclei of cork oak somatic embryos, while peaks 2 and 4 correspond to the internal soy standard (adapted from Fernandes et al., 2008). PI: intensity of propidium fluorescence.

Tissue	DI	SD	2c DNA (pg)	SD	1c DNA (Mbp)[b]
Fresh	0.774[a]	0.0059	1.93	0.015	946
25% Cry[c]	0.777[a]	0.0073	1.94	0.018	950
35% Cry[c]	0.772[a]	0.0138	1.93	0.034	944

Table 1. Nuclear DNA content of *Quercus suber* L. fresh and cryopreserved embryos. The results are given as mean and standard deviation (SD) of the 2C DNA content in mass values (pg). Nuclear DNA content in Mbp is also given. [a] Mean values followed by the same letter are not significantly different according to the Tukey-Kramer multiple comparison test at $P \leq 0.01$. Note: [b]1 pg DNA = 978 Mbp. [c]Cryopreserved embryos (25% and 35 % water content) (adapted from Fernandes et al 2008).

Both works supported therefore that the protocols for somatic embryogenesis and cryopreservation developed so far for cork oak led to "true-to-typeness" (for ploidy and DNA amount parameters) and were worthy of use in cork oak breeding programs.

As explored above, molecular markers provide information on sequence mutation. This information is therefore complementary to the information provided by flow cytometry (or even by chromosome counting).

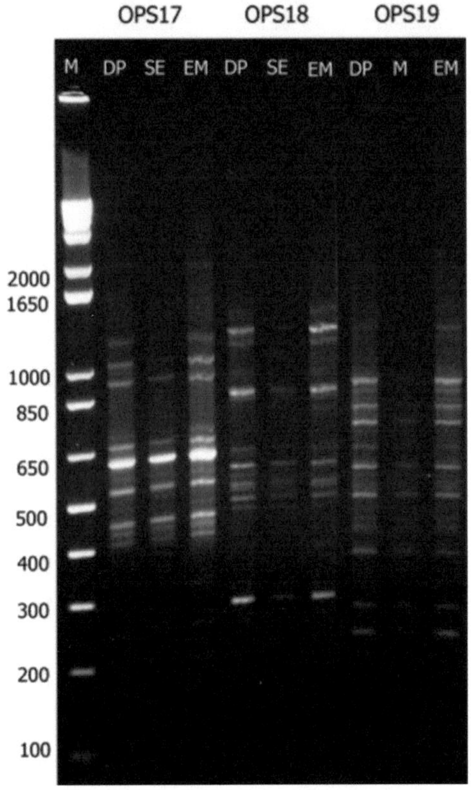

Fig. 6. DNA profiles generated by the RAPD primers OPS 17, 18 and 19, in the three different stages of the somatic embryogenesis process: donor plant (DP), somatic embryo (SE) and embling (EM). M, size marker (1Kb Plus DNA Ladder) (adapted from Fernandes et al., 2011).

In micropropagated *Q. serrate* no aberrations in the banding patterns were detected by RAPD markers (Thakur et al., 1999). Also, RAPDS were used to assess putative occurrence somaclonal variation in *Q. suber* embryogenic lines, but no molecular changes were found (Gallego et al., 1997, Sanchez et al., 2003). More recently, RAPDS were also used to evaluate genetic instability of somatic embryos of *Q. suber* obtained by the above described protocols (Figure 6; Fernandes et al., 2011).

Techniques for RAPD analyses however pose several problems of reproducibility, and give restricted information. So, other molecular analyses can provide complementary information to RAPDS. Microsatellites and AFLP are among the most used markers in *Quercus*. AFLP markers detected changes in cork oak embryogenic lines (Hornero et al., 2001). We also used AFLP to test putative genetic instability during the developed cryopreservation/somatic embryogenesis processes of cork oak (Fernandes et al., 2008). It was used six primer sets that revealed an overall high proximity value between the two vitrification-encapsulation cryopreserved (CRY25 and CRY35) and control samples. Occasionally, few extra AFLP-bands in CRY25 samples were detected and the occurrence of putative small mutations, or DNA methylation or even to subpopulation cryo-selection should not be excluded (Fernandes et al., 2008).

Locus	Allele size (bp)			
	QsG0	QsG5	QsGM1	QsGM2
QM58TGT	201	185/211	203/210	203/210
QM50-3M	276/286	285/287/292	278	282/288
QpZAG9	223/238	223/233	218/223	223/233
QpZAG15	108/123	106/120	nd	nd
QpZAG36	209/219	216/220	207/209	207/209
QpZAG110	223/233	220/238	221	221/233
QrZAG7	106/119	115/127	106/121	106/121
QrZAG11	229/263	261/273	nd	nd

Table 2. Characteristics of the microsatellite loci amplified in *Q. suber*. Allele size found in this study and allele size range and number of alleles (in parenthesis) found in other publications are also given: values for QrZAG7 and QrZAG11 are from Hornero et al. (2001*b*), values for the 6 remaining loci are from Lopes et al. (2006).

Finally, Wilhelm et al. (2005) found that that during somatic embryogenesis process of *Q. robur*, genetic instability occurred. Also, Lopes et al. (2006) confirmed genetic stability of somatic embryos and emblings derived from our somatic embryogenesis protocol (Lopes et al., 2006; Pinto et al., 2002) and after, Fernandes et al. (2008) also confirmed this stability using SSRs in material after recovering from cryopreservation (Table 2). Together with this overall genetic stability, the regenerated cork oak plants looked normal, healthy and well developed shoot. The authors concluded that the encapsulation-dehydration cryopreservation protocol used in cork oak somatic embryos was an efficient method of storage, regarding several parameters: recovery and survival rates and genetic/morphologic stability. To support the battery of protocols for molecular and genetic analyses of cork oak, Santos et al. (2007) published in detail the reliable protocol for analysis of cork oak material by SSRs and ploidy/nDNA quantification, where technical aspects and potential troubleshooting that may occur during analysis of this material are deeply discussed.

4. Concluding remarks

The utility of plant biotechnology tools in woody forest species propagation and preservations has been recognised decades ago, but only recently, it has been effectively incorporated in industrial breeding programs. Despite no robust and efficient protocol for cork oak regeneration by somatic embryogenesis is available yet, the advances observed in the last decade, together with the already available protocol for cryopreservation of this species, open perspectives for the incorporation of these two approaches in future breeding program of this species. Moreover, with the available protocols stable genotypes were obtained.

Genetic and molecular stability was assessed using complementary genetic and molecular techniques such as flow cytometry, RAPDS, AFLP and SSRs. Finally, an interesting research field will focus on the control of cell cycle progression order to control different stages of somatic embryogenesis and preservation. It is our believe that by manipulating proteins that control cell cycle phases transition (which are at the basis of differentiated/ undifferentiated cells) we´ll be able to manipulate the reversion phenomena between NEC and EC and also to better control the developmental somatic embryogenesis stages.

5. References

Benson E (2008) Cryopreservation Theory. In: Plant Cryopreservation - A practical guide (B Reed, Ed.) Springer Science, ISBN 978-0-387-72275-7, Corvallis, OR, USA pp. 15-32

Bouman H & G Klerk (1990) Cryopreservation of lily meristems. Acta Horticulturae. 266: 331-337

Brison M, V Paulus, T Boucaud & F Dosba (1992) Cryopreservation of walnut and plum shoot tips. Cryobiology 29: 738-742

Brito G, A Costa, C Coelho & C Santos (2009) Large scale field acclimatization of *Olea maderensis* micropropagated plants: morphological and physiological survey. Trees 23: 1019–1031

Bueno M, A Gómez & J Manzanera (2000) Somatic and gametic embryogenesis in *Quercus suber* L. In: Somatic embryogenesis in woody plants (Jain S, Gupta P & Newton R, Eds) Kluwer Academic Publishers, ISBN 0-7923-3070-6, Netherlands, pp 479-508

Bueno M, A Gómez, O Vicente & J Manzanera (1996) Stability in ploidy level during somatic embryogenesis in *Quercus canariensis*. In: Somatic Cell Genetics and Molecular Genetics of Trees (Ajuha M, W Bourjan & Neale D, Eds). Kluwer Academic Publishers, Dordrecht, Netherlands, pp 23-28

Chalupa V (2005) Plant regeneration by somatic embryogenesis from cultured immature embryos of oak (*Quercus robur* L.) and linden (*Tilia cordata* Mill.) Plant Cell Reports 9: 398-401

Cools T & L Veylder (2009) DNA stress checkpoint control and plant development. Current Opinion in Plant Biology 12 (1): 23-28

Cuenca B, M San-José, M Martínez, A Ballester & A Vieitez (1998) Somatic embryogenesis from stem and leaf explants of *Quercus robur* L. Plant Cell Reports 18 (7-8): 538-543

Dewitte W, C Khamlichi, S Scofield, J Healy, A Jacqmard, Kilby & J Murray (2003) Altered Cell Cycle Distribution, Hyperplasia, and Inhibited Differentiation in Arabidopsis Caused by the D-Type Cyclin CYCD3. Plant Cell 15(1): 79–92

Dewitte W & J Murray (2003) The Plant Cell Cycle. Annu. Rev. Plant Biology 54: 235–64

Dissmeyer N, A Weimer, A Pusch, K Schutter, C Kamei, N Nowack, G Dua, Zhu Y, De Veylder L & A Schnittger (2009) Control of Cell Proliferation, Organ Growth and DNA Damage Response Operate Independently of Dephosphorylation of the *Arabidopsis* Cdk1 Homolog CDKA;1. The Plant Cell 21: 3641-3654.

Endemann M, K Hristoforoglu, T Stauber & E Wilhelm (2002) Assessment of age-related polyploidy in *Quercus robur* L. somatic embryos and regenerated plants using DNA flow cytometry. Biologia Plantarum, 44(3): 339-345.

Engelmann F, M Arnao, Y Wu & Escobar (2008) Development of encapsulation dehydration. In: Plant Cryopreservation - A practical guide (B Reed, Ed.) Springer Science, Corvallis OR USA, ISBN 978-0-387-72275-7, pp 59-76

Engelmann F (2000) Importance of cryopreservation for the conservation of plant genetic resources. In: Cryopreservation of tropical germplasm. Current research progress and application (Engelmann F & Takagi H, Eds) JIRCAS, ISBN Rome, Italy pp. 8-20

Feng C, Z Yin, Y Ma, Z Zhang, L Chen, B Wang, B Li, Y Huang & Q Wang (2011) Cryopreservation of sweetpotato (*Ipomoea batatas*) and its pathogen eradication by cryotherapy. Biotechnology Advances 29: 84–93

Fernandes P, C Rocha, A Costa & C Santos (2011) genetic stability evaluation of quercus suber l. Somatic embryogenesis by RAPD analysis. Pakistan Journal Botany, in press.

Fernandes P, E Rodriguez, G Pinto, I Roldán-Ruiz, M DeLoose & C Santos (2008) Cryopreservation of *Quercus suber* somatic embryos by encapsulation-dehydration and evaluation of genetic stability. Tree Physiology 28: 1841-1850.

Fernandes P (2011) Strategies of *Quercus suber* micropropagation. PhD Thesis, University Aveiro, Aveiro Portugal, pp 155

Fernández-Guijarro B, C Celestino & M Toribio (1995) Influence of external factors on secondary embryogenesis and germination in somatic embryos from leaves of *Quercus suber* Plant Cell, Tissue and Organ Culture 41 (2): 99-106

Fourré J, P Berger, L Niquet & P Andre (1997) Somatic embryogenesis and somaclonal variation in Norway spruce: Morphogenetic, cytogenetic and molecular approaches. Theoretical Applied Genetics 94(2): 159-169

Gallego F, I Martinez, C Celestino & M Toribio (1997) Testing somaclonal variation using RAPD's in Quercus suber L. somatic embryos. International Journal Plant Sciences 158(5): 563-567

Gahnn P (2007) Totipotency in the cell cycle. In: protocols for micropropagation of woody trees and fruits. (M Jain & H Haggman Ed) Springer, Dordrecht Netherlands. Pp 3-15.

Gonzalez-Benito ME, RM Prieto, E Herradon & C Martin (2002) Cryopreservation of Quercus suber and Quercus ilex embryonic axes: In vitro culture, desiccation and cooling factors. Cryoletters 23:283-290

Hernández I, C Celestino & M Toribio (2003) Vegetative propagation of Quercus suber L. by somatic embryogenesis. I. Factors affecting the induction in leaves from mature cork oak trees. Plant Cell Rep. 21: 759- 64

Hernández I, C Celestino, J Martinez, J Hornero, J Gallego & M Toribio (1999) Induction of somatic embryogenesis in leaves from mature Quercus suber trees. Paper presented at the Physiology and control of plant propagation in vitro. Report of activities, Eur Cost Action 822, Krakow, Poland.

Hirai D & A Sakai (2002) Simplified cryopreservation of sweet potato [Ipomoea batatas (L.) Lam.] by optimizing conditions for osmoprotection Plant Cell ReportsVolume 21, Number 10

Hornero J, I Martinez, C Celestino, FJ Gallego, V Torres & M Toribio (2001) Early checking of genetic stability of cork oak somatic embryos by AFLP analysis. Int. J. Pl. Sci., 162: 827-833.

Inzé D (2000) (ed) The plant cell cycle Klluwer Academic Press. Dordrecht, Netherlands

Isabel N, R Boivin, C Levasseur, PM Charest, J Bousquet & FM Tremblay (1996) Occurrence of somaclonal variation among somatic embryo-derived white spruces (Picea glauca, Pinaceae). Am J Bot 83(9):1121-1130

K Ishii, R Thakur & S Jain (1999) Somatic embryogenesis and evaluation of variability in somatic seedlings of Quercus serrata by RAPD markers. In: Somatic embryogenesis in woody plants (Mohan Jain S, Gupta P & Newton R, Eds), Kluwer Academic Publishers, Dordrecht, Netherlands pp 403-414

Jain SM (1999) An overview on progress of somatic embryogenesis in forest trees. In: Plant Biotechnology and In Vitro Biology in the 21st Century (Altman A, Ziv M & Izhar S, Eds), Kluwer Academic Publishers, Dordrecht, The Netherlands. pp. 57–63

Kim Y (2000) Somatic embryogenesis in *Quercus acutissima*. In: Somatic embryogenesis in woody plants (Jain S, Gupta P & Newton R, Eds). Kluwer Academic Publishers, Dordrecht, Netherlands pp 671-686

Li Z & H Pritchard (2009) The science and economics of ex situ plant conservation. Plant Cell 14(11) 614-622

Lloyd G & B McCown (1981) Commercially-feasible micropropagation of Mountain Laurel, Kalmia latifolia, by shoot tip culture. Proceeding International Plant Prop. Soc. 30: 421-427

Lopes T, G Pinto, J Loureiro, A Costa & C Santos (2006) Determination of genetic stability in long-term somatic embryogenic cultures and derived plantlets of cork oak using microsatellite markers. Tree Physiology 26: 1145-1152

Loureiro J, G Pinto, T Lopes, J Doležel & C Santos (2005) Assessment of ploidy stability of the somatic embryogenesis process in *Quercus suber* L. using flow cytometry. Planta 221(6): 815-822

Martinez M, A Ballester & A Vieitez (2003) Cryopreservation of embryogenic cultures of Quercus robur using desiccation and vitrification procedures. Cryobiology 46: 182-189

Matsumoto T, K Mochida, H Itamura & A Sakai (2001) Cryopreservation of persimmon (Diospyros kaki Thunb.) by vitrification of dormant shoot tips. Plant Cell Reports 20: 398-402

Murashige T & F Skoog (1962) A revised medium for rapid growth and bio assays with tobacco tissue cultures. Physiologia Plantarum 15: 473-497

Niino T & A Sakai (1992) Cryopreservation of alginate-coated in vitro-grown shoot tips of apple, pear and mulberry Plant Science 87: 199-206

Panis B & M Lambardi (2005) Status of cryopreservation technologies in plants (crops and forest trees). FAO: The role of biotechnology for the characterisation and conservation of crop, forest, animal and fishery genetic resources in developing countries.

Park Y (2002) Implementation of conifer somatic embryogenesis in clonal forestry: technical requirements and deployment considerations Ann. Forest Sci. 59: 651-656

Park Y, J Barrett & J Bonga (1998) Application of somatic embryogenesis in high-value clonal forestry: deployment, genetic control, and stability of cryopreserved clones. In Vitro Cell Dev Biol - Plant 34: 231–239.

Park Y, K Klimaszewska , J Park & S Mansfield (2010) Lodgepole pine: the first evidence of seed-based somatic embryogenesis and the expression of embryogenesis marker genes in shoot bud cultures of adult trees. Tree Physiol. 30(11):1469-78.

Paul H, G Daigny & B Sangwan-Noreel (2000) Cryopreservation of apple (Malus x domestica Borkh.) shoot tips following encapsulation-dehydration or encapsulation-vitrification. Plant Cell Reports, 19, 768-774

Pinto G, Y-S Park, S Silva, C Araújo, L Neves & C Santos (2008) Factors affecting maintenance, proliferation, and germination of secondary somatic embryos of Eucalyptus globulus Labill. Plant Cell Tissue and Organ Culture 95(1):69-78.

Pinto G, R Amaral, C Santos & O Carnide (2001) Somatic embryogenesis in calluses of leaves from three year old *Quercus suber* L. plants. In Meeting of the Cost 843 WG3 - II. *Quality enhancement of plant production through tissue culture*, Carcavelos, Portugal.

Pinto G, H Valentim, A Costa, S Castro, C Santos (2002) Somatic embryogenesis in leaf callus from a mature *Quercus suber* L. tree. In Vitro Cellular & Developmental Biology-Plant 38(6): 569-572

Pires I, G Pinto, J Loureiro, C Santos (2003) Study of anatomical and physiological changes in vitro cork oak plants during acclimation. XXXVIII Annual Meeting of the Society of Electron Microscopy and Cell Biology. Ponta Delgada, Portugal (5-7 de Dec) pp 29

Sakai A, D Hirai & T Niino (2008) Development of PVS-based vitrification and encapsulation-vitrification protocols. In Plant Cryopreservation - A practical guide (B Reed Ed.) Springer Science, Corvallis, OR, USA, ISBN 978-0-387-72275-7, pp 33-58

Sanchez M, M Martinez, S Valladares & A Vieitez (2003). Maturation and germination of oak somatic embryos originated from leaf and stem explants: RAPD markers for genetic analysis of regenerants. Journal of Plant Physiology 160: 699-707

Santos C (2008) De que forma é que a embriogénese somática em lenhosas é condicionada pela morte cellular. POCI/AGR/ 60672/2004. MCT-FC -Final Report, Portugal, 59 pp

Santos C, J Loureiro, T Lopes & G Pinto (2007) Genetic fidelity analyses of in vitro propagated cork oak (*Quercus suber* L.) In: Protocols for micropropagation of woody and fruit trees (Mohan Jain S., Häggman H. Eds) Springer, ISBN 978-1-4020-6351-0, Dordrecht, The Netherlands. pp 67-84

Scottez C, E Chevreau, N Godard , Y Arnaud, M Duron & J Dereuddre (1992) Cryopreservation of cold-acclimated shoot tips of pear in vitro cultures after encapsulation-dehydration. Cryobiology 29 (6): 691-700

Thakur R, S Gota, K Ishii & S Jain (1999) Monitoring genetic stability in *Quercus serrata* Thunb. somatic embryogenesis using RAPD markers. Journal of Forestry Research, 4: 157-160

Valladares S, M Toribio, C Celestino, A Vieitez (2004) Cryopreservation of embryogenic cultures from mature *Quercus suber* trees using vitrification. Cryoletters 25: 177-186

Vengadesan G & P Pijut (2009) Somatic embryogenesis and plant regeneration of northern oak (*Quercus rubra*). Plant cell Tissue Organ culture 97(4): 1-9

Verleysen H, G Samym, E Van Bockstaele & P Debergh (2004) Evaluation of analytical techniques to predict viability after cryopreservation. Plant Cell Tissue and Organ Culture 77: 11-21

Volk G (2010) Application of functional genomics and proteomics to plant rryopreservation. Current Genomics 11: 24-29

Wang Q & A Perl (2006) Cryopreservation of embryogenic cell suspensions by encapsulation-vitrification. Methods Molecular BioloGy 318: 77-86

Wilhelm E, M Endemann, K Hristoforoglu, C Prewein & M Tutkova (1999) Somatic embryogenesis in oak (*Quercus robur L.*) and production of artificial seeds. In: Proceedings of application of biotechnology to forest genetics. Biofor 99, Vitoria-Gasteiz

Wilhelm E, K Hristoforoglu, S Fluch & K Burg (2005) Detection of microsatellite instability during somatic embryogenesis of oak. Plant Cell Reports, 23: 790-795

20

Proline and the Cryopreservation of Plant Tissues: Functions and Practical Applications

David J. Burritt
The Department of Botany,
The University of Otago,Dunedin
New Zealand

1. Introduction

Cryopreservation has been proven to be an effective technology for the cost-effective, long-term preservation of genetic material. A wide range of plant material including cultured cells, tissues, embryos, meristems, pollen and seeds can be effectively preserved for extended periods of time and, when thawed, can be used to rapidly produce stock plants, with good preservation of genetic and physiological characteristics. Numerous protocols including controlled rate cooling, PVS2 vitrification, encapsulation-vitrification, and encapsulation-dehydration have been developed that allow the cryopreservation of a wide range of plant germplasm (Burritt, 2008), but irrespective of the protocol used each step in a cryopreservation protocol has the potential to impose a stress on plant cells. Low temperatures that lead to freezing can impose stress on cells and tissues in two ways, by the direct effects of low temperatures on cell function and integrity or by the cellular dehydration that occurs when the cells water freezes. Several of the mechanisms by which these two forms of stress can damage plant cells are shown in Figure 1.

Numerous studies have shown that cold temperatures induce the accumulation of metabolites, including low-molecular-weight carbohydrates such as fructose, glucose, maltose and raffinose, and amino acids such as proline and glutamine (Taji et al., 2002; Cook et al., 2004). These metabolites play important protective roles in freezing tolerance in whole plants (Kaplan and Guy, 2004) and this has lead to their extensive use in the protocols developed for the cryopreservation of isolated plant cells and tissues (Burritt, 2008). In particular, the amino acid proline has been found to help confer freezing tolerance in a wide variety of both animal and plant cells, and is often added to cryoprotective solutions or is used for preconditioning plants or pretreating isolated cells or tissues prior to cryopreservation (Burritt, 2008). Despite its widespread use, little is known of the mechanisms via which proline protects cells during cryopreservation.

This chapter gives an overview of proline synthesis and metabolic regulation in plants and the changes in proline metabolism associated with desiccation and freezing tolerance, which are both of importance for the successful cryopreservation of plants cells and tissues. The use of proline as a cryoprotectant or pre-growth additive for the cryopreservation of plant cells and tissues is then overviewed and the potential mechanisms via which proline can protect plant cells is critically evaluated. Future research needs are then discussed.

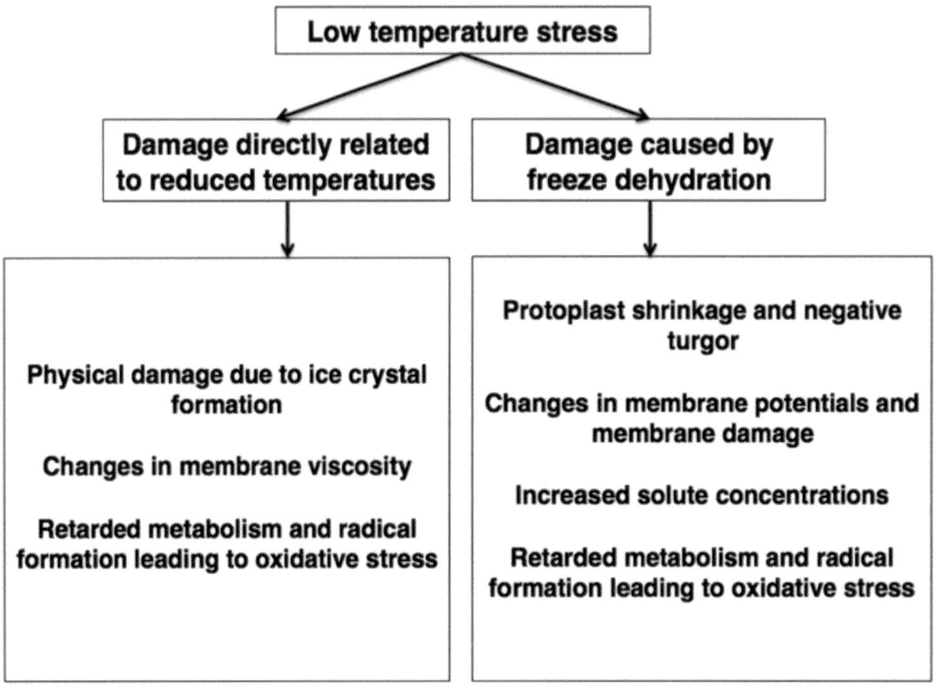

Fig. 1. Potential damage caused by the stresses associated with exposure of plant cells to low temperatures.

2. Proline and plants

2.1 The function of proline in plants

Essential for primary metabolism, both as a free amino acid and as a component of proteins, proline is distinctive among the proteinogenic amino acids as it contains a secondary amino group and a distinctive cyclic structure (Lehmann et al., 2010). The cyclic structure of proline causes exceptional conformational rigidity, compared to other amino acids, as proline's side chain locks its φ backbone dihedral angle at approximately -75° and this determines the arrangement of the peptide chain and can lead to the stabilization or destabilization of secondary protein structures.

As well as being important for primary metabolism proline appears to have numerous other functions in plants. Research has clearly demonstrated that proline levels show significant fluctuations in response to environmental stress (Bohnert et al., 1995), but the precise mode of action of proline remains largely a matter of speculation. In addition to its role in environmental stress tolerance, recent research has provided evidence that proline may also play important roles in plant development both as a metabolite and as a signal molecule (Mattioli et al., 2009). Studies have shown that proline could play important roles in embryo and seed development, stem elongation, and the transition from vegetative growth to flowering (Mattioli et al., 2008; Mattioli et al., 2009)

2.3 Proline biosynthesis and catabolism

The proposed pathways for proline biosynthesis and catabolism in plants are outlined in Figure 2. In plants proline can be synthesized from glutamate or ornithine, however under most conditions proline is mainly synthesized from glutamate rather than from ornithine, as the enzyme ornithine-δ -aminotransferase (dOAT) is down regulated (Szabados & Savoure, 2009). Two enzymes are required for the synthesis of proline from glutamate. The first enzyme, Δ1- pyrroline-5-carboxylate synthase (P5CS) is a bifunctional enzyme that phosphorylates and reduces glutamate to glutamyl-5- semialdehyde (G5SA) that then spontaneously converts to Δ1- pyrroline-carboxylate (P5C). The second enzyme, Δ1-pyrroline-carboxylate reductase (P5CR) further reduces the P5C intermediate to proline (Delauney & Verma, 1993). P5CS has been found to be encoded by 2 genes in most plants, while P5C is encoded by only a single gene (Szekely et al. 2008; Strizhov et al., 1997) The rate-limiting step in the above pathway is the γ -glutamyl kinase activity of P5CS, which is sensitive to feedback inhibition by the presence of relatively low cellular proline levels (Zhang et al., 1995). Alternatively proline can be synthesized from ornithine by dOAT, which converts ornithine and α-ketoglutarate to P5C and glutamate by transamination (Stranska et al., 2008). Funck et al., 2008, in a study of *Arabidopsis thaliana,* found that mutant plants which lacked dOAT activity could not mobilize nitrogen from arginine or ornithine, but could accumulate proline and so suggested the main role for dOAT was arginine degradation. They also suggested that as dOAT is localized in the mitochondria and that it would be unlikely that P5CR could directly utilize dOAT-generated P5C, as P5CR is localized in the cytosol or in plastids.

Proline degradation in plants takes place in mitochondria and so is by in large separated from the biosynthetic pathway. The first step in proline catabolism is the oxidation of proline to P5C by proline dehydrogenase (PDH), which in Arabidopsis and tobacco is encoded by two homologous genes (Mani et al., 2002; Ribarits et al., 2007; Verbruggen & Hermans 2008). The P5C generated is then converted to glutamate by pyrroline-5-carboxylate dehydrogenase (P5CDH), which is thought to be encoded by a single gene in all of the plant species analysed to date (Ayliffe et al. 2005; Mitchell et al. 2006). However, biochemical analysis P5CDH in *Nicotiana plumbaginifolia* and *Zea mays* has revealed two slightly different enzyme activities that may arise from a single gene, or a second P5CDH gene may be present (Elthon & Stewart 1982; Forlani et al. 1997). In plants under stress, the accumulation of proline is thought to be due not only to increased synthesis, but also to inactivation of degradation pathways (Delauney & Verma, 1993)).

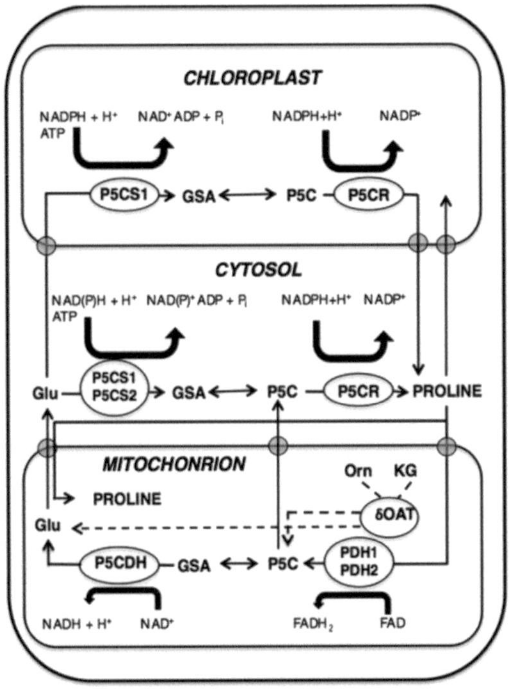

Fig. 2. Proposed model for proline metabolism in higher plants (adapted from Lehmann et al. 2010). Glu glutamate, Orn ornithine, P5C pyrroline-5-carboxylate, GSA glutamic-γ-semialdehyde, KG α-ketoglutarate. P5CS P5C synthetase, P5CR P5C reductase, PDH proline dehydrogenase, P5CDH P5C dehydrogenase, δOAT ornithine-δdaminotransferase. Transporters and potential transporters are shown as grey circles.

3. Proline in plant cells under stress

3.1 Proline accumulation in higher plants

Stress has been shown to induce proline accumulation in a wide range of organisms including eubacteria, protozoa, invertebrates and plants (Verbruggen & Hermans, 2008; Kostal et al., 2011) and proline accumulation is believed to be very important as part of the physiological adaptation of plants to stress. In plants a wide range of abiotic and biotic stressors have been shown to induce proline accumulation including, salt, drought, high temperatures, low temperatures, heavy metals, anaerobiosis, nutrient deficiency, organic pollutants, ultraviolet (UV) radiation and pathogen infection (Chu et al., 1978; Alia & Saradhi, 1991; Saradhi et al., 1995; Hare et al., 1999; Siripornadulsil et al., 2002). The level of proline that accumulates in plants in response to stress varies greatly and is highly dependent on the plant species, with increase of up to 100 fold compared to controls reported in the literature (Verbruggen & Hermans, 2008).

With respect to cryopreservation, numerous studies have demonstrated the importance of proline for plant cold tolerance (Swaaij, Jacobsen & Feenstra 1985, Swaaij et al. 1986; Duncan & Widholm 1987; Ait-Barka & Audran 1997; Hoffman et al., 2010; Javadian et al., 2010; Burbulis et al., 2011). Studies on plants relatively insensitive to chilling, such as barley (Chu et al. 1978), rye (Koster & Lynch 1992), winter wheat (Dorffling et al. 1997), and *Arabidopsis thaliana* (Xin & Browse 1998; Nanjo et al. 1999) have demonstrated significant positive correlations between cellular proline accumulation and improved cold tolerance.

In addition, plant cells under dehydrating conditions, which are often a consequnce of cryopreservation, undergo osmotic adjustment by accumulating one or several low molecular weight organic solutes, which are often referred to as compatible osmolytes and/or osmoprotectants. These molecules play a critical role in counteracting the effect of osmotic stress in plants at the cellular level (Yoshiba et al., 1997). In plants under dehydrating conditions such as drought or high salinity, proline is one of the most common compatible osmolytes and while several amino acids are known to accumulate in response to osmotic stress, proline appears to be the preferred organic osmoticum in many plants and may have a specific protective role in the adaptation of plant cells to dehydration. For example, in a study of *Triticum aestivum* L. (durum wheat) under salinity stress, Poustini et al. (2007) found a positive correlation between proline levels and osmotic potential, and concluded that proline is an important osmolyte for osmotic adjustment in wheat under water stress. In addition, it has been demonstrated that transgenic tobacco plants with elevated levels of proline biosynthesis show increased tolerance to hyperosmotic stress (Kavi Kishot et. al., 1995), providing further evidence of a cause-and-effect relationship between proline levels and osmotic tolerance. Proline normally accumulates in the cytosol, where it contributes to the cytoplasmic osmotic adjustment in response to water loss without interfering with normal cellular processes and biochemical reactions (Ashraf & Foolad, 2007).

3.2 Proline and cryopreservation

During cryopreservation, plant cells encounter similar problems to those they encounter under freezing conditions in the field. They under go changes in the spatial organization of biological membranes, biochemical and chemical reactions can be retarded, and the status and availability of water can be altered. For these reasons proline is likely to be an effective cryoprotectant for cryopreserved plant cells and tissues.

4. The use of proline as a cryoprotectant

Proline has been used for many years in numerous cryoprotection protocals for the preservation of a wide range of both animal and plant cells and tissues. For example, Li et. al. (2003) investigated the effects of addition of proline, glutamine, and glycine to the Tes-Tris-egg yolk (TTE) freezing medium used for cryopreservation of cynomolgus monkey (*Macaca fascicularis*) spermatozoa. They found that the addition of 5 mM proline, 10 mM glutamine, and 10 or 20 mM glycine to TTE significantly improved post-thaw sperm motility and membrane integrity compared to controls without an amino acid. Of the three amino acids tested proline was effective at the lowest concentration.

Proline has also been found to be useful for the cryopreservation of plant cells, meristems and embryos. Jain et al. (1996) included proline in the cryoprotectant solution as part of a protocol that was used to successfully cryopreserve embryogenic suspension cells of two commercially cultivated aromatic Indica rice varieties using a simple one-step freezing procedure that did not require a controlled-rate freezer. Brison et al. (1995) used a preculture medium enriched with dimethylsulfoxide and proline prior to the cropreservation of *in vitro* grown interspecific Prunus rootstock, Fereley-Jaspi (R). In a study to develop a cryoprotection protocol for highly freezing sensitive *Begonia* species, Burritt (2008) found that adventitious shoots of the rhizomatous begonia, *Begonia x erythrophylla* were sensitive to dehydration and very sensitive to freezing. While pre-treatment with 0.75 M sucrose significantly increased the percentage of encapsulated shoots surviving dehydration, pre-treatment with sucrose did not afford cryoprotection without prior dehydration.

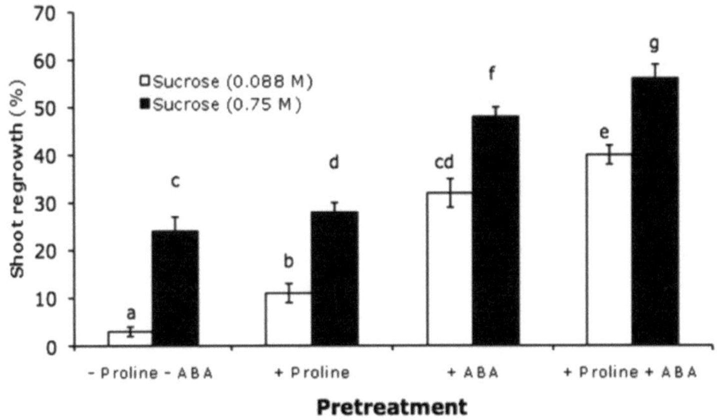

Fig. 3. The percentage of *Begonia x erythrophylla* shoots surviving pre-treatment with ABA (3.8 uM) and/or proline (2.15 mM). The different letters indicate statistically different values at $p < 0.05$ (modified from Burritt 2008).

Addition of abscisic acid (ABA) and proline to the pre-treatment medium significantly improved the percentage of shoots surviving freezing. Pre-treatment of shoots with a medium containing, 0.75M sucrose, 3.8 µM ABA and 2.15 mM proline resulted in greater than 50% of shoots surviving freezing (Figure 3).

Christianson (1998) used a 3-4 day preconditioning treatment using a tissue culture medium supplemented with 10^{-5} M ABA and 100 mM proline to greatly increase survival rates and simplify a protocol for moss cryopreservation. Pretreatment with the combination of proline and ABA was used as part of a cryopreservation protocol that could be used for *Ceratodon purpureus, Funaria hygrometrica, Physcomitrella patens,* and two species of *Sphagnum*. Cryopreserved cultures remained viable at least one year at -80°C.

In addition to both animal and plant cells, proline has been found to be particularily useful for the cryopreservation of algal cells. Kuwano et al. (2004) found the gametophytic cells of six species of Laminariales, *Laminaria japonica* Areschoug, *L. longissima* Miyabe, *Kjellmaniella crassifolia* Miyabe, *Ecklonia stolonifera* Okamura, *E. kurome* Okamura, and *Undaria pinnatifida* (Harvey) Suringar could be cropreserved using a cryoprotective solution containing ethylene glycol and proline. The cells were suspended in a mixture of ethylene glycol and proline, and slowly cooled to -40°C over a period of 4 h. After a cooling step, the cells were immediately immersed in liquid nitrogen. Viabilities ranged from 36.2% to 67.2%. Nanb et al. (2009) developed a cryopreservation protocol for gametophyte strains of the edible macroalgae *Undaria pinnatifida* (Harvey). Following a pretreatment involving exposure of male and female gametophytes to low levels of light, they used a two-step cooling method with a mixture of cryoprotectants including 10% L-proline and 10% glycerol, before freezing in liquid nitrogen. Gametophyte survival rates were high, ranging from 43-60% for females and 64-100% for males. The morphology of the sporophytes formed from the cryopreserved gametophytes appeared normal and the authors suggested that this cryopreservation method could be used to preserve culture stocks of *U. pinnatifida* for mariculture.

5. Proline a multifunctional cryoprotectant?

5.1 Possible mechanisms of protection

Because of its ability to act as an osmoprotectant without interfering with normal cellular processes and biochemical reactions proline has been used in a range of different cryopreservation protocols both for animal and plants cells and tissues, however the exact mode by which protection is achieved is still a matter of considerable debate in the scientific literature. Proline could potentially acting as storage reserve of carbon and nitrogen, a compatible osmolyte, a buffer for cytosolic pH, a scavenger of reactive oxygen species (ROS) and as an aid to balancing cellular redox status (Smirnoff & Cumbes 1989; Hare & Cress, 1997). It has also been proposed that proline could act as a molecular chaperone, helping to stabilize the structure of proteins, and as part of the signal transduction chain alerting plant cells to the presence of a stressor and hence triggering adaptive responses (Maggio et al. 2002).

5.2 Proline as an osmolite

The osmoregulatory role of proline in plant cells exposed to hyperosmotic stress has been the subject of numerous studies and under environmental conditions that result in cellular dehydration such as drought, freezing or extreme salinity, it is widely accepted that proline accumulates and acts as a compatible solute helping to protect cells from damage (Heur, 1994). Accumulation of cytoplasmic osmolytes, such as proline, is thought to aid in reducing the cellular water potential to a level below the external water potential, this enables water to move into the cell and be maintained there, while at the same time minimising potentially deleteriously increases in ionic strength. However, there is some debate in the published literature as to whether increased cytosolic levels of free proline has any direct adaptive value (Heur, 1994). While there are many reports of positive correlations between the capacity for proline accumulation and dehydration and cold tolerance (see section 3.1), some researchers still challenge the value of the ability of plant cells to accumulate proline as a positive index for osmotic stress resistance (Heur, 1994 & references therein).

5.3 Proline as precursor for other molecules

It has been suggested that stress-induced accumulation of amino acids like proline may not only have an osmoregulatory role, but that they could also be a mechanism to provide cells with a pool of the precursors required to synthesis other molecules known to be involved in biotic and abiotic stress responses (Sanchez et al., 2008). For example polyamines can be synthesized from arginine or ornithine and ornithine from glutamate, hence the pathways for proline and polyamine biosynthesis are interlinked, and both groups of molecules are important in plant stress responses (Groppa & Benavides 2008). Little is known about the roles of polyamine metabolism in the process of cryopreservation, but Ramon et al. (2002) reported that an increase in putrescine content was positively correlated with the survival rate after simple freezing or after vitrification of banana meristem cultures.

Stored amino acids could also be useful during the recovery process following stress. The accumulation of large cellular pools of amino acids could allow the rapid synthesis of enzymes and the repair of structural proteins, allowing a more rapid recovery of cells following cryopreservation, but this possibility has yet to investigated.

5.4 Proline as an antioxidant

Reactive oxygen species, such as the superoxide anion (O_2•), hydrogen peroxide (H_2O_2), and the extremely reactive hydroxyl radical (•OH) are produced within cells as a consequence of normal metabolic processes, but the production of ROS often increases when cells are under stress (Smirnoff, 1993; Halliwell & Gutteridge, 1999). When ROS are produced at levels high enough to overcome the antioxidant defences that normally control cellular ROS levels, oxidation of DNA, proteins and membrane fatty acids occurs, the latter can result in lipid peroxidation and loss of membrane function (Halliwell & Gutteridge, 1999). Such damage is commonly referred to as oxidative stress (Lesser, 2006; Burritt & MacKenzie 2003; Burritt, 2008). Cryopreservation protocols comprise a number of steps, each of which has the potential to cause stress that could increase ROS production. Recent studies have shown that dehydration and freezing can both lead to increased ROS production and lead to oxidative stress (Feck et al., 2000; Roach et al., 2008). A recent study on oxidative stress and antioxidant metabolism during the cryopreservation of olive somatic embryos demonstrated the importance of oxidative stress and antioxidant metabolism for the successful cryopreservation of plant cells (Lynch et al., 2011).

As mentioned in Section 4 Burritt (2008) found that addition of ABA and proline to the pre-treatment medium significantly improved the percentage of *Begonia x erythrophylla* shoots surviving freezing, this increase in percentage survival was accompanied by a decrease in levels of hydrogen peroxide (Figure 4) and oxidative damage, measured as the levels of lipid peroxides, observed in the shoots immediately following thawing (Figure 5).

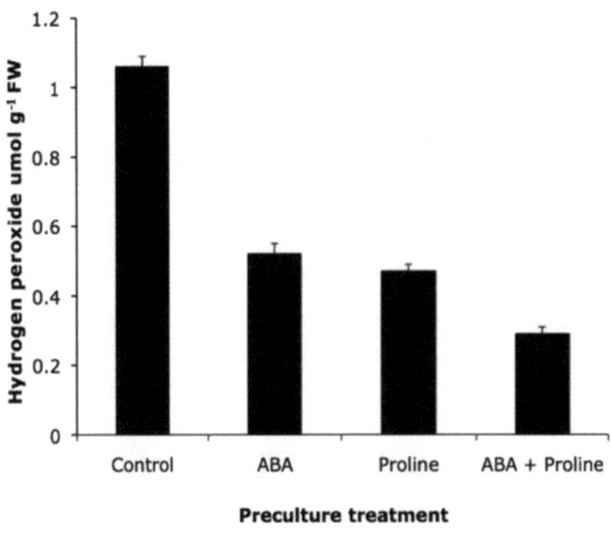

Fig. 4. The influence of pre-treatment with ABA (3.8 uM) and/or proline (2.15 mM) on hydrogen peroxide levels, as determined by Chesseman (2006), in post thaw *Begonia x erythrophylla* shoots cryopreserved as described by Burritt (2008).

Because of its chemical properties proline has a high capacity to quench singlet oxygen and hydroxyl radicals. Pyrrolidine, which forms the 5- membered ring of proline, has a low IP and so proline is able to form a charge-transfer complex, enabling it to quench singlet oxygen effectively. Proline can also react with hydroxyl radicals under hydrogen abstraction forming a stable radical (Matylsik, 2002). Therefore the accumulation of proline to high levels in plant cells under stress or plants cells treated with exogenous proline as part of a cryopreservation protocol could greatly increase the ROS scavenging capacity of said cells and reduce the potential for oxidative damage. In particular, as proline has the potential to reduce ROS levels it could help reduce oxidative damage to vital cellular macromolecules and hence stabilize proteins (Anjum, 2000),) DNA (Iakobashvil, 1999) and lipid membranes (Alia, 1991). The accumulation of proline-rich proteins and particularly proline residues in cellular proteins is thought to provide additional protection against oxidative stress (Matylsik, 2002). The increase in ROS scavenging capacity brought about by increased intracellular proline levels could be a key mechanism by which proline helps reduce the freezing and dehydration associated cellular damage associated with most cryopreservation protocols.

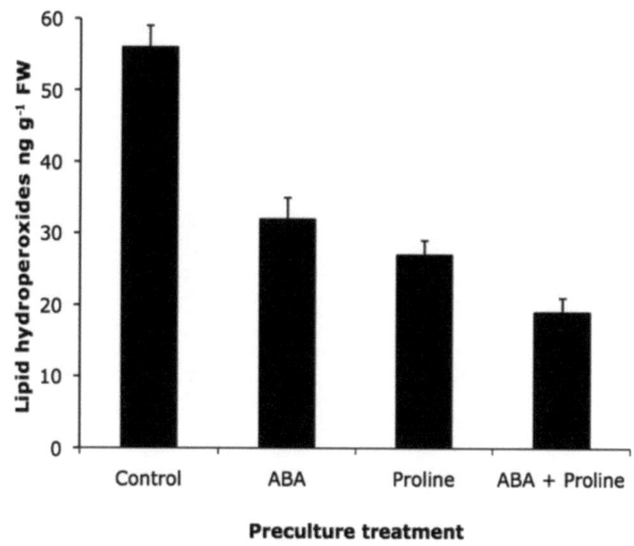

Fig. 5. The influence of pre-treatment with ABA (3.8 uM) and/or proline (2.15 mM) on lipid hydroperoxide levels, as determined by Mihaljevic et al. (1996), in post thaw *Begonia x erythrophylla* shoots cryopreserved as described by Burritt (2008).

Interestingly ABA combined with proline reduced hydrogen peroxide production and oxidative damage, measured as lipid peroxidation, more effectively in post thaw *Begonia x erythrophylla* shoots than ABA or proline alone. Christianson (1998) also found that ABA and proline in combination improved the survival of moss gametophytes following cryopreservation. These results suggest a possible interaction between ABA and proline may exist.

5.5 Is there an inaction between ABA and proline?

Studies have shown a relationship between proline and ABA with respect to cold tolerance (Xi & Li, 1993; Lou & Reid, 1997). In a recent study of maize suspension-cultured cells Chen and Li (2002) showed that an ABA treatment at warm temperatures improved the tolerance of cells to subsequent chilling, and that both ABA-treated and untreated maize cells accumulated proline in response to chilling. Chen and Li also found that ABA-treated cells showed less lipid peroxidation during chilling and unlike untreated cells were able to retain the accumulated proline intracellularly.

In post thaw *Begonia x erythrophylla* shoots ABA combined with proline resulted in much higher shoot survival than pretreatment with ABA or proline alone. Interestingly ABA combined with proline resulted in far higher intracellular proline concentrations (Figure 6). The greater concentrations of proline seen in the combined treatment could be due increased endogenous synthesis of proline, induced by exposure to ABA, combined with uptake of exogenous proline during the pretreatment phase and/or to an ABA induced mechanism that helps reduce proline leakage, but further investigations are required to determine how the combined application of ABA and proline increase shoot survival after cryopreservation.

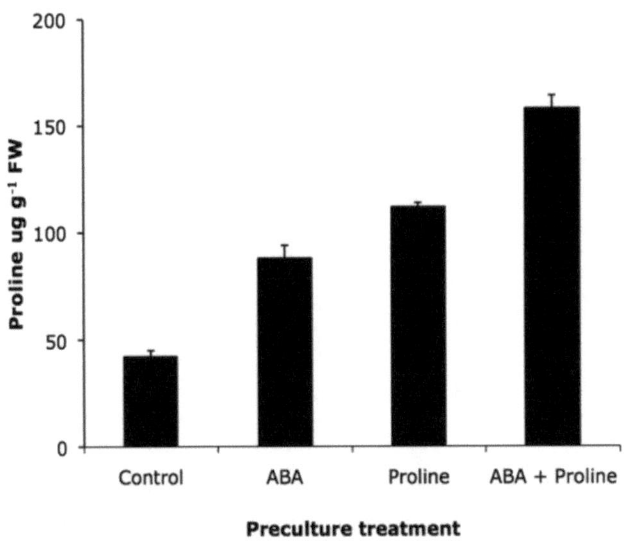

Fig. 6. The influence of pre-treatment with ABA (3.8 uM) and/or proline (2.15 mM) on proline levels, in post thaw *Begonia x erythrophylla* shoots cryopreserved as described by Burritt (2008).

5.5 Proline and direct macromolecule protection

As well as the potential protective mechanisms detailed above, proline has been shown to directly protect key cellular macromolecules, in particular lipid membranes and proteins such as enzymes (Verbruggen & Hermans, 2008). Proline molecules can intercalate between the head groups of membrane phospholipids during freeze-dehydration helping to reduce mechanical stresses in the membranes, or alter the physical properties of membranes making them less prone to a liquid crystalline-to-gel transition (Hoekstra et al., 2001). It has also been suggested that proline molecules can directly replace missing water molecules between the phospholipids headgroups (Rudolph et al., 1986).

In addition, according to the preferential exclusion hypothesis, proline is one of a group of solutes that, when in aqueous solution, are excluded from contact with the surfaces of proteins and phospholipid bilayers (Arakawa & Timasheff, 1983). Accordingly addition of proline to a solution stabilizes the native structure of protein monomers and protects oligomeric protein complexes from denaturation and dissociation. Rudolph et al. (1986) demonstrated that the activity of the enzyme lactate dehydrogenase could be protected in part during freeze-thaw cycles by increasing the concentration of proline from 0 to 200 mM in the buffer in which the enzyme was solubilised.

5.6 Other mechanisms

There are several other mechanisms via which proline could contribute to over coming the stresses associated with cryopreservation. For example, the accumulation of proline could also be a mechanism to store energy as the oxidation of a single proline molecule can produce up to 30 ATP equivalents (Atkinson, 1971). Replenishment of NADP+ and redox cycling have also been sugested as potential mechanisms associated with stress tolerance, (Hare & Cress 1999), as has a role in stress signal transdution (Hare et al., 1997).

6. Conclusions

While numerous studies have demonstrated that proline can be used to improve the survival of plant cells and organs following cryopreservation, there is little definitive evidence as to the mode of action of proline. More research is required to determine how proline protects plant cells at the cellular level and to determine how other treatments that confer cryotolerance, such as ABA pretreatments interact with proline metabolism and could hence improve the cryotolerance of plant cells. However, despite our lack of knowledge with respect to the mode of action of proline, this amino acid continues to be of great value as a cryoprotectant that can be used with a wide range of cell types from many different organisms.

6. References

Ait-Barka, E.; Audran, J.C. (1997) Response of champenoise grapevine to low temperatures: changes of shoot and bud proline concentrations in response to low temperatures and correlations with freezing tolerance. *Journal of Horticultural Science*, Vol.72,577–582.

Alia; Saradhi, P.P. (1991) Proline accumulation under heavy-metal stress. *Journal of Plant Physiology*, Vol.138,554-558.

Anjum, F.; Rishi, V.; Ahmad, F. (2000) Compatibility of osmolytes with Gibbs energy of stabilization of proteins. *Biochimica et biophysica acta-protein structure and molecular enzymology*, Vol.1476,75-84.

Arakawa, T.; Timasheff, S.N. (1985) The stabilisation of proteins by osmolytes. *Biophysical Journal*, Vol.47,411–414.

Ashraf M.; Foolad M. R. (2007) Roles of glycine betaine and proline in improving plant abiotic stress resistance. *Environmental and Experimental Botany*, Vol.59,206-216.

Atkinson, D,E, (1977) Cellular Energy Metabolism and Its Regulation. New York: Academic Press.

Ayliffe, M.A.; Mitchell, H.J.; Deuschle, K.; Pryor, A.J.; (2005) Comparative analysis in cereals of a key proline catabolism gene. *Molecular Genetics & Genomics*, Vol.274,494–505.

Bohnert, H.J.; Nelson, D.E.; Jensen, R.G. (1995) Adaptations to environmental stresses. *The Plant Cell*, Vol.7,1099–1111.

Brison, M.; Deboucaud, M.T.; Dosba, F. (1995) Cryopreservation of in-vitro grown shoot tips of 2 interspecific prunus rootstocks. *Plant Science*, Vol.105,235-242.

Burbulis, N.; Jonytiene, V.; Kupriene,R.; Blinstrubiene, A. (2011) Changes in proline and soluble sugars content during cold acclimation of winter rapeseed shoots in vitro. *Journal of Food Agriculture & Environment*, Vol.9,371-374.

Burritt, D.J. (2008) Efficient cryopreservation of adventitious shoots of *Begonia x erythrophylla* using encapsulation-dehydration requires pretreatment with both ABA and proline. *Plant Cell Tissue and Organ Culture*, Vol.95,209-215.

Burritt, D.J.; MacKenzie, S. (2003). Antioxidant metabolism during the acclimation of *Begonia x erythrophylla* to high-light. *Annals of Botany*, Vol.91, 783-794.

Cheeseman, J.M. (2006) Hydrogen peroxide concentrations in leaves under natural conditions. *Journal of Experimental Botany*, Vol.57,2435-2444.

Chen, W.P.; Li, P.H. (2002) Membrane stabilization by abscisic acid under cold aids proline in alleviating chilling injury in maize (*Zea mays* L.) cultured cells. *Plant Cell and Environment*, Vol.25,955-962.

Christianson, M.L. (1998) A simple protocol for cryopreservation of mosses. *The Bryologist*, Vol.101,32-35.

Chu, T.M.; Jusaitis, M.; Aspinall, D.; Paleg, L.G. (1978) Accumulation of free proline at low temperatures. *Physiologia Plantarum*, Vol.43, 254–260.

Cook, D.; Fowler, S.; Fiehn, O.; (2004) A prominent role for the CBF cold response pathway in configuring the low-temperature metabolome of Arabidopsis. *Proceedings of the National Academy of Sciences of the United States of America*, Vol.101,15243-15248. *Crop Science*, Vol.50,1037-1047.

Delauney, A.J.; Verma, D.P.S. (1993) Proline biosynthesis and osmoregulation in plants. *Plant Journal*, Vol4,215–223.

Dorffling, K.; Dorffling, H.; Lesselich, G.; Luck, E.; Zimmermann, C.; Melz, G.; Jurgens, H.U. (1997) Heritable improvement of frost tolerance in winter wheat by in vitro selection of hydroxyproline-resistant proline overproducing mutants. *Euphytica*, Vol.93,1-10.

Duncan, D.R.; Widholm, J.M. (1987) Proline accumulation and its implication in cold tolerance of regenerable maize callus. *Plant Physiology*, Vol.83,703-708.

Elthon, T.E.; Stewart, C.R.; (1982) Proline oxidation in corn mitochondria: involvement of NAD, relationship to ornithine metabolism, and sidedness on the inner membrane. *Plant Physiology*, Vol.70,567–572.

Fleck, R.A.; Benson, E.E.; Bremner, D.H.; Day, J.G. (2000) Studies of free radical-mediated cryoinjury in the unicellular green alga *Euglena gracilis* using a non-destructive hydroxyl radical assay: A novel approach for developing protistan cryopreservation strategies. *Free Radical Research*, Vol.32,157-170.

Forlani, G.; Scainelli, D.; Nielsen, E. (1997) Two D1-pyrroline-5- carboxylate dehydrogenase isoforms are expressed in cultured *Nicotiana plumbaginifolia* cells and are differentially modulated during the culture growth cycle. *Planta,* Vol. 202,242.

Funck, D.; Stadelhofer, B.; Koch, W. (2008) Ornithine-d-aminotransferase is essential for arginine catabolism but not for proline biosynthesis. *BMC Plant Biology,* Vol.8,40.

Groppa M. D.; Benavides M. P. (2008). Polyamines and abiotic stress: Recent advances. Amino Acids Vol.34, 35-45.

Halliwell, B.; Gutteridge, J.M.C. (1999). Free Radicals in Medicine and Biology, Edition 3. Oxford: Oxford University Press.

Hare, P.D.; & Cress, W.A. (1997) Metabolic implications of stress-induced proline accumulation in plants. *Plant Growth Regulation,* Vol.21,79–102.

Hare, P.D.; Cress, W.; van Staden, J. (1999) Proline synthesis and degradation: a model system for elucidating stress-related signal transduction. *Journal of Experimental Botany,* Vol.50,413–434.

Heuer, B. (1994) Osmoregulatory role of proline in water and salt-stressed plants. In: Pessarakli, M. (ed) Handbook of Plant and Crop Stress, pp 363–381. New York: Marcel Dekker, Inc.

Hoekstra, F.A.; Golovina, E.A.; Buitink, J. (2001) Mechanisms of plant desiccation tolerance. *Trends in Plant Science,* Vol.6,431-438.

Hoffman, L.; DaCosta, M.; Ebdon, S.J.; Watkins, E. (2010) Physiological Changes during Cold Acclimation of Perennial Ryegrass Accessions Differing in Freeze Tolerance.

Iakobashvil, R.; Lapidot, A. (1999) Low temperature cycled PCR protocol for Klenow fragment of DNA polymerase I in the presence of proline. Nucleic Acids Research Vol.27,1566-1568.

Jain, S.; Jain, R.K.; Wu, R. (1996) A simple and efficient procedure for cryopreservation of embryogenic cells of aromatic Indica rice varieties. *Plant Cell Reports,* Vol.15,712-717.

Javadian, N.; Karimzadeh, G.; Mahfoozi, S.; Ghanati, F. (2010) Cold-induced changes of enzymes, proline, carbohydrates, and chlorophyll in wheat. *Russian Journal of Plant Physiology,* Vol.57,540-547.

Kaplan F.; Guy C.L. (2004) Beta-amylase induction and the protective role of maltose during temperature shock. *Plant Physiology,* Vol.135,1674-1684.

Kavi Kishor, P.B.; Hong, Z.; Miao, G.-H.; Hu, C.-A.A.; Verma D.P.S. (1995) Overexpression of Δ1-pyrroline-5-carboxylic acid synthetase increases proline production and confers osmotolerance in transgenic plants. *Plant Physiology,* Vol.108,1387–1394.

Kostal, V.; Zahradnickova, H.; Simek, P. (2011) Hyperprolinemic larvae of the drosophilid fly, *Chymomyza costata,* survive cryopreservation in liquid nitrogen. *Proceedings of the National Academy of Sciences of the United States of America,* Vol.108,13041-13046.

Koster, K.L.; Lynch, D.V. (1992) Solute accumulation and compartmentation during the cold acclimation of Puma rye. *Plant Physiology,* Vol.98,108–113.

Kuwano, K.; Kono, S.; Jo, Y.H.; Shin, J.A.; Saga, N. (2004) Cryopreservation of the gametophytic cells of Laminariales (Phaeophyta) in liquid nitrogen. *Journal of Phycology,* Vol.40,606-610.

Lehmann Silke; Funck Dietmar; Szabados Laszlo; (2010) Proline metabolism and transport in plant development. *Amino Acids,* Vol.39,949-962.

Lesser, M.P (2006). Oxidative stress in marine environments: biochemistry and physiological ecology. *Annual Review of Physiology,* Vol.68, 253-278.

Li, Y.H.; Si, W.; Zhang, X.Z.; Dinnyes, A.; Ji, W.Z. (2003) Effect of amino acids on cryopreservation of cynomolgus monkey (*Macaca fascicularis*) sperm. *American Journal of Primatology,* Vol.59,159-165.

Luo, J. and B.M. Reed. 1997. Abscisic-acid responsive protein, bovine serum albumin,and proline pretreatments improve recovery of in vitro currant shoot-tip meristems and callus preserved by vitrification. *Cryobiology,* Vol.34,240-250.

Lynch, P.T.; Siddika, A.; Johnston, J.W.; Trigwell, S.M.; Mehra, A.; Benelli, C.; Lambardi, M.; Benson, E.E. (2011) Effects of osmotic pretreatments on oxidative stress, antioxidant profiles and cryopreservation of olive somatic embryos. *Plant Science,* Vol.181,47-56.

Maggio, A.; Miyazaki, S.; Veronese, P.; Does proline accumulation play an active role in stress-induced growth reduction? *Plant Journal,* Vol.31,699-712.

Mani, S.; Van de Cotte, B.; Van Montagu, M.; Verbruggen, N. (2002) Altered levels of proline dehydrogenase cause hypersensitivity to proline and its analogs in Arabidopsis. *Plant Physiology,* Vol.128,73–83.

Mattioli, R.; Costantino, P.; Trovato, M.; (2009) Proline accumulation in plants: not only stress. *Plant Signaling & Behavior* Vol.4,1016–1018

Mattioli, R.; Marchese. D.; D'Angeli, S.; Altamura, M.; Costantino, P.; Trovato, M.; (2008) Modulation of intracellular proline levels affects flowering time and inflorescence architecture in *Arabidopsis. Plant Molecular Biology,* Vol.66,277–288 .

Matysik J; Alia; Bhalu B; Mohanty, P. (2002) Molecular mechanisms of quenching of reactive oxygen species by proline under stress in plants. *Current Science,* Vol.82,525-532.

Mihaljevic, B.; Katusin-Razem, B.; Razem, D. (1996) The reevaluation of the ferric thiocyanate assay for lipid hydroperoxides with special considerations of the mechanistic aspects of the response. *Free Radical Biology and Medicine,* Vol.21,53-63.

Mitchell, H.J.; Ayliffe, M.A.; Rashid, K.Y.; Pryor, A.J. (2006) A rust- inducible gene from flax (*fis1*) is involved in proline catabolism. *Planta* Vol.223,213–222.

Nanb, N.; Fujiwara, T.; Kuwano, K.; Ishikawa, Y.; Ogawa, H.; Kado, R. (2009) Effect of pre-incubation irradiance on survival of cryopreserved gametophytes of *Undaria pinnatifida* (Phaeophyta) and morphology of sporophytes formed from the gametophytes. *Aquatic Botany,* Vol.90,101-104.

Nanjo, T.; Kobayashi, M.; Yoshiba, Y.; Kakubari, Y.; Yamaguchi-Shi-nozaki, K.; Shinozaki, K. (1999) Antisense suppression of proline degradation improves tolerance to freezing and salinity in *Arabidopsis thaliana. FEBS Letters,* Vol.461,205–210.

Poustini K.; Siosemardeh A.; Ranjbar M. (2007) Proline accumulation as a response to salt stress in 30 wheat (*Triticum aestivum* L.) cultivars differing in salt tolerance. *Genetic Resources and Crop Evolution,* Vol.54,925-934.

Ramon, M.; Geuns, J.M.C.; Swennen, R.; Panis, B. (2002) Polyamines and fatty acids in sucrose precultured banana meristems and correlation with survival rate after cryopreservation. *Cryoletters,* Vol.23,345-352.

Ribarits, A.; Abdullaev, A.; Tashpulatov, A.; Richter, A.; Heberle-Bors, E.; Touraev, A. (2007) Two tobacco proline dehydrogenases are differentially regulated and play a role in early plant develop- ment. *Planta,* Vol.225,1313–1324.

Roach, T.; Ivanova, M.; Beckett, R.P.; Minibayeva, F.V.; Green, I.; Pritchard, H.W.; Kranner, I. (2008) An oxidative burst of superoxide in embryonic axes of recalcitrant sweet chestnut seeds as induced by excision and desiccation. *Physiologia Plantarum,* Vol.133,131-139.

Rudolph, A.S.; Crowe, J.H.; Crowe, L.M. (1986) Effects of three stabilising agents - proline, betaine and trehalose - on membrane phospholipids. *Archives of Biochemistry and Biophysics,* Vol.245: 134–143.

Sanchez, D.H.; Siahpoosh, M.R.; Roessner, U.; Udvardi, M.; Kopka, J. (2008) Plant metabolomics reveals conserved and divergent metabolic responses to salinity. *Physiologia Plantarum,* Vol.132,209-219.

Saradhi, P.P.; Alia; Arora S; Prasad, KVSK (1995) Proline accumulates in plants exposed to UV-radiation and protects them against UV induced peroxidation. *Biochemical and Biophysical Research Communications,* Vol.209,1-5.

Siripornadulsil, S.; Traina, S.; Verma, D.P.S.; (2002) Molecular mechanisms of proline-mediated tolerance to toxic heavy metals in transgenic microalgae. *The Plant Cell,* Vol.14,2837-2847.

Smirnoff, N. (1993) The role of active oxygen in the response of plants to water deficit and desiccation. *New Phytologist,* Vol.125: 27–58.

Smirnoff, N.; Cumbes. Q.J. (1989) Hydroxyl radical scavenging activity of compatible solutes. *Phytochemistry,* Vol.28:1057–1060.

Stranska, J.; Kopecny, D.; Tylichowa, M.; Snegaroff, J.; Sebela, M. (2008) Ornithine d-aminotransferase: an enzyme implicated in salt tolerance in higher plants. *Plant Signaling & Behavior,* Vol.3,929–935.

Strizhov, N.; Abraham, E.; Okresz, L.; Blickling, S.; Zilberstein, A.; Schell, J.; Koncz, C.; Szabados, L. (1997) Differential expression of two P5CS genes controlling proline accumulation during salt-stress requires ABA and is regulated by ABA1, ABI1 and AXR2 in Arabidopsis. *The Plant Journal,* Vol.12,557–569.

Swaaij, A.C.V.; Jacobsen, E.; Feenstra ,W.J. (1985) Effect of cold hardening, wilting and exogenously applied proline on leaf pro- line content and frost tolerance of several genotypes of Solanum. *Physiologia Plantarum,* Vol.64,230–236.

Swaaij, A.C.V.; Jacobsen, E.; Kiel, J.A.K.W.; Feenstra, W.J.; (1986) Selection, characterization and regeneration of hydroxyproline- resistant cell lines of *Solanum tuberosum*: tolerance to NaCl and freezing stress. *Physiologia Plantarum,* Vol.68, 359–366.

Szabados, L.; Savoure, A. (2010) Proline: a multifunctional amino acid. *Trends in Plant Science,* Vol.15,89–97

Szekely, G.; Abraham, E.; Cseplo, A.; Rigo, G.; Zsigmond, L.; Csiszar, J.; Ayaydin, F.; Strizhov, N.; Jasik, J.; Schmelzer, E.; Koncz, C.; Szabados, L. (2008) Duplicated P5CS genes of Arabidopsis play distinct roles in stress regulation and developmental control of proline biosynthesis. *The Plant Journal,* Vol.53,11–28.

Taji, T.; Ohsumi, C.; Iuchi, S.; Seki, M.; Kasuga, M.; Kobayashi, M.; Yamaguchi-Shinozaki, K.; Shinozaki, K. (2002) Important roles of drought- and cold-inducible genes for galactinol synthase in stress tolerance in *Arabidopsis thaliana. The Plant Journal,* Vol.294,417-426.

Verbruggen, N.; Hermans, C. (2008) Proline accumulation in plants: a review. *Amino Acids,* Vol.35,753–759.

Xin, Z.; Browse, J. (1998) eskimo1 mutants of Arabidopsis are constitutively freezing-tolerant. *Proceedings of National Academy of Science USA,* Vol.95, 7799–7804.

Xin, Z.; Li, P.H. (1993) Relationship between proline and abscisic acid in the induction of chilling tolerance in maize suspension- cultured cells. *Plant Physiology*, Vol.103,607–613.

Yoshiba, Y.; Kiyosue, T.; Nakashima, K.; Yamaguchi-Shinozaki, K; Shinozaki, K. (1997) Regulation of levels of proline as an osmolyte in plants under water stress. *Plant Cell and Physiology*, Vol.38,1095–1102.

Zhang, L.P.; Mehta, S.K.; Liu, Z.P.; Yang, Z.M. (2008) Copper-induced proline synthesis is associated with nitric oxide generation in *Chlamydomonas reinhardtii*. *Plant and Cell Physiology*, Vol.49,411–419.

21

Plant Cryopreservation

R.K. Radha, William S. Decruse and P.N. Krishnan
Plant Biotechnology and Bioinformatics Division
Tropical Botanic Garden and Research Institute
Palode, Thiruvananthapuram, Kerala,
India

1. Introduction

Two basic approaches to conservation of plant genetic resources are *ex situ* and *in situ* conservation. *Ex situ* conservation includes seed storage, *in vitro* storage, DNA storage, pollen storage, field genebanks and botanical gardens while the *in situ* approach encompasses genetic reserves, on farm and home garden conservation.

Cryopreservation is a part of biotechnology. Biotechnology plays an important role in international plant conservation programs and in preservation of the world's genetic resources (Bajaj, 1995; Benson, 1999). Advances in biotechnology provide new methods for plant genetic resources and evaluation (Paunesca, 2009). Cryopreservation, developed during the last 25 years, is an important and the most valuable method for long-term conservation of biological materials. The main advantages in cryopreservation are simplicity and the applicability to a wide range of genotypes (Engelmann, 2004). This can be achieved using different procedures, such as pre-growth, desiccation, pregrowth-desiccation, vitrification, encapsulation-vitrification and droplet-freezing (Engelmann, 2004). Cryopreservation involves storage of plant material (such as seed, shoot tip, zygotic and somatic embryos and pollen) at ultra-low temperatures in LN (-196°C) or its vapor phase (-150°C). To avoid the genetic alterations that may occur in long tissue cultures storage, cryopreservation has been developed (Martin *et al.*, 1998). At this temperature, cell division, metabolic, and biochemical activities remain suspended and the material can be stored without changes and deterioration for long time. Walters *et al.* (2009) proposed that this assumption, based on extrapolations of temperature-reaction kinetic relationships, is not completely supported by accumulating evidence that dried seeds can deteriorate during cryogenic storage. After 30 years of cryogenic storage, seeds of some species exhibited quantitatively lower viability and vigor. In cryopreservation method, subcultures are not required and somaclonal variation is reduced. Advantages of cryopreservation are that germplasm can be kept for theoretically indefinite time with low costs and little space. Besides its use for the conservation of genetic resources, cryopreservation can also be applied for the safe storage of plant tissues with specific characteristics. Different types of plant cell, tissues and organs can be cryopreserved. Cryopreservation is the most suitable long-term storage method for genetic resources of vegetatively maintained crops (Kaczmarczyk *et al.*, 2008). For vegetatively propagated species, the best organs are shoot apices excised from *in vitro* plants. Shoot apices or meristems cultures are suitable because of virus-free plant production, clonal propagation, improving health status, easier recovery and less mutation (Scowcroft, 1984). Seed and field collections have been the only proper for the long-term germplasm conservation of woody species, while a large number of forest angiosperms have recalcitrant seeds with a very limited period of conservability. The species, which are mainly vegetatively propagated, require the conservation of huge number of accessions (Panis and Lambardi, 2005). The storage of this huge number needs large areas of land and high running costs. Preservation of plant germplasm is part of any plant breeding program. The most efficient and economical way of germplasm storage is the form of seeds. However, this kind of storage is not always feasible because 1) some seeds deteriorate due to invasion of pathogens and insects, 2) some plants do not produce seeds and they are propagated vegetatively, 3) some seeds are very heterozygous thus, not proper for maintaining true-to-type genotype, 4) seeds remain viable for a limited time, and 5) clonally propagated crops such as fruit, nut, and many root and tuber vegetables cannot be stored as seed (Chang and Reed, 2001; Bekheet *et al.*, 2007). Cryopreservation offers a good method for conservation of the species, especially woody plant germplasm (Panis and Lambardi, 2005). Cryostorage of seeds in LN was initially developed for the conservation of genetic resources of agriculturally important species (Rajasekharan, 2006). The development of simple cryostorage protocols for orthodox seeds has allowed cryopreservation of a large number of species at low cost, significantly reducing seed deterioration in storage (Stanwood, 1987). Only few reports are available on the application of cryopreservation on seeds of wild and endangered species and medicinal plants (Rajasekharan, 2006). New cryobiological studies of plant materials has made cryopreservation a realistic tool for long-term storage, for tropical species, which are not intrinsically tolerant to low temperature and desiccation, has been less extensively investigated (Rajasekharan, 2006). Cryopreservation has been applied to more than 80 plant species (Zhao *et al.*, 2005). Number of species, which can be cryopreserved has rapidly increased over the last several years because of the new techniques and progress of cryopreservation research (Rajasekharan, 2006). The vitrification/one-step freezing and encapsulation dehydration methods have been applied to an increasing number of species (Panis and Lambardi, 2005). A new method, named encapsulation- vitrification is noteworthy (Sakai, 2000). These techniques have produced high levels of post-thaw and minor modifications (Rajasekharan, 2006). In cryopreservation, information recording such as type and size of explants, pretreatment and the correct type and concentration of cryoprotectants, explants water content, cryopreservation method, rate of freezing and thawing, thawing method, recovery medium and incubation conditions is very important (Reed, 2001 ; González-Benito *et al.*, 2004; Bekheet *et al.*, 2007). All germplasm requires safe storage because even exotic germplasm without obvious economic merit may contain genes or alleles that may be needed as new disease, insect, environmental, or crop production problems arise (Westwood, 1989). It is important to record also the recovery percentage after a short conservation period. A major concern is the genetic stability of conserved material.

For many plant species which produce orthodox seeds, i.e. which can be dehydrated extensively and stored dry at low temperature, the emphasis for genetic resource conservation will be on seed/embryo storage. Recalcitrant seeds cannot tolerate desiccation to moisture content that would permit exposure to low temperature. They are often large with considerable quantities of fleshy endosperm. Therefore, recent investigations have identified species displaying an intermediate form of seed/embryo storage. As regards the balance of techniques employed within complementary strategies developed for conserving the genetic resources of these problems species, the emphasis in the case of non-orthodox (intermediate/ recalcitrant) forest tree species will be on *in situ* conservation in genetic reserves, while for species which are propagated vegetatively the emphasis will be on *ex situ* conservation techniques, including field genebank and *in vitro* storage. However it is essential to recognize that owing to various problems and limitations encountered with both genetic reserves and field genebanks, cryopreservation currently offers the only safe and cost effective option for the long-term conservation of genetic resources of these problem species. Significant progress has been made during the past 10 years in the area of plant cryopreservation with the development of various efficient cryopreservation protocols. An important advantage of these new techniques is their operational simplicity, since they will be applied mainly in developing tropical countries where the largest part of genetic resources of problem species is located. Encouraging results in medicinal plants have been published in recent years which present extensive list of plant species whose embryos and or embryonic axes have been successfully cryopreserved (Kartha and Engelmann 1994, Pence 1995, Engelmann *et al* 1995).

In comparison with results obtained with vegetatively propagated species, it is clear that research is still at a very preliminary stage for recalcitrant seeds. The desiccation technique is mainly employed for freezing embryos and embryonic axes, the survival achieved are extremely uneven. And also survival is often limited and regeneration often restricted to callusing or incomplete development of plantlets. In only a limited number of cases, the whole plants have been regenerated from cryopreserved material (Chin and Pritchard 1988, Assy Bah and Engelmann 1992). Seeds and embryos of recalcitrant species also display various characteristics which make their cryopreservation difficult. One of the characteristics of recalcitrant seeds is that there is no arrest in their development, as with orthodox seeds. It is very difficult to select seeds at a precise developmental stage, even though this parameter is often of critical importance to achieve successful cryopreservation. Seeds of many species are too large to be frozen directly and embryos or embryonic axes have to be employed. However, embryos are often very complex tissue composition which display differential sensitivity to desiccation and freezing, the root pole seeming more resistant than the shoot pole (Pence 1995). In some species, embryos are extremely sensitive to desiccation and even minor reduction in their moisture content down to levels much too high to obtain survival after freezing leads to irreparable structural damage. It should be emphasized that selecting embryos at the right developmental stage is of critical importance for the success of any cryopreservation experiment (Engelmann *et al.*, 1995) However, in these cases basic protocols for disinfection, *in vitro* germination of embryos or embryonic axes, plantlet development and possibly limited propagation will have to be established prior to any cryopreservation experiment.

Cryostorage of seed was initially developed for the preservation of genetic resources of agriculturally important species for breeding and selection. The development of comparatively simple cryostorage protocols allowed seeds of over 155 agricultural species (Stanwood, 1985) to be stored at low cost, in an environment without obvious problems of seed ageing, genetic variations and predation common to many conventional seed storage methods. With the regular use of cryostorage system for seeds of agri-crops, the same process is now viewed as having important application for preserving seeds of medicinal plants (Decruse *et al.*, 1999), endangered species (Touchel and Dixon 1994) and other native plant species (Pence 1991, Touchel and Dixon 1993, Decruse and Seeni 2002). For the long-term preservation of species producing recalcitrant seeds, zygotic embryos were used for cryopreservation. Incidentally excised zygotic embryos or embryonic axes were successfully employed for the cryopreservation of coconut (Assy-Bah and Engelman, 1992 a,b, Chin *et al.*,

1989), cocoa (Pence, 1991, Chandel *et al.*, 1995) oil palm (Chabrillange *et al.*, 1997), walnut (de Boucaud *et al.*, 1991), jack fruit (Chandel *et al.*, 1995, Thammasiri, 1999), rubber (Normann, 1986), tea (Chauduryi *et al.*, 1991) and neem (Berjack and Dumet, 1996).

National Gene bank for Medicinal and Aromatic Plants at Tropical Botanic Garden and Research Institute (TBGRI) is one among the four (CIMAP, Lucknow, NBPGR, New Delhi and RRL, Jammu) having the mandate of conserving the medicinal and aromatic plants (MAPs) of Peninsular India through biotechnological intervention including collection, ex *situ* conservation and characterization of the precious taxa that are rare, endangered, threatened, endemic, vulnerable or over exploited as the case may be. TBGRI has significantly developed cryopreservation protocol on rare and endangered medicinal plants of India (Decruse *et al.*, 1999, Decruse and Seeni, 2002, Radha *et al.*, 2006). A cryobank was also established which now holds more than 25 accessions of medcinal and aromatic plants (Decruse *et al.*, 1999b, Decruse and Seeni 2002b, Radha *et al.*, 2010).

2. Cryopreservation of excised embryonic axes of *Nothapodytes nimmoniana* (Graham) Mebberly, a vulnerable medicinal tree species of the Western Ghats

Nothapodytes nimmoniana (Graham) Mebberly, of family Icacinaceae is a small vulnerable medicinal tree distributed in India, Sri Lanka, Myanmar, Thailand, Malaysia and China. In India it is distributed in upper ranges of the Western Ghats particularly in the Nilgiris and Palni hills of southern peninsula. The stem and roots are an important source of the anti-tumour quinoline alkaloid camptothecin (Hsiang etal.,1985) and also find applications against retrovirus and human immunodeficiency virus. Consequently natural population of this species in the Western Ghats are severely depleted owing to habitat destruction and over exploitation (Cragg *et al.*,1993, Ravikumar and Ved, 2000) and hence conservation efforts are undertaken by certain agencies in the region.

Seeds of N. nimmoniana are large intermediate type showed 100% germination under controlled conditions. Embryonic axes with cotyledons having moisture content of 55.7% presumed to be intermediate in nature, lose their viability within a short period after maturity. Cryopreservaton of zygotic embryos is recognized as an effective tool for the long-term preservation of such plant species those produce recalcitrant/large seeds (Engelmann, 1997).

Desiccation and cryopreservation. The seeds were separated from the fruits (drupe), rinsed in running tap water for one hour to remove the mucilage and washed in commercial detergent (1% Teepol, Godrej, India Ltd., Mumbai) for 10 min. followed by thorough washing in running tap water for 10-20 min. Seeds were then surface decontaminated by immersion in 0.01% (w/v) $HgCl_2$ for 5-10 min. followed by 3-5 rinses in sterile distilled water. Seed coat was broken and embryos with cotyledons were dissected out free of the endosperm in aseptic condition in the laminar air flow cabinet. Immediately after dissection, batches of 20 embryos each were subjected to dehydration under laminar airflow for 30, 60, 90, 120,150,180 and 210 min. period. A sample of 10 embryos was inoculated into MS medium (Murashige and Skoog, 1962) devoid of PGR as fresh control and cultured under 10/14h light/dark periods (30 - 50 µmol m^{-2} s^{-2}) at 25±2 ^{0}C for 8 weeks. After desiccation at 30 min intervals, equally divided samples of 10 embryos were transferred to germination medium and another 10 packed in 2ml cryovial and transferred to LN (at -196 ^{0}C) After 24h storage, the vials were retrieved from LN and rewarmed in a water bath at 40°C for 1-2 min. The rewarmed embryos were also transferred to germination medium and cultured under stated conditions for recovery. The whole experiment was repeated three times.

Observations on the germination of embryos were made after 8 weeks and results analyzed statistically in a completely randomized model. Survival rate was assessed as the percentage of embryonic axes that exhibited any kind of growth, including seedling development; shoot growth and root growth.

Moisture content determination. Moisture content (MC) of the embryos was determined by constant temperature oven method (103 °C) for 17h.

The embryonic axes with cotyledons (Fig 1a) freshly dissected from the seeds possessed 55.7% MC and exhibited 86.67% germination and normal growth in MS medium devoid of PGRs within a week of culture. Dehydration under laminar airflow reduced the MC to 43.7% after 30min and 31.3% after 60min. without appreciable reduction in viability so that 76-77% of them germinated (Fig.1). Dehydration for 120min reduced MC to 19.6% and germination to 66.67% and was the optimum dehydration period (Fig.2) to get maximum germination (60%) after LN treatment (Fig.2). Root and shoot emergence was observed after one week of culture (Fig.1b) in 60% of the desiccated (120 min.) and LN treated embryonic axes and well developed seedlings were obtained within 20 days of culture (Fig.1c). Dehydration beyond 120 min. gradually reduced MC and drastically reduced viability. The MC came down to 12.1% after 210 min. when none of the embryos survived. Prolonged dehydration (150-180min) not only reduced survival down to 16.67-10% but also caused abnormal growth with only radicle development in the survived embryos (Fig.1d).

Research in the past two decades has shown that most orthodox seeds remain viable for long periods of storage after attaining appropriate desiccation levels of about 3-5% moisture content (Roberts, 1973). Contrary to this, recalcitrant seeds of several tropical and temperate species are desiccation sensitive, eg. Tea, Cocoa, Citrus, Jack fruit (Chin and Roberts, 1980). There are various options available to improve storage of non-orthodox seeds/embryos. Desiccation is the simplest procedure since it consists of dehydrating explants, and then freezing them rapidly by direct immersion in LN has been applied to embryonic axes extracted from recalcitrant and intermediate seeds (Engelmann, 1997). It should also be noted that selection of embryos at the right developmental stage is of critical importance for the success of any cryopreservation experiment (Engelmann *et al.*, 1995). The conservation efforts of *N. nimmoniana* are hampered mainly due to relatively large and intermediate type of seeds with desiccation sensitivity. The viability of the embryos was not much affected when the embryos were desiccated from 55.7% to 43.7% (i.e. 30 min. desiccation). Significant loss of viability due to further reduction of moisture content shows the intermediate nature of the embryos is in line with the report of Dussert *et al* (1995). Safe moisture content of the embryonic axes as obtained in the present study is 19.6% (60% survival). Damage to plumule rather than radicle occurred due to excessive dehydration of *N. nimmoniana* embryos is as observed earlier in *Auracaria hunstenni* where desiccation damage is reported to be more serious in the plumule (Pritchard and Prendergast, 1986). The exact causes of embryonic death and its relationship with moisture content are not fully understood. Chin *et al* stated that seed death could be due either to the moisture content falling below a critical value or simply a general physiological deterioration with time. If embryonic axes have been desiccated to around 20% moisture content without loss of viability, it is possible that

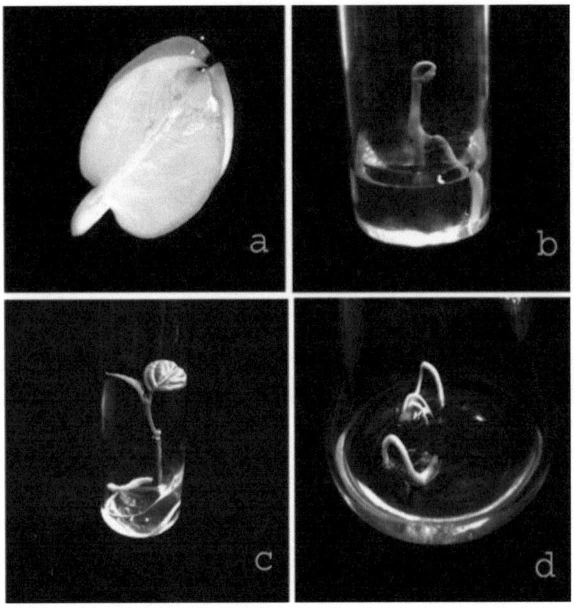

Fig..1 a. Isolated embryonic axes, b. Cryopreserved embryo showing germination after 30 days of culture on MS basal medium, c. Seedling from Cryopreserved embryo after 60 days of culture on MS basal medium and d. radicle development and degeneration of plumule in embryo subjected to desiccation for 180 min.

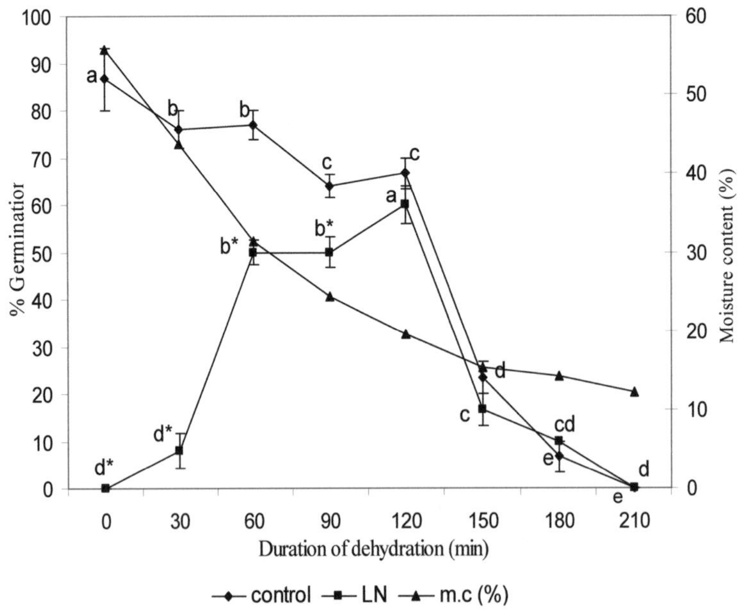

Fig. 2. Effect of cryopreservation on germination of *N. nimmoniana* zygotic embryos. Different letter (s) in a data series shows significant difference at 5% level based on LSD multiple 't' test. *Control and LN treated values differ significantly at 5% level based on Student 't' test. The bars represent SEM.

cooling and storage in LN will be progressed more easily. In most of the reports of successful cryopreservation, excised embryos or embryonic axes have been used for desiccation sensitive species, i.e. zygotic embryos of Citrus (Mumford and Grout, 1979) Oil palm (Grout *et al.*, 1983) Coconut (Chin *et al.*, 1989) Hevea (Normah *et al.*,1986) where the embryos withstand freezing after being subjected to partial desiccation. The desiccated embryonic axes do not lose viability after rapid cooling and storage at the temperature of LN. At such temperature there should be no change in the tissue either genetic or developmental, over a period of decades (Ashwood *et al.*, 1977). This situation together with the ease to develop independent plants *in vitro* (Satheeshkumar and Seeni, 2000, Ravishankar Rai, 2002) from embryonic axes suggest cryopreservation is an effective technique for the long-term conservation of *N. nimmoniana,* a medicinal tree species producing large intermediate type of seeds.

3. References

Ashwood Smith MJ and Grant E (1977). The freezing of mammalian embryos, at *CIBA symposium*, (eds) Elliot K, Whelan J, Elsevier, Holland, 251-271.

Assy-Bah B and Engelman F (1992a). *Cryo-Letters* 13: 67-74

Assy-Bah B and Engelman F (1992b). *Cryo-Letters* 13: 117-26

Bajaj YPS (1995). Cryopreservation of plant cell, tissue and organ culture for the conservation of germplasm and biodiversity. In: Bajaj YPS (ed) *Biotechnology in Agriculture and Forestry Cryopreservation of Plant Germplasm I*, New York, Springer-Verlage, pp. 3-18.

Bekheet SA, Taha HS, Saker MM, Solliman ME (2007). *J Appl Sci Res.*, 3 (9): 859-866.

Benson EE (1999). Cryopreservation. In: Benson EE (ed) *Plant Conservation Biotechnology*, Taylor and Francis, London, pp. 83-95.

Berjack P and Dumet D (1996). *Cryo-Letters* 17: 99-104

Chabrillange N, Aberlenc-Bertossi F, Engelman F and Duval Y (1997). *Cryo-Letters* 18: 68-74.

Chandel KPS, Chaudhury R, Radhamani J and Malik SK (1995). *Ann.Bot.*, 76: 443-450.

Chang Y and Reed BM (2001. *HortSci* 36 (7): 1329-1333.

Chaudury R, Radhamani J and Chandel KPS (1991). *Cryo-Letters* 12: 31-6

Chin HF, Krishnapillay B and Hor YL (1989). *Pertanika* 12: 183-86

Chin HF, Krishnapillay B and Standwood PC (1989). In: *Seed Moisture* (eds. P C Standwood and M B MC Donald). *Crop Science Society of America,* Madison, WI, USA, 15-22.

Chin HF and Pritchard HW (1988). Recalcitrant seeds, A Status Report. *IBPGR*, Rome.

Chin HF and Roberts EH (1980). *Recalcitrant crop seeds,* Tropical press SDN, BHD.

Cragg GM, Schepartz SA, Suffness M, Grever MR (1993). *J Nat. Prod.* 56:1657-1668.

De Boucaud M, Brison M, Ledoux C, Germain E and Lutz A (1991). *Cryo-Letters* 12:163-166

Decruse SW, Seeni S and Pushpangadan P (1999a) *Seed Sci and Technol.* 27: 501- 505

Decruse SW Seeni S and Pushpangadan P (1999b). *Cryo-Letters* 20: 243-250

Decruse SW and Seeni S (2002b). *Cryo-Letters* 23:55-60

Decruse SWand Seeni S (2002a). *Seed Sci and Technol* (In Press)

Engelmann F (1997). *Plant Genetic Resources Newsletter* 112: 9-18.

Engelmann F (2004). *In Vitro Cell Dev Biol Plant* 40: 427-433.

Engelmann FD, Dumet N, Chabrillange A, Abdelnour Esquivel B, Assy-Bah J Dereuddre and Duval Y (1995). *Plant Genetic Resources Newsletter* 103: 27-31.

González-Benito ME, Clavero-Ramirez I, López-Aranda JM (2004). *Spanish J Agric Res* 2 (3): 341-351.

Hsiang YH, Herzberg R, Hecht S and Liu LF (1985). *J Biol Chem* 260: 14873-14878.

Kaczmarczyk A, Shvachko N, Lupysheva Y, Hajirezaei MR, Keller ERJ (2008). *Plant Cell Rep* 27: 1551-1558.

Kartha KK, Engelmann F (1994). Cryopreservation and germplasm storage. In: Vasil, I.K. and Thorpe, T.A. (eds) *Plant Cell Tiss Cult* Springer 195-230.

Martin C, Iridono JM, Benito-Gonzales E, Perez C (1998). *Agro-Food-Ind Hi-Tech* 9 (1): 37-40.

Mumford PM, Grout BWW (1979). *Seed Science Tech* 7: 407-410.

Murashige T and Skoog F (1962). *Physiol Plant* 15: 473-497.

Normah MN, Chin HF and Hor LY (1986). *Pertanika* 9: 299-303.

Panis B, Lambardi M (2005). Status of cryopreservation technologies in plants (Crops and Forest trees).At: *The Role of Biotechnology*, Villa Gualino, Turin, Italy.

Paunesca A (2009). *Romanian Biotech Letters* 14 (1): 4095-4104.

Pence VC (1991). *Plant Cell Rep.* 10: 144-7

Pence VC (1991). *Seed Sci. and Technol.* 19: 235-251

Pence VC (1995). In: Bajaj YPS (ed) *Cryopreservation of Plant Germplasm, Biotechnology in Agriculture and Forestry*, Springer, Berlin Heidelberg, pp. 29-50.

Pritchard HW, and Prendergast, FG (1986). *J. Exp. Bot* 37: 1388-1397

Radha RK, Decruse SW, Seeni S (2006). Cryopreservation of embryonic axes of recalcitrant seed species *Myristica malabarica* Lam., a rare medicinal plant of the Southern Western Ghats. Presented in the *National Seminar on Plant Resources of the Western Ghats* organized by Karnataka Biodiversity Board at IISc. Bangalore on 7th and 8th December.

Radha RK, Decruse SW, Krishnan PN (2010). *Ind. J. Biotechnology* 9: 435-437.

Rajasekharan PE (2006). Prospects of new cryopreservation techniques for conservation oftropical horticultural species. Paper presented at the *ICAR Short Course on In Vitro Conservation and Cryopreservation-New Options to Conserve Horticultural Genetic Resources*, Banglore, India, 21-30 September.

Ravikumar K, and Ved,DK (2000). Illustrated field guide of 100 Red listed medicinal plants of Conservation concern in Southern India. Foundation for Revitalization of Local Health traditions (FRLHT), Bangalore, pp 465.

Ravishankar Rai V (2002). *In Vitro Cellular & Developmental Biology. Plant* 38 (4): 347-351.

Reed BM (2001). *Cryo-Letters* 22:97-104

Roberts EH (1973). *Seed Sci Tech,* 1: 499-514

Sakai A (2000). Development of cryopreservation techniques. In: Engelmann F Takagi H (eds) *Cryopreservation of tropical plant germplasm.* International of Plant Genetic Resources Institute, Rome, pp. 1-7.

Satheeshkumar K and Seeni S (2000). *Ind J Expt Biol* 38: 273-277.

Stanwood PC (1985). In: Kartha KK ed. *Cryopreservation of plant cells and organs* (Boca Raton, Florida: CRC press) pp 199-226

Stanwood PC (1987). *Crop Sci* 27: 327-331.

Scowcroft WR (1984). Genetic variability in tissue culture: Impact on germplasm conservation and utilization. *International Board for Plant Genetic Resources* Secretariat, Rome, p. 42.

Touchel DH and Dixon KW (1993). *Biodiversity and Conservation* 2:594-602

Touchel DH and Dixon KW (1994). *Annals of Bot.* 74: 541-546

Walters C, Volk GM, Towill LE, Forsline P (2009). Survival of cryogenically-stored dormant apple buds: a 20 year assessment. Paper presented at the *1st International Symposium on Cryopreservation in Horticultural Species, Leuven, Belgium,* 5-9 April.

Westwood MN (1989). *Plant Breeding Rev* 7: 111-128.

Zhao MA, Dhital SP, Fang YL, Khu DM, Song YS, Park EJ, Kang CW and Lim HT (2005). *J Plant Biotech* 7 (3):183-186

Part 6

Equipment and Assays

22

X Ray Diffraction: An Approach to Structural Quality of Biological Preserved Tissues in Tissue Banks

H. Perez Campos et al.*
Instituto Nacional de Donación y Trasplante (INDT), Ministerio de Salud Pública - Fac. de Medicina, Uruguay

1. Introduction

The purpose of this chapter is to introduce new methods of analysis, to evaluate the final quality of human origin bio therapeutics products generated in Tissue Banks (TB), using well developed and known techniques in various fields of Physics, Chemistry and Biology as applied X - Ray diffraction (XRD), and Raman Scattering (RS).

Cryopreservation techniques are fundamental supports in the conservation procedures of biological materials in TB work. However, controversial views remain on the effects at the molecular level that cryogenic temperatures and thawing could produce on the functional structures of tissues. The same concept can be sustained to glycerolized tissue preservation. Taking into account this scope, we implemented a methodological scheme to analyze tissue specimens before and after programmed cryopreservation, or glycerolization in order to find structural differences in the basic material constitutive collagen, using the techniques formerly mentioned: diffractive and scattering.

It is noteworthy that both methods of analysis can be applied to any type of tissue preserved for the aforesaid purposes, with other conservation techniques, such as freeze drying or un programmed freezing.

2. The tissue banks and the "viability" of it's therapeutically products

The TB are technical establishments whose main institutional objectives are collection, preservation, storage, release and distribution of biological tissues for therapeutic use in transplantation medicine. These objectives is met according to scientific criteria from agreed international protocols (Spanish Association of Tissue Banks: AEBT, International Atomic Energy Agency, IAEA, European Association of Tissue Banks: EATB; American Association of Tissue Banks: AATB) and according to the legal frameworks of the different countries and

* Mc. Saldias[1], G. Sanchez[1], P. Martucci[1], Mc. Acosta[1], I. Alvarez[1], R. Faccio[2], L. Suescun[2], M. Romero[2] and A. Mombru[2]
[1] *Instituto Nacional de Donación y Trasplante (INDT), Ministerio de Salud Pública - Fac. de Medicina, Uruguay*
[2] *Laboratorio de Cristalografía, Estado Sólido y Materiales (DETEMA) Fac. de Química, Uruguay*

their ethical rules. On the other hand, tissue banking activities are carried out following strict bio-security criteria by the selection of suitable donors, permanent quality control and continuous management of physical plant, equipment, supplies procurement procedures, and end products, which must comply criteria for "viability" therapeutic apply.

According to (Pegg 2006) this concept of "viability," applies whenever a graft of tissue obtained as final product, meets the leading natural-biological function for which it was preserved, and that is pathologically affected in the recipient. This explains why there are different procedures and methods of conservation in TB, according to the type of tissues and its expected restored function.

In some cases, we must preserve, as the main function, a biological synthesis, which requires mandatory of vitality in cellular functions (eg, parathyroid glands, or ovarian tissue). In other cases, the objective is to preserve static and mechanical functions as a segmental allograft bone support, or biodynamic behavior, as in cases of preservation of blood vessels. This implies that "viability" is not necessarily synonymous with "cell vitality" and therefore the requirement is to preserve elements of the Extra Cellular Matrix (ECM).

Hence, the importance of assessing possible changes on the components of the ECM that TB procedures may be generate on processing and stocking biological tissues.

3. The concept of functional structure of ECM

The ECM was considered early in the twentieth century as filler material and mechanical support of cell structures, which was thought as the only protagonists of tissues functionality. In the 50´s, (Grobstein, 1953) proposed that the induction in the development of a tissue depended on the presence of ECM.

In the 60´s, (Hauschka & Königsberg, 1966) establish that pig embryonic muscle cell cultures proliferate more properly in a media with presence of metabolic products from fibroblasts, and identify the collagen to induce the development of them. This protagonist role of collagen in the processes of induction and cell proliferation, is corroborated by (Meier & There, 1974), by testing the inductive capacity of collagen, for the synthesis of ECM in the corneal epithelium of pig embryos (5 days old development). (Sanders 1988) working on neural crest and sclerotome cells of early chicken cells embryos, prove that the presence of type I collagen is necessary for cellular migration and *de novo* synthesis of ECM is a prerequisite for normal cell migration and attachment in earliest stages of embryogenesis.

Additionaly, since the early 80´s (Bissell 1982), work in the field of cancer biology emphasizes the importance of "micro" immediate cellular environment and posits the hypothesis of 'Dynamic Reciprocity' by which the ECM contact trans membrane receptors, influence gene expression through signals transmitted via cytoskeleton, generating so "new" products for the ECM. Thus, the cell and the ECM, form a binomial reciprocal exchange interaction, which has vital importance in the early stages of embryonic morphogenesis and later in postnatal life, the physiological mechanisms of growth and development as well as in response to injury. (Bissell, 1982; Davis, 2010; Kelleher, 2004; Nadiarnykh, 2010; Schultz, 2005; Schwinn, 2010). The ECM consists of a complex variety of macro molecules that can be summarized schematically as follows: 1) protein collagen, 2) structural glycoproteins, 3) proteoglycans and glucosamine glycans, and 4) elastin. These complex and diverse molecular groups, organized into super families, are shown with a dynamic distribution, and functional modulated behavior, with variations between different tissues. This bio plasticity is observed even within the same tissue type, as homeostatic biochemical, and bio mechanical requirements, including interactions with various molecules: growth factors, cytokines, enzymes and other inducers synthesis products as well as lytic and degradative matrix one.

The surface receptors of the cell membrane, in close contact with this complex and dynamic molecular ECM set, interact through the cytoskeleton to the genome, which modulates the different stages of ontogeny, growth and postnatal development sequences as well as molecular structural and functional physiology and patho physiology biological requirements of tissues. (Abraham, 2007; Bowers, 2010; Worthley, 2010). Related to the own collagen structure there are ligands and functional domains that take contact with other ECM molecules (fibronectin, proteoglicans, and collagen - collagen interactions) and with the neighbor cells microenvironment (cell integrin receptors). Several poly peptid sequences, (eg.: GFPGER: glycine - phenylalanine - hyroxyproline - glycine - glutamic acid - arginine) and ligands domains (eg.: Matrix Metalloproteinase Interaction Domain or MMP ID; Colagen V Cross-link site or Col V X-link), have been identified and play an important role in regulation of migration, proliferation, adhesion and apoptosis in biology cell tissues. (Orgel, 2011; Sottile, 2007; Sweeney, 2008).

It is obvious therefore that the extra cellular medium, forms a molecular complex of plastic, in both dynamic up regulator as down regulator in constant cross talk with cell pole whose points of contact and mutual information imply the presence of binding sites at ECM structure related to cell surface receptors. Fibril collagen constitutes approximately 25% of tissues for all species of mammals and is the main component of the total molecules that make up the ECM. (Kielty, & Grant, 2002). To date, they have been described up to 29 different types of collagens with the corresponding genetic determinants. "Structural" Collagens in the ECM are called fibrils (Fibril Collagen) consisting of types I, II and III, V and XI. Type I constitute 90% of body collagen and mainly, perform mechanical resistance functions. In addition it provides three - dimensional modeling formation of tissues. An important bio molecular feature of our study is it hierarchical and sequenced shaping showing collagen. Taking collagen I model, pro collagen, amino acid primary structure have a intracellular synthesis (endoplasmic reticulum), with repeated tripeptides, design whose residues are Gly-Pro- Hyp or Gly-X- Hyp which Pro and Hyp are near to 17% and 33% respectively. Therefore 50% of the average 1000 residues of the total composition of the molecule, pro collagen, are other amino acids. Its length of 300 nm and width of 1.5 nm, is organized in a left-handed secondary structure of three amino acids per turn, with Gly residues central and peripheral Pro and Hyp out of the spiral. Three assembly helical pro collagen monomers (2 $\alpha 1$ and 1 $\alpha 2$) in right-handed configuration, determine the tropo collagen structure in the extra cellular space. In this space the molecule is arranged in staggered bundles with a gap of 67 nm by inter molecular bonds tropo collagen units, which gives to the collagen fibril new product design, a repetitive sequence which observed in the ME identified a characteristic D - banding. The final design shows collagen fibers arranged, spatially distributed in regular packages along the lines of force of the biomechanical characteristics of each tissue. This last aspect brings an added dimension of ordering design given by the spatial distribution of fiber bundles, and their inter reciprocal space.

These three characteristics: a) repetitive and periodic sequencing of the D - banding, b) structuring hierarchically ordered by supra molecular complexes, and c) nano-scale dimensions of the structures, made of collagen complex an para crystalline super molecule, liable to be analyzed by techniques diffractive as discussed later. (Sweeney, 2008; Berenger, 2009).

In this context becomes important the analysis of the changes that for preservation purposes, can be induced in the molecular components of the ECM. Particularly taking account that collagen is the main structural component of the extra cellular microenvironment and has a proven role in the functional biological mechanisms, developmental, physiological homeostatic and physio - pathological tissues behavior. (Kielty & Grant, 2002; Orgel, 2011).

These reasons justify work TB, to design tissue preservation processing models, with conservation of the collagen component, from both, structural and biochemical characteristics. It must be take in mind, that allografts should meet a homeostatic interaction with cell biology recipient patient, through its membrane receptors in functional contacts with molecular ligands and domains of preserved ECM, to improve physio pathological situations.

Note therefore that the biological behavior of a suitable allograft depend on the presence and indemnity of molecular epitopes or ligands, which can be eventually altered in its stereo chemical distribution during cryogenic or glycerolized procedures.

4. The interaction between ECM components, and the physic-chemical phenomena preservation procedures

About the cryogenic effects on ECM, there are several references in different disciplines, about the ultra cold temperatures on biological material. Indeed, at the molecular level have been observed different types alterations generated by freezing / thawing phenomenon, on biological structures. As earlier in the 60´s, (Levit, 1962; 1966), had postulated irreversible changes in the tertiary structures of soluble vegetal proteins, with loss of its biological capacity, such as the rearrangement of disulfidric functional bonds to non-functional disulfide covalent configurations.

The concept of "repulsion hydration forces" refers to cell membranes, was developed by Wolfe, J. (1999). These phenomena is induced by water efflux through such semi-permeable membranes during cooling process, promoting large mechanical stress and strain in the biological structures, and generating physical deformations and changes in the molecular functional membrane behavior. Such condition is done under the observed ground of the structure of the crystalline ice mass in the extra cellular space. This, results in intra cellular dehydration of the tissue and the extra cellular hyper osmolarity of super cooling liquid. So, it generates a displacement of inter atomic and molecular chemical equilibrium that, explains changes in the stereochemistry and molecular architecture of biological structures. This scope would agree to Levit postulates. (Levit, 1962; 1966).

The effect that the conventional criopreservation exerts on the ECM structure is controversial information. (Gerson, 2009), comparing morpho structural collagen mesh from fresh and cryopreserved human heart valves by second harmonic generation, sets no changes between both categories. However, it was found extensive damage in collagen structure in porcine frozen leaflets related to fresh control one, using laser-induced auto fluorescence imaging , (Schenke Layland, 2006), and second-harmonic generation. (Schenke Layland, 2007).

In other field, there are many studies showing that the biomechanical behavior of tissue collagen framework, is not altered by effect of cryopreservation / thaw cycle, in vascular (Armentano, 2006; Bia, 2006; Langerak, 2001, 2007; Pukacki, 2000), tendon , (Woo, 1986; Park, 2009) or bone tissues. (Hamer, 1996).

However, controversial literature is also observed for biomechanical variables. (Rosset, 1996) observed in vitro, decreased compliance and hysteresis an increase of modulus of elasticity in thawed cryopreserved human carotid arteries, related to fresh one. (Gianni, 2008) found that the freezing of human posterior tibial tendons significantly affected behavior in vitro biomechanical performance. Finally, either way, under many point of view is possible highlight that the functional character of fibril collagen depending to the particular structural and biochemical preservation, which may be damaged during cryopreservation defrosted process in TB.

About the interaction between alcohols and polymerized amino acid, in early 70´s (Frushour, & Koenig, 1975) postulated, in Raman Scattering field, that methanol modified an aqueous poly-DL-alanine (PDLA) solution, by disruptions of the helical regions by breaking the hydrophobic bonds.

5. Diffractometry: a tool for analysis of structural ordering of collagen

5.1 The matter and its organization

It defines that the spatial arrangement of ions and atoms of matter, have a crystalline profile when its design shows a repeating sequence. The frequency of repeated and symmetrical distribution of the atomic stereo chemical units constituents of matter, determine the solid crystalline character. It is understood that a substance is "homogeneous" when each constituent unit of the solid is linked by chemical bonds to another identical unit in any sense of space, and is identified as the ideal "homogeneous crystal" model when, theoretically, is infinitely extended in space. The sequential nature of the repetitive and symmetric atomic elements, defines the spatial network model, under the so called "cells ordering". The three dimensional symmetric distribution of elemental units of the complex let likened to an orderly succession of planes separated by a distance "d". This design is easily identifiable in crystalline substances of inorganic chemistry: quartz, diamond, graphite, etc. Diffracted analysis of inorganic or organic crystalline matter can provide detailed information about molecular design, related to intermolecular distance and stereo chemical angle conformation. In the world of bio molecular chemistry, matter is organized by more complex models, through an extensive variety of atomic molecular combined structures. In this picture certain combinations become repetitive units, consisting of several basic types of atoms links together by different kinds of bonds. Nevertheless, one can observe the character of certain spatially ordered molecular configurations, which are equally repetitive. This setting defines the so called "molecular crystals", despite not showing the perfection system of "homogeneous crystals". So, no bond lengths and angular atomic positions can be determined, but an approximate view about relative ordering structure is given applying XRD techniques. These structures are typical of biological substances such as proteins or DNA molecules whose functional design depends on the molecular arrangement, and type of chemical bond established between its molecules.

5.2 The diffractive phenomenon and elastic scattering of x-rays

X-rays are a form of electromagnetic energy produced by a source to be impacted by electrons of high kinetic energy supplied (usually a tungsten filament named cathode). The incident electron impact, destabilizes the internal atomic orbital of a target material (an anode built with pure copper), generating atoms in electronically excited state. The movement of electrons from outer orbital to balance the impact generates heat energy and emission of x-rays, in a spectrum of wavelengths (λ) measurable in Angstrom units (**1 Å = 10^{-10} m**). The band spectral x-rays emission is "filtered" through mono chromator to obtain a single wavelength that corresponds to the "characteristic radiation" used in the x-rays crystallography: K-alpha line (Kα) for each material property anode of the device. To our work, XRD is defined by the interference between monochromatic characteristic emissions (Kα) with the ordered material in crystalline form. When a photon interferes with an orderly molecular structure without loss of energy (elastic scattering), produces a deviation from its original direction which is the diffraction phenomenon. The condition for diffraction to be possible is that the distance between the periodic and ordered structure elements (atoms or molecules), fall in the wavelength range of incident ray. The condition to be detectable is that some degree of ordering is present in the material to be studied, in order that the interference could be constructive. As previously mentioned, collagen shows particular characteristics like a biomolecular arrangement built by: a) the hierarchical supra molecular order fibers, b) the observed crystalline structure of the spaces sequenced "D" of fibrillar collagen, c) the nano scale distance of repetitive molecular units, all of them allows to analyze biological stroma collagen by the XRD. (Aspden, 1987; Berenguer, 2009; Connon, 2007; Hickey & Hukins, 1980; Horton, 1958; Pauling & Corey, 1951; Pérez Campos, 2008).

The equation that allows the practical application of this technique is Bragg's Law.

$$n\lambda = 2d \sin\theta \tag{1}$$

Formula 1 Bragg´s Law: **n** is an integer, **λ** is the x - ray monochromatic wavelength, d is de distance between the planes of the crystalline net, and **θ** is the angular value between the incidental x -ray and the considered crystalline plane.

Given a certain incidence angle of a monochromatic beam on the material structure, sequencing crystalline spacing "d" determines a dispersive interference when the rays are emerging in construction phase. Emerging rays can be recorded in a Cartesian coordinate system where the independent variable "x" records the range of 2θ values and the dependent variable "y", the relative values (R.V.) of the ordering lattice system. The recordable diffraction graphics or diffractogram depend of the content and the atomic distribution within the repetitive units that define the three dimensional arrangement. (See Figure 1)

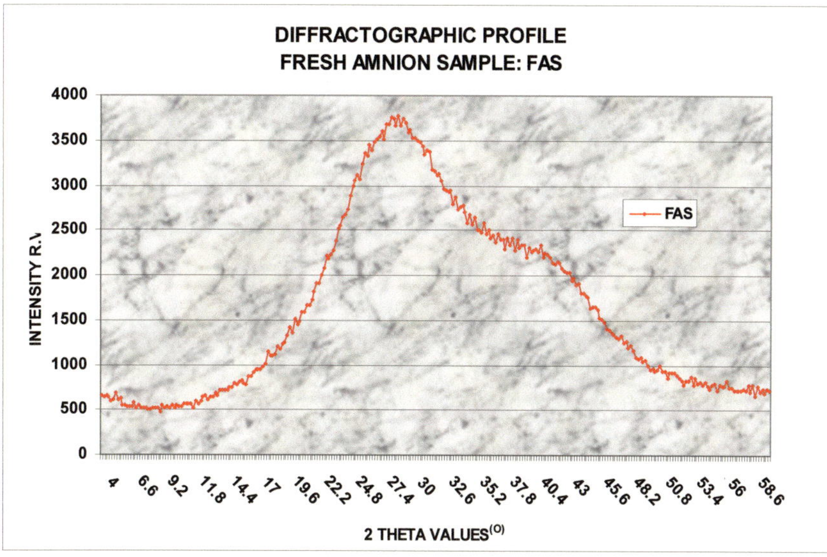

Fig. 1.

5.3 The Raman effect, or inelastic scattering of light

Given a monochromatic light beam incident on a material, there will be a phenomenon of elastic light scattering as a result of the interaction of photons and electron atomic elements of the network links. The elastic scattering implies that the frequency of the incident light beam and the scattered light emerging is the same, so that no changes have occurred in the respective energy levels. This phenomenon known as Rayleigh scattering is highly significant from a statistical point of view. However an extremely low intensity incident light (in the order of 1 photon in 10^7 to 10^{10}) shows inelastic behavior in the interaction with the structure determining slight changes in the emerging wave frequency, which depends on the characteristics of matter incised. This phenomenon or Raman Effect was discovered by (Raman & Krishnan, 1928), and allows the chemical structure analysis of biological material.

Atomic particles mass and its energy states (vibration and/or rotational), maintain chemical bonds that define crystal structure, sustaining dynamic design stability by neighbor interaction. The movements of both vibration and rotation of the particulates: (υ = frequency), defined the dynamically stable energy level where they are. If the interaction of a beam of photons of frequency (υ_0), print a change of unstable frequency at the particle network link, they scatter photons with a different frequency (υ_r), define an inelastic scattering. The energy can be dispersed in a model $\upsilon_r > \upsilon_0$ called scattering Raman - Stokes or $\upsilon r < \upsilon_0$ know as scattering Raman anti Stokes. υ_r values are characteristic of each design structures and define atomic molecular matter. Approximately 99% of the output assays is Stokes Raman scattering hence, those are the models profiles recorded. In the Cartesian co ordinate system the independent variable x records the difference υ_r-υ_0 in cm^{-1}, and the ordinates y, the scattering intensities for each differential rate, in relative units (R.U.) (See Figure 2)

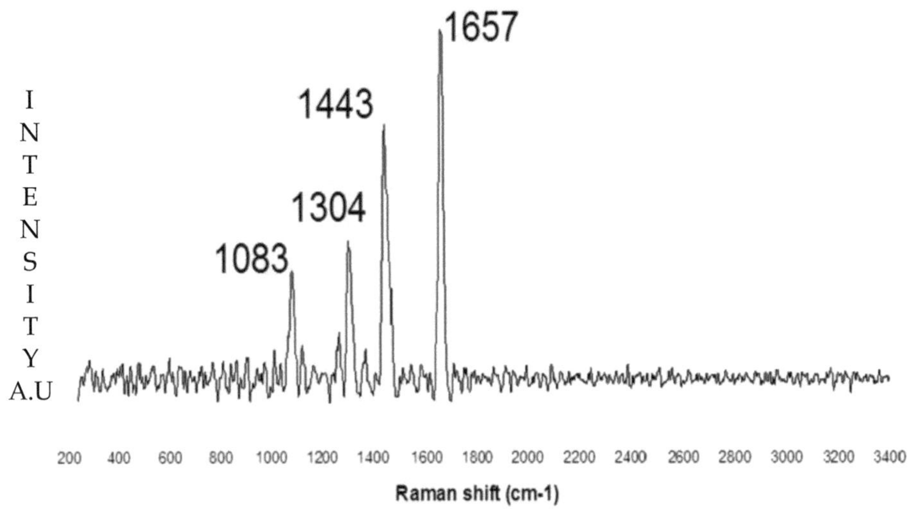

Fig. 2. Raman Spectum oleic acid

5.4 Interaction x-rays - Collagen: A model analysis

The x-rays diffractive analysis of the collagen structure was earlier studied (Pauling & Corey, 1951). Defined diffraction peak was described, in the range d = 2.86 Å (2 θ = 31.3 ^{o} using CuK α radiation λ = 1.5418 Å value), based on the criteria of the Bragg Law. This phenomenon was interpreted as the expression of cis configurations for the amide groups of the polypeptide chain.

Using this background, our group analyzed tissue banking allograft, in order to compare the diffraction profiles obtained before and after cryo-preservation method (vascular tissues and amnion tissue), and glycerolized preserved method (amniotic membrane). The working hypothesis states that the preservation methods can modify the stereochemistry molecular structures, determining changes in collagen and the consequent differentiation of diffractive or dispersive profile.

6. Materials and methods

6.1 Donors and biological samples procedures

The applied procurement protocol to vascular tissues was made on cadaver multi organ donors through informed consent and in accordance with standard operating manuals in the National Institute for Donation and Transplantation (INDT) of Uruguay. They were likewise applied legal and ethics regulations (Law 14005/1971 - 17668/2003) valid in our country. The exclusion criteria and biological safety were applied, in accordance with the Standards for Tissue Banking: International Atomic EnergyAgency (IAEA - 2005) and the

Spanish Association of Tissue Banks (AEBT - 2005). The same selection criteria, exclusion, tissue procurement and processing were applied to living donor placenta with clinical controlled normal pregnancy and delivery, by the parameters set by the Ministry of Public Health of Uruguay.

Donors were selected according to the protocol in a range between 18 and 60 years (35.5 ± 11.8 years mean age, 47% M 53% F) obtained by aseptic dissection, 10 aortic arterial segments, and 8 carotid. It proceeded under a laminar flow cabinet to cleaning, package, and storage in physiological saline at 4°C. Fresh Vascular Samples (FVS) were shipped to DETEMA within 24 hs, for XRD. The same process aseptic protocol was applied to 6 amnion obtained by manual dissection from donor placenta. Fresh Amnion Samples, (FAS) were stored in saline solution at 4ºC and shipping to DETEMA within 24 hs for XRD analysis.

The Cryopreserved Vascular Samples (CVS), segments of each contra lateral carotid donor, and hemi ring segments of thoracic descendent aorta, were processed for cryopreservation in a Controlled Rate Freezing System (Model 9000, Gordinier Electronics, Inc. Michigan). Stored CVS were maintained up to 30 days at -142ºC. The cryopreservation media was: RPMI 1640, 85 cc; Human Albumin (20%), 5 cc; DMSO 10%. Cryopreservation was made into termal sealed double cryo resistant bag (Joisten and Kettenbaum D51429, Bereisch Gladbach, Mod.011342). The mean cooling rate applied was -1ºC/min from 4ºC to -90ºC and then quickly stored at -142ºC during 30 days in steam liquid nitrogen.

Same procedures were applied to obtain 2 Cryopreserved Amnion Samples (CAS), stored up to 30 days at -142ºC.

Defrost protocol applied for vascular and amniotic tissues, was according with Pegg et al. (1997), and defrosted samples were shipped to DETEMA for XRD.

6 Glycerolized Amnion Samples (GAS) were obtained by soaked in screw cap flask in Glycerol (95%) and stored at 4º C for 30 days. Glycerol from amnion was removed by three sequential shaking washing for 15 min. each, in saline solution and then shipped to DETEMA for x-rays diffraction, and Raman Scattering analysis. The comparative assays were done with FVS vs CVS; FAS vs CAS; and FAS vs GAS.

6.2 Diffractographic and Raman scattering technical procedures

XRD measurements were conducted at the Laboratory of Crystallography, Solid State and Materials, School of Chemistry (DETEMA), with a CuKα radiation source of wavelength λ = 1.5418 Å, using a Rigaku Ultima IV diffraction system. The incident ray is calibrated to arterial vessels in a range for 2θ between 5 ° and 60 °, step scan of 0.1 ° for 10 sec. each. The respective diffraction profiles (FVS vs. CVS) were filed for later analysis. Comparative profiles for amnion (FAS vs. GAS; and FAS vs CAS) were treated the same way as having been calibrated for 2θ between 5º and 60 º ranges scanning with steps of 0.2° for 10 sec each.

Raman spectra were recorded using a Raman DeltaNu Advance 532 spectrometer with a laser frequency doubled Nd: YAG, 100mW, with a 532 nm wavelength, scanning in the 200 and 3400 cm^{-1} region.

6.3 Planimetric analysis: Obtaining the order coefficients for XRD (Perez Campos, 2008)

Given the diffraction profiles of two tissues A and B to compare tests, we can define the respective planimetric surfaces, defined under the corresponding diffraction curve, that are a function of the degree of molecular arrangement of the studied tissue.

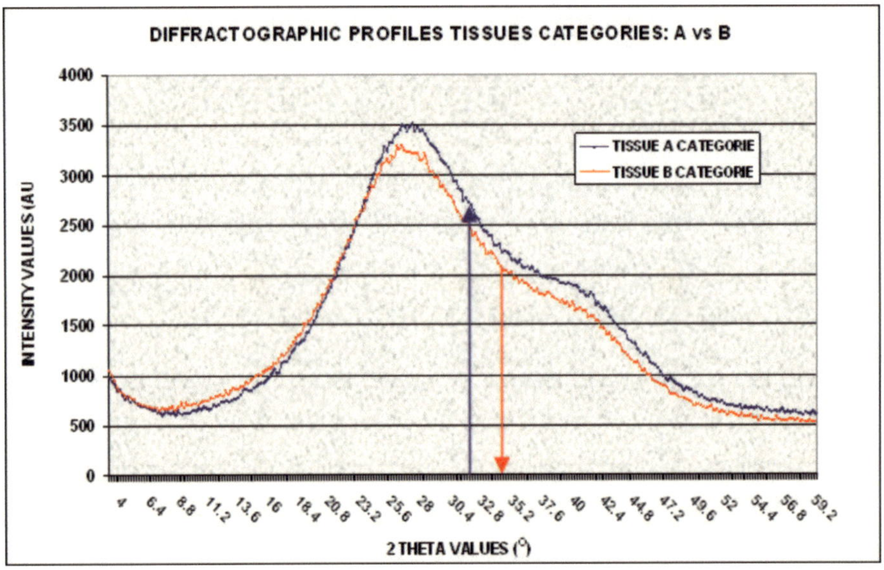

Fig. 3. Diffractive profiles from two different tissues categories A vs B.

Relative Differential Intensity values (RDIV) established for each 2 θ point between 5° and 60°, will produce a result that: if absolute value tissue A > tissue B will have a resulting positive (+) value; but if tissue A < tissue B will have a resulting negative (-) value.

With those values obtained in each point from 2θ it can be developed Differential Planimetric Surfaces (DPS) that represent the sum of every relative intensity value.

Its mathematical expression is given by the equation:

$$\Sigma\ (\uparrow - \downarrow)\ \mathrm{x}\ I_{A-B}\,[2\,\theta\,(^{\circ})] = \mathrm{DPS} \tag{2}$$

Where ($\uparrow$) represents ordering diffractometric intensity for *y* axes values of tissue A in one point in 2θ; ($\downarrow$) represents ordering diffractometric intensity for *y* axes values of tissue B, in the same point in 2 θ; I_{A-B} is the difference between each respective ordering diffractometric intensity for *y* axes value at the same point in 2θ. Finally, [2θ (°)] is each point in **x** axes between 5° and 60° angular incidence.

DPS can be edited in a Cartesian model too, where *x* axis is 2θ values and *y* axes is the Intensity Relative Differential Values (IRDV) between both comparative samples for each point 2θ values. (See figure 4)

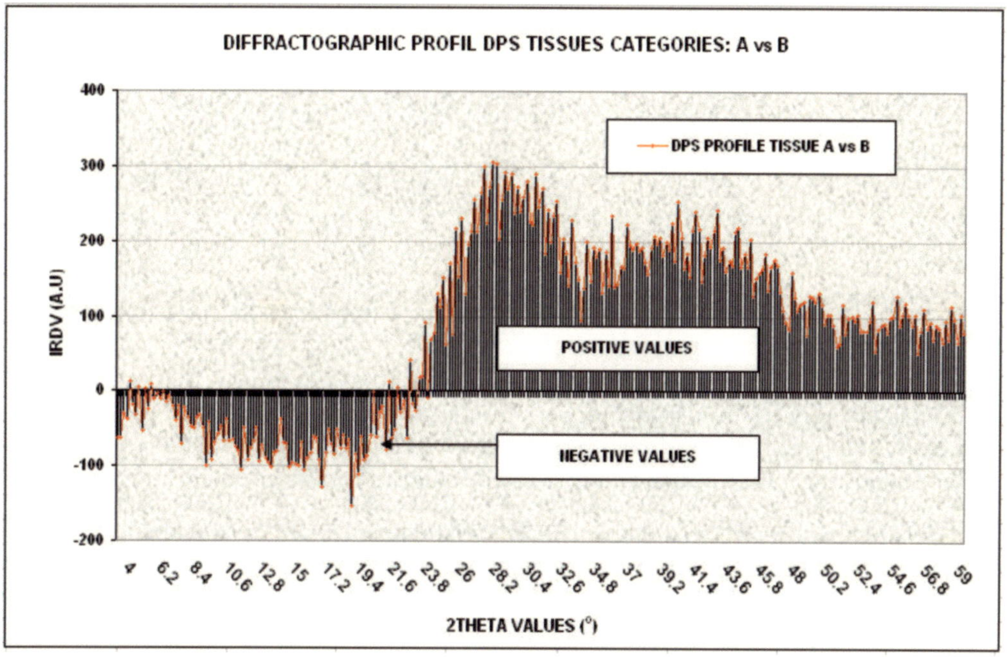

Fig. 4. DPS from two tissues categories A vs B to be studied by X-ray diffraction.

The ratio DPS (+) values vs DPS (-) values define the Ordering Profile Coefficient (OPC) according to the following formula:

$$OPC = \frac{DPS + VALUES}{DPS - VALUES} \tag{3}$$

OPC absolute values are always above 0 and they are greater than 1 when +DPS values > -DPS values. When +DPS values < -DPS, OPC falls into an interval greater than 0 and lower of 1.

7. Results

7.1 Cryopreserved vascular tissues results

Analysis of the diffraction curves shows that regardless of the condition FVS or CVS, the same design with a peak of maximum intensity to 31.3 ° and another lower, at 42 ° in 2θ is kept, whether there are noticeable differences in design profiles between the two categories, even for a single donor. (Perez Campos, 2008). Comparatives diffraction profiles shows the confirmation of a peak intensity for 2θ = 31.3 ° corresponding a d - spacing = 2.86 Å. The lower peak intensity, obtain d - spacing = 2.15 Å applying Bragg Low calculus. This behavior is independent of the vessel (aorta or carotid) and the processed sample (FVS or CVS). See Figure 5

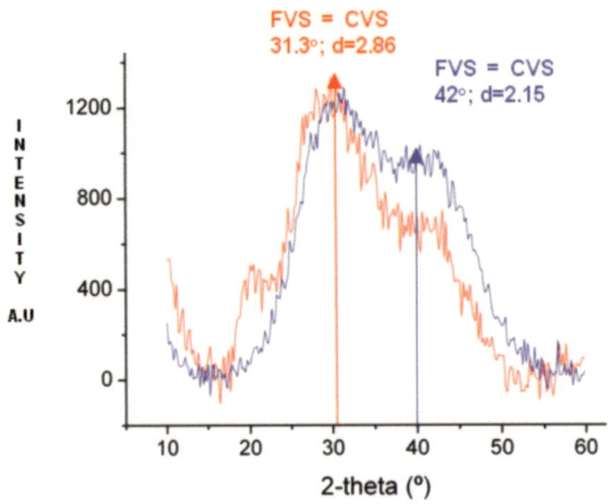

Fig. 5. Diffractografic profiles from thoracic descending aorta. Code color: FVS, Red; CVS, Blue. Note: The 1st and 2nd maximum peak labels of each respective diffractive curve indicate; Category of tissue: 2 Theta value; and calculated d spacing.

Calculated OPC values in respective analyzed vessels, shows a greater crystalline framework for CVS vs FVS, regardless the kind of arterial segment: aorta or carotid. 75% of aortic samples showed OPC values > 1 and 62,5% of carotid samples had the same behavior. Perez Campos et al (2008). Figure 6 show DPS defined from diffractografic FVS and CVS of Figure 5:

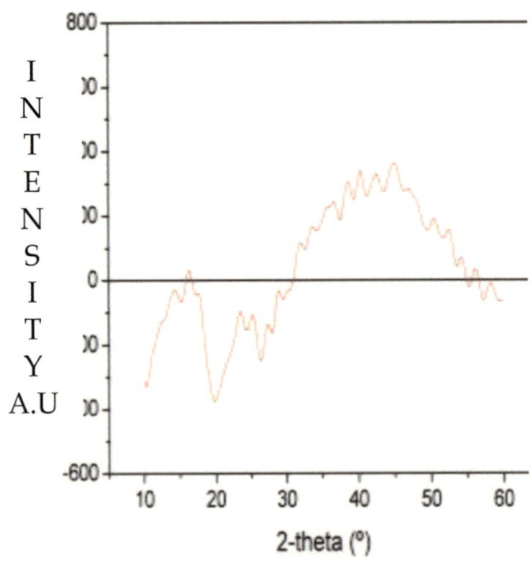

Fig. 6. DPS profile FVS vs CVS in a descending thoracic aorta from a male donor 50 years old.

Note the great difference of design shape of DPS curve related to the same one obtained from amnion tissues; (see below).

7.2 Glycerolized amnion tissues results

The diffraction curves of the glycerolized amniotic membrane, also shows the same kind of form and design for both the FAS and GAS. Notwithstanding the maximum diffractive peak in FAS d spacing = 3.24 (28.4 ° in 2θ) and in GAS, d spacing = 3.28 (28 ° in 2θ). A second peak is shown to both kinds of samples FAS and GAS for same d spacing = 2.35 (40.4° in 2θ). Contrary to the notable profiles differences showing in the two categories of vascular tissues (FVS and CVS), both amnion profiles -fresh and glycerolized- have almost the same design curve. (See Figure 7)

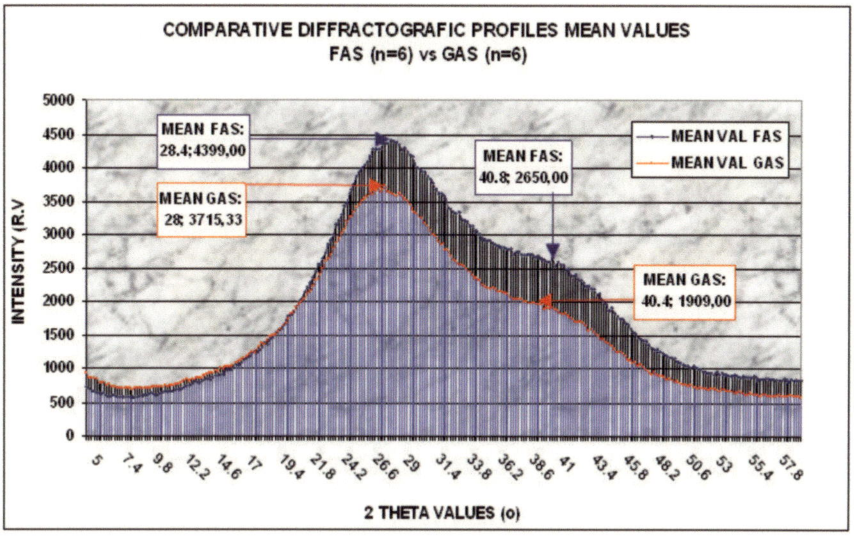

Fig. 7. Diffractive curves FAS and GAS profiles. Note: The 1st and 2nd maximun peak labels of each respective diffractive curve indicate; Categories of tissue: 2 Theta value; and Intensity RV.

Mean diffractographic profiles for 6 FAS vs 6 GAS let us obtain DPS picture and calculate OPC values = 14.76 (See respective Figure: 7 and Table: 1)

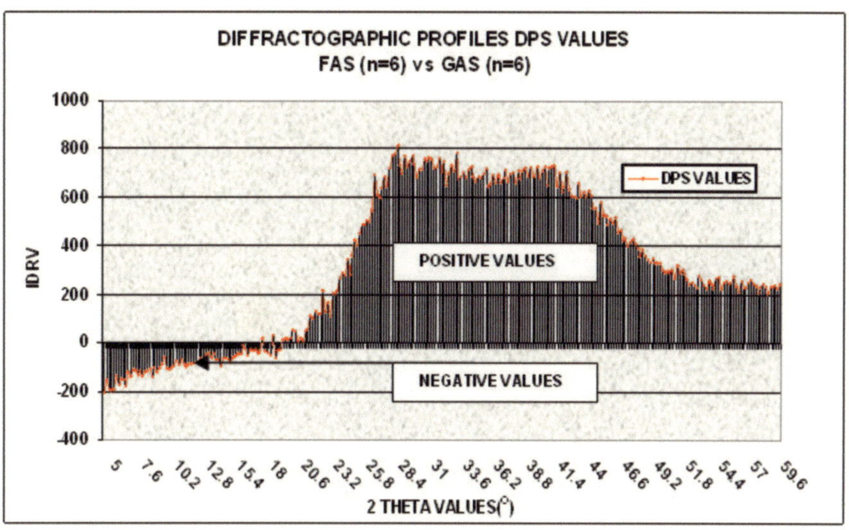

Fig. 8. DPS profile by FAS vs GAS analysis obtained.

OPC CALCULATION FAS vs GAS		
OPERATION		ABS. VALUES
∑ DPS + VALUES		94228,17
∑ DPS - VALUES		-6385,33
OPC VALUE = (+DPS) / (-DPS) = 14,76		

Table 1. OPC value from FAS vs GAS.

7.3 Cryopreserved amnion tissues results: X-ray diffraction

4 FAS vs 2 CAS was analysed. The same phenomena of Glycerolized amnion about form and design maintenance, was observed between FAS and CAS. Equally, there are a maximum peak in FAS d spacing = 3.26 (28.2° in 2θ) and in CAS, the same d spacing. Also a second peak is detectable to FAS in d spacing = 2.36 (40.8° in 2θ) and to CAS with equal values. (See Fig 9)

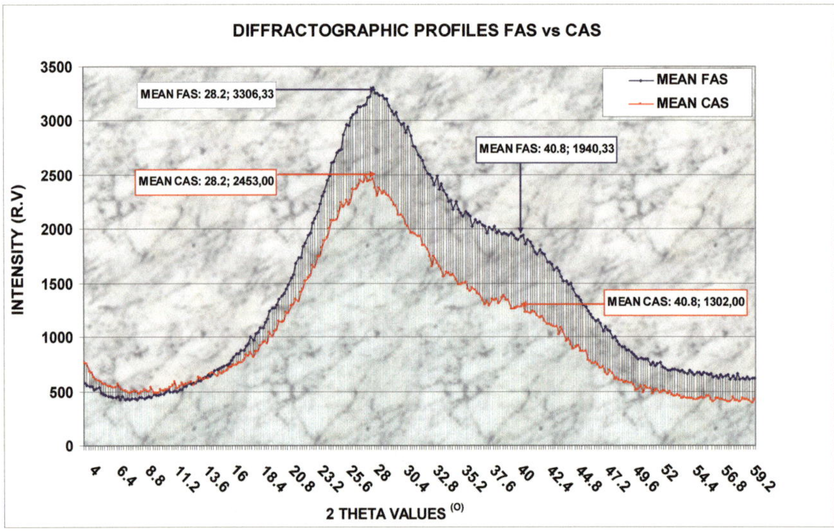

Fig. 9. Diffractive curves FAS and GAS profiles. Note: The 1st and 2nd maximun peak labels of each respective diffractive curve indicate; Categories of tissue: 2 Theta value; and Intensity RV.

Figure 10: show DPS profile obtained from operative subtractions analysis between FAS Vs CAS respective diffractive curves:

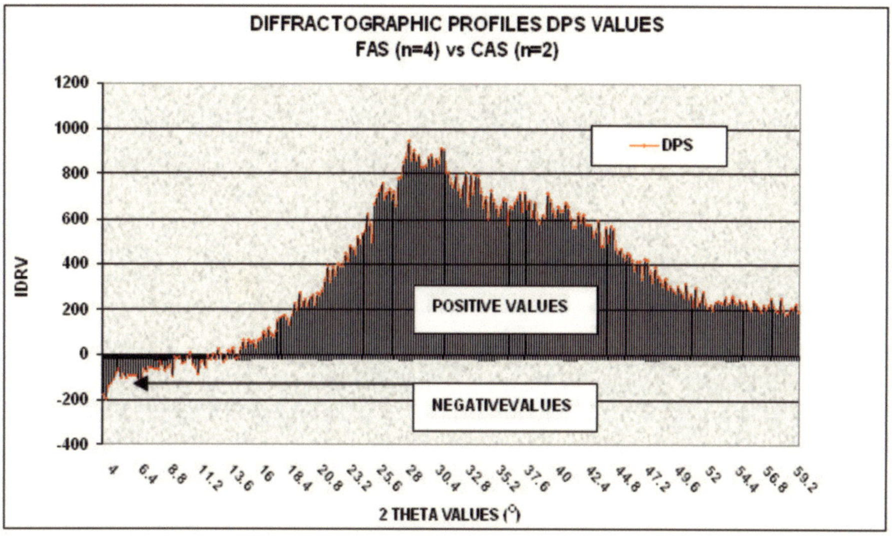

Fig. 10. DPS planimetric picture FAS vs CAS.

Operative planimetric values DPS obtained showed an OPC = 42.02 (See Table 2).

Newly highlight the notable differences in shape and form between vascular and amnion DPS pictures.

OPC CALCULATION FAS vs GAS		
OPERATION		ABS. VALUES
∑ DPS + VALUES		102887.67
∑ DPS - VALUES		-2448,33
OPC VALUE = (+DPS) / (-DPS) = 42.02		

Table 2. OPC values FAS vs GAS

7.4 Cryopreserved amnion tissues results: Raman Spectra

Figure 11 Show Raman Spectra profile from FAS vs CAS assays. Note that having regard to the best imaging definition, and taking account the meaningful change area of Raman Shift between both categories, the showed range is 1000 - 2000 cm^{-1} in x.

It should be highlighted that arrows points marked correspond to noticeable differences in positive Intensity values (AU) between FAS (red line) and CAS (blue line).

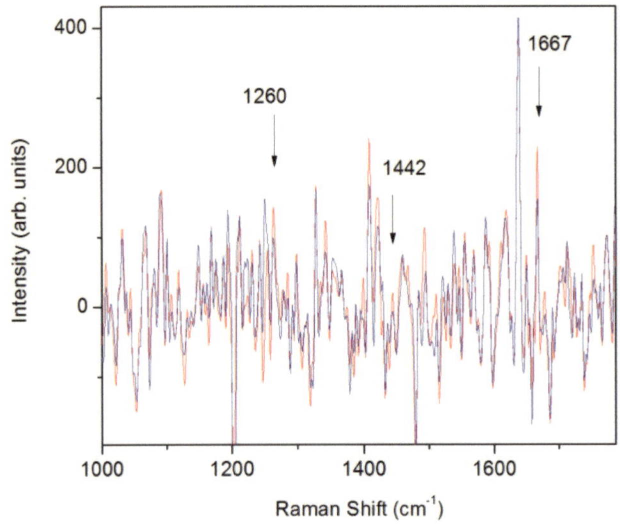

Fig. 11. Raman Spectra FAS vs CAS to area 1000 – 1800 Raman Shift (cm^{-1}). Color code; Red: FAS, Blue: CAS.

8. Comments

8.1 X-ray diffraction

The application of XRD on final quality of stromal collagen tissues analyzed, show differential results according tissue type and / or method of preservation applied. Indeed, relative to the observed results in cryopreservation of arterial vessels we see that regardless of the vessel -carotid or descending thoracic aorta- and the condition of FVA or CVS, a common diffractive peak at d spacing 2.86 Å is seen. The same phenomenon is shown for a second peak at d spacing 2.15 Å. These results show that vascular cryopreservation -defrost procedures did not alter the sequential structure of vascular fresh collagen. In reference to the results of amniotic membrane processing under the same preservation procedures, we see that for both varieties, fresh and cryopreserved, the maximum diffractive profiles remain unchanged: FAS, with d spacing 3.262 Å (28.2 ° in 2 θ) and equal values are checked for CAS. Other 2nd peak at d - spacing 2.359 (40.8° in 2 θ) is verified for both study categories. This confirms that the collagen is resistant to the cryopreservation defrost technique in regard to its sequential molecular structure independently of type of tissue.

Contrary, when both amniotic membrane categories study values are observed we found that FAS variant shows a peak for d spacing of 3.241 Å. (28.4 ° in 2θ), while GAS is expressed in the maximum deflection for spacing d 3.284 (28 ° in 2θ). This lag is not verified for the 2nd peak in both categories that match at d - spacing = 2.359 (40.8° in 2 θ). (See figure 7). These findings showed significant data in the sense that the chemical preservation of amniotic membrane with glycerol, modifies sequencing molecular design of collagen, while this variable is not changed under the cryopreserved defrosted condition.

These would be in according with aforementioned work from (Frushour & Koenig 1975).

Additionally, by analyzing the profiles of ordering by OPC values -in relative terms- we see that both, the cryopreservation defrosted and glicerolización procedures, down modify profiles designs, defined as "molecular crystals". (FAS vs GAS, OPC value = 14.76; FAS vs. CAS, OPC value = 42.02 respectively).

The main conclusion from these data is: amnion chemical glycerolized procedures, change sequencing molecular design, while physical cryopreserved method does not. But, physical cryopreservation method, and chemical glycerolized, modifies OPC values related to both tissue categories: cryopreserved amnion and vascular tissue.

It must be noted the observed differences in the profiles of diffractive curves between fresh and cryopreserved arterial vessels. Indeed, there are a disparity between those varieties, which were not verified by the corresponding samples fresh and cryopreserved amniotic membrane, that maintain substantially similar profiles. This aspect is independent of the OPC values for both tissues and categories of each study. Our hypothesis is that these differences are related to the anatomical and functional collagen distribution in different tissues. Indeed, the hierarchical order of collagen mesh reach a final design bundles arranged following the lines of force according to bio mechanical requirements.

In this sense, arterial wall of large conduit vessels such as aorta and carotid, are under pulsate hemodynamic regimens alternating expansion and elastic contraction states. Under these conditions, the main loads acting on the vessel wall are pressure and blood flow. The blood pressure acts directly on the inner wall of the vessel in normal direction, and flow proactively work generating a pressure proportional to the square of blood velocity, Fung YC (1997). There are therefore two preferred directions in the distribution of the charges: one circumferential and other longitudinal. This results in a complex morphological organization of the cellular components of the middle layer, composed of smooth muscle cells and collagen mesh ECM, whose design will trace circumferential and longitudinal lines, giving to vascular tissue an anisotropic mechanical behavior condition. (Rodriguez, 2007). The primary ice spontaneous nucleation happen at random in many sites of ECM, Muldrew. 1999). The crystallization growth front, follows the preferred direction lines according to design collagen mesh. According with OPC values, the new organizational picture of the cryopreserved defrosted vascular tissues would be the result of a complex sequence of physic chemical events through preservation procedures, applying changing a complex vascular structural tissue.

This is not the organizational situation of amnion collagen ECM. Amnion membrane has a laminar design, and is not subject to biomechanical pulsate regimen. Its function as an external fetal covering meets fundamentally amniotic liquid metabolic regulations, more than biomechanical functions. In spite of, both kind of tissue studied, (vascular, and amnion membrane) have basically the same fibril collagen composition, namely: Collagen I, III, and V. Additionally, Colagen IV in structural Basements Membranes.

Then, our hypothesis is that the morphologic compositions, and the architectural organizations, according to specific functional requirements to each tissue, define the proper molecular assembly, and therefore its own diffractografic profiles.

8.2 Raman scattering

Our preliminary results obtained on amnion membrane (FAM vs CAS) show punctual differences between both categories in three ranges of Raman Sepectra: 1260, 1442, and 1667 cm^{-1} band (See figure 11) where is observed an increased Intensity (AU) values in FAS related to CAS. According to references Frank, C. et al (1995), these Raman Spectra range areas aforementioned belong to fibril collagen I, III and V from human placenta. The defined corresponding chemical residues assignments, (Frushour & Koenig 1975). are respectively: Amide III; CH3, CH2 (deform); and Amide I for each range Raman Spectra recorded.

These findings support our work hypothesis about the potential power of cryopreservation procedure to change collagen structure at molecular level.

9. Conclusions

The aim of our study was the analysis of tissues produced in TB for therapeutic purposes, by known techniques, XRD and RS, able to approach the study of structures at the molecular level, mainly in reference to collagen, the fundamental component of ECM,. Taking account the hypothesis that preservation techniques introduce changes in the matrix elements of the tissues, we subjected amniotic membrane and arterial vascular samples under two types of procedures: a) cryopreservation-defrost as physical process, and glycerolization - deglycerolization as chemical process. Regarding to the results of our tests, we accept our hypothesis and concluded that while cryopreservation modifies the structural arrangement of collagen at the level of ECM, glycerolization changes molecular d spacing of biological polymers, besides the aforementioned order. However, the changes between the processed vascular tissues and amniotic membrane are different, because while the vascular cryopreservation increases the molecular order of the crystalline structure measured by OPC value, the amniotic membrane glycerolization and cryopreservation decrease the referred molecular order. These differences are interpreted as the result of complex physicochemical phenomena that occur during preservation procedures on molecular structures and its designs. These phenomena promote variations in the tissue molecular complexity and order distribution. Preliminary data from the Raman tests corroborate the hypothesis of specific modifications in the molecular structures. The consequences of these findings on the allograft biological behavior applied to clinical purposes is a challenge to research and development. Both types of tissues studied are widely applied in the world with beneficial results for the restoration of altered structures and functions in the recipients. But the gold standard allograft is not yet produced, so it is necessary to obtain the allograft that better reproduce the structural and functional conformation of the patient implanted. This objective will be achieved through the best inter relation between recipient structures and the preserved tissues applied, at molecular level to obtain the better possible allograft behavior and patency. Mainly, taking account the advanced development of applied bio engineering, and the design of complex products (composites) that combine different types of cells and artificial, biological, or modified scaffolds.

10. References

Abraham LC, Dice JF, Finn PF, Mesires NT, Lee K, & Kaplan DL., (2007) Extracellular matrix remodeling--methods to quantify cell-matrix interactions. Biomaterials.. 28(2):151-61. PMID: 16893566

Armentano RL, Santana DB, Cabrera Fischer EI, Graf S, Perez Cámpos H, Germán YZ, Carmen Saldías MD, Alvarez I. (2006). An in vitro study of cryopreserved and fresh human arteries: a comparison with ePTFE prostheses and human arteries studied non-invasively in vivo. Cryobiology. Feb;52(1):17-26. ISSN 0011-2240

Aspden, RM., Bornstein NH., & Hukins DW., (1987) Collagen organization in the interspinous ligament and its relationship to tissue function. *J. Anat.*. 155: 141 -151. PMCID: PMC1261882

Berenguer F., Wenger MPE., Bean RJ., Boze L., Horton MA., & Robinson IK. (2009).Coherent X-ray diffraction from collagenous soft tissues. *PNAS.* (106) 15297 - 15301. PNAS: 0905151106

Bia D., Pessana F., Armentano R., Pérez Campos H., Graf S., Zócalo Y., Saldías M., Perez N., Alvarez O., Silva W., Machin D., Sueta P., Ferrin S., Acosta M.,& Alvarez I. (2006). Cryopreservation procedure does not modify human carotid homografts mechanical properties: an isobaric and dynamic analysis. *Cell Tissue Bank.* 7(3):183-94. PMID:16933040

Bia D., Armentano RL., Zócalo Y., Pérez Cámpos H., Cabrera FE., Graf S., Saldías M., Silva W., & Alvarez I. (2007). Functional properties of fresh and cryopreserved carotid and femoral arteries, and of venous and synthetic grafts: comparison with arteries from normotensive and hypertensive patients. *Cell Tissue Bank.*; 8(1):43-57. PMID:16826454

Bissell MJ., Hall HG., & Parry G. (1982). How does the extracellular matrix direct gene expression? *J. Theor. Biol.* 99:31–68 PMID: 6892044

Bowers SL., Banerjee I., & Baudino TA. (2010). The extracellular matrix: at the center of it all. J Mol Cell Cardiol. Mar;48(3):474-82. PMID:19729019

Brewster L., Brey EM., & Greisler HP. (2007). Blood Vessels. In: *Principles of Tissue Enginieering.* R. Lanza, R. Langer and J. Vacanti Ed. 568 - 584. ISBN: 978-0-12-370615-7

Davis GE. (2010). The development of the vasculature and its extracellular matrix: a gradual process defined by sequential cellular and matrix remodeling events. *Am J Physiol Heart Circ Physiol.* 299 (2): H245-247. ISSN: 0363-6135

Frushour, B. G. and Koenig, J. L. (1975), Raman spectra of D and L amino acid copolymers. Poly-DL-alanine, poly-DL-leucine, and poly-DL-lysine. *Biopolymers,* 14: 363–377. doi: 10.1002/bip.1975.360140210

Fung, Y.C. (1997). Blood flow in arteries. In: Boold Flow in Arteries. *Biomechanics Circulation..* Springer-Verlag Ed. N. York. Pp.: 108 - 200.- ISBN: 0-387-94386-6

Grobstein, C. (1953) Morphogenetic Interaction between Embryonic Mouse Tissues separated by a Membrane Filter. *Nature.* (172): 869 - 871. ISSN: 0028-0836

Kelleher CM., McLean SE.,& Mecham RP. (2004) Vascular Extra cellular Matrix and Aortic Development. Current Topics in Developmental Biology. 62: 153-188. PMID:15522742

Gerson C.J., Goldstein S., & Heacox, A.E. (2009) Retained structural integrity of collagen and elastin within cryopreserved human heart valve tissue as detected by two-photon laser scanning confocal microscopy. Cryobiology. 59(2):171-179. PMID:19591817

Giannini S, Buda R, Di Caprio F., Agati P., Bigi A., De Pasquale V. & Ruggeri A.. (2008). Effect of freezing on the biomechanical and structural properties of human posterior tibial tendons. *Int Orthop* 32(2):145–151. PMCID: PMC2269010

Hamer, AJ., Strachan JR., Black MM., Ibboston CJ. Stockley I., & Elson RA. (1996). Biomechanical properties of cortical allograft bone using a new method of bone strength measurement: a comparison of fresh, fresh-frozen and irradiated bone. J Bone Joint Surg Br. 78(3): 363 – 368. PMID: 8636167

Hauschka S. & Konigsberg I.R. (1966) The influence of collagen on the development of muscle clones. *Zoology;* 55: 119 - 126. PMCID: PMC285764

Hickey, DS. And Hukins DWL. (1980) X-ray diffraction studies of the arrangement of collagenous fibres in human fetal intervertebral disc. *J. Anat.* 131 (1): 81 - 90. PMCID: PMC1233288

Horton, WG. (1958) Further observations on the elastic mechanism of the inter vertebral disc. *J. Bone and Joint Surg.* 40 (3) 552 - 558. PMID: 13575471

Kielty, C. M., & Grant, M. E. (2002). The collagen family: structure, assembly, and organization in the extracellular matrix. In "*Connective Tissue and Its Heritable Disorders: Molecular, Genetic, and Medical Aspects*" (P. M. Royce and B. Steinmann, eds.), pp. 159–221. Wiley-Liss, New York. ISBN: 9780471251859

Langerak S. E. , Groenink M., van der Wall E. E., Wassenaar C., Vanbavel E., van Baal M. C. & Spaan J.A. E. (2001). Impact of current cryopreservation procedures on mechanical and functional properties of human aortic homografts. *Transpl Int.* 14(4):248-255. PMID:11512058

Levit, J. (1962). A sulfhydryl-disulfide Hipótesis of Frost Injury and Resístanse in Plants. *J. Theoret. Biol..* 3: 355 - 391.-

Levit, J. (1966). Cryochemistry of Plant Tissue. Protein Interactions. *Cryobiology.* 3. 243 - 251. PMID: 5970348

Meier S. & Hay E. (1974). Control of corneal differentiation by extracellular materials. Collagen as a promoter and stabilizer of epithelial stroma production. Developmental Biology. Volume 38, (2): 249-270 PMID: 4275424

Muldrew, & K., McGann, L. (1999). Cryobiology - A Short Course In: *http://www.ucalgary.ca/~kmuldrew/cryo_course/cryo_*

Nadiarnykh O., LaComb R. B., Brewer M. A., & Campagnola P. J. (2010). Alterations of the extracellular matrix in ovarian cancer studied by Second Harmonic Generation imaging microscopy. BMC Cancer, 10:94. In: *www.ncbi.nlm.nih.gov/pmc/articles/PMC2841668/* PMCID: PMC2841668

Orgel J.P., San Antonio J.D., & Antipova O. (2011) Molecular and structural mapping of collagen fibril interactions. *Connective Tissue Research.* 52 (1): 2-17 PMID: 21182410

Park H. J., Urabe K., Naruse K., Onuma K., Nemoto N. & Itoman M. (2009).The effect of cryopreservation or heating on the mechanical properties and histomorphology of rat bone-patellar tendon-bone. *Cell Tissue Bank.* 10:11–18 PMID: 18830689.

Pauling, L & Corey, R. B. (1951) The structure of fibrous proteins of the collagen gelatin group. *Proc. Nat.Acad. Sci.* 37:272 - 281. ISSN 0027-8424

Pegg D.E., Wusteman M.C., & Boylan S. (1997). Fractures in Cryopreserved Elastic Arteries *Cryobiology* 34, 183–192. PMID: 9130389

Pegg, D. The preservation of tissues for transplantation. 2006. *Cell and Tissue Banking.* 7: 349 - 358. PMID: 16957871

Perez Campos, H., Saldias, MC., Silva, W., Machin, D., Suescun, L., Faccio, R., Mombru, A., & Alvarez I. (2008). Control of Cryopreservation Procedures on Blood Vessels

Using Fiber X-Ray Diffraction. *Transplantation Proceedings.* 40(3):668-674.PMID:18454983

Pukacki F., Jankowski T., Gabriel M., Oszkinis G., Krasinski Z., & Zapalski S. The mechanical properties of fresh and cryopreserved arterial homografts. *Eur J Vasc Endovasc Surg.* 2000. 20 (1): 21-24. PMID: 10906292

Raman C.V. & Krishnan, KS. A New Type of Secondary Radiation. *Nature.* 1928. *121(3048),* 501 - 502. ISSN: 0028-0836

Rodriguez, J. Goicolea J. M., & Gabaldón F. (2007). A volumetric model for growth of arterial walls, with arbitrary geometry and loads. *J. Biomechanics.* 40: 961 - 971. PMID: 16797020

Rosset E., Friggi A., Novakovitch G., Rolland P.H., Rieu R., Pellissier J.F., Magnan P.E. & Branchereau A. (1996). Effects of cryopreservation on the viscoelastic properties of human arteries. Ann Vasc Surg..10 (3): 262-72. PMID:8792995

Sanders E.J, Prasad S. & Cheung E. (1988). Extracellular matrix synthesis is required for the movement of sclerotome and neural crest cells on collagen. Differentiation.Volume 39, Issue 1, November, Pages 34-41. PMID: 3246291

Schenke-Layland K, Madershahian N, Riemann I, Starcher B, Halbhuber KJ, König K, & Stock UA. (2006) Impact of cryopreservation on extracellular matrix structures of heart valve leaflets. Ann Thorac Surg. 81 (3): 918-926. PMID:16488695

Schenke-Layland K., Xie J., Heydarkhan-Hagvall S. Hamm-Alvarez, S. F., Stock U A., Brockbank K.G.M, & MacLellan W. R. (2007). Optimized preservation of extracellular matrix in cardiac tissues: implications for long-term graft durability. Ann Thorac Surg. 83 (5): 1641-1650. PMID:17462373

Schultz G.S., Ladwig G., & Wysocki A. (2005). Extracellular matrix: review of its roles in acute and chronic wounds. In: *http://www.worldwidewounds.com/2005/august/Schultz/Extrace-Matric-Acute-Chronic-Wounds.html*

Schwinn, M.K., Faralli, J.A., Filla, M.S., Peters D.M., (2010). The Fibrillar Extracellular Matrix of the Trabecular Meshwork. *Encyclopedia of the Eye.* Edited By Joseph Besharse, Reza Dana, & Darlene A. Dart. *Pages 135-141.* Elsevier.ISBN: 978-0-12-374198-1 USA.

Sottile J., Shi, F., Rublyevska, I.,Chiang, HY., Lust, J & Chandler, J. (2007). Fibronectin-dependent collagen I deposition modulates the cell response to fibronectin. *Am J Physiol Cell Physiol.* 293: C1934 - C1943. PMID: 17928541

Standards for Tissue Banking: International Atomic EnergyAgency (IAEA - 2005); *www.iaea.org*.Spanish Association of Tissue Banks (AEBT - 2005). www.aebt.org

Sweeney S., Orgel JP., Fertala A., McAuliffe JD., Turner KR., Di Lullo GA., Chen S., Antipova O., Perumal S., Ala-Kokko L., Forlino A., Cabral WA., Barnes AM., Marini JC., & San Antonio JD. (2008). Candidate Cell and Matrix Interaction Domains on the Collagen Fibril, the Predominant Protein of Vertebrates J. Biol. Chem. 2008. 283: 21187-21197. ISSN 0021-9258

Wolfe, J. & Bryant, G. (1999). Freezing, drying and/or vitrification of membrane-solute-water systems.. Cryobiology 39: 103 – 129. PMID: 10529304

Woo S.L., Orlando C.A., Camp J.F., & Akeson W.H. (1986). Effects of postmortem storage by freezing on ligament tensile behavior. J Biomech. 19:399–404. PMID:3733765

Worthley, DL. Giraud, A.S. & Wang, T.C.(2010) The Extracellular Matrix in Digestive Cancer

Cancer microenvironment official journal of the International Cancer Microenvironment Society. 3 (1) p. 177-185. PMCID: PMC2990481